Foodborne Disease Handbook

FOODBORNE DISEASE HANDBOOK

Editor-in-Chief
Y. H. Hui

Volume Editors
**J. Richard Gorham
David Kitts
K. D. Murrell
Wai-Kit Nip
Merle D. Pierson
Syed A. Sattar
R. A. Smith
David G. Spoerke, Jr.
Peggy S. Stanfield**

Volume 1   *Bacterial Pathogens*

Volume 2   *Viruses, Parasites,
    Pathogens, and HACCP*

Volume 3   *Plant Toxicants*

Volume 4   *Seafood and
    Environmental Toxins*

Foodborne Disease Handbook

Second Edition, Revised and Expanded

Volume 2: Viruses, Parasites, Pathogens, and HACCP

edited by

Y. H. Hui
Science Technology System
West Sacramento, California

Syed A. Sattar
University of Ottawa
Ottawa, Ontario, Canada

K. D. Murrell
U.S. Department of Agriculture
Beltsville, Maryland

Wai-Kit Nip
University of Hawaii at Manoa
Honolulu, Hawaii

Peggy S. Stanfield
Dietetics Resources
Twin Falls, Idaho

MARCEL DEKKER, INC. NEW YORK · BASEL

ISBN: 0-8247-0338-3

This book is printed on acid-free paper.

Headquarters
Marcel Dekker, Inc.
270 Madison Avenue, New York, NY 10016
tel: 212-696-9000; fax: 212-685-4540

Eastern Hemisphere Distribution
Marcel Dekker AG
Hutgasse 4, Postfach 812, CH-4001 Basel, Switzerland
tel: 41-61-261-8482; fax: 41-61-261-8896

World Wide Web
http://www.dekker.com

The publisher offers discounts on this book when ordered in bulk quantities. For more information, write to Special Sales/Professional Marketing at the headquarters address above.

Introduction to the Handbook

The *Foodborne Disease Handbook, Second Edition, Revised and Expanded*, could not be appearing at a more auspicious time. Never before has the campaign for food safety been pursued so intensely on so many fronts in virtually every country around the world. This new edition reflects at least one of the many aspects of that intense and multifaceted campaign: namely, that research on food safety has been very productive in the years since the first edition appeared. The *Handbook* is now presented in four volumes instead of the three of the 1994 edition. The four volumes are composed of 86 chapters, a 22% increase over the 67 chapters of the first edition. Much of the information in the first edition has been carried forward to this new edition because that information is still as reliable and pertinent as it was in 1994. This integration of the older data with the latest research findings gives the reader a secure scientific foundation on which to base important decisions affecting the public's health.

We are not so naive as to think that only scientific facts influence decisions affecting food safety. Political and economic factors and compelling national interests may carry greater weight in the minds of decision-makers than the scientific findings offered in this new edition. However, if persons in the higher levels of national governments and international agencies, such as the Codex Alimentarius Commission, the World Trade Organization, the World Health Organization, and the Food and Agriculture Organization, who must bear the burden of decision-making need and are willing to entertain scientific findings, then the information in these four volumes will serve them well indeed.

During the last decade of the previous century, we witnessed an unprecedentedly intense and varied program of research on food safety, as we have already noted. There are compelling forces driving these research efforts. The traditional food-associated pathogens, parasites, and toxins of forty years ago still continue to cause problems today, and newer or less well-known species and strains present extraordinary challenges to human health.

These newer threats may be serious even for the immunocompetent, but for the immunocompromised they can be devastating. The relative numbers of the immunocompromised in the world population are increasing daily. We include here not just those affected by the human immunodeficiency virus (HIV), but also the elderly; the very young; the recipients of radiation treatments, chemotherapy, and immunosuppressive drugs; patients undergoing major invasive diagnostic or surgical procedures; and sufferers of debilitating diseases such as diabetes. To this daunting list of challenges must be added numerous instances of microbial resistance to antibiotics.

Moreover, it is not yet clear how the great HACCP experiment will play out on the worldwide stage of food safety. Altruism and profit motivation have always made strange bedfellows in the food industry. It remains to be seen whether HACCP will succeed in wedding these two disparate motives into a unifying force for the benefit of all concerned—producers, manufacturers, retailers, and consumers. That HACCP shows great promise is thoroughly discussed in Volume 2, with an emphasis on sanitation in a public eating place.

All the foregoing factors lend a sense of urgency to the task of rapidly identifying toxins, species, and strains of pathogens and parasites as etiologic agents, and of determining their roles in the epidemiology and epizootiology of disease outbreaks, which are described in detail throughout the *Foodborne Disease Handbook*.

It is very fortunate for the consumer that there exists in the food industry a dedicated cadre of scientific specialists who scrutinize all aspects of food production and bring their expertise to bear on the potential hazards they know best. A good sampling of the kinds of work they do is contained in these four new volumes of the *Handbook*. And the benefits of their research are obvious to the scientific specialist who wants to learn even more about food hazards, to the scientific generalist who is curious about everything and who will be delighted to find a good source of accurate, up-to-date information, and to consumers who care about what they eat.

We are confident that these four volumes will provide competent, trustworthy, and timely information to inquiring readers, no matter what roles they may play in the global campaign to achieve food safety.

Y. H. Hui
J. Richard Gorham
David Kitts
K. D. Murrell
Wai-Kit Nip
Merle D. Pierson
Syed A. Sattar
R. A. Smith
David G. Spoerke, Jr.
Peggy S. Stanfield

Preface

Much of the thought and action surrounding food safety deals with preventing foodborne bacteria from causing disease—and that is as it should be. But there are other threats that command our attention. These—the viruses and parasites—take center stage in the second volume of the *Foodborne Disease Handbook, Second Edition, Revised and Expanded*.

Viruses play an important role as agents of human diseases, and indeed their relative significance is increasing as we try to prevent and control the spread of common bacterial pathogens. The potential for foodborne spread of viruses is also stronger now than ever because of a combination of many current societal changes. Ongoing changes in demographics, changing lifestyles, faster and more frequent movement of peoples and goods, and rapidly expanding global trade in produce have already had a profound impact on the potential of viruses and other pathogens to spread through foods.

Eight chapters on foodborne viruses, contributed by internationally recognized experts in their respective fields, represent an overview of the most up-to-date information in this area. They cover well-known viral pathogens, as well as those that are less well understood but may acquire greater significance if left unheeded. The chapters, when considered together, represent a valuable resource on the biology of foodborne viruses, clinical diagnosis, and medical management as well as laboratory-based identification of viral infections transmitted through foods, and the epidemiology, prevention, and control of foodborne spread of viral pathogens. Wherever appropriate, the challenges and difficulties of detecting viruses in foods are highlighted and research needs identified. The guidelines for reducing the risk of spread of hepatitis A through foods should be applicable to many other foodborne pathogens.

It is anticipated that the information presented here will assist researchers, epidemiologists, physicians, public health officials and government regulators, and those in the food production and marketing business, to become better informed on the human health impact of foodborne viral infections and to work collectively in making foods safer.

All chapters on parasites from the first edition have been revised and updated. The addition of a chapter on the occurrence of parasites in seafood completes the overall subject of foodborne and waterborne diseases transmitted by parasites.

Although Americans are relatively unfamiliar with parasitic infection, three examples of areas in which diseases have been transmitted by parasites will help us to remember this important subject: northern Taiwan (from consuming raw or undercooked beef), Asian countries such as Thailand (undercooked pork), and Japan and Hawaii (undercooked seafood). Various chapters in this volume provide detailed description of these types of diseases transmitted by parasites.

There is a gradual movement—sometimes voluntary, sometimes mandated—toward implementation of HACCP principles in all aspects of the food industry, beginning with production and harvesting and continuing through the various stages of manufacturing, warehousing, wholesaling, retailing, and serving, until the food eventually reaches the consumer's plate. However, for the moment, at least, it is in the area of food service that the application of HACCP principles has reached the highest level of achievement. While much remains to be done even in the food service sector with regard to the implementation of HACCP, it will be seen from this second volume of the *Foodborne Disease Handbook* that the mechanics of implementation have been worked out and fine-tuned, and that this accomplishment may now serve as a model and a guide for other facets of the food industry.

The food industry in general seems to be buying into the HACCP system with an apparent high level of commitment, and this bodes well for the consumer. But rather than relaxing vigilance and trusting that HACCP will solve all food safety problems, the food industry must heighten vigilance and redouble efforts to ensure that HACCP programs are protected on all sides by a secure fortress of state-of-the-art environmental sanitation.

The editors and contributors to this volume have given researchers, microbiologists, parasitologists, food-industry managers, food analysts, and HACCP managers a wealth of information on how to detect and identify foodborne parasites and viral pathogens, how to investigate the disease outbreaks they cause, and, most importantly, how to prevent foodborne diseases by the practical application of HACCP principles.

Y. H. Hui
Syed A. Sattar
K. D. Murrell
Wai-Kit Nip
Peggy S. Stanfield

Contents

Contributors

Ann M. Adams   Seafood Products Research Center, U.S. Food and Drug Administration, Bothell, Washington

Hazel Appleton   Virus Reference Division, Public Health Laboratory Service, Central Public Health Laboratory, London, England

Sabah Bidawid   Food Directorate, Health Canada, Ottawa, Ontario, Canada

William C. Campbell   Research Institute for Scientists Emeriti, Drew University, Madison, New Jersey

Ramsey C. Cheung   Department of Medicine (Gastroenterology), Stanford University School of Medicine, Stanford, and VA Palo Alto Health Care System, Palo Alto, California

Theresa L. Cromeans   Hepatitis Branch, Division of Viral and Rickettsial Diseases, National Center for Infectious Diseases, Centers for Disease Control and Prevention, Atlanta, Georgia, and Department of Environmental Sciences and Engineering, School of Public Health, University of North Carolina, Chapel Hill, North Carolina

John H. Cross   Preventive Medicine and Biometrics, Uniformed Services University of the Health Sciences, Bethesda, Maryland

Debra D. DeVlieger   Division of Emergency and Investigational Operations, U.S. Food and Drug Administration, Washington, D.C.

J. P. Dubey   BARC, Parasite Biology and Epidemiology Laboratory, LPSI, Agricultural Research Service, U.S. Department of Agriculture, Beltsville, Maryland

Michael O. Favorov   Hepatitis Branch, Division of Viral and Rickettsial Diseases, National Center of Infectious Diseases, Centers for Disease Control and Prevention, Atlanta, Georgia

Ronald Fayer   Agricultural Research Service, U.S. Department of Agriculture, Beltsville, Maryland

H. Ray Gamble   Parasite Biology and Epidemiology Laboratory, Agricultural Research Service, U.S. Department of Agriculture, Beltsville, Maryland

Bruno Gottstein   Institute of Parasitology, University of Berne, Berne, Switzerland

Y. H. Hui   Science Technology System, West Sacramento, California

Lee-Ann Jaykus   Department of Food Science, North Carolina State University, Raleigh, North Carolina

Alan M. Johnson   Professor of Microbiology, University of Technology, Sydney, New South Wales, Australia

Thomas M. Lüthi   Official Food Control Authority of the Canton of Solothurn, Solothurn, Switzerland

Harold S. Margolis   Hepatitis Branch, Division of Viral and Rickettsial Diseases, National Center for Infectious Diseases, Centers for Disease Control and Prevention, Atlanta, Georgia

Suzanne M. Matsui   Department of Medicine (Gastroenterology), Stanford University School of Medicine, Stanford, and VA Palo Alto Health Care System, Palo Alto, California

K. D. Murrell   Agricultural Research Service, U.S. Department of Agriculture, Beltsville, Maryland

Omana V. Nainan   Hepatitis Branch, Division of Viral and Rickettsial Diseases, National Center for Infectious Diseases, Centers for Disease Control and Prevention, Atlanta, Georgia

Zbigniew S. Pawlowski   University of Medical Sciences, Poznan, Poland

Paul Prociv   Department of Microbiology and Parasitology, The University of Queensland, Brisbane, Queensland, Australia

Syed A. Sattar   Department of Biochemistry, Microbiology, and Immunology, Faculty of Medicine, University of Ottawa, Ottawa, Ontario, Canada

O. Peter Snyder, Jr.   Hospitality Institute of Technology and Management, St. Paul, Minnesota

David G. Spoerke, Jr.   Bristlecone Enterprises, Denver, Colorado

V. Susan Springthorpe   Centre for Research on Environmental Microbiology, Department of Biochemistry, Microbiology, and Immunology, Faculty of Medicine, University of Ottawa, Ottawa, Ontario, Canada

Jason A. Tetro   Centre for Research on Environmental Microbiology, Department of Biochemistry, Microbiology and Immunology, Faculty of Medicine, University of Ottawa, Ottawa, Ontario, Canada

Contents of Other Volumes

VOLUME 3: PLANT TOXICANTS

VOLUME 4: SEAFOOD AND ENVIRONMENTAL TOXINS

I. Poison Centers

II. Seafood Toxins

1

The Role of Poison Centers in the United States

David G. Spoerke, Jr.
Bristlecone Enterprises, Denver, Colorado

I. EPIDEMIOLOGY

Epidemiological studies aid treatment facilities in determining risk factors, determining who becomes exposed, and establishing the probable outcomes with various treatments. A few toxicology organizations have attempted to gather such information and organize it into yearly reports. The American Association of Poison Control Centers (AAPCC) and some federal agencies work toward obtaining epidemiological information, but the AAPCC also has an active role in assisting with the treatment of potentially toxic exposures. Epidemiological studies assist government and industry in determining package safety, effective treatment measures, conditions of exposure, and frequency of exposure.

Studies on viral exposures provide information on the type of people most commonly involved in exposures. Are they children, adults at home, outdoorsmen, industrial workers, or blue collar workers? Studies can also tell us which viral species are most commonly involved. What symptoms are seen first, what the onset of symptoms is like, and if there are any sequelae may also be determined and compared with current norms.

A. AAPCC

1. What Are Poison Centers and the AAPCC?

The group in the United States that is most concerned on a daily basis with potential poisonings due to household agents, industrial agents, and biologics is the American Association of Poison Control Centers (AAPCC). This is an affiliation of local and regional centers that provides information concerning all aspects of poisoning and refers patients to treatment centers. This group of affiliated centers is often supported by local government, private funds, and industrial sources.

Poison centers were started in the late 1950s; the first were in the Chicago area. The idea caught on quickly and at the peak of the movement there were hundreds of centers throughout the United States. Unfortunately, there were few or no standards as to what might be called a poison center, the type of staff, hours of operation, or information resources. One center may have had a dedicated staff of doctors, pharmacist, and nurses trained specifically in handling poison cases; the next center may just have a book on toxicology in the emergency room or hospital library. In 1993, the Health and Safety Code (Section 777.002) specified that a poison center must provide a 24-hour service for public and health care professionals and meet requirements established by the AAPCC. This action helped the AAPCC to standardize activities and staffs of the various centers.

The federal government does not fund poison centers, even though for every dollar spent on poison centers there is a savings of $2 to $9 in unnecessary medical expenses (1,2). The federal agency responsible for the Poison Prevention Packaging Act is the U.S. Consumer Product Safety Commission (CPSC). The National Clearinghouse for Poison Control Centers initially collected data on poisonings and provided information on commercial product ingredients and biologic toxic agents. For several years the National Clearinghouse provided product and treatment information to the poison centers that handled the day-to-day calls.

At first, most poison centers were funded by the hospital in which they were located. As the centers grew in size and number of calls handled, both city and state governments look on the responsibility of contributing funds. In recent years the local governments have found it difficult to fund such operations and centers have had to look to private industry for additional funding. Government funding may take several forms, either as a line item on a state budget, as a direct grant, or as moneys distributed on a per-call basis. Some states with fewer residents may contract with a neighboring state to provide services to its residents. Some states are so populous that more than one center is funded by the state. Industrial funding also varies—sometimes as a grant, sometimes as payment for handling the company's poison or drug information–related calls, sometimes as payment for collection of data regarding exposure to the company's product.

Every year the AAPCC reports a summary of all kinds of exposures.

2. Regional Centers

As the cost of providing this service has risen, the number of listed centers has dropped significantly since its peak of 600-plus. Many centers have been combined into regional organizations. These regional poison centers provide poison information, offer telephone management and consultation, collect pertinent data, and deliver professional and public education. Cooperation between regional poison centers and poison treatment facilities is crucial. The regional poison information center, assisted by local hospitals, should determine the capabilities of the treatment facilities of the region. They should also have a working relationship with their analytical toxicology, emergency and critical care, medical transportation, and extracorporeal elimination services. This should be true for both adults and children.

A "region" is usually determined by state authorities in conjunction with local health agencies and health care providers. Documentation of these state designations must be in writing unless a state chooses (in writing) not to designate any poison center or accepts a designation by other political or health jurisdictions. Regional poison information centers should serve a population base of greater than 1 million people and must receive at least 10,000 human exposure calls per year.

The number of certified regional centers in the United States is now under 50. Certification as a regional center requires the following.

1. Maintenance of a 24 hour-per-day, 365-days-per-year service.
2. Provision of service to both health care professionals and the public.
3. Availability of at least one specialist in poison information in the center at all times.
4. Having a medical director or qualified designee on call by telephone at all times.
5. Service should be readily accessible by telephone from all areas in the region.
6. Comprehensive poison information resources and comprehensive toxicology information covering both general and specific aspects of acute and chronic poisoning should be available.
7. The center is required to have a list of on-call poison center specialty consultants.
8. Written operational guidelines that provide a consistent approach to evaluation, follow-up, and management of toxic exposures should be obtained and maintained. These guidelines must be approved in writing by the medical director of the program.
9. There should be a staff of certified professionals manning the phones (at least one of the individuals on the phone has to be a pharmacist or nurse with 2000 hours and 2000 cases of supervised experience)
10. There should be a 24 hour-per-day physician (board-certified) consultation service.
11. The regional poison center shall have an ongoing quality assurance program.
12. Other criteria, determined by the AAPCC, may be established with membership approval.
13. The regional poison information center must be an institutional member in good standing of the AAPCC. Many hospital emergency rooms still maintain a toxicology reference such as the POISINDEX® system to handle routine exposure cases but rely on regional centers to handle most of the calls in their area.

B. Poison Center Staff

The staffing of poison centers varies considerably from center to center. The three professional groups most often involved are physicians, nurses, and pharmacists. Who answers the phones is somewhat dependent on the local labor pool, monies available, and the types of calls being received. Others personnel used to answer the phone include students in medically related fields, toxicologists, and biologists. Persons responsible for answering the phones are either certified by the AAPCC or are in the process of obtaining certification. Passage of an extensive examination on toxicology is required for initial certification, with periodic recertification required.

Regardless of who takes the initial call, there is a medical director and other physician back-up available. These physicians have specialized training or experience in toxicology, and are able to provide in-depth consultations for health care professionals calling a center.

1. Medical Director

A poison center medical director should be board-certified in medical toxicology or be board-certified in internal medicine, pediatrics, family medicine, or emergency medicine. The medical director should be able to demonstrate ongoing interest and expertise in toxicology as evidenced by publications, research, and meeting attendance. The medical director must have a medical staff appointment at a comprehensive poison treatment facility and be involved in the management of poisoned patients.

2. Managing Director

The managing director must be registered nurse, pharmacist, physician, or hold a degree in a health science discipline. The individual should be certified by the American Board of Medical Toxicology (for physicians) or by the American Board of Applied Toxicology (for nonphysicians). He or she must be able to demonstrate ongoing interest and expertise in toxicology.

3. Specialists in Poison Information

These individuals must be registered nurses, pharmacists, or physicians, or be currently certified by the AAPCC as specialists in poison information. Specialists in poison information must complete a training program approved by the medical director and must be certified by the AAPCC as specialists in poison information within two examination administrations of their initial eligibility. Specialists not currently certified by the Association must spend an annual average of no fewer than 16 hours per week in poison center–related activities. Specialists currently certified by the AAPCC must spend an annual average of no less than 8 hours per week. Other poison information providers must have sufficient background to understand and interpret standard poison information resources and to transmit that information understandably to both health professionals and the public.

4. Consultants

In addition to physicians specializing in toxicology, most centers also have lists of experts in many other fields as well. Poison center specialty consultants should be qualified by training or experience to provide sophisticated toxicology or patient care information in their area(s) of expertise. In regard to viral exposures, the names and phone numbers of persons in infectious disease at nearby hospitals might be helpful. Funding is usually a crucial issue, so these experts should be willing to donate their expertise in identification and handling cases within their specialty. Poison centers usually do not have specific specialists in viral diseases. Local hospital departments are most often contacted and the patient referred to the specialist.

C. What Types of Calls Are Received?

All types of calls are received by poison centers, most of which are handled immediately while others are referred to more appropriate agencies. Which calls are referred depends on the center, its expertise, its consultants, and the appropriateness of a referral. Below are lists of calls that generally fall into each group. Remember there is considerable variation between poison centers; if there is doubt, call the poison center and they will tell you if your case is more appropriately referred. Poison centers do best on calls regarding acute exposures. Complicated calls regarding exposure to several agents over a long period of time, which produces nonspecific symptoms, are often referred to another medical specialist, to the toxicologist associated with the center, or to an appropriate government agency. The poison center will often follow up on these cases to track outcome and type of service given.

Types of Calls Usually Accepted
Drug identification
Actual acute exposure to a drug or chemical
Actual acute exposure to a biologic agent (e.g., plants, mushrooms, various animals)
Information regarding the toxic potential of an agent
Possible food poisonings

Types of Calls Often Referred
Questions regarding treatment of a medical condition (not poisoning)
Questions on common bacterial, viral, or parasitic infections
General psychiatric questions
Proper disposal of household agents such as batteries, bleach, insecticides
Use of insecticides (which insecticide to use, how to use it) unless related to a health issue, e.g., a person allergic to pyrethrins wanting to know which product does not contain pyrethrins)

Records of all calls/cases handled by the center should be kept in a form that is acceptable as a medical record. The regional poison information center should submit all its human exposure data to the Association's National Data Collection System. The regional poison information center shall tabulate its experience for regional program evaluation on at least an annual basis.

1. AAPCC Toxic Exposure Surveillance System

In 1983 the AAPCC formed the Toxic Exposure Surveillance System (TESS) from the former National Data Collection System. Currently, TESS contains nearly 16.2 million human poison exposure cases. Sixty-five poison centers, representing 181.3 million people, participate in the data collection. The information has various uses to both governmental agencies and industry, providing data for product reformulations, repackaging, recalls, bans, injury potential, and epidemiology.

The summation of each year's surveillance is published in the *American Journal of Emergency Medicine* late each summer or fall.

D. How Calls Are Handled

Most poison centers receive requests for information via the telephone. Calls come from both health care professionals and consumers. Only a few requests are received by mail or in person, and these are often medicolegal or complex cases. Most centers can be reached by a toll-free phone number in the areas they serve, as well as a local number.

Busy centers will have a single number that will ring on several lines. Calls are often direct referrals from the 911 system. In most cases, poison center specialists are unable to determine the virus involved, so the caller is referred to a infectious disease physician.

Poison information specialists listen to the caller, recording the history of the case on a standardized form developed by AAPCC. Basic information such as the agent involved, amount ingested, time of ingestion, symptoms, previous treatment, and current condition are recorded, as well as patient information such as sex, age, phone number, who is with the patient, relevant medical history, and sometimes patient address. All information is considered a medical record and is therefore confidential.

The case is evaluated (using various references) as:

1. Information only, no patient involved
2. Harmless and not requiring follow-up
3. Slightly toxic, no treatment necessary but a follow-up call is given
4. Potentially toxic, treatment given at home and follow-up given to case resolution
5. Potentially toxic, treatment may or may not be given at home, but it is necessary for the patient to be referred to a medical facility
6. Emergency—an ambulance and/or paramedics are dispatched to the scene

Cases are usually followed until symptoms have resolved. In cases where the patient is referred to a health care facility, the hospital is notified, the history relayed, toxic potential discussed, and suggestions for treatment given.

E. What References Are Used?

References used also vary from center to center, but virtually all centers use a toxicology system called POISINDEX®, which contains lists of products, their ingredients, and suggestions for treatment. The system is compiled using medical literature and editors throughout the world. Biologic products such as plants, insects, mushrooms, animal bites, and so forth are handled similarly. Viral infections are not listed in the system. An entry for an individual household product or plant might contain a description, potentially toxic agent, potential toxic amounts, and so forth. The physician or poison information specialist is then referred to a treatment protocol that may apply to a general class of agents. Using plants as an example, an exposure to a philodendron would be referred to a protocol on oxalate-containing plants. An unknown skin irritation or potential infection would deserve a consult with an infectious disease specialist. POISINDEX is available on microfiche, a CD-ROM, over a network, or on a mainframe. It is updated every 3 months.

Various texts are also used, especially when the exposure agents, like viruses, are usually not in POISINDEX. It is very difficult to identify infectious cases over the phone, so often the assistance of an epidemiologist and an infectious disease specialist is used. Some poison centers have more experience with certain types of poisonings, so often one center will consult another on an interesting case. These are often more complex cases, or cases involving areas within both centers' regions.

A recent trend has been for various manufacturers not to provide product information to all centers via POISINDEX but to contract with one poison center to provide for poison information services for the whole country. Product information is given to only that center and cases throughout the country are referred to that one center.

F. How Poison Centers Are Monitored for Quality

Most poison centers have a system of peer review in place. One person takes a call, another reviews it. Periodic spot review is done by supervisor and physician staff. General competence is assured by certification and recertification via examination of physicians and poison information specialists. The review process helps the poison control specialists to be consistent.

G. Professional and Public Education Programs

The regional poison information center is required to provide information on the management of poisoning to the health professionals throughout the region who care for poisoned patients. Public education programs aimed at educating both children and adults about poisoning concerns and dangers should be provided.

In the past, several centers provided stickers or logos such as Officer Ugh, Safety Sadie, and Mr. Yuck that could be placed on or near potentially toxic substances. While the intent was to identify potentially toxic substances from which children should keep away, the practice has been much curtailed on the new assumption that in some cases the stickers actually attracted the children to the products.

In the spring of every year there is a poison prevention week. National attention is focused on the problem of potentially toxic exposures. During this week many centers run special programs for the public. This may include lectures on prevention, potentially toxic agents in the home, potentially toxic biologic agents, or general first-aid methods using during a poisoning. Although an important time for poison centers, public and professional education is a year-round commitment. Physicians are involved with medical toxicology rounds, journal clubs, and lectures by specialty consultants. Health fairs, school programs, and various men's and women's clubs are used to educate the public. The extent of these activities is often determined by the amount of funding from government, private organizations, and public donations.

H. Related Professional Toxicology Organizations

> **ACGIH** American Conference of Governmental and Industrial Hygienists
> Address: Kemper Woods Center, Cincinnati, OH, 45240
> Phone: 513-742-2020
> FAX: 513-742-3355
> **ABAT** American Board of Applied Toxicology
> Address: Truman Medical Center, West, 2301 Holmes St., Kansas City, MO, 64108
> Phone: 816-556-3112
> FAX: 816-881-6282
> **AACT** American Association of Clinical Toxicologists
> Address: c/o Medical Toxicology Consultants, Four Columbia Drive, Suite 810, Tampa FL, 33606
> **AAPCC** American Association of Poison Control Centers
> Address: 3201 New Mexico Avenue NW, Washington, DC, 20016
> Phone: 202-362-7217
> FAX: 202-362-8377

ABEM   American Board of Emergency Medicine
   Address: 300 Coolidge Road, East Lansing, MI, 48823
   Phone: 517-332-4800
   FAX: 517-332-2234
ACEP   American College of Emergency Physicians (Toxicology Section)
   Address: P.O. Box 619911, Dallas, TX, 75261-9911
   Phone: 800-798-1822
   FAX: 214-580-2816
ACMT   American College of Medical Toxicology (formerly ABMT)
   Address: 777 E. Park Drive, P.O. Box 8820, Harrisburg, PA, 17105-8820
   Phone: 717-558-7846
   FAX: 717-558-7841E-mail: lkoval@pamedsoc.org (Linda L. Koval)
ACOEM   American College of Occupational and Environmental Medicine
   Address: 55 West Seegers Road, Arlington Heights, IL, 60005
   Phone: 708-228-6850
   FAX: 708-228-1856
ACS   Association of Clinical Scientists, Dept. of Laboratory Medicine, University
   of Connecticut Medical School
   Address: 263 Farmington Ave., Farmington, CT, 06030-2225
   Phone: 203-679-2328
   FAX: 203-679-2328
ACT   American College of Toxicology
   Address: 9650 Rockville Pike, Bethesda, MD, 20814
   Phone: 301-571-1840
   FAX: 301-571-1852
AOEC   Association of Occupational and Environmental Clinics
   Address: 1010 Vermont Ave., NW, #513, Washington, DC, 20005
   Phone: 202-347-4976
   FAX: 202-347-4950
   E-mail: lo478x@gwis.circ.gwu.edu
ASCEPT   Australian Society of Clinical and Experimental Pharmacologists and
   Toxicologists
   Address: 145 Macquarie St., Sydney N.S.W. 2000, Australia
   Phone: 61-2-256-5456
   FAX: 61-2-252-3310
BTS   British Toxicology Society, MJ Tucker, Zeneca Pharmaceuticals
   Address: 22B11 Mareside; Alderley Park, Macclesfield, Cheshire, Sk10 4TG,
   United Kingdom
   Phone: 0428 65 5041
CAPCC   Canadian Association of Poison Control Centers, Hopital Sainte-Justine
   Address: 3175 Cote Sainte-Catherine; Montreal, Quebec, H3T1C5
   Phone: 514-345-4675
   FAX: 514-345-4822
CSVVA (CEVAP)   Center for the Study of Venoms and Venomous Animals
   Address: UNESP, Alameda Santos, N 647, CEP 01419-901, Sao Paulo, SP;
   Brazil
   Phone: 55 011 252 0233
   FAX: 55 011 252 0200

EAPCCT   European Association of Poison Control Centers
Address: J. Vale, National Poisons Information Centre, P.O. Box 81898 Dep;
N-0034 Oslo, Norway
Phone: 47-260-8460

HPS   Hungarian Pharmacological Society, Central Research Institute for
Chemistry
Address: Hungarian Academy of Sciences, H-1525 Budapest, P.O. Box 17,
Pusztaszeri ut 59-67
Phone: 36-1-135-2112

ISOMT   International Society of Occupational Medicine and Toxicology
Address: USC School of Medicine, 222 Oceanview Ave., Suite 100, Los
Angeles, CA, 90057
Phone: 213-365-4000

JSTS   Japanese Society of Toxicological Sciences
Address: Gakkai Center Building, 4-16, Yayoi 2-chome, Bunkyo-ku, Tokyo
113, Japan
Phone: 3-3812-3093
FAX: 3-3812-3552

SOT   Society of Toxicology
Address: 1101 14th Street, Suite 1100, Washington, DC., 20005-5601
Phone: 202-371-1393
FAX: 202-371-1090
E-mail: sothq@toxicology.org

SOTC   Society of Toxicology of Canada
Address: P.O. Box 517, Beaconsfield, Quebec, H9W 5V1, Canada
Phone: 514-428-2676
FAX: 514-482-8648

STP   Society of Toxicologic Pathologists
Address: 875 Kings Highway, Suite 200, Woodbury, NJ, 08096-3172
Phone: 609-845-7220
FAX: 609-853-0411

SSPT   Swiss Society of Pharmacology and Toxicology
Address: Peter Donatsch, Sandoz Pharma AG, Toxicologtie 881/130, CH-
4132 Muttenz, Switzerland
Phone: 41-61-469-5371
FAX: 41-61-469-6565

WFCT   World Federation of Associations of Clinical Toxicology Centers and Poi-
son Control Centers
Address: Centre Anti-Poisons, Hopital Edonard Herriot, 5 pl d, Arsonval,
69003 Lyon, France
Phone: 33 72 54 80 22
FAX: 33 72 34 55 67

I. International AAPPC Affiliations

The AAPCC and its members attend various world conferences to learn of toxicology
problems and new methods used by these agencies. An especially close relationship has
formed between the American and Canadian poison center associations. Once a year the

AAPCC and CAPCC hold a joint scientific meeting and invite speakers and other toxicology specialists from around the world. Some international affiliated organizations are listed with the North American groups above.

J. Toxicology and Poison Center Web sites

Association of Occupational and Environmental Clinics This group is dedicated to higher standards of patient-centered, multidisciplinary care emphasizing prevention and total health through information sharing, quality service, and collaborative research.
Address: lo478x@gwis.circ.gwu.edu

Finger Lakes Regional Poison Center
Address: pwax@ed.urmc.rochester.edu

Medical/Clinical/Occupational Toxicology Professional Groups A list of primarily U.S. professional groups interested in toxicology. There is a description of each group, the address, phone numbers, and contact names.
Keyword: poison centers, toxicology
Address: http://www.pitt.edu/~martint/pages/motoxorg.htm

Poison Net A mailing list dedicated to sharing information, problem solving, and networking in the areas of poisoning, poison control centers, hazardous materials, and related topics. The list is intended for health care professionals, not the lay public. The moderators do not encourage responses to individual poisoning cases from the public:
Keyword(s): poisoning, poison control centers

II. U.S. POISON INFORMATION CENTERS

The following poison control center telephone numbers and addresses are thought to be accurate as of the date of publication. Poison control center telephone numbers or addresses may change. The address and phone number of the poison control center nearest you should be checked frequently. If the number listed does not reach the poison center, contact the nearest emergency service, such as 911 or local hospital emergency rooms. The authors disclaims any liability resulting from or relating to any inaccuracies or changes in the phone numbers provided below. An asterisk indicates a regional center designated by the American Association of Poison Control Centers. This information should NOT be used as a substitute for seeking professional medical diagnosis, treatment, and care.

ALABAMA

Birmingham
Regional Poison Control Center*
Children's Hospital of Alabama
1600 Seventh Avenue, South Birmingham,
    AL 35233-1711
(800) 292-6678 (AL only)
(205) 933-4050

Tuscaloosa
Alabama Poison Control System, Inc.
408 A Paul Bryant Drive, East
Tuscaloosa, AL 35401
(800) 462-0800 (AL only)
(205) 345-0600

ALASKA

Anchorage
Anchorage Poison Center
Providence Hospital
P.O. Box 196604
3200 Providence Drive
Anchorage, AK 99519-6604
(800) 478-3193 (AK only)

Fairbanks
Fairbanks Poison Center
Fairbanks Memorial Hospital
1650 Cowles St.
Fairbanks, AK 99701
(907) 456-7182

ARIZONA

Phoenix
Samaritan Regional Poison Center*
Good Samaritan Medical Center
1130 East McDowell Road, Suite A-5
Phoenix, AZ 85006
(602) 253-3334

Tucson
Arizona Poison and Drug Information
    Center*
Arizona Health Sciences Center, Room
    1156
1501 N. Campbell Ave
Tucson, AZ 85724
(800) 362-0101 (AZ only)
(602) 626-6016

ARKANSAS

Little Rock
Arkansas Poison & Drug Information
    Center
University of Arkansas
College of Pharmacy
4301 Wes Markham, Slot 522
Little Rock, AR 77205
(800) 482-8948 (AR only)
(501) 661-6161

CALIFORNIA

Fresno
Fresno Regional Poison Control Center*
Fresno Community Hospital & Medical
    Center
2823 Fresno Street
Fresno, CA 93721
(800) 346-5922 (CA only)
(209) 445-1222

Los Angeles
Los Angeles County
University of Southern California Regional
    Poison Center*
1200 North State, Room 1107
Los Angeles, CA 90033
(800) 825-2722
(213) 222-3212

Orange
University of California
Irvine Medical Center Regional Poison
    Center*
101 The City Drive, South
Route 78
Orange, CA 92668-3298
(800) 544-4404 (CA only)
(714) 634-5988

Richmond
Chevron Emergency Information Center
15299 San Pablo Avenue
P.O. Box 4054
Richmond, CA 94804-0054
(800) 457-2202
(510) 233-3737 or 3738

Sacramento
Regional Poison Control Center*
University of California at Davis Medical
    Center
2315 Stockton Boulevard Rm HSF-124
Sacramento, CA 95817
(800) 342-3293 (northern CA only)
(916) 734-3692

San Diego
San Diego Regional Poison Center*
University of California at San Diego
    Medical Center
225 West Dickinson Street
San Diego, CA 92013-8925
(800) 876-4766 (CA only)
(619) 543-6000

San Francisco
San Francisco Bay Area Poison Center*
San Francisco General Hospital
1001 Potrero Avenue Rm 1E86
San Francisco, CA 94122
(800) 523-2222
(415) 476-6600

San Jose
Regional Poison Center
Santa Clara Valley Medical Center
751 South Bascom Avenue
San Jose, CA 95128
(800) 662-9886, 9887 (CA only)
(408) 299-5112, 5113, 5114

COLORADO

Denver
Rocky Mountain Poison Center*
1010 Yosemite Circle
Denver, CO 80230
(800) 332-3073 (CO only)
(303) 629-1123

CONNECTICUT

Farmington
Connecticut Poison Control Center
University of Connecticut Health Center
263 Farmington Avenue
Farmington, CT 06030
(800) 343-2722 (CT only)
(203) 679-3456

DELAWARE

Wilmington
Poison Information Center
Medical Center of Delaware
Wilmington Hospital
501 West 14th Street
Wilmington, DE 19899
(302) 655-3389

DISTRICT OF COLUMBIA

Washington
National Capital Poison Center*
Georgetown University Hospital
3800 Reservoir Road, North West
Washington, DC 20007
(202) 625-3333

FLORIDA

Jacksonville
Florida Poison Information Center
University Medical Center
655 West Eighth Street
Jacksonville, FL 32209
(904) 549-4465 or 764-7667

Tallahassee
Tallahassee Memorial Regional Medical
    Center
1300 Miccosukk Road
Tallahassee, FL 32308
(904) 681-5411

Tampa
Tampa Poison Information Center*
Tampa General Hospital
Davis Islands
P.O. Box 1289
Tampa, FL 33601
(800) 282-3171 (FL only)
(813) 253-4444

GEORGIA

Atlanta
Georgia Regional Poison Control Center*
Cerady Memorial Hospital
80 Butler Street South East
Box 26066
Atlanta, GA 30335-3801
(800) 282-5846 (GA only)
(404) 616-9000

Macon
Regional Poison Control Center
Medical Center of Central Georgia
777 Hemlock Street
Macon, GA 31208
(912) 744-1146, 1100 or 1427

Savannah
Savannah Regional Poison Control Center
Memorial Medical Center Inc.
4700 Waters Avenue
Savannah, GA 31403
(912) 355-5228 or 356-5228

HAWAII

Honolulu
Kapiolani Women's and Children's
    Medical Center
1319 Punahou Street
Honolulu, HI 96826
(800) 362-3585, 3586 (HI only)
(808) 941-4411

IDAHO

Boise
Idaho Poison Center
St. Alphonsus Regional Medical Center
1055 North Curtis Road
Boise, ID 83706
(800) 632-8000 (ID only)
(208) 378-2707

ILLINOIS

Chicago
Chicago and NE Illinois Regional Poison
    Control Center
Rush Presbyterian—St. Luke's Medical
    Center
1653 West Congress Parkway
Chicago, IL 60612
(800) 942-5969 (Northeast IL only)
(312) 942-5969

Normal
Bromenn Hospital Poison Center
Virginia at Franklin
Normal, IL 61761
(309) 454-6666

Springfield
Central and Southern Illinois Poison
    Resource Center

St. John's Hospital
800 East Carpenter Street
Springfield, IL 62769
(800) 252-2022 (IL only)
(217) 753-3330

Urbana
National Animal Poison Control Center
University of Illinois Department of
    Veterinary Biosciences
2001 South Lincoln Avenue, 1220 VMBSB
Urbana, IL 61801
(800) 548-2423 (Subscribers only)
(217) 333-2053

INDIANA

Indianapolis
Indiana Poison Center*
Methodist Hospital
1701 North Senate Boulevard
Indianapolis, IN 46202-1367
(800) 382-9097
(317) 929-2323

IOWA

Des Moines
Variety Club Drug and Poison Information
    Center
Iowa Methodist Medical Center
1200 Pleasant Street
Des Moines, IA 50309
(800) 362-2327
(515) 241-6254

Iowa City
University of Iowa Hospitals and Clinics
200 Hawkins Drive
Iowa City, IA 52246
(800) 272-6477 or (800) 362-2327 (IA
    only)
(319) 356-2922

Sioux City
St. Luke's Poison Center
St. Luke's Regional Medical Center
2720 Stone Park Boulevard
Sioux City, IA 51104
(800) 352-2222 (IA, NE, SD)
(712) 277-2222

KANSAS

Kansas City
Mid America Poison Center
Kansas University Medical Center
39th and Rainbow Boulevard Room B-400
Kansas City, KS 66160-7231
(800) 332-6633 (KS only)
(913) 588-6633

Topeka
Stormont Vail Regional Medical Center
Emergency Department
1500 West 10th
Topeka, KS 66604
(913) 354-6100

Wichita
Wesley Medical Center
550 North Hillside Avenue
Wichita, KS 67214
(316) 688-2222

KENTUCKY

Ft. Thomas
Northern Kentucky Poison Information
    Center
St. Luke Hospital
85 North Grand Avenue
Ft. Thomas, KY 41075
(513) 872-5111

Louisville
Kentucky Poison Control Center of Kosair
    Children's Hospital
315 East Broadway
P.O. Box 35070
Louisville, KY 40232
(800) 722-5725 (KY only)
(502) 589-8222

LOUISIANA

Houma
Terrebonne General Medical Center Drug
    and Poison Information Center
936 East Main Street
Houma, LA 70360
(504) 873-4069

Monroe
Louisiana Drug and Poison Information
    Center
Northeast Louisiana University School of
    Pharmacy, Sugar Hall
Monroe, LA 71209-6430
(800) 256-9822 (LA only)
(318) 362-5393

MAINE

Portland
Maine Poison Control Center
Maine Medical Center
22 Bramhall Street
Portland, ME 04102
(800) 442-6305 (ME only)
(207) 871-2950

MARYLAND

Baltimore
Maryland Poison Center*
University of Maryland School of
    Pharmacy
20 North Pine Street
Baltimore, MD 21201
(800) 492-2414 (MD only)
(410) 528-7701

MASSACHUSETTS

Boston
Masschusetts Poison Control System*
The Children's Hospital
300 Longwood Avenue
Boston, MA 02115
(800) 682-9211 (MA only)
(617) 232-2120 or 735-6607

MICHIGAN

Adrian
Bixby Hospital Poison Center
Emma L. Bixby Hospital
818 Riverside Avenue

Adrian, MI 49221
(517) 263-2412

Detroit
Poison Control Center
Children's Hospital of Michigan
3901 Beaubien Boulevard
Detroit, MI 48201
Outside metropolitan Detroit; (800) 462-
    6642 (MI only)
(313) 745-5711

Grand Rapids
Blodgett Regional Poison Center
1840 Wealthy Street, South East
Grand Rapids, MI 49506
Within MI: (800) 632-2727

Kalamazoo
Bronson Poison Information Center
252 East Lovell Street
Kalamazoo, MI 49007
(800) 442-4112 616 (MI only)
(616) 341-6409

MINNESOTA

Minneapolis
Hennepin Regional Poison Center*
701 Park Avenue South
Minneapolis, MN 55415
(612) 347-3144
(612) 347-3141 (Petline)

St. Paul
Minnesota Regional Poison Center*
St. Paul-Ramsey Medical Center
640 Jackson Street
St. Paul, MN 55101
(800) 222-1222 (MN only)
(612) 221-2113

MISSISSIPPI

Jackson
University of Mississippi Medical Center
2500 North State Street
Jackson, MS 39216
(601) 354-7660

Hattiesburg
Forrest General Hospital
400 S. 28th Avenue
Hattiesburg, MS#39402
(601) 288-4235

MISSOURI

Kansas City
Poison Control Center
Children's Mercy Hospital
2401 Gillham Road
Kansas City, MO 64108-9898
(816) 234-3000 or 234-3430

St. Louis
Regional Poison Center*
Cardinal Glennon Children's Hospital
1465 South Grand Boulevard
St. Louis, MO 63104
(800) 392-9111 (MO only)
(800) 366-8888 (MO, West IL)
(314) 772-5200

MONTANA

Denver
Rocky Mountain Poison and Drug Center
Denver, CO 80204
(800) 525-5042 (MT only)

NEBRASKA

Omaha
The Poison Center*
Children's Memorial Hospital
8301 Dodge Street
Omaha, NE 68114
(800) 955-9119 (WY, NE)
(402) 390-5400, 5555

NEVADA

Las Vegas
Humana Hospital—Sunrise*
3186 Maryland Parkway
Las Vegas, NV 89109
(800) 446-6179 (NV only)

Reno
Washoe Medical Center
77 Pringle Way
Reno, NV 89520
(702) 328-4144

NEW HAMPSHIRE

Lebanon
New Hampshire Poison Center
Dartmouth-Hitchcock Medical Center
1 Medical Center Drive
Lebanon, NH 03756
(800) 562-8236 (NH only)
(603) 650-5000

NEW JERSEY

Newark
New Jersey Poison Information and
    Education Systems*
201 Lyons Avenue
Newark, NJ 07112
(800) 962-1253 (NJ only)
(973) 923-0764

Phillipsburg
Warren Hospital Poison Control Center
185 Rosberg Street
Phillipsburg, NJ 08865
(800) 962-1253
(908) 859-6768

NEW MEXICO

Albuquerque
New Mexico Poison and Drug Information
    Center*
University of New Mexico
Albuquerque, NM 87131
(800) 432-6866 (NM only)
(505) 843-2551

NEW YORK

Buffalo
Western New York Poison Control Center

Children's Hospital of Buffalo
219 Bryant Street
Buffalo, NY 14222
(800) 888-7655 (NY only)
(716) 878-7654

Mineola
Long Island Regional Poison Control
    Center*
Winthrop University Hospital
259 First Street
Mineola, NY 11501
(516) 542-2323, 2324, 2325

New York City
New York City Poison Control Center*
455 First Avenue, Room 123
New York, NY 10016
(212) 340-4494
(212) 764-7667

Nyack
Hudson Valley Regional Poison Center
Nyack Hospital
160 North Midland Avenue
Nyack, NY 10920
(800) 336-6997 (NY only)
(914) 353-1000

Rochester
Finger Lakes Regional Poison Control
    Center
University of Rochester Medical Center
601 Elmwood Avenue
Rochester, NY 14642
(800) 333-0542 (NY only)
(716) 275-5151

Syracuse
Central New York Poison Control Center
SUNY Health Science Center
750 E Adams Street
Syracuse, NY 13210
(800) 252-5655
(315) 476-4766

NORTH CAROLINA

Ashville
Western North Carolina Poison Control
    Center
Memorial Mission Hospital

509 Biltmore Avenue
Ashville, NC 28801
(800) 542-4225 (NC only)
(704) 255-4490 or 258-9907

Charlotte
Carolinas Poison Center
Carolinas Medical Center
100 Blythe Boulevard
Charlotte, NC 28232-2861
(800) 848-6946
(704) 355-4000

Durham
Duke Regional Poison Control Center
P.O. Box 3007
Durham, NC 27710
(800) 672-1697 (NC only)
(919) 684-8111

Greensboro
Triad Poison Center
Moses H. Cone Memorial Hospital
1200 North Elm Street
Greensboro, NC 27401-1020
(800) 953-4001 (NC only)
(919) 574-8105

Hickory
Catawba Memorial Hospital Poison Control
    Center
810 Fairgrove Church Road, South East
Hickory, NC 28602
(704) 322-6649

NORTH DAKOTA

Fargo
North Dakota Poison Center
St. Luke's Hospital
720 North 4th Street
Fargo, ND 58122
(800) 732-2200 (ND only)
(701) 234-5575

OHIO

Akron
Akron Regional Poison Center
281 Locust Street

Akron, OH 44308
(800) 362-9922 (OH only)
(216) 379-8562

Canton
Stark County Poison Control Center
Timken Mercy Medical Center
1320 Timken Mercy Drive, North West
Canton, OH 44667
(800) 722-8662 (OH only)
(216) 489-1304

Cincinnati
South West Ohio Regional Poison Control
    System and Cincinnati Drug and
    Poison Information Center*
University of Cincinnati College of
    Medicine
231 Bethesda Avenue ML #144
Cincinnati, OH 45267-0144
(800) 872-5111 (Southwest OH only)
(513) 558-5111

Cleveland
Greater Cleveland Poison Control Center
2074 Abington Road
Cleveland, OH 44106
(216) 231-4455

Columbus
Central Ohio Poison Center*
700 Children's Drive
Columbus, OH 43205
(800) 682-7625 (OH only)
(614) 228-1323

Dayton
West Ohio Regional Poison and Drug
    Information Center
Children's Medical Center
One Children's Plaza
Dayton, OH 45404-1815
(800) 762-0727 (OH only)
(513) 222-2227

Lorain
County Poison Control Center
Lorain Community Hospital
3700 Kolbe Road
Lorain, OH 44053
(800) 821-8972 (OH only)
(216) 282-2220

Sandusky
Firelands Community Hospital Poison
    Information Center
1101 Decatur Street
Sandusky, OH 44870
(419) 626-7423

Toledo
Poison Information Center of Northwest
    Ohio
Medical College of Ohio Hospital
3000 Arlington Avenue
Toledo, OH 49614
(800) 589-3897 (OH only)
(419) 381-3897

Youngtown
Mahoning Valley Poison Center
St. Elizabeth Hospital Medical Center
1044 Belmont Avenue
Youngstown, OH 44501
(800) 426-2348 (OH only)
(216) 746-2222

Zanesville
Bethesda Poison Control Center
Bethesda Hospital
2951 Maple Ave
Zanesville, OH 43701
(800) 686-4221 (OH only)
(614) 454-4221

OKLAHOMA

Oklahoma City
Oklahoma Poison Control Center
Children's Memorial Hospital
940 Northeast 13th Street
Oklahoma City, OK 73104
(800) 522-4611 (OK only)
(405) 271-5454

OREGON

Portland
Oregon Poison Center
Oregon Health Sciences University
3181 South West Sam Jackson Park Road
Portland, OR 97201

(800) 452-7165 (OR only)
(503) 494-8968

PENNSYLVANIA

Hershey
Central Pennsylvania Poison Center*
Milton Hershey Medical Center
Pennsylvania State University
P.O. Box 850
Hershey, PA 17033
(800) 521-6110
(717) 531-6111

Lancaster
Poison Control Center
St. Joseph Hospital and Health Care Center
250 College Avenue
Lancaster, PA 17604
(717) 299-4546

Philadelphia
Philadelphia Poison Control Center*
One Children's Center
34th and Civic Center Boulevard
Philadelphia, PA 19104
(215) 386-2100

Pittsburgh
Pittsburgh Poison Center*
One Children's Place
3705 Fifth Avenue at DeSoto Street
Pittsburgh, PA 15213
(412) 681-6669

Williamsport
The Williamsport Hospital Poison Control
    Center
777 Rural Avenue
Williamsport, PA 17701
(717) 321-2000

RHODE ISLAND

Providence
Rhode Island Poison Center*
593 Eddy Street
Providence, RI 02903
(401) 444-5727

SOUTH CAROLINA

Charlotte
Carolinas Poison Center
Carolinas Medical Center
1000 Blythe Boulevard
Charlotte, NC 28232-2861
(800) 848-6946

Columbia
Palmetto Poison Center
University of South Carolina
College of Pharmacy
Columbia, SC 29208
(800) 922-1117 (SC only)
(803) 765-7359

SOUTH DAKOTA

Aberdeen
Poison Control Center
St. Luke's Midland Regional Medical
 Center
305 S. State Street
Aberdeen, SD 57401
(800) 592-1889 (SD, MN, ND, WY)
(605) 622-5678

Rapid City
Rapid City Regional Poison Control Center
835 Fairmont Boulevard
P.O. Box 6000
Rapid City, SD 57709
(605) 341-3333

Sioux Falls
McKennan Poison Center
McKennan Hospital
800 East 21st Street
P.O. Box 5045
Sioux Falls, SD 57117-5045
(800) 952-0123 (SD only)
(800) 843-0505 (IA, MN, NE)
(605) 336-3894

TENNESSEE

Knoxville
Knoxville Poison Control Center

University of Tennessee Memorial
 Research Center and Hospital
1924 Alcoa Highway
Knoxville, TN 37920
(615) 544-9400

Memphis
Southern Poison Center, Inc.
Lebanheur Children's Medical Center
848 Adams Avenue
Memphis, TN 38103-2821
(901) 528-6048

Nashville
Middle Tennessee Regional Poison Center,
 Inc.
501 Oxford House
1161 21st Avenue South B-101VUII
Nashville, TN 37232-4632
(800) 288-9999 (TN only)
(615) 322-6435

TEXAS

Conroe
Montgomery County Poison Information
 Center
Medical Center Hospital
504 Medical Center Blvd.
Conroe, TX 77304
(409) 539-7700

Dallas
North Central Texas Poison Center*
Parkland Memorial Hospital
5201 Harry Hines Boulevard
P.O. Box 35926
Dallas, TX 75235
(800) 441-0040 (TX only)
(214) 590-5000

El Paso
El Paso Poison Control Center
Thomas General Hospital
4815 Alameda Avenue
El Paso, TX 79905
(915) 533-1244

Galveston
Texas State Poison Control Center
University of Texas Medical Branch
8th and Mechanic Street

Galveston, TX 77550-2780
(800) 392-8548 (TX only)
(713) 654-1701 (Houston)
(409) 765-1420 (Galveston)

Lubbock
Methodist Hospital Poison Control
3615 19th Street
Lubbock, TX 79413
(806) 793-4366

UTAH

Salt Lake City
Utah Poison Control Center*
Intermountain Regional Poison Control
   Center
410 Chipeta Way, Suite 230
Salt Lake City, UT 84108
(800) 456-7707 (UT only)
(801) 581-2151

VERMONT

Burlington
Vermont Poison Center
Medical Center Hospital of Vermont
111 Colchester Avenue
Burlington, VT 05401
(802) 658-3456

VIRGINIA

Charlottesville
Blue Ridge Poison Center*
University of Virginia Health Sciences
   Center
Box 67
Charlottesville, VA 22901
(800) 451-1428 (VA only)
(804) 924-5543

Richmond
Virginia Poison Center
Virginia Commonwealth University
MCV Station Box 522
Richmond, VA 23298-0522

(800) 552-6337 (VA only)
(804) 786-9123

WASHINGTON

Washington
Washington Poison Center
P.O. Box 5371
Seattle, WA 98105-0371
Within WA: (800) 732-6985
(206) 526-2121

WEST VIRGINIA

Charleston
West Virginia Poison Center*
West Virginia University
3110 MacCorkle Avenue, South East
Charleston, WV 25304
(304) 348-4211
(800) 642-3625 (WV only)

Parkersburg
St. Joseph's Hospital Center
19th Street and Murdoch Avenue
Parkersburg, WV 26101
(304) 424-4222

WISCONSIN

Madison
Regional Poison Control Center
University of Wisconsin Hospital
600 Highland Avenue
Madison, WI 53792
(608) 262-3702

Milwaukee
Poison Center of Eastern Wisconsin
Children's Hospital of Wisconsin
9000 West Wisconsin Avenue
P.O. Box 1997
Milwaukee, WI 53201
(414) 266-2222

WYOMING

Omaha
The Poison Center*
Children's Memorial Hospital
8301 Dodge Street
Omaha, NE 68114
(800) 955-9119 (WY, NE)
(402) 390-5400, 5555

REFERENCES

1. DL Harrison, JR Draugalis, MK Slack, PC Langly. Cost effectiveness of Regional Poison Control Centers. Arch Intern Med 156:2601–2608, 1996.
2. CPSC: CPSC Chairman Ann Brown Suggests Information Technology Study to Support Work of Poison Centers. News Release #94-047, Tuesday March 15, 1994.
3. TG Martin. Summarization of the American Association of Poison Control Centers Certification Criteria for Regional Poison Information Centers. Internet Address: motoxorg.htm.

2
Hepatitis A and E Viruses

Theresa L. Cromeans
Centers for Disease Control and Prevention, Atlanta, Georgia, and University of North Carolina, Chapel Hill, North Carolina

Michael O. Favorov, Omana V. Nainan, and Harold S. Margolis
Centers for Disease Control and Prevention, Atlanta, Georgia

I. INTRODUCTION

Our knowledge of the disease that was initially called infectious hepatitis has increased greatly since the 1950s when hepatitis A became reportable in the United States (1). Studies in human volunteers showed that hepatitis A was transmitted primarily by the fecal-oral route, and distinguished this disease from serum hepatitis, or hepatitis B (2). However, it was not until after hepatitis A virus (HAV) was identified and diagnostic tests were developed that the full epidemiological spectrum of the disease was appreciated, and this has led to the development of an approach to its control and potential eradication.

    In the United States, the most common source of infection is contact with an HAV-infected person. The importance of food and water in the transmission of hepatitis A has been well established; however, food and waterborne outbreaks contribute only 3–8% of the total cases reported in any year (3). Hepatitis A virus infection is highly endemic in developing countries where children become infected at a young age (4). Detection of HAV by polymerase chain reaction (PCR) in clinical and environmental samples combined with nucleotide sequence analysis has expanded our understanding of the epidemiology of virus transmission (5–12). New approaches and methods for detection of HAV in clinical and environmental samples will be important for further understanding of the epidemiology of HAV transmission (13–18). The ability to grow HAV in cell culture has culminated in the development of inactivated vaccines that have been shown to effectively prevent hepatitis A (19–21). Vaccine use in selected populations may reduce disease, but only routine childhood vaccination will have significant and long-term impact on total disease burden in developed countries. Genetic engineering is being used to develop an attenuated vaccine (22,23), but it remains to be determined as to whether such a vaccine would be better than the current inactivated ones.

Following the introduction of serological tests for acute HAV and hepatitis B virus (HBV) infection, epidemics of viral hepatitis were identified that were not due to HAV but that had a fecal-oral route of transmission. This second type of viral hepatitis was initially designated enterically transmitted non-A, non-B (ETNANB) hepatitis. In the 1980s, virus-like particles were identified in patients, and this new virus, named hepatitis E virus (HEV), was cloned and partially sequenced in 1990 (24). Reagents for the serological diagnosis of HEV infection have been developed, and epidemiological studies have suggested that a substantial proportion of acute viral hepatitis in developing countries may be due to HEV (25,26). Detection of HEV requires amplification by PCR since antigen detection assays with adequate sensitivity have not been developed. Hepatitis E virus replication in several cell culture systems has been reported; however, the degree of replication appears limited and does not produce enough antigen for detection, and detailed replication studies have not been performed in these systems (27–30). In most outbreaks of this disease, fecally contaminated water has been identified as the source. However, several aspects of the epidemiology of hepatitis E could be explained by zoonotic transmission. Recently, HEV infection has been identified in several species (26,31–33). In addition, a swine HEV has been identified in the United States (34), which is similar to HEV isolated from several cases not associated with travel (35). A number of important questions concerning the epidemiology and natural history of HEV infections must be answered before prevention strategies can be devised. Evaluation of a recombinant vaccine in animal models indicated protection against disease although virus excretion occurred (36).

II. HEPATITIS A

A. Clinical Features

1. Disease Pattern

Acute hepatitis caused by infection with HAV is generally self-limited and infrequently causes fulminant disease that results in death. The degree of clinically evident illness is best predicted by the age of the infected individual (37,38). Among children younger than 6 years, less than 10% develop jaundice and fewer than 50% have any symptoms associated with acute viral hepatitis (37). In older children and adults, icterus is present in 40–80% and almost all have some signs or symptoms associated with viral hepatitis (38).

Typically a symptomatic infection begins as a mild illness characterized by malaise, nausea, low-grade fever, and headache and progresses to more severe symptoms that include vomiting, diarrhea, right upper quadrant discomfort, maculopapular rashes, arthritis, and pruritus. In most cases these symptoms are accompanied by jaundice, and the illness usually lasts for 2–6 weeks. Death from fulminant hepatitis A is a rare event. In the United States, the reported case-fatality rate is 0.3% and varies from 0.004% in persons 5–14 years old to 2.7% in persons older than 49 years (4). Underlying chronic liver disease or concomitant infection with HBV has often been found in persons dying from hepatitis A and may predispose to a more severe outcome from this infection.

Hepatitis A virus has not been shown to cause a persistent infection and has not been associated with chronic liver disease. However, 3–20% of persons with hepatitis A have been shown to have a relapse of symptoms, liver enzyme elevations, or jaundice that occurs 1–3 months after resolution of their initial liver enzyme abnormalities (39,40). These relapses have been associated with reactivation of viral shedding, and in some instances extrahepatic manifestations of disease have been observed (39–41).

The average incubation period for HAV infection is 28 days, but can range from 2 to 6 weeks (2,42–44). The patterns of virus replication and excretion have been best defined using experimental infection of nonhuman primates (chimpanzees and tamarins) by the oral-gastric route (44,45) (Fig. 1A). Viremia, as measured by extraction of RNA from chimpanzee serum followed by RT-PCR, begins approximately 1 week after inoculation and continues up to 12 weeks (W. Bower, et al., manuscript in preparation). The duration of HAV RNA in serum was dependent on dose; RNA was detected up to 5 weeks when one infectious dose was administered and up to 13 weeks with 10^6 infectious doses. Approximately 1 week prior to the onset of liver enzyme elevations, HAV antigen is detected in hepatocytes surrounding the portal region. At the peak of liver enzyme elevations, most hepatocytes have become infected, and several weeks after the resolution of disease HAV antigen is no longer detected. Hepatitis A virus detected by enzyme immunoassay is excreted in feces soon after it can be detected in the liver. Cyclical shedding may occur, and excretion measured by antigen detection usually ceases once liver enzymes return to normal.

In humans, the duration of virus shedding may be age-related. In adults, 22–50% have detectable HAV antigen in stool specimens within a week of the onset of jaundice, and virus shedding measured by immunoassay may continue for another 2 weeks in about 20% of patients (46,47). Infants and children appear to shed virus for longer periods than adults, perhaps accounting for the high rates of infection observed among young children and their contacts (7,47,48). Detection of HAV RNA in human stool samples from outbreaks has demonstrated shedding for as long as 2–3 months after the onset of symptoms (11,49) (B. H. Robertson et al., in preparation). Whether HAV RNA is excreted for prolonged periods has not been evaluated fully in adults. However, the presence of HAV RNA does not prove the presence of infectious virus in persons with prolonged excretion.

The diagnosis of acute HAV infection can only be confirmed through serological testing; no combinations of signs or symptoms are predictive of the diagnosis. IgM antibody to HAV (anti-HAV IgM) is first detected 7–10 days after infection and is usually present at the time patients present with clinical illness. An IgM capture assay is commercially available for the detection of anti-HAV IgM as either a radioimmunoassay (RIA) or an enzyme immunoassay (EIA) (50,51). Although anti-HAV IgM may be present for long periods of time, the diagnostic assays are configured in such a manner that they usually do not produce a positive result 4–6 months after acute infection. Anti-HAV IgM

(A)

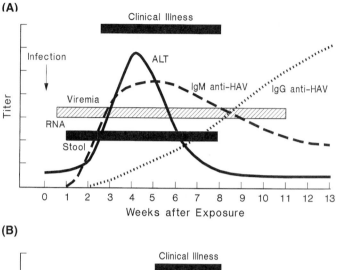

(B)

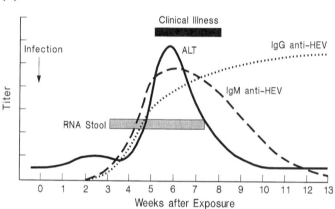

Figure 1 Viral replication and excretion: (A) virological, immunological, and biochemical events during the course of experimental hepatitis A virus (HAV) infection in chimpanzees inoculated intravenously with human HAV, strain HLD-2. (B) Virological, immunological, and biochemical events during the course of hepatitis E virus (HEV) infection in experimentally infected cynomolgus macaques and humans. ALT, alanine aminotransferase; IgG, immunoglobulin G; IgG, immunoglobulin M. (Fig. 1A adapted from Ref. 59; Fig. 1B adapted from Refs. 392, 449, 455, 457).

is quickly followed by the appearance of IgG antibody (anti-HAV IgG) that appears to confer lifelong immunity. Competitive inhibition immunoassays that detect both IgG and IgM anti-HAV (total anti-HAV) are commercially available. The presence of total anti-HAV and the absence of anti-HAV IgM indicates that the person had a prior HAV infection and is not currently infected.

There is no specific therapy for hepatitis A. In the case of fulminant disease, orthotopic liver transplantation has been used successfully, although persistent infection has been reported infrequently (52,53).

2. Pathogenesis

Hepatitis A virus is primarily hepatotropic, although limited data suggest that the virus may replicate in cells other than hepatocytes. Whether limited infection of some intestinal cells occurs is unclear. Using antigen and nucleic acid detection methods, tamarins inoculated intragastrically had no evidence of intestinal replication (54), whereas nucleic acid detection methods suggested that there is intestinal replication in both experimentally infected monkeys and tamarins (55,56). In experimentally infected chimpanzees, saliva obtained 18 days after inoculation was found to contain viral RNA (57), and transmission by oropharyngeal secretions has been suggested in one foodborne outbreak (58).

In the chimpanzee model of HAV infection, the virus can be recovered from blood during a viremic phase that begins early and lasts until up to 13 weeks after inoculation of chimpanzees (Fig. 1A)(W. Bower et al., in preparation). Circulating virus is probably contained in immune complexes (59). During infection, virus is released from hepatocytes into the bile and is then shed in feces (60). The exact routes by which virus enters the liver after ingestion and passage into the small intestine are not known. Experimental infections can be initiated by intravenous inoculation of fecal material (44), and hepatitis A has been transmitted by blood transfusion (7,61) showing that the gastrointestinal tract and/or oral ingestion are not required for HAV infection.

Hepatitis A virus antigen can be detected in hepatocytes by immunofluorescence microscopy within a week of experimental inoculation, and its distribution in cells near the portal tract during the initial phase of infection suggests delivery by the portal blood supply. The histopathological picture of acute hepatitis A does not differ significantly from acute hepatitis B in terms of the cellular infiltrates and pathomorphological changes. Unlike hepatitis B, HAV can be identified in Kupffer cells (liver macrophages) and other sinusoidal cells from the period of peak liver enzyme elevation onward (44,62).

Vesicular structures containing virus particles have been seen in hepatocytes and Kupffer cells of infected animals (63,64), and similar structures have been identified in cell culture where HAV produces no cytopathic effect (65). In cell culture, HAV can replicate and be released without cell damage, and the virus does not appear to produce the hepatocellular injury observed in human or nonhuman primates. Several studies have suggested that cytotoxic T cells are involved in the resolution of HAV infection and may contribute to the observed hepatocellular injury (66,67). Clonal analysis has shown that at the site of inflammation in the liver, CD8 interferon-producing cytotoxic T lymphocytes are enriched (68). Peripheral blood lymphocytes have been shown to produce interferon in response to exposure to HAV-infected cells (69), and γ-interferon production by T lymphocytes has been found during acute infection (70). Although serum complement levels drop during infection and complement has been shown to bind to HAV capsid proteins (44,59), it is not clear as to whether complement-mediated cellular injury occurs (71).

B. Epidemiology

1. Distribution of Infection and Disease

Worldwide, infection with HAV is very common. In many developing countries (e.g., most of Africa and much of Asia) more than 90% of children may be infected by 6 years of age, with most infections being asymptomatic (4). However, in countries where there has been significant improvement in the standard of living, a noticeable decrease has been reported in early childhood infections (72–76). These changes in age-specific infection rates have resulted in a large proportion of the childhood and adult population being susceptible to infection. When HAV is introduced into these populations (e.g., through contaminated food and/or water, or from person-to-person contact) large outbreaks of disease have occurred.

An example of the impact of the changing epidemiology of HAV infection occurred in 1988 in Shanghai, China. Seroprevalence studies had shown low rates of infection among young children and young adults (74). However, in 1988, a very large epidemic of hepatitis A occurred over several months (more than 300,000 cases) because adults ate partially cooked clams that were harvested from areas contaminated with raw sewage (77,78).

Cyclical epidemics of hepatitis A have been shown to occur in populations where HAV infection is considered highly endemic (79,80). In many Native American populations, epidemics of hepatitis A occur every 5–7 years. Prior to one such epidemic, 40% of very young children and over 80% of adults had serological evidence of previous infection (80). However, inapparent person-to-person transmission among younger children was the most likely source of the epidemic, and resulted in a 10–40% attack rate of hepatitis A among older children and adolescents.

In the United States, epidemics of hepatitis A have occurred with almost regular periodicity until the 1980s (Fig. 2) and are probably related to the age-specific proportion of the population that is susceptible to infection and the degree to which HAV is circulating in the general population or in selected risk groups. The highest rates of hepatitis A have consistently been in the western part of the country and are primarily in counties with a low population density and large Hispanic or Native American populations (3). In addition, these epidemic periods have been associated with extended, community-wide epidemics that occur among children and young adults of low socioeconomic status via person-to-person transmission (81,82).

In 1988, approximately 28,500 cases of hepatitis A were reported to the Centers for Disease Control (CDC) and this number increased to 35,800 in 1989, reflecting a country-

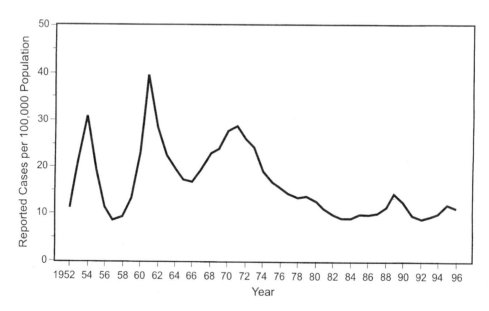

Figure 2   Rates of hepatitis A (per 100,000 population) by year in the United States, 1952–1996. (Adapted from Refs. 458 and 459.)

wide epidemic that lasted through 1991 (Fig. 2). Although the actual number of cases was probably several times higher because of underreporting, it has been shown that health care providers report hepatitis A better than other types of viral hepatitis (84,88). Among reported cases, the primary risk factors associated with hepatitis A include contact with an infected person (24–30% of cases); attending, working in, or having a child in a day care center (15% of cases); international travel (5% of cases); injection drug use (2% of cases); and food-or waterborne outbreaks (5–7%). In addition, approximately 15% of cases are associated with the presence of a child less than 5 years of age in the household who is not in day care (CDC, unpublished data). However, in all surveillance studies, 20–35% of the cases of hepatitis A have no known source for their infection (3,8) (CDC, unpublished data).

The distribution of risk factors associated with hepatitis A since 1983 is shown in Fig. 3. Injection drug use as a risk factor for infection has only been recognized since the early 1980s, and a number of outbreaks of hepatitis A have been observed in this risk group (86). Whether transmission occurs primarily through needle sharing or whether this risk factor is a surrogate for poor hygiene is not clear; both factors probably play a role in disease transmission.

The risk of infection among persons traveling to countries where HAV is highly endemic has been well documented (87). Risk factors include the region of the world being visited and the length of stay in that country. When traveling in these areas, it is recommended that hepatitis A vaccine be administered 4 weeks prior to travel; immune globulin (IG) should be administered if travel is to commence within 4 weeks of immunization. It is recommended for travelers under the age of 2 (88). Although vaccination will provide protection from HAV, travelers should take precautions to minimize ingestion of other enteric pathogens.

Epidemiological studies have also identified other settings in which HAV infection has been transmitted. These include outbreaks among persons working with nonhuman primates (89), high rates of infection among personnel during military operations in countries with a high endemicity of infection (21,90), rare instances of transmission by blood transfusion with secondary transmission to health care workers or other caregivers

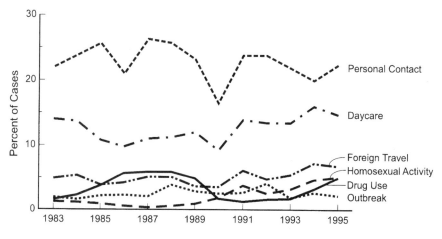

Figure 3 Distribution of risk factors for hepatitis A in the United States since 1983, mutually exclusive groups. (Adapted from Refs. 458 and 459.)

(7,91,92), and high rates of infection among homosexual men (93–95). Transmission by clotting factor concentrates has also been described (96). Some investigators have suggested that sewage workers are at risk of HAV infection; however, more evaluation of this potential risk factor in different geographic locations with appropriate control groups is needed (97).

2. Modes of Transmission

Hepatitis A virus is transmitted by the ingestion of feces on contaminated objects, as fomites, or in food or water. The most important source of infection is transmission from person to person. This is shown by the high rates of transmission among young children in developing countries, in populations with a high degree of crowding or poor sanitation, as well as within households, day care settings, and institutions for the developmentally disabled (4). In almost all settings, secondary infections occur among 4–20% of household or sexual contacts of persons involved in outbreaks of hepatitis (58,81,98). Investigation of transmission in the hospital setting has confirmed that poor handwashing practices and eating in an environment where HAV is present were risk factors for infection (7).

Compared with other picornaviruses, HAV has been shown to be more resistant to low pH, heat (99–101) and drying (102), and probably allows the virus to persist in the environment for months. Hepatitis A virus can survive on human hands and be transmitted to environmental surfaces (103), a characteristic that may allow transmission from sources that are not apparent or readily traceable.

Common source outbreaks due to contaminated food or water have been well documented. When they occur they attract a great deal of public attention and concern. However, in countries such as the United States they contribute to a very small proportion of the total disease burden. Even in developing countries, foodborne transmission is probably not the primary source of HAV infection. Outbreaks may be due to contamination of the food where it is grown or harvested, or to contamination from infected food handlers. In both cases the food is usually not cooked adequately, has been eaten raw, or has been contaminated after being cooked. The food most often contaminated at its origin has been shellfish. Filter-feeding shellfish may concentrate virus up to 100-fold from large volumes of water (104,105), thus permitting the accumulation of HAV from fecally contaminated water. Although outbreaks from contaminated shellfish are usually small, an outbreak in China resulted in over 300,000 cases (77), and the degree of low-level endemic transmission from this source has not been quantified. One study has suggested that in Italy hepatitis A is mainly foodborne and that shellfish consumption is the most frequently reported risk factor (106). Other foods contaminated at the source include fruits and vegetables that may be consumed raw or fresh-frozen. These agricultural products are most likely contaminated by infected agricultural workers. Reclaimed wastewater used for irrigation is a potential source of contamination if sprayed on the plants or if contaminated soil remains on produce. Contamination at the processing and packaging plants is also possible, and it can be difficult to determine whether the contamination occurred in the agricultural field or in a processing plant. Certain preparative processes including freezing do not inactivate HAV and the usual washing procedures do not always remove virus from the produce. Some authors have suggested that foodborne outbreaks could be an increasing problem in populations with low immunity due to the increased use of imported fresh food from countries where disease is endemic (10).

Uncooked food or food handled after cooking has been the usual source of infection transmitted by food handlers. The major risk factor in these outbreaks has been poor personal hygiene prior to the recognition that the food handler has hepatitis A (58,107). The attack rate among patrons of the involved eating establishment has usually been low, although some notable exceptions have been reported (58). In addition, the potential exists for widespread transmission when commercial facilities prepare food that is distributed to geographically distant locations. Recently, outbreaks possibly due to contamination by agricultural field workers prior to retail food distribution have been associated with fresh lettuce (108) and frozen strawberries (5,12).

Hepatitis A virus has been shown to be transmitted by water, and fecally polluted water used in food processing can be a source of contamination. Epidemiological studies have shown an association between HAV infection and consumption of defined groundwater sources that were subsequently shown to have been fecally contaminated using bacterial indicators (109–111). In addition, outbreaks of hepatitis A and viral or bacterial gastroenteritis have been associated with contaminated groundwater. Ingestion of water while swimming in a chlorinated pool or in a lake has been associated with disease (112,113). Most convincing has been the isolation of HAV from groundwater sources associated with outbreaks of hepatitis (110,111). In one recent outbreak HAV was isolated from the source of the groundwater contamination, the implicated groundwater, and the infected individuals, with molecular confirmation of the chain of transmission (11). In a recent analysis of 139 groundwater concentrates from different geographic locations in the United States, 8.6% of samples were HAV RNA positive (114). If confirmed, this would suggest there has been a relatively high level of contamination of groundwater sources, although RNA detection does not necessarily reflect presence of infectious virus.

Viable HAV has been detected in groundwater as long as 17 months after the original contamination of wells associated with outbreaks of HAV (11,110,111) (Cromeans et al., in preparation). The significance of HAV RNA detection in groundwater needs to be established, including its potential infectivity.

C. Hepatitis A Virus Biology

1. Isolation and Characterization

For decades it was known that hepatitis was an infectious disease, but it was not until World War II that it became apparent that two epidemiologically distinct forms existed (115). Early studies in human volunteers demonstrated the intestinal-oral route of infection for the disease that has now become known as hepatitis A (42). Later, studies in human volunteers distinguished infectious hepatitis (hepatitis A) and serum hepatitis (hepatitis B) by their incubation period, primary source of the virus in the patient, and route of transmission. For hepatitis A, these studies identified fecal shedding of the virus, the viremic phase of the infection, and the effect of antibody on disease expression (42,116). However, attempts to culture the virus associated with hepatitis A were unsuccessful for almost three decades.

In 1973, Feinstone and co-workers used immune electron microscopy to visualize, in feces of patients, the 27-nm virus particles that reacted with serum of persons known to have recovered from infectious hepatitis (117). After an experimental model of infection was established in chimpanzees and tamarins (43,118), the virus was characterized further, along with the events that occurred in the host from time of infection to the resolution of disease. Initially, cell culture of HAV was only achieved after virus was passaged multiple times in marmosets (tamarins) (119), but subsequently HAV was cultivated directly from human stools without animal passage (120–124). Whether the inoculum was human- or animal-derived, long periods of growth ranging from 4 to 10 weeks have been required for detection of significant amounts of HAV antigen in infected cells. However, no cytopathic effect (CPE) typical of many picornaviral infections was observed in any of these culture systems.

Hepatitis A virus has an RNA genome (125), is a 27- to 32-nm icosahedral particle with 32 capsomeres on the surface, has a buoyant density of 1.33–1.34 g/mL, a sedimentation coefficient of 156–160S, and contains four polypeptides based on genomic sequence, although the fourth has not been isolated (126,127). Based on its early characterization, HAV was classified as an enterovirus in the Picornavirus family (128,129). However, when compared to other enteroviruses, HAV replicated much more slowly, had greater heat stability (100), had essentially no nucleotide and amino acid sequence identity, and did not appear to have a primary intestinal tract phase of replication. Thus, HAV has been reclassified in a separate genus, the *Hepatovirus* in the Picornavirus family (131,132).

2. Replication in Cell Culture

Since no CPE was observed in HAV-infected cultures, virus was detected by immunofluorescent localization of HAV antigen in the cytoplasm (119,124). Upon initial isolation, HAV antigen accumulates slowly, but with increasing passage the time required for maximum antigen production can be shortened (133,134). However, in all instances maximum yields of virus were not obtained before 3 weeks post infection. The lack of CPE resulted in the establishment of persistently infected cell cultures which have been used to produce large quantities of virus (135–139). Some investigators found HAV to be only cell-associated (119,124,134), whereas others have found infectious virus and/or antigen in the cell culture fluid (133,135,140–142). HAV isolate and passage level and type of cell substrate and passage level (134,138,141,142) influence the degree of viral replication and may influence virus release.

A radioimmunofocus assay (RIFA) for the detection of infectious virus developed by Lemon and co-workers (143) made analysis of HAV replication kinetics possible. The maximum virus titer under one-cycle conditions occurred 5 days after infection, and the block in HAV replication was postulated to be at the uncoating phase (144). The synthesis of HAV RNA and depletion of the RNA pool required for replication due to encapsidation have also been suggested as rate-limiting steps for strains HM-175 and GBM (145–147). In addition, asynchronous virus replication has been suggested as a reason for the slow rate of growth (148,149). Asynchronous replication of HAV as a rate-limiting factor can also be inferred from kinetic studies with cytopathic HM-175 (142). Recent studies have also suggested that the inefficient translation initiation due to the unusual internal ribosome entry site in the 5′-NTR contributes to the slow replication (150,151). The reason(s) for the slow replication cycle of HAV as compared to other picornaviruses is not yet well defined.

Cytopathic variants of HAV have been isolated in several laboratories (152–154), and the specificity of this effect has been shown by neutralization with antibody to HAV, thus eliminating the possibility of adventitious viruses. Generally, these variants have been obtained from persistently infected cells and were subsequently serially (or acutely) passaged (152–154). Cytopathic isolates have also been obtained from serially passaged virus (141,155–157), although HAV was not shown by neutralization to be the only cause of CPE in all of these reports (141,155). The CPE is similar to other picornavirus-induced CPE. The replication cycle of these cytopathic isolates is shortened to 2–3 days and the yield is 20–560 infectious virions per cell, depending on the cell type employed (152,158). In studies of morphogenesis, replication has been shown to be in close association with cytoplasmic membranes for both cytopathic (157) and noncytopathic strains (65,159). A cytopathic variant has been shown to increase the number of annulate lamellae in infected cells examined by electron microscopy (160). In addition, studies of noncytopathic (161) as well as cytopathic HAV (162) show no overall effect on host cell metabolism.

Strains of HAV that have been adapted to grow efficiently in cell culture appear to have mutations in the 2B and 2C areas of the P2 region (Fig. 4) that codes for nonstructural proteins (163). Genomic analysis of cytopathic HAV variants has revealed mutations in the P2 region as well as in the P3 region and the 5′- and 3′-NTR regions (164). Although the specific changes associated with the cytopathic effect were not identified, these studies suggested that genetic recombination of HAV strain HM-175 occurred during passage in persistently infected cells. The involvement of multiple genomic regions was shown using infectious cDNA clones of a rapidly replicating variant (165). In another cytopathic strain, mutations in the 5′-NTR, P2, and P3 appear to contribute to the cytopathic phenotype (166). In addition, a cytopathogenic variant with mutation sites similar to those described in other cytopathic variants has produced an apoptotic reaction in infected cells (167). Mutations in the 5′-NTR, 2B, 2C, 3A, 3B, 3D, and 3′-NTR regions of noncytopathic HM-175 have been shown to enhance replication in cell culture (168–170). No consistent pattern of mutation has been observed following adaptation to different cell lines or among different HAV isolates, indicating a role of host factors in replication and adaptation of HAV to cell culture (22,142,168–172).

Hepatitis A virus has been shown to replicate in nonprimate as well as primate cell lines, indicating the presence of cell surface receptor(s) and other host factor(s) required for HAV replication (173). Viral attachment has been shown to be pH-and calcium ion–dependent, and inhibited by fetal calf serum (174–176). A surface glycoprotein on two different cell types has been identified as a receptor for HAV (177,178). Further studies

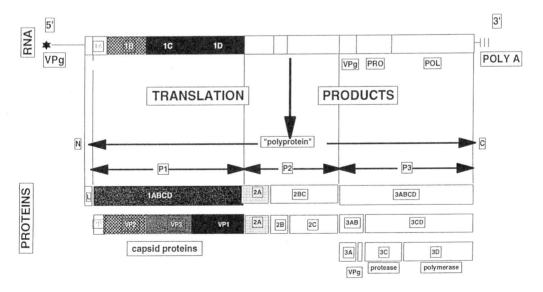

Figure 4 HAV genomic organization and proposed cleavage products. (Adapted from Ref. 455.)

have shown that this receptor is expressed in several human tissues including liver, and that the receptor is HAV binding and possibly functional in replication (179,180).

3. Molecular Biology

a. Genomic Organization

Hepatitis A virus has a single-stranded RNA genome of positive polarity that is approximately 7500 nucleotides long (Fig. 4). Genomic structure and replication scheme of HAV are similar to that of poliovirus and other members of the Picornaviridae family. HAV genome contains a single open reading frame (ORF) of 6500 nucleotides encoding a single polyprotein that is cleaved by virus-specific proteases, has a poly-A tract at its 3′ terminus, and has no 5′-cap structure (181). VPg, a viral protein linked to the 5′ end of the genome, has a lower isoelectric point than that of other picornaviruses (182). Although the genomic organization of HAV and other picornaviruses are similar, there is little similarity in their nucleotide sequences (131,183,184), or amino acid composition (131).

Hepatitis A virus appears to have the same translational strategy as that of other picornaviruses and proteins identified following in vitro translation were related immunobiologically to the structural proteins (185). Locarnini et al. (186) detected 11 virus-specific proteins in cells infected with wild-type HAV, a pattern similar to that found in cells infected with poliovirus type 1. However, other investigators found only structural proteins of the mature virion in infected cells (185,187), which may be due to low protein stability or the slow growth rate of HAV and its failure to inhibit host protein synthesis. Using various recombinant in vivo and in vitro expression systems, the 3C proteinase (3C^{pro}) was shown to cleave all structural and nonstructural proteins from the primary translation product (188–194). P1-2A has been proposed as the functional precursor of structural proteins (192,195,196). Transient expression of a nested set of 3C^{pro}-containing proteins also has been reported (197,198) and suggests that these intermediate products may be involved in promoting viral assembly.

The HAV capsid is composed of three major structural proteins—VP1, VP2 and VP3—with sizes of 33, 27, and 29 kDa, respectively (137). Based on the nucleotide sequence, a fourth capsid protein VP4, of approximately 2.5 kDa, should be encoded (199) but has not been identified in mature virions.

The entire nucleotide sequence and derived amino acid sequence has been obtained from cDNA clones for strains HM-175, HAS-15, LA, and MBB (183,200–202). When the sequence of wild-type human HAV (HM-175) in marmosets was compared to the partial sequence of three other cell culture–adapted human HAV strains (HM-175, LAV, CR326 and HAS-15), the majority of amino acid changes were located in the capsid region, mostly in VP1 (184). Nucleotide changes have also been shown to occur following cell culture adaptation and attenuation of HM-175 (203). Twenty-four nucleotide changes distributed throughout the genome were present in the cell culture–adapted virus compared with its wild-type parent. The changes found in the cell culture–adapted isolates do not appear specific for the adaptative process, and some may be cell-specific mutations (170). Studies by Emerson et al. (163) indicate that mutations in the P2 region, specifically in the 2B and 2C areas, resulted in increased growth of the virus in cell culture.

b. Antigenic Sites

Members of the Picornavirus family are distinct from each other antigenically (204,205). These differences are reflected, to a large extent, in the detailed molecular topography of the virion with most of their antigenic sites contained on the exposed outer surface of the capsid structure (206,207). Because x-ray crystallographic data are not yet available for HAV, information concerning its surface structure and potential antigenic sites comes from several sources. Hepatitis A virus exists as a single serotype, and neutralizing monoclonal antibodies that identify overlapping epitopes of cell culture–adapted strains of HAV generally recognize wild-type isolates from all parts of the world (208). Surface labeling studies have identified VP1 as the major exposed protein (209–211), as is the case for other picornaviruses. Alignments of VP1 sequences of poliovirus type 1 and HAV have identified three peptides with potential cross-reactive antigenicity, and a synthetic peptide from one of these sequences has been used to generate antibodies in rabbits that neutralized HAV in cell culture (212). Ostermayr et al. (213) produced a recombinant VP1 fusion protein, expressed in *Escherichia coli*, that reacted with rabbit anti-HAV serum. However, chimpanzees immunized with this recombinant protein produced antibodies that only reacted with VP1 of denatured virus and not with intact virus. Similarly, chimpanzees immunized with recombinant VP1 produced high titer antibody, but were not protected when challenged with wild-type HAV (Margolis et al., unpublished data).

Virus mutants selected for resistance to neutralization by monoclonal antibodies, as well as competitive antibody binding studies, suggest that HAV has a single conformational immunogenic epitope composed of several sites located on VP1 and VP3 (208,214–216). Under antibody selection, amino acid changes occur in both sites simultaneously, which further supports the concept that these two sites interact to form a single neutralization epitope (216). However, findings from HAV strains isolated from Old World monkeys suggest that sites other than those identified by the escape mutants may be involved in protecting the host from infection. These viruses have been shown to be genetically distinct from human HAV (see section below); they generally are not recognized by monoclonal antibodies produced against human HAV (217, 218, Nainan et al., unpublished data), but they bind polyclonal antibody to human HAV. In these viruses, the specific amino acids in VP1 and VP3 associated with the binding of neutralizing monoclonal antibodies differ from those found in human HAV, and probably explains their poor binding by monoclonal antibodies to human HAV (216). However, chimpanzees immunized with an HAV isolated from an Old World monkey (*Cynomolgus macaques*) induced an antibody response that usually protected them from human HAV infection, suggesting that additional antibody sites may be involved in aborting HAV infection (Nainan et al., unpublished data). In a recent study, a continuous epitope of HAV in VP3, 12 amino acids long, has been identified and antibodies generated against this peptide were found to be capable of binding intact HAV and neutralizing its infectivity (219).

A three-dimensional structure of HAV has been developed using amino acid sequence alignments of the known structures of Mengo virus and human rhinovirus (220). However, these models have not always accurately predicted the structure of other viruses (221). The escape mutant data (214–216) suggest that this may be the case for HAV, since only 6 of the 11 amino acids associated with antibody binding are located on the outer surface. Although the data suggest that the sites in VP1 and VP3 interact to form a single antigenic determinant, they are located too far apart to fit under a single immunoglobulin binding site. However, it is possible that amino acid changes at one site affect primary antibody binding at a distant site, as has been found with foot-and-mouth disease virus (FMDV) (221a). Thus, the true nature of the neutralization epitope(s) of HAV probably will not be resolved without a crystallographic structure.

c. Molecular Epidemiology

A high degree of nucleic acid conservation exists among human HAV isolates. Nucleotide variation has averaged 1–4%, a few have differed by 10%, and one group differs from the majority of human isolates by 16–24%. Polymerase chain reaction (PCR) has been used to amplify selected regions of VP1 and VP3, and HAV has been classified into seven genotypes based on the nucleotide sequence patterns from these areas (222,223). In three of the genotypes all of the viruses came from human cases; the viruses in another genotype were isolated from humans and from captive New World monkeys; and the viruses from the remaining three genotypes were isolated from recently captured Old World monkeys that had serological evidence of acute HAV infection. Most of the HAV strains isolated from North America were of a common genotype, while the strains isolated from Japan and Western Europe were of multiple genotypes and may reflect viruses imported as a result of international travel (223,224).

Genomic analysis has been applied in the epidemiological investigation of hepatitis A outbreaks. Two geographically separated foodborne outbreaks in which frozen strawberries were the likely vehicle of transmission (5) were shown to be linked to this single food source by the identical HAV sequences obtained from patient specimens. The strawberries implicated in both outbreaks came from the same processor, were packed during the same shift, and had been frozen for several years before being distributed. In a recent multistate foodborne outbreak of hepatitis A, genetic identity of virus isolated from case patients confirmed the epidemiological evidence of a common source outbreak. Genetic relatedness of HAV isolates may prove useful in determining the extent of foodborne transmission of HAV in both epidemic and nonepidemic settings (12).

4. General Virus Detection Methods

a. Cell Culture Propagation and Antigen Detection

Hepatitis A Virus has been propagated in several cell types of human and nonhuman primate origin, including primary and secondary African green monkey kidney cells (AGMK), fetal rhesus kidney (FRhK-4 and FRhK-6) cells, Alexander hepatoma cells, and human fibroblasts (120,122,124,140,225) in addition to recent propagation in nonprimate cells (173). Cell lines such as MRC-5 (human lung fibroblast) (141) and HFS (human embryo fibroblast) (226) have been used to propagate adapted virus for vaccine production. BSC-1 (African green monkey kidney) and FRhK have been primarily used for the propagation of HAV, and these cell lines or primary monkey kidney cells have been the most useful in the initial isolation of virus from clinical or environmental samples (110,111,121,133). Adventitious agents contained in primary cell cultures have been a potential problem in the primary isolation of HAV due to its slow growth. Because HAV does not usually produce CPE and is very stable in the environment, contamination of cell lines has been and can be a problem in those laboratories that also work with noncytopathic cell culture–adapted viruses (224). Appropriate separation of procedures and, where feasible, use of cytopathic strains in laboratory modeling experiments will reduce this problem.

The lack of CPE required the use of immunological assays to detect HAV antigen in cells, including immunofluorescence microscopy (62,119,134) and solid phase radioimmunoassay (RIA) or enzyme immunoassay (EIA) (137,227–229). Methods used to quantitate infectivity in vitro have included the RIFA (143), the fluorescent focus assay, the in situ RIA (134), and in situ hybridization (230), which has shortened the assay time to 3–4 days as compared to 7 days for in situ RIA and 6–21 days for the RIFA.

The standard for infectious virus identification has been the RIFA, an assay that utilizes radioactively labeled antibody to detect foci in cell culture, analogous to the standard viral plaque assay (143). The RIFA has been used for virus quantitation, neutralization studies, studies of virus stability and disinfection, as well as clonal isolation of virus populations (139,142,144,216,231). Hepatitis A Virus recently isolated from clinical or environmental samples forms smaller foci in the RIFA and may require longer incubation time to be detected (up to 3–4 weeks) than cell culture–adapted strains. With most cell culture–adapted strains, radioimmunofoci can be visualized within 2 weeks, whereas rapidly replicating variants of HAV form larger radioimmunofoci that are detectable in 5–7 days (142). An in situ enzyme immunoassay that is instrument-independent may be useful in high-throughput situations, such as screening compounds for inhibition or assessing neutralizing ability (232).

Cytopathic strains of HAV (described above) provide a classic viral plaque assay for studies of neutralization, environmental stability and disinfection (103,233), replication, and clonal isolation (142,164). However, the assay is only useful in laboratory modeling experiments and is not applicable to the isolation of HAV directly from clinical or environmental samples.

b. Nucleic Acid Detection

Nucleic acid hybridization with radioactive or colorimetric nucleic acid probes (DNA or RNA) has been shown to be more sensitive than detection of HAV antigen by immunoassay, but less sensitive than RIFA or infectivity in susceptible nonhuman primates (47,234–237). Hybridization using single-stranded RNA (ssRNA) or cDNA probes have been shown to detect 500–1000 infectious units of HAV (234–236). When fecal specimens have been analyzed by hybridization, the period of virus excretion following infection has been longer than that detected by antigen detection with RIA (47). The sensitivity of nucleic acid detection has been greatly increased by amplification of viral RNA by reverse transcription followed by the polymerase chain reaction (RT-PCR). A selected segment of the viral RNA genome is reverse-transcribed using the enzyme reverse transcriptase and the resulting cDNA amplified by a heat-stable polymerase (8,238–240). Primers used for the RT-PCR amplification of HAV RNA have been selected from these capsid regions of the genome that are highly conserved in the majority of HAV strains (12,224). Hepatitis A Virus RNA has been extracted from a variety of sources by proteinase K digestion, phenol-chloroform extractions, and ethanol precipitations (59), and many commercial kits are available for RNA extraction of clinical and environmental samples. Hepatitis A Virus is efficiently recovered from most specimens (except serum) following immunocapture on a solid phase coated with polyclonal anti-HAV (IC-RT-PCR) (8,11,223). The specificity of the amplified nucleic acid products has been determined by hybridization with radiolabeled or nonisotopically labeled oligonucleotide probes (8,18,59,223,241). Low levels of HAV RNA have been detected by RT-PCR in a variety of clinical and environmental samples (5–8,11,12,223,224,242). Nested PCR using a second round of primers internal to the first set has been used to increase the sensitivity of detection by an order of magnitude and provide the same degree of specificity observed with hybridization of amplification products. Fecal excretion can be detected with IC-RT-PCR up to several months after onset of symptoms in humans (11,49) (B. H. Robertson et al., in preparation); however, the infectivity of virus excreted at the later time points has not been determined. The level of detection of HAV by IC-RT-PCR has been as low as 0.1–0.01 radioimmunofocus units or plaque-forming units (PFUs) using laboratory preparations of virus (8,18). Laboratory contamination of RT-PCR can occur easily and yield false-positive results unless precautions have been taken (243). Furthermore, although detection of nucleic acid serves as an indicator that HAV is present in the specimen, additional information (usually epidemiological) must be used to determine if it is associated with infection (6,7,11).

D. Hepatitis A Detection in Foods

1. Isolation from Foods

Foods may become contaminated with HAV in the growing environment (e.g., shellfish), or at the time of harvest or preparation by infected agricultural workers or food handlers, respectively. Shellfish become contaminated from fecal material in growing waters, since filter-feeding mollusks concentrate viruses very efficiently. These waters usually become contaminated due to the flow of untreated sewage into coastal areas; however, treatment of sewage might not adequately remove viable virus contaminants (244). Potentially, foods can become contaminated through irrigation with untreated or treated sewage effluent. This could occur through external contamination of fruits and vegetables (245), or possibly through virus entry via damage to the root system and translocation to the plant (246). However, uptake of animal viruses by plants has not been a consistent finding (247), and these potential sources of infection have not been linked epidemiologically with HAV infection (74,245). In areas of China such as Shanghai, where "night-soil" (untreated human excrement) is routinely used for fertilizer, background rates of HAV infection are low (74). Presumed external contamination of fruits and vegetables at the time of harvest or processing has been associated with HAV infection. Outbreaks have been recognized due to contaminated lettuce (7), strawberries (5,12), and raspberries (248–250). Epidemiologically, the time of contamination was shown to be during growing, during harvest, or possibly during processing and packing, rather than at the time of food preparation. However, in none of these outbreaks was virus isolated from the implicated food.

Since HAV is shed prior to the onset of illness, persons handling food (e.g., field workers, cooks, and food handlers) may contaminate food without knowledge of their potential infectivity. Fecal material may contain as much as 10^8 infectious particles/mL (251). Thus, the likelihood of contamination from infected food handlers who do not practice good hygiene would appear high. However, of the estimated 1000 food handlers per year who may become infected with HAV, only 2 would be likely to become the source of an outbreak of hepatitis A (107).

Outbreaks due to ingestion of contaminated shellfish (including oysters, clams, mussels, and cockles) have been documented for more than 30 years (6,78,252–259), and are associated with the consumption of raw shellfish or those cooked for times and temperatures that allow survival of HAV (101,139,260,261). In addition, outbreaks have been traced to contamination of pastries (262,263), bread (264), sandwich meats (265), green salad, spaghetti, and hamburgers (10,266,267). The contribution of contaminated food to sporadic HAV infections is difficult to determine because of the long incubation period. A particular food eaten weeks before the onset of a disease is less likely to be recognized than food eaten only hours or days before onset of disease (268). However, in coastal areas where seafood is consumed regularly, cases of hepatitis A due to inadequately cooked seafood have been found to be more common than was initially recognized from disease reporting (257,269,270).

Isolation of HAV from food is difficult for several reasons. Due to the long incubation period, the food or foods involved in the outbreak have usually been completely consumed or discarded. If samples of the food are still available, detection of low levels of virus is difficult, since the contaminating virus may not be uniformly distributed in the food and the portions remaining for analysis may not be contaminated. Methods for isolation of viruses from other environmental samples are applicable to foods with modifications that may be food-specific. Modeling experiments have demonstrated isolation of HAV and detection of HAV RNA by RT-PCR in frozen strawberries (271,272). Percentage recovery and sensitivity of the method may be adequate for detection of low levels present on field-contaminated fruit. Hepatitis A virus has been detected using RT-PCR from market lettuce harvested during months of high incidence of diarrhea (273). Concentrates of ultracentrifuged lettuce washes contained fecal coliform bacteria in addition to HAV. Shellfish are the only food from which HAV has been isolated after an outbreak (6,274). Hepatitis A virus was identified in clams associated with the 1988 Shanghai epidemic by isolation in cell culture and by nucleic acid hybridization (274), and it was detected, following RT-PCR amplification, in oysters and other shellfish harvested from waters implicated in a large outbreak in Florida (6).

2. Detection Methods

Detection of HAV in foods linked to outbreaks has been difficult because of the low concentrations of virus and the difficulty in growing HAV in cell culture. For a complete review of the parameters of isolation of viruses in food, see Cliver et al. (275) and Larkin (276). Bacterial counts have been used as indicators of fecal contamination in food. Currently, the approval of oyster harvesting waters is based on bacterial examination, as well as examination of harvested shellfish (277). However, in shellfish, bacterial counts have not been valid indicators for enteric contamination (278–282). Outbreaks of HAV from consumption of raw oysters taken from approved beds suggest that the use of bacterial indicators for shellfish may not be adequate (256,283), although in some instances shellfish harvested from unapproved waters had been illegally marketed as ''approved'' (6).

Methods for the quantitative recovery of poliovirus and other enteroviruses from laboratory-contaminated shellfish have been developed by numerous investigators over the past three decades (16,284–298). Current methods for recovery of cytopathic enteric viruses (e.g., polio) from shellfish have been compared, critically reviewed, and evaluated by Sobsey (299). Virus particles can be recovered either by extraction, by extraction-concentration, or by precipitation of virus following adsorption, elution, and concentration. Recovery of virus ranges from 50% to 100%, large amounts of meat (50–100 g) must be used to ensure accurate quantitation of low levels of virus, and recovery varies according to the method and the type of shellfish being tested. Comparative studies suggest that poliovirus is a poor model for predicting recovery of HAV (139); thus prediction of HAV recovery from other Picornavirus data may not be appropriate. When modifications of each recovery method have been evaluated using oysters experimentally contaminated with HAV, the adsorption-elution-precipitation method yielded better virus recovery (46%) than the elution-precipitation method (6%) (299). In another study, polyethylene glycol precipitation recovered 98% of HAV from eluates of oysters (292). These investigators used infectivity (299) or an antigen detection by RIA (292) to detect laboratory strains of HAV in the final eluate or concentrate.

Because of the difficulty in cultivating wild-type HAV and the high sensitivity provided by nucleic acid detection methods, these have been investigated extensively

(16,18,295–298). Although nucleic acid detection by gene amplification circumvents the cultivation difficulty, the detection of nucleic acid is not a measure of infectious virus; rather, it is an indicator of potential infectivity. Hybridization with cDNA has detected 10^4 physical particles (300), while ssRNA probes have been shown to detect as few as 500–1000 infectious units of HAV in estuarine samples (236), and 1000 infectious particles in oyster and clam meat (294). To achieve the sensitivity required for detection of low-level HAV in food and water, RT-PCR amplification was used to examine shellfish implicated in one outbreak (6). In this instance, some of the shellfish (oysters) had HAV antigen detectable by EIA, indicating high levels of contamination, but HAV RNA was also detected in EIA-negative specimens.

In the last decade, numerous investigators have published RT-PCR methods for detection of HAV in shellfish, and most investigators have analyzed whole shellfish. In laboratory experiments, HAV RNA has been detected in experimentally contaminated shellfish using RT-PCR (18,242,295,298,301) with sensitivity in the range of 0.2–20 PFU/g of meat. Analysis of potentially contaminated field samples has also demonstrated the utility of RT-PCR to indicate the presence of viruses in shellfish (296,297,302); cell culture infectivity was also observed in some samples (296,302). In addition, in situ detection of HAV in shellfish tissues by transcription of viral RNA has shown the virus to be localized in stomach and digestive diverticulum (303), and a method has been developed for assay of this tissue specifically (295).

Approaches to overcome inhibition of assay enzymes by substances in food or environmental samples have been examined and include concentration and extensive purification prior to RT-PCR or extraction of RNA from whole shellfish or shellfish concentrate (see above). Comparison of seven RNA extraction methods for HAV-contaminated shellfish demonstrated that all methods detected RNA in 90% of cases (304).

Another approach for detection of HAV in environmental samples has been capture of virus particles using virus-specific antibody-coated magnetic beads or microfuge tubes prior to RT-PCR (11,13,305,306). These methods have the potential for greater sensitivity because of larger sample volume per assay and removal of RT-PCR inhibitors. Although this detection method selects immunoreactive virus particles, these particles may or may not be infectious.

Currently, the sensitivity of detection is defined in units per gram of meat (e.g., PFU); however, quantification of HAV RNA in environmental samples is needed. Several investigators have used internal standard RNA in RT-PCR to quantify HAV in shellfish concentrates (307–309).

E. Prevention of Hepatitis A

A number of prevention strategies can be used to reduce the risk of HAV infection from consumption of contaminated food. These include (a) prevention or identification of contamination where food is grown, harvested, processed, and prepared; (b) virus inactivation or disinfection of potentially contaminated foods to render them safe for consumption; and (c) immunization to lower the general risk of infection or disease transmission.

Prevention of HAV transmission is primarily related to foods that are eaten raw or have the potential for contamination after they have been cooked. Foods implicated in HAV transmission include shellfish, fresh vegetables, fresh-frozen fruits, and glazed pastries.

1. Prevention or Identification of Food Contamination

Probably the most extensive efforts have been those to protect consumers who eat raw shellfish. The primary strategy has been development and enforcement of regulations to ensure that shellfish are harvested from waters that are free from fecal contamination. Harvesting has been restricted in waters that have more than 70 total coliforms or 14 fecal coliforms per 100 mL (277). Unapproved areas are patrolled to prevent harvesting, and identification of the harvesting site is required for each lot of shellfish shipped to the retailer. In spite of these regulations and precautions, outbreaks of hepatitis A from contaminated shellfish continue to occur, including ones from distribution of illegally harvested shellfish (6). In addition, HAV RNA has been found in oysters from approved beds that were possibly involved in an outbreak (6). All of this suggests that the absence of fecal coliforms may not ensure that shellfish are free of pathogenic viruses. Contaminating bacteria may have been eliminated or may not be infectious at the time of sampling, while viruses may be present and retain infectivity. Elimination of bacteria has not reflected the elimination of HAV (105,310) or other viruses (310a) in laboratory depuration experiments.

Since concentrations of coliform bacteria in harvesting waters may not reliably predict virus contamination, virus detection in shellfish themselves could be used to ensure that contaminated shellfish would not be eaten raw. One approach would be to use ''indicator viruses'' such as those commonly found in sewage—rotaviruses, enteroviruses, or adenoviruses (311,312). However, each of these viral indicators has its disadvantages, and the optimum solution would be detection of HAV itself.

The RT-PCR amplification of viral RNA coupled with sensitive nonisotopic detection methods could provide sensitive and rapid screening of selected food samples (18,242). A disadvantage of RT-PCR amplification is that it cannot differentiate between infectious and noninfectious HAV because no component of the genome has been solely associated with infectivity. In cell culture, the ratio of noninfectious to infectious particles can range from 3:1 to 1000:1 (138,294,313); for virus that has not been adapted to cell culture, the ratio was 2×10^5:1 (313), and for a cytopathic cell culture strain the ratio was 79:1 (13). Because HAV can survive in the environment for prolonged periods, the detection of HAV RNA must be equated with the presence of infectious virus until such time that an ''infectivity indicator'' is identified. However, food potentially contaminated with HAV can be consumed if cooked (see section below). Probably the best available method to prevent HAV infection is public education concerning the risks of eating raw or partially cooked seafood and improved enforcement of current regulations for harvest of shellfish.

Another source of contamination is from infected persons who harvest or handle food that is eaten raw, or from persons who handle food after it has been cooked. Although not a major source of infection in the United States, prevention of HAV transmission by infected food handlers poses a public health and management problem and relies primarily on good hygiene, most importantly thorough handwashing (7). In addition, the use of disposable gloves might be considered a simple solution to the problem of inadequate handwashing, and some jurisdictions have enacted laws requiring food handlers not to have ''bare hand contact'' with food that will be eaten without further cooking (314). However, transmission of foodborne infections other than HAV has occurred in spite of glove use, suggesting that this might not be an ideal prevention strategy.

Outbreaks of hepatitis A due to contamination of food at or near the time of harvest (e.g., frozen strawberries and raspberries, lettuce, green onions) and the widespread use of uncooked foods in our diets underscores the need to examine public health practices of agricultural workers. Field workers who harvest foods that are not subsequently cooked or processed in a way that would inactivate HAV should be considered foodhandlers with respect to their personal hygiene. This includes provision of adequate toilet and handwashing facilities and attention to eliminating those activities that might contaminate the harvested food with human fecal material. Fresh vegetables imported from areas with high endemic rates of HAV infection and where infected persons are likely to be field workers (e.g., children, adolescents, young adults) could become the source for hepatitis A outbreaks.

2. Making Potentially Contaminated Foods Safe for Consumption

Depuration, or the transfer to clean water to allow the natural pumping mechanisms to remove impurities, is one approach to making potentially contaminated shellfish safe for raw consumption. However, HAV has been recovered from contaminated oysters after depuration under commercial conditions, which included 96 h of immersion in a continuous flow of ozonated marine water (261).

Thorough cooking will inactivate HAV and render potentially contaminated food safe for consumption. Internal temperatures of 85–95°C for one minute completely inactivate HAV in shellfish (260). However, HAV and rotaviruses have been recovered from experimentally contaminated mussels that were steamed for 5 min (261), indicating that they had not reached a temperature adequate to inactivate these viruses.

Hepatitis A virus stability in the environment facilitates transmission from contaminated foodstuffs. It has been shown to survive frozen for many years, to survive in dried feces for a month (102), to resist pH 1.0 at 38°C for 90 min (99), and to be more resistant to temperatures known to reduce cell culture infectivity of poliovirus by 50% (139). When suspended in milk and treated for 30 min at 62.8°C, 0.1% infectivity remains (101), suggesting that pasteurization methods may not completely inactivate HAV. Poliovirus cannot be used to predict survival of HAV as shown by comparative studies in cookies and oysters, on polystyrene, and in water and sediments (139). Hepatitis A virus has been shown to survive on human hands and be transferred to surfaces (103), which probably facilitates its spread. In addition, the ability of HAV to survive for extended periods in the environment is probably facilitated by the presence of biological material, such as feces. However, HAV was inactivated more rapidly in mixed human and animal wastes than in septic tank effluent alone or in phosphate-buffered saline (PBS) (315).

While disinfectants are not commonly used directly on food products, they should be considered an adjunct to the prevention of HAV infection. Disinfectants are used to ensure a clean environment in which food is prepared or stored, and to ensure a safe water supply. Fresh foods are often washed in disinfectant solutions before commercial processing. However, strawberries washed in 12 ppm chlorine before preparation and freezing were the source of a hepatitis A outbreak (12). Recently, the potential for adequate disinfection of fresh foods by disinfectant washes has been evaluated (316,317). With a dose of 10 mg/L and a contact time of 5 min, 90–99% inactivation of HAV was produced in a model system; however, peroxyacetic acid at concentrations of 40 and 80 ppm was less effective with only 86% inactivation.

Hepatitis A virus has been shown to be relatively resistant to free chlorine in human and animal infectivity studies (318,319), whereas one study suggests that HAV is more sensitive to chlorine than poliovirus when evaluated by cell culture infectivity (320). The differing conclusions drawn from these studies are most likely related to the use of dispersed HAV (320) versus aggregated or cell-associated virus (318,319). Cell-associated virus is inactivated by chlorine more slowly than dispersed virus. Concerns about the health effects of chlorine byproducts in water have lead to studies of alternative disinfectants (321). Hepatitis A virus contamination in the environment is most often associated with organic matter (e.g., feces), which results in slower virus inactivation by chlorine. An outbreak of hepatitis A among swimmers associated with fecal contamination of pool water (112) indicated that if a prescribed free chlorine level of 0.3–0.5 ppm existed, it failed to inactivate HAV. In a comparative study of the efficacy of a number of chemical disinfectants to inactivate HAV on contaminated surfaces, only sodium hypochlorite, 2% glutaraldehyde, and quaternary ammonia compound (QAC) with 23% HCl were effective in reducing the titer of HAV by more than 10^4 (103). The current recommendation for general purpose disinfection using a 1:11 dilution of 6% sodium hypochlorite (household bleach is 5% in the United States) or 5000 µg/mL free chlorine with 1 min contact time should inactivate HAV in most situations (103).

3. Immunization to Lower Risk of Infection

The ability to provide long-lasting immunity against HAV infection through active immunization has the potential to further lower the already low risk of foodborne infection found in the United States. However, a decrease in risk of foodborne HAV infection would not be expected to result from the specific vaccination of food handlers or others involved in the food harvest or processing, but rather from a general increase in the proportion of persons immune to HAV infection due to widespread immunization. Only one situation exists in which hepatitis A immunization is recommended to protect against foodborne transmission and that is for persons traveling to countries with a high endemicity of infection, where there is a high risk of foodborne infection (83).

a. Immunizing Agents

Passive transfer of anti-HAV through the administration of various therapeutic preparations of IG has been shown to prevent infection if given prior to exposure to HAV (83,116,322). In addition, when given within 1–2 weeks following exposure, IG has been shown to prevent the development of hepatitis A in a high proportion of children and adults (116,322). With postexposure prophylaxis, IG alters the expression of clinical illness (hepatitis A) but does not always prevent HAV infection, thus providing "passive-active" immunization (38,323). Although IG has been used widely for both preexposure and postexposure protection, its practical effectiveness is limited by the half-life of IgG. For persons living or working in areas endemic for HAV infection, compliance with multiple injections of IG has been poor and has resulted in high rates of infection (324,325). The availability of hepatitis A vaccine generally limits use of IG to postexposure immunization, except in a limited number of preexposure situations (see below).

Postexposure prophylaxis has been highly effective in preventing secondary cases following the early identification of outbreaks of hepatitis A in day care centers or in other relatively closed populations. Because of the late identification of the source, its use following foodborne outbreaks has been generally limited to the prevention of secondary cases. The widespread use of IG has generally not stopped extended community-wide epidemics of hepatitis A.

Inactivated hepatitis A vaccines, prepared by methods similar to inactivated polio vaccine (326), have proven to be highly immunogenic in adults and children. However, passively acquired maternal anti-HAV blunts the immune response to vaccine among infants born to antibody-positive mothers (327–330, CDC, unpublished data). For persons of all ages, antibody levels after vaccination may be below that detected with commercially available assays for anti-HAV. However, these vaccines have been shown to protect against disease and infection in animal studies and placebo-controlled clinical trials (19,327,331–334), including after a single vaccine dose (19). The duration of protective efficacy and whether booster doses are needed for these vaccines is not known. However, estimates of antibody persistence derived from kinetic models indicate that the protective level of anti-HAV could be present for at least 20 years after a primary immunization series (335). Currently, vaccines produced by two manufacturers have been licensed in the United States.

Although several live, attenuated vaccines have been developed (331,336,337), and some have been shown to have good immunogenicity when injected intramuscularly (338–340), their efficacy has not been evaluated in controlled clinical trials. At this time it is unclear as to whether these vaccines will offer any advantages over the currently licensed inactivated vaccines.

Vaccines produced by recombinant expressed proteins have not been developed commercially. Most recombinant expressed HAV capsid proteins have failed to elicit a protective antibody response (213, 341, Margolis, unpublished data), probably due to the conformational nature of the epitopes that induce a protective antibody response. However, tamarins immunized with vaccinia-vectored capsid proteins were protected upon challenge with wild-type virus (342), indicating that a conformationally correct epitope can be produced.

b. Passive Immunization

Passive immunization with IG is indicated as soon as possible for persons exposed to HAV (88). However, IG does not provide protection when given more than 2 weeks after exposure (88,343). Persons who should receive Ig following exposure to hepatitis A include household and sexual contacts of an infected person; staff and attendees of a day care center where one or more cases are diagnosed, or where cases are recognized in two or more households of attendees of the center; individuals in institutional settings if epidemiological investigation shows HAV transmission between individuals; and any common source exposures if the exposure is recognized in time. If persons have received one dose of hepatitis A vaccine at least 1 month prior to HAV exposure, Ig is not needed (88).

When a foodborne outbreak occurs or when hepatitis A is recognized in a food handler, it is often too late to give Ig to prevent disease in those who have been exposed. When hepatitis A is recognized in a food handler, transmission to patrons in uncommon (88,107). In this situation, use of Ig is based on the hygiene of the food handler, the type of food handled, and whether patrons can be located and treated within 2 weeks of exposure (88). It is recommended that other food handlers in the establishment be given Ig because of the secondary cases that are generally identified in this situation (88).

Rarely does transmission of HAV occur among contacts of persons with hepatitis A in the classroom setting, in offices or factories, or in the hospital (88). The use of Ig is not routinely indicated in these situations unless warranted by the epidemiological circumstances.

Immune globulin is not recommended for preexposure protection except for travelers in the following situations (88): (a) for children <2 years of age because the vaccine is not licensed for this age groups; (b) persons traveling to a country with a high endemicity of HAV infection who were not vaccinated at least 2–4 weeks prior to departure should receive both vaccine and IG; and (c) a person who requires preexposure protection and in whom vaccination is contraindicated.

c. Active Immunization

Hepatitis A vaccination is recommended for the limited groups at high risk of infection, for persons with chronic liver disease in whom infection is known to produce an adverse outcome, and for persons living in those communities in the United States with the highest rates of disease (88). The ultimate goal of hepatitis A vaccination is to eliminate transmission of infection and HAV (88). Recommendations have evolved based on the epidemiology of disease patterns and transmission, feasibility of implementation, prevention effectiveness, and vaccine performance.

Vaccination is recommended for persons at high risk of infection, including travelers to countries with high endemic rates of infection, men who have sex with men, and intravenous drug abusers. Routine vaccination of food handlers is not recommended since they do not have a higher rate of infection than the general population. In addition, food handlers have not been shown to contribute significantly to infection within community-wide outbreaks and routine vaccination of this group has not been shown cost-effective (CDC, unpublished results). Vaccination is also recommended for persons with chronic liver disease because they are more likely to have acute liver failure if they have hepatitis A (88).

However, vaccination of persons in risk groups is unlikely to lower national disease incidence because most hepatitis A occurs in community-wide outbreaks. Children account for at least one-third of cases and are often a major source of infection, and routine vaccination of children is the best means to control this disease (88). However, because hepatitis A vaccine is not immunogenic in all infants, immunization of children must wait until after the second year of life (88).

Routine vaccination of children in communities with high rates of HAV infection and recurrent outbreaks of hepatitis A has been shown to prevent new epidemics and hastened the control of an ongoing epidemic (19). In communities with intermediate rates of HAV infection, widespread vaccination has been used to attempt to control epidemics with limited success (344).

In the United States, the highest rates of hepatitis A occur primarily in western states (88). The success of sustained routine childhood vaccination in preventing hepatitis in communities with high rates of infection indicates the feasibility and effectiveness of such a prevention strategy. In addition, routine vaccination of children and adolescents in communities with rates of infection higher than the average for the United States has been shown to be cost-effective (88). Recently, the Advisory Committee on Immunization Practices recommended routine vaccination of children living in states and communities with reported rates of hepatitis A that exceeded the national average rate of infection (10/100,000 population) over the 10-year period 1986–1996 (88). These states contribute 50% of reported cases of hepatitis A in the United States, and routine vaccination of children should ultimately result in a reduction of national disease incidence.

III. HEPATITIS E

A. Epidemiology and Clinical Disease

Studies indicate that a substantial proportion of acute viral hepatitis that occurs in young adults in Asia, the Indian subcontinent, and possibly Africa is caused by HEV (345–350).

The disease has been shown to occur in both epidemic and sporadic forms and is primarily associated with the ingestion of fecally contaminated drinking water (351,352). In 1955, an epidemic occurred in New Delhi, India, with 29,000 reported cases of icteric hepatitis following widespread fecal contamination of the city's drinking water (130). A similar epidemic occurred in 1975–1976 in Ahmedabad City, India, also associated with fecally contaminated water (353). At the time they occurred, both outbreaks were thought to be caused by HAV. However, retrospective serological analysis showed that acute phase markers of infection with either HAV or HBV were not present, indicating the presence of a new agent of enterically transmitted hepatitis (345).

In Asia, epidemics and acute cases of HEV have been reported in the republics of Kirghiz (354) and Tajikistan (355), northern India (356–358), Nepal (359–361), Burma (362), Pakistan (361,363), China (364,365), Hong Kong (366), and Vietnam (352). In Africa, hepatitis E has been reported in Algeria (367), Egypt (368), Ivory Coast (369,370), eastern Sudan (370), Ethiopia (371), and Senegal (372). In the Americas, outbreaks have been reported in two rural villages in Mexico (373), and cases have been imported to the United States from areas where HEV is known to be endemic, but there has been no evidence of spread (361). In the United States, one patient with no history of travel to endemic areas has been shown to have hepatitis E infection (374). Nucleotide sequence analysis of HEV RNA from this individual demonstrated significant divergence from other human HEV isolates and may represent a new strain of HEV along with the U.S. swine isolate (34,35) (see below).

The precise worldwide distribution of HEV infections has not been determined due to the lack of well-standardized and readily available serodiagnostic tests. However, the geographic distribution in countries where outbreaks and sporadic cases have been reported indicates that this disease may be endemic in developing countries. The endemicity of this infection is further suggested by the occurrence of 7- to 10-year epidemic cycles that have been documented in central Asia (375,376), similar to that observed for hepatitis A. In areas where long-term surveillance has been conducted, outbreaks of hepatitis E have shown a pronounced seasonel distribution most often associated with the local rainy season (130,352,353). The rapid onset of these epidemics has suggested a common source such as contaminated water (351,376–379).

The age-specific prevalence of HEV infection in countries where the disease is endemic indicated the highest rates of infection/disease in older children and young adults (359,380–383), an epidemiological feature that was inconsistent with the high rates of other enteric infections among children in these countries. However, studies have shown that hepatitis E occurs in children (366,368,382,383), although the true prevalence of infection in this age group is not known.

Among persons with symptomatic infection, the male-to-female ratio has ranged from 1:1 to 3:1 (130,346,359,375). However, epidemiological studies have shown that sociological factors contributed to the predominance of men in some outbreaks, including the temporary residence of men seeking work (375). In outbreaks in Central Asia, Africa, and Mexico there has not been a predominance of males (373,384,385,387). The mean age for persons with symptomatic infection has been 29 years, and the highest age-specific attack rates have been in the 20- to 30-year age group (358,362,375,387,388). Clinical attack rates among adults have ranged from 3% to 10%, whereas among children less than 15 years of age the attack rates have ranged from 0.2% to 3% (346,370,373,375,387). A high case fatality rate among pregnant women has been a consistent feature of hepatitis E. Mortality has ranged from 17% to 33% in pregnant women with hepatitis E, while case fatality rates among nonpregnant women have been the same as observed for men

(375,388,389). The highest rates of fulminant hepatitis and death have occurred during the 20th to 32nd week of gestation, as well as during labor (M. Favorov, personal communication).

The incubation period for hepatitis E has ranged from 22 to 60 days with a mode of 40 days (388,390). Symptoms have included nausea (46–85%), dark urine (92–100%), abdominal pain (41–85%), vomiting (50%), pruritus (13–55%), joint pain (3–81%), rash (3%), and diarrhea (3%). Fever is present in approximately 20%, and almost all patients have hepatomegaly (346,358,377). Among patients with fulminant hepatitis, renal failure secondary to hemoglobinuria or a hemorrhagic diathesis has been the most common cause of death (391).

Transmission of HEV in several nonhuman primate models has provided information concerning the relationship of virological, immunological, clinical, and histopathological events during infection. The most widely used model is the cynomolgus macaque (390,392,393), although experimental infections have been produced in African green monkeys (394), tamarins (392), squirrel monkeys (395), and chimpanzees (396). In all animal models, inoculation has been done intravenously and has been followed by liver enzyme elevations after 24–38 days (see Fig. 1B). Hepatitis E virus has been detected in the stool of humans and experimentally infected animals approximately 2 weeks prior to the onset of liver enzyme elevations, and virus shedding appears to have ended when the level of liver enzymes returned to normal (26,397). In one study, the infectious dose per gram of feces was 10^6–10^7 when inoculated intravenously in cynomolgus macaques and the dose responsible for infection corresponded to the endpoint detection by RT-PCR (393). An antigen associated with HEV (HEV ag) appears in the cytoplasm of hepatocytes of experimentally infected animals as early as 10 days post inoculation. The presentation of HEV ag in hepatocytes may persist for as long as 21 days and most often preceded the period of ALT elevations. In some cases, however, the presentation of antigen and alanine aminotransferase elevations were coincident (45,397).

An IgM antibody response to HEV begins during the early phase of the infection, is generally detectable at the time of onset of symptoms or enzyme elevations, and persists for 5–6 months after the acute infection (366,398,399). IgG antibody to HEV has generally been detectable almost simultaneously with IgM antibody (366,398,399), but its duration of persistence is not known. It has been suggested that IgG is relatively short lived (368); however, IgG anti-HEV activity has been found for at least a year after acute infection in most patients (399,400), and for at least 4 years in experimentally infected primates. The long-term persistence of IgG anti-HEV is further demonstrated by its detection among 85% of adults 10 years after they were jaundiced during an HEV epidemic and in asymptomatic persons living in areas where HEV infection has been shown to be endemic.

Pathomorphological changes observed in patients with hepatitis E include colestasis with glandlike transformation of bile ducts with preservation of the lobular structure, portal inflammation, ballooning degeneration, Kupffer cell hyperplasia, and liver cell necrosis that has varied from single-cell degeneration to bridging necrosis (346,358). Follow-up studies of persons with hepatitis E indicate that chronic liver disease has not been a consequence of this infection.

B. Reservoirs of Hepatitis E Virus

Fecally contaminated water has been identified as the source of HEV transmission in most outbreaks, and simple means of water purification such as the addition of chlorine or boiling has interrupted disease transmission. Transmission occasionally occurs through person-to-person contact, as shown in case-control studies and by secondary attack rates ranging from 0.7% to 8.0% in households of cases (362,401,402). Several reports have suggested that hepatitis E may be foodborne (403–406). However, these studies have not had adequate control groups, have lacked serological testing, and have not determined whether HEV can be transmitted by food. Sewage contamination of shellfish areas could be a source of infection, since HEV has been detected in sewage (407,408); however, this has not been demonstrated to be a source of transmission, nor has HEV been isolated from shellfish.

The reservoir of HEV between epidemics is unknown. One possible reservoir for HEV is serial transmission among susceptible individuals by occasional failures of water plants or sporadic contamination of individual untreated water supplies. In many countries in which hepatitis E outbreaks have been reported, sporadic cases of hepatitis E account for a substantial proportion of acute viral hepatitis, and these sporadic infections may maintain transmission in endemic regions during interepidemic periods. Contamination could also occur when migrant populations introduce the virus into an area. Although epidemics of HEV have always been associated with contaminated water, the origin of the fecal contamination of the water has frequently not been determined. Hepatitis E virus RNA has been detected in wastewater by RT-PCR (407) and in a community water supply in India (409). In one epidemic associated with river and well water in Somalia, the attack rate was higher in those using river water than well water, although cases occurred in association with both (378).

A second possible reservoir of HEV could be animals. In endemic regions, HEV has been detected in feces of wild-caught pigs by RT-PCR and anti-HEV has been detected in serum of pigs, cattle, and sheep (26,31–33, Favorov, unpublished data). Contamination of natural water sources by regional domestic and/or free-roaming animals is conceivable, especially during rainy seasons and floods. However, cross-species transmission of HEV has not been demonstrated.

Recently, a novel virus, designated swine hepatitis E virus, was identified in pigs from the midwestern United States (34). Swine HEV cross-reacts with antibody to the human HEV capsid antigen. The entire ORFs 2 and 3 were amplified by reverse transcription-PCR from sera of naturally infected pigs. The putative capsid gene (ORF-2) of swine HEV shared about 79–80% sequence identity at the nucleotide level and 90–92% identity at the amino acid level with human HEV strains. The small ORF-3 of swine HEV had 83–85% nucleotide sequence identity and 77–82% amino acid identity with human HEV strains. Phylogenetic analysis showed that swine HEV is closely related to, but distinct from human HEV strains (34). This discovery may explain the detection of anti-HEV antibody in regions where the disease is not endemic. In Baltimore, Maryland (USA), 900 individuals were tested for anti-HEV by EIA, and anti-HEV was found in an unexpectedly high percentage of homosexual men (15.9%) and injection drug users (23.0%). However, anti-HEV was present in a similar proportion of blood donors (21.3%) ($p > 0.05$) (410). Hepatitis E virus has been isolated by RT-PCR from acute hepatitis patients in Taiwan and shown to be different from HEV isolated from other parts of the world, therefore constituting a novel group (411). Since the genomic region examined is not the same as for the U.S. swine HEV (34), no comparison of the two has been made. Interestingly, pork is the primary source of meat protein in Taiwan. Sequence analysis of a U.S. isolate of HEV (35,374) has shown that the U.S. human and swine HEV compose a separate group distinct from the Asian and Mexican strains.

Further research is needed to clarify the potential relationships of animal HEV to human HEV in developing and developed countries. The transmission of swine HEV to humans could be of concern in developed countries in xenotransplantation (34). In addition, the low prevalence of antibody to HEV may indicate exposure and/or infection without clinical disease in individuals in developed countries. More studies of animal populations in developing countries are needed to investigate the potential for an animal reservoir of HEV that is the source of human disease.

C. Hepatitis E Virus Characterization

Hepatitis E virus is a 32- to 34-nm virus that appears to be unstable upon exposure to high concentrations of salt. Pelleting from stool suspensions frequently results in complete loss of detectable virus by immunoelectron microscopy (IEM), while banding in linear, preformed sucrose gradients has been found to yield virus suitable for IEM studies (394). The computed sedimentation coefficient was approximately 183S, and the buoyant density as determined in a potassium tartrate/glycerol gradient was 1.29 g/mL. Hepatitis E virus that sediments at 165S has been presumed to be defective (394). Although HEV particles were labile in certain types of laboratory experiments, recent experiments demonstrated infectious HEV remained after drying and storage up to 28 days (McCaustland et al., unpublished results).

The absence of a cell culture system for virus propagation required the use of either nonhuman primate or human material for characterization. Since sufficient quantities of virus could not be obtained from fecal specimens (392), bile that contained large quantities of HEV became the starting material for molecular cloning (412–414). Molecular cloning was accomplished by differential hybridization to heterogeneous cDNA obtained from the bile of infected and uninfected animals using "plus–minus" screening after expression in λgt10 (415,416), and immunoscreening of expressed antigens from cDNA libraries using convalescent antibodies from an outbreak in Mexico (401,417,418). Virus-specific clones have been shown to hybridize specifically to HEV-infected source cDNA (415), to HEV-infected liver RNA (416), and to cDNAs prepared from HEV-infected stools collected from outbreak-related cases in Borneo, Mexico, Pakistan, Somalia, and the former USSR (419).

Although a system for propagation of HEV has been described with evidence for replication based on HEV RNA detection by PCR, the yield of HEV particles is inadequate for biochemical characterization or virus production (28). This system has been used to demonstrate that antibodies elicited during HEV infection have broad HEV neutralizing activity (420). Other culture systems have been reported, but none have yielded sustained production of virus particles or a system adequate for characterization of replication in vitro (29,30). Production of HEV in vitro has been demonstrated by cultivation of in vivo–infected primary macaque hepatocytes (421), but this system is only useful for limited replication studies.

The genomic organization of HEV is substantially different from picornaviruses; the HEV genome codes for structural and nonstructural proteins through discontinuous, partially overlapping ORFs (Fig. 5). The HEV genome has been shown to be a single-stranded positive sense 3′-polyadenylated (poly-A) RNA molecule containing three ORFs. The nonstructural genes are located at the 5′ end and the structural genes at the 3′ end of the molecule. ORF-1 is approximately 5 kb in length and appears to contain several consensus sequences, including one associated with an NTP-binding, and another with a helicase-

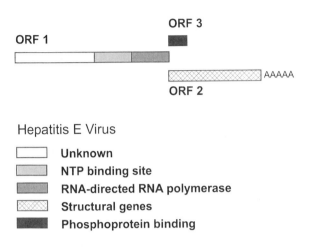

Figure 5   Open reading frames of hepatitis E virus.

like function that is located 5′ to the putative RNA-dependent RNA polymerase (422). ORF-2 is approximately 2 kb in size and contains the major structural proteins. A third ORF (ORF-3) of 328 bp overlaps ORF-1 by one nucleotide and ORF-2 by 328 nucleotides, and encodes a phosphoprotein that is cytoskeleton-associated (423). Studies suggested that HEV could belong to a larger family of single-stranded, poly-A RNA viruses that possess three ORFs and at least two subgenomic transcripts ranging in size from 2 to 4 kb (424). However, HEV could be classified in the Caliciviridae based on morphology, absence of a lipid envelope, nucleic acid and general genome organization (425). Official classification has not been made at this time.

Geographically distinct isolates of HEV have been demonstrated, although definitive comparisons cannot be made across all isolates since the same genomic regions have not been analyzed for all. The Mexican isolate was shown to vary from the Asian isolates in the structural as well as nonstructural regions (417,426). More recently, geographic isolates from all epidemic and endemic areas have been analyzed and compared with each other to an extent to establish geographic origin of divergence. Isolates from Nepal have been shown to have greater identity (>96–95%) with the Indian and Burmese isolates and less with the African (>94%) and the Chinese isolates (>91%), although all of these differ from the Mexican genotype (427,428). African strains have been shown to be distinct from but more like Asian strains than the Mexican strain, whereas Central Asian isolates were almost completely identical (429,430). Genotyping of HEV by restriction endonuclease analysis suggested that there are three, possibly four Asian subgenotypes (431,432). Isolation of a novel HEV from swine in the United States and comparison of that sequence with a U.S. human isolate has demonstrated a genotype unique among all previously described and continued to establish the geographic identity of isolates (34,35). Analysis of genetic variability within and between outbreaks in India demonstrated significantly higher variation between than within (433). Further genomic analysis of all geographic isolates and outbreaks should answer significant remaining questions about the natural history of hepatitis E disease.

D. Detection of Hepatitis E Virus Infection

Early detection of antibody to HEV (anti-HEV) was accomplished by IEM using virus isolates from geographically distinct regions of the world (434). In 1990, isolation of a partial cDNA clone from the virus responsible for ET-NANB hepatitis was reported (415). The authors designated the newly identified agent as HEV. This clone, derived from the Burmese HEV isolate, hybridized with cDNA isolates from five other outbreaks in Asia, Africa, and Mexico. These findings strongly suggested that a single agent was responsible for the majority of ETNANB hepatitis disease seen worldwide (391, 399, 417, Favorov, unpublished data).

Recombinant expressed proteins from ORF-2 and ORF-3 have been used in the development of a specific immunoblot (IB) and an EIA for the detection of anti-HEV activity (366,368,382,399). Studies regarding immunoreactivity of synthetic peptides prepared from ORF-2 and ORF-3 have been published elsewhere (435–438).

Several EIAs have been developed using recombinant expressed antigens for the detection of IgG or IgM activity (366,368,439). One assay utilized four recombinant antigens representing two distinct antigenic domains from two different HEV strains (439). Both isotypes were detected in seven of eight pedigreed specimens from known outbreaks of HEV in Mexico, Burma, Somalia, and Pakistan.

An EIA utilizing four recombinant antigens was used to diagnose acute sporadic hepatitis E in rural Egyptian children (368,417). Of 36 children with non-A, non-B hepatitis, 15 (42%) were positive for IgG anti-HEV, whereas of 20 healthy children, only 5 (25%) were positive for IgG anti-HEV (417). When tested for IgM anti-HEV, 6 of the 15 IgG anti-HEV-positive children, but none of the controls, were positive and were regarded as having an acute HEV infection. An EIA composed of two recombinant antigens from the C-terminal region of ORF-3 of the Mexican and Burmese strains, detected IgG and IgM anti-HEV activity in 16.5% and 5.3%, respectively, of patients with acute viral hepatitis in Hong Kong (366). Of 18 patients diagnosed as having acute non-A, non-B, non-C hepatitis, 6 were IgM anti-HEV positive. Several studies of antibody persistence following acute hepatitis E suggest that IgG anti-HEV may disappear in some patients after a year (366,440); however, IgG anti-HEV was found 14 years after an outbreak in Kashmir, India (441).

A diagnostic EIA has been developed using a mosaic protein composed of linear antigenic epitopes from HEV structural proteins (438) and was used in conjunction with a blocking neutralization confirmation test of a mixture of individual synthetic peptides (442). This assay identified most (89%, range 77–100%) patients with acute hepatitis from 10 geographically distinct outbreaks. A subset of patients tested for IgM anti-HEV (97%) were positive (442). Of patients tested 6–8 months after onset of jaundice, 92% were anti-HEV IgG-positive and 40–50% were IgM-positive (442). Similar data were obtained with EIA based on complete ORF-2 protein expression in insect cells (443). In another study, IgM anti-HEV was not detected in any of 33 patients 20 months after acute infection, although IgG was found in all patients at that time (382). At the present time, it is not known as to whether the IgG anti-HEV detected by these assays is a neutralizing antibody.

In seroprevalence studies conducted in HEV-endemic countries, anti-HEV has been detected in ≤5% of children under 10 years of age, increasing to 10–40 among adults 25 years of age (381,444,445). By comparison, the prevalence of antibody to HAV among children 10 years of age in most HEV-endemic countries is >90% (446). These findings suggest that HEV, unlike other enterically transmitted agents, is infrequently transmitted among young children in developing countries. However, until the natural history of long-term anti-HEV persistence is defined, the true frequency of infection cannot be determined. In nonendemic regions anti-HEV was evaluated among 5000 U.S. blood donors by two EIAs. Overall, 1.2–1.4% were seropositive. The concordance among reactive sera by either test was only 27%. In a case-control study, seroreactive persons were more likely than seronegative persons to have traveled to countries in which HEV is endemic (447). Panel evaluation studies have been done to compare different anti-HEV detection tests (447). In several tests results were highly concordant when acute and convalescent sera were tested. However, when 12 different tests were compared in a panel evaluation study, highly discrepant results were found among U.S. blood donors' sera. Anti-HEV seroprevalence data in non-HEV-endemic countries should be interpreted with caution (447).

Reverse transcriptase PCR has been used to detect HEV RNA in experimentally infected animals as well as in humans (426, 448, 449, Sheng-li Bi, unpublished data). In these studies HEV RNA is extracted from stool suspensions prior to RT-PCR amplification, or virus is immunoprecipitated for analysis (449); however, immunocapture, as developed for HAV RT-PCR, has been successful (450,451). Several methods of RNA extraction from stool suspensions and from serum have proven successful for preparation of RNA suitable for RT-PCR (407,431,452). HEV RNA detected in the serum of experimentally infected animals prior to elevation of liver enzymes coincided with virus shedding into the bile and remained positive until the appearance of anti-HEV (426,448). Hepatitis E viral RNA has been detected, using nested PCR, in feces and bile as early as day 6 and 7, respectively, confirming that excretion of virus was an early indicator of infection (426). Viral excretion in feces began on day 6 and ended by day 35 post inoculation. However, three samples taken between days 10 and 15 were repeatedly negative for viral RNA, suggesting a level of viral excretion less than 10^3 infectious particles. In one study of patients with acute hepatitis E, HEV RNA was detected in stools by RT-PCR for up to 52 days after illness onset (453); however, the correlation of HEV RNA detection with infectivity in stools has not been demonstrated.

E. Prevention and Control

The available epidemiological data indicate that HEV infection is primarily transmitted by water that has become contaminated with human (or animal) fecal material, and less so by person-to-person spread. The disease attack rate has been directly correlated to the quantity of contaminated water used by households in one study (351). Thus, measures to prevent and control hepatitis E infections have been directed at improving the quality of the water supply in areas that have experienced outbreaks. If a zoonotic infection exists, prevention will require protection of water supplies from fecal matter of the implicated animals. The relative importance of other modes of transmission, such as food, must be determined through appropriate epidemiological studies that confirm infections through serological and nucleic acid detection methods. Only then can an appropriate comprehensive prevention strategy be developed for the prevention of this infection.

Passive protection through the use of IG has been suggested as a means of preventing HEV infection. Attempts have been made to use IG in various outbreaks and endemic settings; however, no conclusions can be drawn as to the efficacy of passive immunoprophylaxis (356,454). Whether IG has a role in the prevention of HEV infection must be determined through appropriately designed clinical trials that evaluate the efficacy of IG having known titers of anti-HEV. Studies indicate that antibodies to recombinant proteins from ORF-2 provide protection from infection in the cynomolgus macaque model (36,455,456). Additional vaccine studies, as well as studies of the natural history and epidemiology of HEV infection, will be required to determine if active immunization might effectively prevent this infection and associated liver disease.

REFERENCES

1. U.S. Public Health Service (1951). Viral Hepatitis: A New Disease, No. 5, Communicable Disease Center, Atlanta, GA.
2. Krugman, S., Giles, J. P., and Hammond, J. (1967). Infectious hepatitis: evidence for two distinctive clinical, epidemiological and immunological types of infection, J. Am. Med. Assoc., 200: 365.
3. Shapiro, C. N., Shaw, F. E., Mendel, E. J., and Hadler, S. C. (1991). Epidemiology of hepatitis A in the United States, Viral Hepatitis and Liver Disease (F. B. Hollinger, S. M. Lemon, and H. S. Margolis, eds.), Williams and Wilkins, Baltimore, p.214.
4. Hadler, S. C. (1991). Global impact of hepatitis A virus infection changing patterns, Viral Hepatitis and Liver Disease (F. B. Hollinger, S. M. Lemon, and H. S. Margolis, eds.), Williams and Wilkins, Baltimore, p.14.
5. Niu, M. T., Polish, L. B., Robertson, B. H., Khanna, B., Woodruff, B. A., Shapiro, C. N., Miller, M. A., Smith, J. D., Gedrose, J. K., Alter, M. J., and Margolis, H. S. (1992). A multistate outbreak of hepatitis A associated with frozen strawberries, J. Infect. Dis., 166: 518.
6. Desenclos, J. A., Klontz, K. C., Wilder, M. H., Nainan, O., V., Margolis, H. S., and Gunn, R. A. (1991). A multistate outbreak of hepatitis A caused by the consumption of raw oysters, Am. J. Pub. Health, 81: 1268.
7. Rosenblum, L. S., Villarino, M. E., Nainan, O., V., Melish, M. E., Hadler, S. C., Pinsky, P. P., Jarvis, W. R., Ott, C. E., and Margolis, H. S. (1991). Hepatitis A outbreak in a neonatal intensive care unit: risk factors for transmission and evidence of prolonged viral excretion among preterm infants, J. Infect. Dis., 164: 476.
8. Jansen, R. W., Siegl, G., and Lemon, S. M. (1990). Molecular epidemiology of human hepatitis A virus defined by an antigen-capture polymerase chain reaction method, Proc. Natl. Acad. Sci. USA, 87: 2867.
9. Boswami, B. B., Burkhardt, W. I. I. I., and Cebula, T. A. (1997). Identification of genetic variants of hepatitis A virus, J. Virol. Meth., 65: 95.
10. Pebody, R. G., Leino, T., Ruutu, P., Kinnunen, L., Davidkin, I., Nohynek, H., and Leinikki, P. (1998). Foodborne outbreaks of hepatitis A in a low endemic country: an emerging problem?, Epidemiol. Infect., 120: 55.
11. De Serres, G., Cromeans, T. L., Levesque, B., Brassard, N., Barthe, C., Dionne, M., Prud'home, H., Paradis, D., Shapiro, C. N., Nainan, O. V., and Margolis, H. S. (1999). Molecular confirmation of hepatitis virus from well water: epidemiology and public health implications, J. Infect. Dis., 179: 37.
12. Hutin, Y. J. F., Pool, V., Cramer, E. H., Nainan, O. V., Weth, J., Williams, I. T., Goldstein, S. T., Gensheimer, K. F., Bell, B. P., Shapiro, C. N., Alter, M. J., and Margolis, H. S. (1999). A multistate foodborne outbreak of hepatitis A, N. Engl. J. Med., 340: 595.
13. Deng, M. Y., Day, S., and Cliver, D. O. (1994). Detection of hepatitis A virus in environmental samples by antigen-capture PCR, Appl. Environ. Microbiol., 60: 1927.
14. Schwab, K. J., De Leon, R., and Sobsey, M. D. (1995). Concentration and purification of beef extract mock eluates from water samples for the detection of enteroviruses, hepatitis A virus, and Norwalk virus by reverse transcription-PCR, Appl. Environ. Microbiol., 61: 531.
15. Shieh, Y.-S. C., Wait, D., Tai, L., and Sobsey, M. D. (1995). Methods to remove inhibitors

in sewage and other fecal wastes for enterovirus detection by the polymerase chain reaction, J. Virol. Meth., 54: 51.

16. Jaykus, L.-A., De Leon, R., and Sobsey, M. D. (1996). A virion concentration method for detection of human enteric viruses in oysters by PCR and oligoprobe hybridization, Appl. Environ. Microbiol., 62: 2074.

17. Nainan, O. V., Cromeans, T. L., and Margolis, H. S. (1996). Sequence-specific, single-primer amplification and detection of PCR products for identification of hepatitis viruses, J. Virol. Meth., 61: 127.

18. Cromeans, T. L., Nainan, O. V., and Margolis, H. S. (1997). Detection of hepatitis A virus RNA in oyster meat, Appl. Environ. Microbiol., 63: 2460.

19. Werzeberger, A., Mensch, B., Kuter, B., Brown, L., Lewis, J., Sitrin, R., Miller, W., Shouval, D., Wiens, B., Calandra, G., Ryan, J., Provost, P., and Nalin, D. (1992). A controlled trial of formalin-inactivated hepatitis A vaccine in healthy children, N. Engl. J. Med., 327: 453.

20. Margolis, H. S., Hadler, S. C., Shapiro, C. N., and Alter, M. J. (1991). Immunization strategies for the control of hepatitis A in the United States, Viral Hepatitis and Liver Disease (F. B. Hollinger, S. M. Lemon, and H. S. Margolis, eds.), Williams and Wilkins, Baltimore, p.724.

21. Hoke, C. H., Binn, L. N., Egan, J. E., DeFraites, R. F., MacArthy, P. O., Innis, B. L., Eckels, K. H., Dubois, D., D'Hondt, E., Sjogren, K. H., Rice, R., Sodoff, J. C., and Bancroft, W. H. (1992). Hepatitis A in the U.S. Army: epidemiology and vaccine development, Vaccine, 10: 575.

22. Funkhouser, A. W., Purcell, R. H., D'Hondt, E., and Emerson, S. U. (1994). Attenuated hepatitis A virus: genetic determinants of adaptation to growth in MRC-5 cells, J. Virol., 68: 148.

23. Funkhouser, A. W., Raychaudhuri, G., Purcell, R. H., Govindarajan, S., Elkins, R., and Emerson, S. U. (1996). Progress toward the development of a genetically engineered attenuated Hepatitis A virus vaccine, J. Virol., 70: 7948.

24. Reyes, G. R., Purdy, M. A., Kim, J. P., Luk, K.-C., Young, LaV. M., Fry, K. E., and Bradley, D. W. (1990). Molecular cloning from the virus responsible for enterically transmitted non-A, non-B hepatitis, Science, 247: 1335.

25. Purcell, R. H. and Ticehurst, J. (1988). Enterically transmitted non-A, non-B hepatitis: epidemiology and clinical characteristics, Viral Hepatitis and Liver Disease (A. J. Zuckerman, eds.), Alan R Liss, New York, p. 131.

26. Ticehurst, J. (1991). Identification and characterization of hepatitis E virus, Viral Hepatitis and Liver Disease (F. B. Hollinger, S. M. Lemon, and H. S. Margolis, eds.), Williams and Wilkins, Baltimore, p. 501.

27. Huang, R. T., Li, D. R., Wei, J., Huang, X. R., Yuan, X. T., and Tian, X. (1992). Isolation and identification of hepatitis E virus in Xinjiang, China, J. Gen. Virol., 73: 1143.

28. Meng, J., Dubreuil, P., and Pillot, J. (1997). A new PCR-based seroneutralization assay in cell culture for diagnosis of hepatitis E, J. Clin. Microbiol., 35: 1373.

29. Kazachkov, Y. A., Balayan, M. S., Ivannikova, T. A., Panina, L. I., Orlova, T. M., Zamyatina, N. A., and Kusov, Y. Y. (1992). Hepatitis E virus in cultivated cells, Arch. Virol., 127: 399.

30. Huang, R., Nakazono, N., Ishii, K., Li, D., Kawamata, O., Kawaguchi, R., and Tsukada, Y. (1995). I. Hepatitis E virus (87A strain) propagated in A549 cells, J. Med. Virol., 47: 299.

31. Balayan, M. S., Usmanov, R. K., Zamyatina, N. A., Djumalieva, D. I., and Karas, F. R. (1990). Brief report: experimental hepatitis E infection in domestic pigs, J. Med. Virol., 32: 58.

32. Usmanov, R. K., Balayan, M. S., Dvoinikova, O. V., Alymbaevam, D. B., Zamiatina, N. A., Kazachkov, IuA., and Belov, V. I. (1994). An experimental infection in lambs by the hepatitis E virus, Vopr. Virusol., 39: 165.

33. Clayson, E. T., Innis, B. L., Myint, K. S. A., Narupiti, S., Vaughn, D. W., Biri, S., Ranabhat, P., and Shrestha, M. P. (1995). Detection of hepatitis E virus infections among domestic swine in the Kathmandu valley of Nepal, Am. J. Trop. Med. Hyg., 53: 228.

34. Meng, X.-J., Purcell, R. H., Halbur, P. G., Lehman, J. R., Webb, D. M., Tsareva, T. S., Haynes, J. S., Thacker, B. J., and Emerson, S. U. (1997). A novel virus in swine is closely related to the human hepatitis E virus, Proc. Natl. Acad. Sci. USA, 94: 9860.

35. Schlauder, G. G., Dawson, G. J., Erker, J. C., Kwo, P. Y., Knigge, M. F., Smalley, D. L., Rosenblatt, J. E., Desai, S. M., and Mushahwar, I. K. (1998). The sequence and phylogenetic analysis of a novel hepatitis E virus isolated form a patient with acute hepatitis reported in the United States, J. Gen. Virol., 79: 447.

36. Tsarev, S. A., Tsareva, T. S., Emerson, S. U., Govindarajan, S., Shapiro, M., Gerin, J. L., and Purcell, R. H. (1997). Recombinant vaccine against hepatitis E: dose response and protection against heterologous challenge, Vaccine, 15: 1832.

37. Hadler, S. C. and McFarland, L. (1986). Hepatitis in day care centers: epidemiology and prevention, Rev. Infect. Dis., 8: 548.

38. Lednar, W. M., Lemon, S. M., Kirkpatrick, J. W., Redfield, R. R., Fields, M. L., and Kelly, P. W. (1985). Frequency of illness associated with epidemic hepatitis A virus infection in adults, Am. J. Epidemiol., 122: 226.

39. Sjogren, M. H., Tanno, H., Fay, O., Sileoni, S., Cohen, B. D., Burke, D. S., and Feighny, R. J. (1987). Hepatitis A virus in stool during clinical relapse, Ann. Intern. Med., 106: 221.

40. Glikson, M., Galun, E., Oren, R., Tur-Kaspa, R., and Shouval, D. (1992). Relapsing hepatitis A, Medicine, 71: 14.

41. Inman, R. D., Hodge, M., Johnston, M. E., Wright, J., and Heathcote, J. (1986). Arthritis, vasculitis and cryoglobulinemia associated with relapsing hepatitis A virus infection, Ann. Intern. Med., 105: 700.

42. Krugman, S., Ward, R., and Giles, W. P. (1962). The natural history of infectious hepatitis, Am. J. Med., 32: 717.

43. Dienstag, J. L., Feinstone, S. M., Purcell, R. H., Hoofnagle, J. H., Barker, L. F., London, W. T., Hopper, H., Peterson, J. M., and Kapikian, A. Z. (1975). Experimental infection of chimpanzees with hepatitis A virus, J. Infect. Dis., 132: 532.

44. Margolis, H. S., Nainan, O., V, Krawczynski, K., Bradley, D. W., Ebert, J. W., Spelbring, J., Fields, H. A., and Maynard, J. E. (1988). Appearance of immune complexes during experimental hepatitis A infection in chimpanzees, J. Med. Virol., 26: 315.

45. Krawczynski, K. and Bradley, D. W. (1989). Enterically transmitted non-A, non-B hepatitis: identification of virus-associated antigen in experimentally infected cynomolgus macaques, J. Infect. Dis., 159: 1042.

46. Carl, M., Kantor, P. J., Webster, H. M., Fields, H. A., and Maynard, J. E. (1982). Excretion of hepatitis A virus in the stools of hospital patients, J. Med. Virol., 9: 125.

47. Tassopoulos, N. C., Papaevangelou, G. J., Ticehurst, J. R., and Purcell, R. H. (1986). Fecal excretion of Greek strains of hepatitis A virus in patients with hepatitis A and in experimentally infected chimpanzees, J. Infect. Dis., 154: 231.

48. Chiriaco, P., Gaudalupi, C, Armigliato, M. K., Bortolotti, F., and Realdi, G. (1986). Polyphasic course of hepatitis type A in children, J. Infect. Dis., 153: 378.

49. Yotsuyanagi, H., Koike, K., Yasuda, K., Moriya, K., Shintani, Y., Fujie, J., Kurodawa, K., and Iino, S. (1995). Prolonged fecal excretion of hepatitis A virus in adult patients with hepatitis A as determined by polymerase chain reaction, Hepatology, 24: 10.

50. Locarnini, S. A., Ferris, A. A., Lehmann, N. I., and Gust, I. (1977). The antibody response following hepatitis A infection, Intervirology, 8: 309.

51. Kao, H. W., Aschavai, M., and Redeker, A. G. (1984). The persistence of hepatitis A IgM antibody after acute clinical hepatitis A, Hepatology, 4: 933.

52. Devictor, D., Desplangues, L., Debray, D., Ozier, Y., Dubousset, A., Valayer, J., Houssin,

D., Bernard, O., and Huavlt, G. (1992). Emergency liver transplantation for fulminant liver failure in infants and children, Hepatology, 16: 1156.

53. Fagan, E., Yousef, G., Brahm, J., Garelick, H., Mann, G., Wjolstenholme, A., Portmann, B., Harrison, T., Mowbray, J. F., Mowat, A., Zuckerman, A., and Williams, R. (1990). Persistence of hepatitis A virus in fulminant hepatitis and after liver transplantation, J. Med. Virol., 30: 131.

54. Krawczynski, K. Z., Bradley, D. W., Murphy, B. L., Ebert, J. W., Anderson, T. A., Doto, I. L., Nowoslawski, A., Duermeyer, W., and Maynard, J. E. (1981). Pathogenetic aspects of hepatitis A virus infection in enterally inoculated marmosets, Am. J. Clin. Pathol., 76: 698.

55. Karayiannis, P., Jowett, T., Enticott, M., Moore, D., Pignatelli, M., Brenes, F., Scheuer, P. J., and Thomas, H. C. (1986). Hepatitis A virus replication in tamarins and host immune response in relation to pathogenesis of liver cell damage, J. Med. Virol., 18: 261.

56. Asher, L. V. S., Binn, L. N., Mensing, T. L., Marchwicki, R. H., Vassell, R., and Young, G. D. (1995). Pathogenesis of hepatitis A in orally inoculated owl monkeys (Aotus trivirgatus), J. Med. Virol., 47: 260.

57. Cohen, J. I., Feinstone, S., and Purcell, R. H. (1989). Hepatitis A virus infection in a chimpanzee: duration of viremia and detection of virus in saliva and throat swabs, J. Infect. Dis., 160: 887.

58. Levy, B. S., Fontaine, R. E., Smith, C. A., Brinda, J., Hirman, G., Nelson, D. B., Johnson, P. M., and Larson, O. (1975). A large food-borne outbreak of hepatitis A, J. Am. Med. Assoc., 234: 289.

59. Margolis, H. S., and Nainan, O. V. (1990). Identification of virus components in circulating immune complexes isolated during hepatitis A infection, Hepatology, 11: 31.

60. Schulman, A. N., Dienstag, J. L., Jackson, D. R., Hoofnagle, J. H., Gerety, R. J., Purcell, R. H., and Barker, L. F. (1976). Hepatitis A antigen particles in liver, bile and stool of chimpanzees, J. Infect. Dis., 134: 80.

61. Hollinger, F. B., Khan, N. C., Oefinger, P. E., Yawn, D. H., Schmulan, A. C., Dressman, G. R., and Melnick, J. L. (1983). Posttransfusion hepatitis type A, JAMA, 250: 2313.

62. Mathiesen, L. R., Drucker, J., Lorenz, D., Wagner, J., Gerety, R. J., and Purcell, R. H. (1978). Localization of hepatitis A antigen in marmoset organs during acute infection with hepatitis A virus, J. Infect. Dis., 138: 369.

63. Shimizu, Y. K., Mathiesen, L. R., Lorenz, D., Drucker, J., Feinstone, S. M., Wagner, J. A., and Purcell, R. H. (1978). Localization of hepatitis A antigen in liver tissue by peroxidase-conjugated antibody method: light and electron microscopic studies, J. Immunol., 121: 1671.

64. Shimizu, Y. K., Shidata, T., Benninger, P. R., Sata, M., Setoyama, H., Abe, H., and Tanikawa, K. (1982). Detection of hepatitis A antigen in human liver, Infect. Immun., 36: 320.

65. Asher, L. V. S., Binn, L. N., and Marchwicki, R. H. (1987). Demonstration of hepatitis A virus in cell culture by electron microscopy with immunoperoxidase staining, J. Virol. Meth., 15: 323.

66. Vallbracht, A., Gabriel, P., Maier, K., Hartman, F., Steinhardt, H. J., Muller, C., Wolf, A., Manncke, K. H., and Flehmig, B. (1986). Cell-mediated cytotoxicity in hepatitis A virus infection, Hepatology, 6: 1308.

67. Vallbracht, A., Maier, K., Stierhof, Y.-D., Wiedmann, K. H., Flehmig, B., and Fleischer, B. (1989). Liver-derived cytotoxic T cells in hepatitis A virus infection, J. Infect. Dis., 160: 209.

68. Fleischer, B., Fleischer, S., Maier, K., Wiedmann, K. H., Sacher, M. K., Thaler, H., and Vallbracht, A. (1990). Clonal analysis of infiltrating T lymphocytes in liver tissue in viral hepatitis A, Immunology, 69: 14.

69. Kurane, I., Binn, L. N., Bancroft, W. H., and Ennis, F. A. (1985). Human lymphocyte responses to hepatitis A virus–infected cells: interferon production and lysis of infected cells, J. Immunol., 135: 2140.

70. Maier, K., Gabriel, P., Koscielniak, E., Stierhof, Y. D., Wiedman, K. H., Flehmig, B., and Vallbracht, A. (1988). Human gamma interferon production by cytotoxic T lymphocytes sensitized during hepatitis A virus infection, J. Virol., 62: 3756.

71. Gabriel, P., Vallbracht, A., and Flehmig, B. (1986). Lack of complement-dependent cytolytic antibodies in hepatitis A virus infection, J. Med. Virol., 20: 23.

72. Innis, B. L., Snitbhan, R., Hoke, C. H., Munindhorn, W., and Laorakponyse, T. (1991). The declining transmission of hepatitis A in Thailand, J. Infect. Dis., 163: 989.

73. Stroffolini, T., De Crescenzo, L., Giammanco, A., Intonazzo, V., LaRosa, G., Cascio, A., Sarzana, A., Chiarini, A., and Dardanoni, L. (1990). Changing patterns of hepatitis A virus infection in children in Palermo, Italy, Eur. J. Epidemiol., 6: 84.

74. Hu, M. D., Schenzle, D., Deinhardt, F., and Scheid, R. (1984). Epidemiology of hepatitis A and B in the Shanghai area: prevalence of serum markers, Am. J. Epidemiol., 120: 404.

75. Hsu, H. Y., Chang, M. H., Chen, D. S., Lee, C. Y., and Sung, J. L. (1985). Changing seroepidemiology of hepatitis A virus infection in Taiwan, J. Med. Virol., 17: 297.

76. Reyes, G. R. and Baroudy, B. M. (1991). Molecular biology of non-A, non-B hepatitis agents: hepatitis C and hepatitis E viruses, Adv. Virus Res., 40: 57.

77. Yao, G. (1991). Clinical spectrum and natural history of viral hepatitis A in the 1988 Shanghai epidemic, Viral Hepatitis and Liver Disease (F. B. Hollinger, S. M. Lemon, and H. S. Margolis, eds.), Williams and Wilkins, Baltimore, p. 76.

78. Halliday, M. L., Kang, L.-Y., Zhou, T.-kui, Hu, M.-D., Pan, Q.-C., Fu, T.-Y., Huang, Y.-S., and Hu, S.-L. (1991). An epidemic of hepatitis A attributed to the ingestion of raw clams in Shanghai, China, J. Infect. Dis., 164: 852.

79. Shaw, F. E., Hadler, S. C., Maynard, J. E., Konfortion, P., Law-min, G., Brissonette, G., and Ramphul, J. (1989). Hepatitis A in Mauritius: an apparent transition from endemic to epidemic transmission patterns, Ann. Trop. Med. Parasitol., 83: 179.

80. Shaw, F. E., Shapiro, C. N., Welty, T. K., Dill, W., Reddington, J., and Hadler, S. C. (1990). Hepatitis transmission among the Sioux Indians of South Dakota, Am. J. Pub. Health, 80: 1091.

81. Shaw, F. E., Jr., Sudman, J. H., Smith, S. M., Williams, D. L., Kapell, L. A., Hadler, S. C., Halpin, T. J., and Maynard, J. E. (1986). A community-wide epidemic of hepatitis A in Ohio, Am. J. Epidemiol., 123: 1057.

82. Crusberg, T. C., Burke, W. M., Reynolds, J. T., Morse, L. E., Reilly, J., and Hoffman, A. H. (1978). The reappearance of classical epidemic infectious hepatitis in Worcester, Massachusetts, Am. J. Epidemiol., 107: 545.

84. Alter, M. J., Mares, A., Hadler, S. C., and Maynard, J. E. (1986). The effect of underreporting on the apparent incidence and epidemiology of acute viral hepatitis, Am. J. Epidemiol., 125: 133.

85. Bell, B. P., Shapiro, C. N., Alter, M. J., Moyer, L. A., Judson, F. N., Mottram, K., Fleenor, M., Ryder, P. L., and Margolis, H. S. (1998). The diverse patterns of hepatitis A epidemiology in the United States—implication for vaccination strategies, J. Infect. Dis., 178: 1579.

86. Schade, C. P. and Komorwska, D. (1988). Continuing outbreak of hepatitis A linked with intravenous drug abuse in Multnomah County, Pub. Health Rep., 103: 452.

87. Steffen, R., Rickenbach, M., Wilhelm, U., Helminger, A., and Schar, M. (1987). Health problems after travel to developing countries, J. Infect. Dis., 156: 84.

88. Centers for Disease Control and Prevention. (1996). Prevention of hepatitis A through active or passive immunization: recommendations of the Advisory Committee on Immunization Practices (ACIP), MMWR 48(RR-12): 1.

89. Krushale, D. H. (1970). Application of preventive health measures to curtail chimpanzee-associated infectious hepatitis in handlers, Lab. Anim. Care. 20: 52.

90. Bancroft, W. H. and Lemon, S. M. (1984). Hepatitis A from the military perspective, Hepatitis A (R. J. Gerety, eds.), Academic Press, New York, p.81.

91. Noble, R. C., Kane, M. A., Reeves, S. A., and Roeckel, I. (1984). Posttransfusion hepatitis A in a neonatal intensive care unit, JAMA, 252: 2711.

92. Hollinger, F. B., Khan, N. C., Oefinger, P. E., Yawn, D. H., and Schmulen, A. C. (1983). Posttransfusion hepatitis type A, JAMA, 250:2313.

93. Corey, L., and Holmes, K. K. (1980). Sexual transmission of hepatitis A in homosexual men, N. Engl. J. Med., 302: 435.

94. Kosatsky, T., and Middaugh, J. P. (1986). Linked outbreaks of hepatitis A in homosexual men and in food service patrons and employees, West. J. Med., 144: 307.

95. Centers for Disease Control (1992). Hepatitis A among homosexual men—United States, Canada, and Australia, MMWR, 41: 155.

96. Soucie, J. M., Robertson, B. H., Bell, B. P., McCaustland, K. A., and Evatt, B. L. (1998). Hepatitis A virus infections associated with clotting factor concentrate in the United States, Transfusion, 38: 573.

97. Brugha, R., Heptonstall, J., Farrington, P., Andren, S., Perry, K., and Parry, J. (1998). Risk of hepatitis A infection in sewage workers, Occup. Environ. Med., 55: 567.

98. Hadler, S. C., Webster, H., Erben, J. J., Swanson, J. E., and Maynard J. E. (1980). Hepatitis A in day-care centers: a community-wide assessment, N. Engl. J. Med., 302: 1222.

99. Scholz, E., Heinricy, U., and Flehmig, B. (1989). Acid stability of hepatitis A virus, J. Gen. Virol., 70: 2481.

100. Siegel, G., Weitz, M., and Kronauer, G. (1984). Stability of hepatitis A virus, Intervirology, 22: 218.

101. Parry, J., V. and Mortimer, P. P. (1984). The heat sensitivity of hepatitis A virus determined by a simple tissue culture method, J. Med. Virol., 14: 277.

102. McCaustland, K. A., Bond, W. W., Bradley, D. W., Ebert, J. W., and Maynard, J. E. (1982). Survival of hepatitis A virus in feces after drying and storage for 1 month, J. Clin. Microbiol., 16: 957.

103. Mbithi, J. N., Springthorpe, V. S., Boulet, J. R., and Sattar, S. A. (1992). Survival of hepatitis A virus on human hands and its transfer on contact with animate and inanimate surfaces, J. Clin. Microbiol., 30: 757.

104. Mitchell, J., Presnell, N., Akin, E., Cummins, J., and Liu, O. (1966). Accumulation and elimination of poliovirus by the eastern oyster, Am. J. Epidemiol., 84: 40.

105. Enriquez, R., Frosner, G. G., Hochstein-Mintzel, V., Riedemann, S., and Reinhardt, G. (1992). Accumulation and persistence of hepatitis A virus in mussels, J. Med. Virol., 37: 174.

106. Mele, A., Stroffolini, T., Palumbo, F., Gallo, G., Ragni, P., Balocchini, E., Tosti, M. E., Corona, R., Marzolini, A., Moiraghi, A., and SIEVA Collaborating Group (1997). Incidence of and risk factors for hepatitis A in Italy: public health indications from a 10-year surveillance, J. Hepatol., 26: 743.

107. Carl, M., Francis, D. P., and Maynard, J. P. (1983). Food-borne hepatitis A: recommendation for control, J. Infect. Dis., 148: 1133.

108. Rosenblum, L. S., Mirkin, I. R., Allen, D. T., Safford, S., and Hadler, S. C. (1990). A multifocal outbreak of hepatitis A traced to commercially distributed lettuce, Am. J. Pub. Health, 80: 1075.

109. Bowen, G. S. and McCarthy, M. (1983). Hepatitis A associated with a hardware store water fountain and a contaminated well in Lancaster County, Pennsylvania, 1980, Am. J. Epidemiol., 117: 695.

110. Sobsey, M. D., Oglesbee, S. E., Wait, D. A., and Cuenca, A. I. (1985). Detection of hepatitis A virus (HAV) in drinking water, Water Sci. Tech., 17: 23.

111. Bloch, A. B., Stramer, S. L., Smith, D., Margolis, H. S., Fields, H. A., McKinley, T. W., Gerba, C. P., Maynard, J. E., and Sikes, K. (1990). Recovery of hepatitis A virus from a water supply responsible for a common source outbreak of hepatitis A, Am. J. Pub. Health, 80: 428.

112. Mahoney, F. J., Farley, T. A., Kelso, K. Y., Wilson, S. A., Horna, J. M., and McFarland,

L. M. (1992). An outbreak of hepatitis A associated with swimming in a public pool, J. Infect. Dis., 165: 613.

113. Bryan, J. A., Lehmann, J. D., Setiady, I. F., and Hatch, M. H. (1974). An outbreak of hepatitis A associated with recreational lake water, Am. J. Epidemiol., 99: 145.

114. Abbaszadegan, M., Stewart, P., and LeChevallier, M. (1999). A strategy for detection of viruses in groundwater by PCR, Appl. Environ. Microbiol., 65: 444.

115. Gust, I. D. and Feinstone, S. M. (1988). History, Hepatitis (I. D. Gust and S. M. Feinstone, eds.), CRC Press, Boca Raton, FL, p.1.

116. Krugman, S., Ward, R., Giles, J. P., and Jacobs, A. M. (1960). Infectious hepatitis, study on effect of gamma globulin and on the incidence of apparent infection, JAMA, 174: 823.

117. Feinstone, S. M., Kapikian, A. Z., and Purcell, R. H. (1973). Hepatitis A: detection by immune electron microscopy of a virus-like antigen association with acute illness, Science, 182: 1026.

118. Maynard, J. E., Bradley, D. W., Gravelle, C. R., Ebert, J. W., and Krushak, D. H. (1975). Preliminary studies of hepatitis A in chimpanzees, J. Infect. Dis., 131: 194.

119. Provost, P. J., and Hilleman, M. R. (1979). Propagation of human hepatitis A virus in cell culture in vitro, Proc. Soc. Exp. Biol. Med, 160: 213.

120. Frosner, G. G., Deinhardt, F., Scheid, R., Gauss-Muller, V., Holmes N., Messelberger, V., Siegl, G., and Alexander, J. J. (1979). Propagation of human A virus in a hepatoma cell line, Infection, 7: 303.

121. Gauss-Muller, V., Frosner, G. G., and Deinhardt, F. (1981). Propagation of hepatitis A virus in human embryo fibroblasts, J. Med. Virol., 7: 233.

122. Flehmig B. (1980). Hepatitis A virus in cell culture: I. Propagation of different hepatitis A virus isolates in a fetal rhesus monkey kidney cell line (FRhK-4), Med. Microbiol. Immunol., 168: 239.

123. Bradley, D. W., Schable, C. A., McCaustland, K. A., Cook, E H., and Maynard, J. E. (1984). Hepatitis A virus: growth characteristics of in vivo and in vitro propagated wild and attenuated virus strains, J. Med. Virol., 14: 373.

124. Daemer, R. J., Feinstone, S. M., Gust, I. D., and Purcell, R. H. (1981). Propagation of human hepatitis A virus in African green monkey kidney cell culture: Primary isolation and serial passage, Infect. Immun., 32: 388.

125. Bradley, D. W., Fields, H. A., McCaustland, K. A., Cook, E. H., Gravelle, C. R., and Maynard, J. E. (1978). Biochemical and biophysical characterization of light and heavy density hepatitis A virus particles: evidence HAV is an RNA virus, J. Med. Virol., 2: 175.

126. Coulepis, A. G., Locarnini, S. A., Westaway, E. G., Tannock, G. A., and Gust, I. D. (1982). Biophysical and biochemical characterization of hepatitis A virus, Intervirology, 18: 107.

127. Tesar, M., Jia, X.-Y., Summers, D. F., and Ehrenfeld, E. (1993). Analysis of a potential myristoylation site in hepatitis A virus capsid protein VP4, Virology, 194: 616.

128. Melnick, J. L. (1982). Classification of hepatitis A virus as enterovirus type 72 and of hepatitis B virus as hepadnavirus type I, Intervirology, 18: 105.

129. Gust, I. D., Coulepis, A. G., Feinstone, S. M., Locarnini, S. A., Moritsugu, Y., Najera, R., and Siegl, G. (1983). Taxonomic classification of hepatitis A virus, Intervirology, 20: 1.

130. Viswanathan, R. (1957). Infectious hepatitis in Delhi (1955–56): a critical study; epidemiology, Indian J. Med. Res., 1.

131. Ticehurst, J. R. (1986). Hepatitis A virus: clones, cultures, and vaccines, Semin. Liver Dis., 6: 46.

132. Minor, P. D. (1991). Picornaviridae, Classification and Nomenclature of Viruses: The Fifth Report of the International Committee on Taxonomy of Viruses (R. I. B. Franki, C. M. Fauquet, D. L. Knudson, and F. Brown, eds.), Springer-Verlag, Wien, p.320.

133. Binn, L. N., Lemon, S. M., Marchwicki, R. H., Redfield, R. R., Gates. N. L., and Bancroft, W. H. (1984). Primary isolation and serial passage of hepatitis A virus strains in primate cell cultures, J. Clin. Microbiol., 20: 28.

134. Siegl, G., De Chastonay, J., and Kronauer, G. (1984). Propagation and assay of hepatitis A virus in vitro, J. Virol. Meth., 9: 53.

135. Vallbracht, A., Hofmann, L., Wurster, K. G., and Flehmig, B. (1984). Persistent infection of human fibroblasts by hepatitis A virus, J. Gen. Virol., 65: 609.

136. Simmonds, R. S., Szucs, G., Metcalf, T. G., and Melnick, J. L. (1985). Persistently infected cultures as a source of hepatitis A virus, Appl. Environ. Microbiol., 49: 749.

137. Wheeler, C. M., Robertson, B. H., Van Nest, G., Dina, D., Bradley, D. W., and Fields, H. A. (1986). Structure of hepatitis A virion: peptide mapping of the capsid region, J. Virol., 58: 307.

138. Robertson, B. H., Khanna, B., Brown, V. K., and Margolis, H. S. (1988). Large scale production of hepatitis A virus in cell culture: effect of type of infection on virus yield and cell integrity, J. Gen. Virol., 69: 2129.

139. Sobsey, M. D., Shields, P. A., Hauchman, F. S., Davis, A. L., Rullman, V. A., and Bosch, A. (1988). Survival and persistence of hepatitis A virus in environmental samples, Viral Hepatitis and Liver Disease (A. J. Zuckerman, eds.), Alan R. Liss, New York, p.121.

140. Flehmig, B. (1981). Hepatitis A virus in cell culture: II. Growth characteristic of hepatitis A virus in FRhK-4/R cells, Med. Microbiol. Immunol., 170: 73.

141. Gregersen, J.-P., Mehdi, S., and Mauler, R. (1988). Adaptation of hepatitis A virus to high titre growth in diploid and permanent cell cultures, Med. Microbiol. Immunol., 177: 91.

142. Cromeans, T., Fields, H. A., and Sobsey, M. D. (1989). Replication kinetics and cytopathic effect of hepatitis A virus, J. Gen. Virol., 70: 2051.

143. Lemon, S. M., Binn, L. N., and Marchwicki, R. H. (1983). Radioimmunofocus assay for quantitation of hepatitis A virus in cell culture, J. Clin. Microbiol., 17: 834.

144. Wheeler, C. M., Fields, H. A., Schable, C. A., Meinke, W. J., and Maynard, J. E. (1986). Adsorption, purification, and growth characteristics of hepatitis A virus strain HAS-15 propagated in fetal Rhesus monkey kidney cells, J. Clin. Microbiol., 23: 434.

145. Anderson, D. A., Locarnini, S. A., Coulepis, A. G., and Gust, I. D. (1985). Restrictive events in the replication of hepatitis A virus in vitro, Intervirology, 24: 26.

146. De Chastonay, J. and Siegl, G. (1987). Replicative events in hepatitis A virus–infected MRC-5 cells, Virology, 157: 268.

147. Anderson, D. A., Ross, B. C., and Locarnini, S. A. (1988). Restricted replication of hepatitis A virus in cell culture: encapsidation of viral RNA depletes the pool of RNA available for replication, J. Virol., 62: 4201.

148. Harmon, S. A., Summers, D. F., and Ehrenfeld, E. (1989). Detection of hepatitis A virus RNA and capsid antigen in individual cells, Virus Res., 12: 361.

149. Cho, M. W. and Ehrenfeld E. (1991). Rapid completion of the replication cycle of hepatitis A virus subsequent to reversal of guanidine inhibition, Virology, 180: 770.

150. Glass, M. J., Jia, S.-Y., and Summers, D. F. (1993). Identification of the hepatitis A virus internal ribosome entry site: in vivo and in vitro analysis of bicistronic RNA's containing the HAV 5′ noncoding region, Virology, 193: 842.

151. Whetter, L. E., Day, S. P., Elroy-Stein, O., Brown, E. A., and Lemon, S. M. (1994). Low efficiency of the 5′ nontranslated region of hepatitis A virus RNA in directing capindependent translation in permissive monkey kidney cells, J. Virol., 68: 5253.

152. Anderson, D. A. (1987). Cytopathology, plaque assay, and heat inactivation of hepatitis A strain HM-175, J. Med. Virol., 22: 35.

153. Cromeans, T., Sobsey, M. D., and Fields, H. A. (1987). Development of a plaque assay for a cytopathic, rapidly replicating isolate of hepatitis A virus, J. Med. Virol., 22: 45.

154. Nasser, A. M. and Metcalf, T. G. (1987). Production of cytopathology in FRhK-4 cells by BS-C-1-passaged hepatitis A virus, Appl. Environ. Microbiol., 53: 2967.

155. Shen, W., Lian-cai, J., Hui-Xun, S., Xi-tan, Z., Qing-hai, M., and Xue-guang, L. (1986). Strain of hepatitis A virus causing cytopathic effects isolated in A549 cell line, Chinese Med. J., 99: 387.

156. Venuti, A., Di Russo, C., del Grosso, N., Patti, A.-M., Ruggery, R., De Stasio, P. R., Martiniello, M. G., Pagnotti, P., Degener, A. M., Midulla, M., Pana, A., and Perez-Bercoff, R. (1985). Isolation and molecular cloning of a fast-growing strain of human hepatitis A virus from its double-stranded replicative form, J. Virol., 56: 579.

157. Tinari, A., Ruggeri, F. M., Divizia, M., Pana, A., and Donelli, G. (1989). Morphological changes in HAV-infected Frp/3 cell and immunolocalization of HAAg, Arch. Virol., 104: 209.

158. Cromeans, T. L., Fields, H. A., and Sobsey, M. D. (1988). Kinetic studies of a rapidly replicating, cytopathic hepatitis A virus, Viral Hepatitis and Liver Disease (A. J. Zuckerman, eds.), Alan R. Liss, New York, p.24.

159. Cromeans, T., Humphrey, C. D., Sobsey, M. D., and Fields, H. A. (1989). Use of immunogold pre-embedding technique to detect hepatitis A viral antigen in infected cells, Am. J. Anat., 185: 314.

160. Marshall, J. A., Borg, J., Coulepis, A. G., and Anderson, D. A. (1996). Annulate lamellae and lytic HAV infection in vitro, Tissue and Cell, 28: 205.

161. Gauss-Muller, V. and Deinhardt, F. (1984). Effect of hepatitis A virus infection on cell metabolism in vitro, Proc. Soc. Exp. Biol. Med, 175: 10.

162. Siegl, G., Nuesch, J. P. F., and deChastonay, J. D. (1989). Defective interfering particles of hepatitis A virus in cell cultures and clinical specimens, New Aspects of Positive Strand RNA Viruses (M. Brinton and F. Heinz, eds.), ASM, Washington, DC, p.102.

163. Emerson, S. U., Huang, Y. K., McRill, C., Lewis, M., and Purcell, R. H. (1992). Mutations in both the 2B and 2C genes of hepatitis A virus are involved in adaptation to growth in cell culture, J. Virol., 66: 650.

164. Lemon, S. M., Murphy, P. C., Shields, P. A., Ping, L.-H., Feinstone, S. M., Cromeans, T., and Jansen, R. W. (1991). Antigenic and genetic variation in cytopathic hepatitis A virus variants arising during persistent infection: evidence for genetic recombination, J. Virol., 65: 2056.

165. Zhang, H., Chao, S.-F., Ping, L.-H., Grace, K., Clarke, B., and Lemon, S. M. (1995). An infectious cDNA clone of a cytopathic hepatitis A virus: genomic regions associated with rapid replication and cytopathic effect, Virology, 212: 686.

166. Morace, G., Pisani, G., Beneduce, F., Divizia, M., and Pana, A. (1993). Mutations in the 3A genomic region of two cytopathic strains of hepatitis A virus isolated in Italy, Virus Res., 28: 187.

167. Brack, K., Frings, W., Dotazuer, A., and Vallbracht, A. (1998). A cytopathogenic, apoptosis-inducing variant of hepatitis A virus, J. Virol., 72: 3370.

168. Day, S. P., Murphy, P., Brown, E. A., and Lemon, S. M. (1992). Mutations within the 5' non-translated region of hepatitis A virus RNA enhance replication in BS-C-1 cells, J. Virol., 66: 6533.

169. Graff, J., Dasang, C., Normann, A., Pfisterer-Hunt, M., Feinstone, S. M., and Flehmig, B. (1994). Mutational events in consecutive passages of hepatitis A virus strain GBM during cell culture adaptation, Virology, 204: 60.

170. Graff, J., Normann, A., Feinstone, S. M., and Flehmig, B. (1994). Nucleotide sequence of wild-type hepatitis A virus GBM in comparison with two cell culture-adapted variants, J. Virol., 68: 548.

171. Chang, K. H., Brown, E. A., and Lemon, S. M. (1993). Cell type-specific proteins which interact with the 5' nontranslated region of hepatitis A virus RNA, J. Virol., 67: 6716.

172. Schultz, D. E., Honda, M., Whetter, L. E., McKnight, K. L., and Lemon, S. M. (1996). Mutations within the 5' nontranslated RNA of cell culture-adapted hepatitis A virus which enhance cap-independent translation in cultured African green monkey kidney cells, J. Virol., 70: 1041.

173. Dotzauer, A., Feinstone, S. M., and Kaplan, G. (1994). Susceptibility of nonprimate cell lines to hepatitis A virus Infection, J. Virol., 68: 6064.

174. Stapleton, J. T., Frederick, J., and Meyer, B. (1991). Hepatitis A attachment to cultured cell lines, J. Infect. Dis., 164: 1098.

175. Azjac, A. J., Amphlett, E. M., Rowlands, D. J., and Sangar, D. V. (1991). Parameters influencing the attachment of hepatitis A virus to a variety of continuous cell lines, J. Gen. Virol., 72: 1667.

176. Bishop, N. E., and Anderson, D. A. (1997). Early interactions of hepatitis A virus with cultured cells: viral elution and the effect of pH and calcium ions, Arch. Virol., 142: 2161.

177. Ashida, M., and Hamada, C. (1997). Molecular cloning of the hepatitis A virus receptor from a simian cell line, J. Gen. Virol., 78: 1565.

178. Kaplan, G., Totsuka, A., Thompson, P., Akatsuka, T., Moritsugu, Y., and Feinstone, S. M. (1996). Identification of a surface glycoprotein on African green monkey kidney cells as a receptor for hepatitis A virus, EMBO, 15: 4282.

179. Kaplan, G., Totsuka, A., Matoo, P., Thompson, P., Akatsuka, T., Moritsugu, Y., and Feinstone, S. (1997). HAV cellular interaction—identification of a cell surface glycoprotein as a cellular receptor for hepatitis A virus infection, in Viral Hepatitis and Liver Disease (M. Rizetto, R. H. Purcell, J. L. Gerin, et al., eds.) Minerva Medica, Turin, Italy, p. 32.

180. Feigelstock, D., Thompson, P., Mattoo, P., Zhang, Y., and Kaplan, G. G. (1998). The human homolog of HAVcr-1 codes for a hepatitis A virus cellular receptor, J. Virol., 72: 6621.

181. Siegl, G., and Frosner, G. G. (1978). Characterization and classification of virus particles associated with hepatitis A: type and configuration of nucleic acids, J. Virol., 26: 48.

182. Weitz, M., Baroudy, B. M., Maloy, W. L., Ticehurst, J. R., and Purcell, R. H. (1986). Detection of a genome-linked protein (VPg) of hepatitis A virus and its comparison with other picornaviral VPgs, J. Virol., 60: 124.

183. Baroudy, B. M., Ticehurst, J. R., Miele, T. A., Maizel, J. V., Purcell, R. H., and Feinstone, S. M. (1985). Sequence analysis of hepatitis A virus cDNA coding for capsid region and RNA polymerase, Proc. Natl. Acad. Sci. USA, 82: 2143.

184. Cohen, J. I., Ticehurst, J. R., Purcell, R. H., Buckler-White, A., and Baroudy, B. M. (1987). Complete nucleotide sequence of wild-type hepatitis A virus: comparison with different strains of hepatitis A virus and other picornaviruses, J. Virol., 61: 50.

185. Gauss-Muller, V., von der Helm, K., and Deinhardt, F. (1984). Translation in vitro of hepatitis A virus RNA, Virology, 137: 182.

186. Locarnini, S. A., Coulepis, A. G., Westaway, E. G., and Gust, I. D. (1981). Restricted replication of human hepatitis A virus in cell culture: intracellular biochemical studies, J. Virol., 37: 216.

187. Siegl, G., Frosner, G. G., Gauss-Muller, V., Tratschin, J., and Deinhardt, F. (1981). The physiochemical properties of infectious hepatitis A virions, J. Gen. Virol., 57: 331.

188. Malcolm, B. A., Chin, S. M., Jewell, D. A., Stratton-Thomas, J. R., Thudium, K. B., Ralston, R., and Rosenberg, S. (1992). Expression and characterization of recombinant hepatitis A virus 3C proteinase, Biochemistry, 31: 3358.

189. Jia, X.-Y., Ehrenfeld, E., and Summers, D. F. (1991). Proteolytic activity of hepatitis A virus 3C protein, J. Virol., 65: 2595.

190. Harmon, S. A., Updike, W., Jia, X.-Y., Summers, D. F., and Ehrenfeld, E. (1992). Polyprotein processing in cis and trans by hepatitis A virus 3C protease cloned and expressed in Escherichia coli, J. Virol., 66: 5242.

191. Jurgensen, D., Kusov, Y. Y., Facke, M., Drausslich, H.-G., and Gauss-Muller, V. (1993). Cell-free translation and proteolytic processing of the hepatitis A virus polyprotein, J. Gen. Virol., 74: 677.

192. Schultheiss, T., Kusov, Y. Y., and Gauss-Muller, V. (1994). Proteinase 3C of hepatitis A virus (HAV) cleaves the HAV polyprotein P2-P3 at all sites including VP1/2A and 2A/2B, Virology, 198: 275.

193. Schultheiss, T., Sommergruber, W., Kusov, Y., and Gauss-Muller, V. (1995). Cleavage specificity of purified recombinant hepatitis A virus 3C proteinase on natural substrates, J. Virol., 69: 1727.

194. Goser, R., Cassinotti, P., Siegl, G., and Weitz, M. (1996). Identification of hepatitis A virus non-structural protein 2B and its release by the major virus protease 3C, J. Gen. Virol., 77: 247.

195. Kusov, Y. Y., Sommergruber, W., Schreiber, M., and Gauss-Muller, V. (1992). Intermolecular cleavage of hepatitis A virus (HAV) precursor protein P1-P2 by recombinant HAV proteinase 3C, J. Virol., 56: 6794.

196. Probst, C., Jecht, M., and Gauss-Muller, V. (1997). Proteinase 3C–mediated processing of VP1-2A of two hepatitis A virus strains: in vivo evidence for cleavage at amino acid position 273/274 of VP1, J. Virol., 71: 3288.

197. Tesar, M., Pak, I., Jia, X.-Y., Richards, O. C., Summers, D. F., and Ehrenfeld, E. (1994). Expression of hepatitis A virus precursor protein P3 in vivo and in vitro: polyprotein processing of the 3CD cleavage site, Virology, 198: 524.

198. Probst, C., Jecht, M., and Gauss-Muller, V. (1998). Processing of proteinase precursors and their effect on hepatitis A virus particle formation, J. Virol., 72: 8013.

199. Wimmer, E. and Murdin, A. D. (1991). Hepatitis A virus and the molecular biology of picornaviruses: a case for a new genus of the family Picornaviridae, Viral Hepatitis and Liver Disease (F. B. Hollinger, S. M. Lemon, and H. Margolis, eds.), Williams & Wilkins, Baltimore, p.31.

200. Ovchinnikov, I. A., Sverdlov, E. D., Tsarev, S. A., Arsenian, S. G., Rokhlina, T. O., Chizhikov, V. E., Petrov, N. A., Prikhod'ko, G. G., Blinov, V. M., Basilenko, S. K., Sandakhchiev, L. S., Kusov, I. I., Grabko, V. I., Fleer, G. P., Balayan, M. S., and Drozdov, S. G. (1985). Sequence of 3372 nucleotide units of RNA of the hepatitis A virus, coding the capsids VP4-VP1 and some nonstructural proteins, Dokl. Akad. Nauk. SSSR, 285: 1014.

201. Najarian, R., Caput, D., Gee, W., Potter, S. J., Renard, A., Merryweather, J., Van Nest, G., and Dina, D. (1985). Primary structure and gene organization of human hepatitis A virus, Proc. Natl. Acad. Sci. USA, 82: 2627.

202. Paul, A. V., Tada, H., von der Helm, K., Wissel, T., Kiehn, R., Wimmer, E., and Deinhardt, F. (1987). The entire nucleotide sequence of the genome of human hepatitis A virus (isolate MBB), Virus Res., 8: 153.

203. Cohen, J. I., Rosenblum, B., Ticehurst, J. R., Daemer, R. J., Feinstone, S. M., and Purcell, R. H. (1987). Complete nucleotide sequence of an attenuated hepatitis A virus: comparison with wild-type virus, Proc. Natl. Acad. Sci. USA, 84: 2497.

204. Crowell, R. L., and Landau, B. J. (1983). Receptors in the initiation of picornavirus infection, Comprehensive Virology, Vol. 18 (H. Fraenkel-Conrat and R. R. Wagner, eds.), Plenum Press, New York, p.1.

205. Rueckert, R. R. (1990). Picornaviridae and their replication, Virology (B. N. Fields and D. M. Knipe, eds.), Raven Press, New York, p.507.

206. Rossman, M. G., Arnold, E., Erickson, J. W., Frankenberger, E. A., Griffith, J. P., Hecht, H. J., Johnson, J. E., Kamer, G., Luo, M., Mosser, A. G., Rueckert, R. R., Sherry, B., and Vriend, G. (1985). Structure of human cold virus and functional relationship to other picornaviruses, Nature, 317: 145.

207. Luo, M., Vriend, G., Kamer, G., Minor, I., Arnold, E., Rossmann, M. G., Boege, U., Scraba, D. G., Duke, G. M., and Palmenberg, A. C. (1987). The atomic structure of mengo virus at 3.0 Å resolution, Science, 235: 182.

208. Stapleton, J. T. and Lemon, S. M. (1987). Neutralization escape mutants define a dominant immunogenic site on hepatitis A virus, J. Virol., 61: 491.

209. Gerlich, W. H. and Frosner, G. G. (1983). Topology and immunoreactivity of capsid proteins in hepatitis A virus, Med. Microbiol. Immunol., 172: 101.

210. Ross, B. C., Anderson, B. N., Coulepis, A. G., Chenoweth, M. P., and Gust, I. D. (1986).

Molecular cloning of cDNA from hepatitis A virus strain HM-175 after multiple passages in vivo and in vitro, J. Gen. Virol., 67: 1741.

211. Hughes, J. V., Stanton, L. W., Tomassini, J. E., Long, W. J., and Scolnick, E. M. (1984). Neutralizing monoclonal antibodies to hepatitis A virus: partial localization of a neutralizing antigenic site, J. Virol., 52: 465.

212. Emini, E. A., Hughes, J. V., Perlow, D. S., and Boger, J. (1985). Induction of hepatitis A virus neutralizing antibody by a virus-specific synthetic peptide, J. Virol., 55: 836.

213. Ostermayr, R., Von Der Helm, K., Gauss-Muller, V., Winnacker, E. L., and Deinhardt, F. (1987). Expression of hepatitis A virus cDNA in Escherichia coli: antigenic VP1 recombinant protein, J. Virol., 61: 3645.

214. Ping, L.-H., Jansen, R. W., Stapleton, J. T., Cohen, J. I., and Lemon, S. M. (1988). Identification of an immunodominant antigenic site involving the capsid protein VP3 of hepatitis A virus, Proc. Natl. Acad. Sci. USA, 85: 8281.

215. Ping, L.-H., and Lemon, S. M. (1992). Antigenic structure of human hepatitis A virus defined by analysis of escape mutants selected against murine monoclonal antibodies, J. Virol., 66: 2208.

216. Nainan, O. V., Brinton, M. A., and Margolis, H. S. (1992). Identification of amino acids located in the antibody binding sites of human hepatitis A virus, Virology, 191: 984.

217. Karetnyi, Y. V., Andjaparidze, A. G., Orlova, T. M., and Balayan, M. S. (1989). Study of hepatitis A virus isolates of human and simian origin by enzyme immunoassay using polyclonal and monoclonal antibodies, Vopr. Virusol., 1: 50.

218. Nainan, O. V., Margolis, H. S., Robertson, B. H., Balayan, M., and Brinton, M. A. (1991). Sequence analysis of a new hepatitis A virus naturally infecting cynomolgus macaques (Macaca fascicularis), J. Gen. Virol., 72: 1685.

219. Bosch, A., Gonzalez-Dankaart, J. F., Haro, I., Gajardo, R., Perez, J. A., and Pinto, R. M. (1998). A new continuous epitope of hepatitis A virus, J. Med. Virol., 54: 95.

220. Luo, M., Rossmann, M. G., and Palmenberg, A. C. (1988). Prediction of three-dimensional models for foot-and-mouth disease virus and hepatitis A virus, Virology, 166: 503.

221. Acharya, R., Fry, E., Stuart, D., Fox, G., and Brown, F. (1989). The three-dimensional structure of foot-and-mouth disease virus at 2.9 Å resolution, Nature, 337: 709.

221a. Parry, N., Fox, G., Rowlands, D., Brown, F., Fry, E., Acharya, R., Logan, D., and Stuart, D. (1990). Structural and serological evidence for a novel mechanism of antigenic variation in foot and mouth disease virus, Nature, 347: 569.

222. Jansen, R. W., Siegl, G., and Lemon, S. M. (1991). Molecular epidemiology of human hepatitis A virus (HAV), Viral Hepatitis and Liver Disease (F. B. Hollinger, S. M. Lemon, and H. S. Margolis, eds.), Williams & Wilkins, Baltimore, p.58.

223. Robertson, B. H., Khanna, B., Nainan, O. V., and Margolis, H. S. (1991). Epidemiologic patterns of wild-type hepatitis A virus determined by genetic variation, J. Infect. Dis., 163: 286.

224. Robertson, B. H., Jansen, R. W., Khanna, B., Totsuka, A., Nainan, O., V, Siegl, G., Widell, A., Margolis, H. S., Isomura, S., Ito, K., Ishizu, T., Moritsugu, Y., and Lemon, S. M. (1992). Genetic relatedness of hepatitis A virus strains recovered from different geographic regions, J. Gen. Virol., 73: 1365.

225. Wang, K.-Q., Nielsen, C. M., and Vestergaard, B. F. (1985). Isolation and adaptation characteristics of hepatitis A virus in primary African green monkey kidney cells: production of antigen useful for ELISA serology, Intervirology, 24: 99.

226. Flehmig, B., Vallbracht, A., and Wurster, G. (1981). Hepatitis A virus in cell culture III. Propagation of hepatitis A virus in human embryo kidney cells and human embryo fibroblast strains, Med. Microbiol. Immunol., 170: 83.

227. Hollinger, F. B., Bradley, D. W., Maynard, J. E., Dreesman, G. R., and Melnick, J. L. (1975). Detection of hepatitis A viral antigen by radioimmunoassy, J. Immunol., 115: 1464.

228. Mathieson, L. R., Feinstone, S. M., Wong, D. C., Skinhoej, P., and Purcell, R. H. (1978).

Enzyme-linked immunosorbent assay for detection of hepatitis A antigen in stools and antibody to hepatitis A antigen in sera: comparison with solid-phase radioimmunoassay, immune electron microscopy and immune adherence hemagglutination assay, J. Clin. Microbiol., 1: 184.

229. Coulepis, A. G., Veale, M. F., MacGregor, A., Kornitschuk, M., and Gust, I. D. (1985). Detection of hepatitis A virus and antibody by solid-phase radioimmunoassay and enzyme-linked immune absorbant assay with monoclonal antibodies, J. Clin. Microbiol., 22: 119.

230. Jiang, X., Estes, M. K., and Metcalf, T. G. (1989). In situ hybridization for quantitative assay of infectious hepatitis A virus, J. Clin. Microbiol., 27: 874.

231. Lemon, S. M., and Jansen, R. W. (1985). A simple method for clonal selection of hepatitis A virus based on recovery of virus from radioimmunofocus overlays, J. Virol. Meth., 11: 171.

232. Borovec, S., and Uren, E. (1997). Single-antibody in situ enzyme immunoassay for infectivity titration of hepatitis A virus, J. Virol. Meth., 68: 81.

233. Mbithi, J., Sringthorpe, S., and Sattar, S. A. (1990). Chemical disinfection of hepatitis A virus on environmental surfaces, Appl. Environ. Microbiol., 56: 3601.

234. Jansen, R. W., Newbold, J. E., and Lemon, S. M. (1985). Combined immunoaffinity cDNA-RNA hybridization assay for detection of hepatitis A virus in clinical specimens, J. Clin. Microbiol., 22: 984.

235. Ticehurst, J. R., Feinstone, S. M., Chestnut, T., Tassopoulos, N. C., Popper, H., and Purcell, R. H. (1987). Detection of hepatitis A virus by extraction of viral RNA and molecular hybridization, J. Clin. Microbiol., 25: 1822.

236. Shieh, Y.-S., Baric, R. S., Sobsey, M. D., Ticehurst, J., Miele, T. A., DeLeon, R., and Walter, R. (1991). Detection of hepatitis A virus and other enteroviruses in water by ssRNA probes, J. Virol. Meth., 31: 119.

237. Dubrou, S., Kopecka, H., Pila, J. M. L., Marechal, J., and Prevot, J. (1991). Detection of hepatitis A virus and other enteroviruses in wastewater and surface water samples by gene probe assay, Water Sci. Tech., 24: 267.

238. Robertson, B. H., Khanna, B., Nainan, O., V., and Margolis, H. S. (1991). Genetic variation of wild-type hepatitis A isolates, Viral Hepatitis and Liver Disease (B. N. Hollinger, S. M. Lemon, and H. S. Margolis, eds.), Harper Graphics, Waldorf, MD, p.54.

239. Saiki, R. K., Gelfand, D. H., Stoffel, S., Schaft, S. J., Higuchi, R., Horn, G. T., Mullis, K. B., and Erlich, H. A. (1988). Primer-directed enzymatic amplification of DNA with a thermostable DNA polymerase, Science, 239: 487.

240. Chapman, N. M., Tracy, S., Gauntt, C. J., and Fortmueller, U. (1990). Molecular detection and identification of enteroviruses using enzymatic amplification and nucleic acid hybridization, J. Clin. Microbiol., 28: 843.

241. DeLeon, R., Shieh, C., Baric, R. B., and Sobsey, M. D. (1990). Detection of enteroviruses and hepatitis A virus in environmental samples by gene probes and polymerase chain reaction, Advances in Water Analysis and Treatment, American Water Works Association Proceedings (AWWA, eds.), AWWA, San Diego, CA, p. 833.

242. Jaykus, L. A., DeLeon, R., and Sobsey, M. D. (1993). Application of RT-PCR for the detection of enteric viruses in oysters, Water Sci. Tech., 27: 49.

243. Kwok, S., and Higuchi, R. (1989). Avoiding false positives with PCR, Nature, 339: 237.

244. Rao, V. C. (1982). Introduction to environmental virology, Methods in Environmental Virology (C. P. Gerba and S. M. Goyal, eds.), Marcel Dekker, New York, p. 1.

245. Fattal, B. (1983). The prevalence of viral hepatitis and other enteric disease in communities utilizing wastewater in agriculture, Water Sci. Tech., 15: 43.

246. Katzenelson, E., and Mills, D. (1984). Contamination of vegetables with animal viruses via the roots, Monog. Virol., 15: 216.

247. Murphy, W. H. and Syverton, J. T. (1958). Absorption and translocation of mammalian viruses by plants, Virology, 6: 623.

248. Reid, T. M. S., and Robinson, H. G. (1987). Frozen raspberries and hepatitis A, Epidemiol. Infect., 98: 109.

249. Ramsay, C. N., and Upton, P. A. (1989). Hepatitis A and frozen raspberries, Lancet, 1: 43.

250. Noah, N. D. (1981). Foodborne outbreaks of hepatitis A, Med. Lab. Sci., 38: 428.

251. Purcell, R. H., Feinstone, S. M., Ticehurst, J. R., Daemer, R. J., and Baroudy, B. M. (1984). Hepatitis A virus, Viral Hepatitis and Liver Disease (G. N. Vyas, J. L. Dienstag, and J. H. Hoofnagle, eds.), Grune and Straton, Orlando, FL, p.9.

252. Roos, B. (1956). Hepatitis epidemic conveyed by oysters, Svensk. Lakartidn., 53: 989.

253. Mason, J. O., and McLean, W. R. (1962). Infectious hepatitis traced to the consumption of raw oysters, Am. J. Hyg., 75: 90.

254. Dismukes, W. E., Bisno, A. L., Katz, S., and Johnson, R. F. (1969). An outbreak of gastroenteritis and infectious hepatitis attributed to raw clams, Am. J. Epidemiol., 89: 555.

255. Ruddy, S. J., Johnson, R. F., Mosley, J. W., Atwater, J. B., Rosetti, M. A., and Hart, J. C. (1969). An epidemic of clam-associated hepatitis, JAMA, 208: 649.

256. Portnoy, B. L., Mackowiak, P. A., Caraway, C. T., Walker, J. A., McKinley, T. W., and Klein, C. A. (1975). Oyster-associated hepatitis failure of shellfish certification programs to prevent outbreaks, JAMA, 233: 1065.

257. O'Mahony, M. C., Gooch, C. D., Smyth, D. A., Thrussell, A. J., Bartlett, C. L. R., and Noah, N. D. (1983). Epidemic hepatitis A from cockles, Lancet, 1: 518.

258. Ohara, H., Naruto, H., Watanabe, W., and Ebisawa, I. (1983). An outbreak of hepatitis A caused by consumption of raw oysters, J. Hyg. Camb., 91: 163.

259. Heller, D., Gill, O. N., Raynham, E., Zadick, P.M., and Stanwell-Smith, R. (1986). An outbreak of gastrointestinal illness associated with consumption of raw depurated oysters, Br. Med. J., 1986: 1726.

260. Millard, J., Appleton, H., and Parry, J. V. (1987). Studies on heat inactivation of hepatitis A virus with special reference to shellfish. 1. Procedures for infection and recovery of virus from laboratory-maintained cockles, Epidemiol. Infect., 98: 397.

261. Abad, F. X., Pinto, Rosa, M., Gafardo, R., and Bosch, A. (1997). Viruses in mussels: public health implications and depuration, J Food Prot., 60: 677.

262. Schoenbaum, S. C., Baker, O., and Jezek, Z. (1976). Common-source epidemic of hepatitis due to glazed and iced pastries, Am. J. Epidemiol., 104: 74.

263. Weltman, A. C., Bennett, N. M., Ackman, D. A., Misage, J. H., Campana, J. J., Fine, L. S., Doniger, A. S., Balzano, G. J., and Birkhead, G. S. (1996). An outbreak of hepatitis A associated with bakery, New York, 1994: The 1968 ''West Branch, Michigan'' outbreak repeated, Epidemiol. Infect., 117: 333.

264. Warburton, A. R. E., Wreghitt, T. G., Rampling, A., Buttery, R., Ward, K. N., Perry, K. R., and Parry, J. V. (1991). Hepatitis A outbreak involving bread, Epidemiol. Infect., 106: 199.

265. Zachoval, R., Frosner, G., Deinhardt, F., and John, I. (1981). Hepatitis A transmission by cold meats, Lancet, 2: 260.

266. Leger, R. T., Boyer, K. M., Pattison, C. P., and Maynard, J. E. (1975). Hepatitis A: report of a common-source outbreak with recovery of a possible etiologic agent. I. Epidemiologic studies, J. Infect. Dis., 131: 163.

267. Mishu, B., Hadler, S., Boza, V. A., Hutcheson, R. H., Horan, J. M., and Schaffner, W. (1990). Foodborne hepatitis A: evidence that microwaving reduces risk?, J. Infect. Dis., 162: 655.

268. Bean, N. H., Griffin, P, M, Goulding, J. S., and Ivey, C. B. (1990). Foodborne disease outbreaks, 5-year summary, 1983–1987, MMWR, 39: 1.

269. Koff, R. S., Grady, G. F., Chalmers, T. C., Mosley, J. W., and Swartz, B. L. (1967). Viral hepatitis in a group of Boston hospitals. III. Importance of exposure to shellfish in a nonepidemic period, N. Engl. J. Med., 276: 703.

270. Stille, W., Kunkel, B., and Nerger, K. (1972). Oyster-transmitted hepatitis, Dtsh. Med. Wschr., 97: 145.

271. Cromeans, T. L., and Margolis, H. S. (1998). Methods for isolation and detection of hepatitis A virus in frozen strawberries, ASM Abstracts, P10, p. 405.

272. Cromeans, T. L., Sobsey, M. D., and Margolis, H. S. (1999). Molecular detection of hepatitis A virus in frozen strawberries, ASM Abstracts, P108.

273. Hernandez, F., Monge, R., Jimenez, C., and Taylor, L. (1997). Rotavirus and hepatitis A virus in market lettuce (Latuca sativa) in Costa Rica, Int. J. Food Microbiol., 37: 221.

274. Hu, M. D., Sun, B. F., Xi, Z., and He, J. F. (1989). Detection of hepatitis A virus from the clams produced by Qi-Dong using cell culture and nucleic acid hybridization assays, Shanghai Med. J., 12: 73.

275. Cliver, D. O., Ellender, R. D., and Sobsey, M. D. (1983). Methods for detecting viruses in foods: background and general principles, J. Food Prot., 46: 248.

276. Larkin, E. P. (1982). Detection of viruses in foods, Methods in Environmental Virology (C. P. Gerba and S. M. Goyal, eds.), Marcel Dekker, New York, p. 221.

277. National Shellfish Sanitation Program (1989). Manual of Operations. Part 1. Sanitation of Shellfish Growing Areas; U.S. Dept. of Health and Human Services, Public Health Service, Food and Drug Administration, Washington, D.C.

278. Gerba, C. P., and Goyal, S. M. (1978). Detection and occurrence of enteric viruses in shellfish: a review, J. Food Prot., 41: 743.

279. Ellender, R. D., Mapp, J. B., Middlebrooks, B. L., Cood, D. W., and Cake, E. W. (1980). Natural enterovirus and fecal coliform contamination of gulf coast oysters, J. Food Prot., 43: 105.

280. Vaughn, J. M., Landry, E. F., Vicale, T. J., and Dahl, M. C. (1980). Isolation of naturally occurring enteroviruses from a variety of shellfish species residing in Long Island and New Jersey marine embayments, J. Food Prot., 43: 95.

281. Wait, D. A., Hackney, C. R., Carrick, R. J., Lovelace, G., and Sobsey, M. D. (1983). Enteric bacterial and viral pathogens and indicator bacteria in hard shell clams, J. Food Prot., 46: 493.

282. Jehl-Pietri, C., Dupont, J., Derve, C., Menard, D., and Munro, J. (1991). Occurrence of faecal bacteria, salmonella and antigens associated with hepatitis A virus in shellfish, Zbl. Hyg., 192: 230.

283. Mackowiak, P. A., Caraway, C. T., and Portnoy, B. L. (1976). Oysters associated hepatitis: lessons from the Louisiana experience, Am. J. Epidemiol., 103: 181.

284. Konowalchuk, J., and Speirs, J. I. (1972). Enterovirus recovery from laboratory-contaminated samples of shellfish, Can. J. Microbiol., 18: 1023.

285. Kostenbader, K. D., and Cliver, D. (1972). Polyelectrolyte flocculation as an aid to recovery of enteroviruses from oysters, Appl. Microbiol., 24: 540.

286. Sobsey, M. D., Wallis, C., and Melnick, J. L. (1975). Development of a simple method for concentrating enteroviruses from oysters, Appl. Environ. Microbiol., 29: 21.

287. Sobsey, M. D., Carrick, R. J., and Jensen, H. R. (1978). Improved methods for detecting enteric viruses in oysters, Appl. Environ. Microbiol., 36: 121.

288. Larkin, E. P., and Metcalf, T. G. (1980). Cooperative study of methods for the recovery of enteric viruses from shellfish, J. Food Prot., 43: 84.

289. Richards, G. P., Goldmintz, D., Green, D. L., and Babinchak, J. A. (1982). Rapid methods for extraction and concentration of poliovirus from oyster tissues, J. Virol. Meth., 5: 285.

290. Sullivan, R, and Peeler, J. T. (1982). Evaluation of a method for recovery of virus from oysters, J. Food Prot., 45: 636.

291. Sullivan, R., Peeler, J. T., Tierney, J. T., and Larkin, E. P. (1984). Evaluation of a method for recovering poliovirus 1 from 100-gram oyster samples, J. Food Prot., 47: 108.

292. Lewis, G. D., and Metcalf, T. G. (1988). Polyethylene glycol precipitation for recovery of pathogenic viruses, including hepatitis A virus and human rotavirus, from oyster, water, and sediment samples, Appl. Environ. Microbiol., 54: 1983.

293. Tierney, J. T., Sullivan, R, Peeler, J. T., and Larkin, E. P. (1985). Detection of low numbers of poliovirus 1 in oysters: collaborative study, J. Assoc. Off. Anal. Chem., 68: 884.

294. Zhou, Y.-J., Estes, M. K., Jiang, X., and Metcalf, T. G. (1991). Concentration and detection of hepatitis A virus and rotavirus from shellfish by hybridization tests, Appl. Environ. Microbiol., 57: 2963.

295. Atmar, R. L., Neil, F. H. and Romalde, J. L., et al. (1995). Detection of Norwalk virus and hepatitis A virus in shellfish tissues with the PCR, Appl. Environ. Microbiol., 61: 3014.

296. Lees, D. N., Henshilwood, K., and Butcher, S. (1995). Development of a PCR-based method for the detection of enteroviruses and hepatitis A virus in molluscan shellfish and its application to polluted field samples, Water Sci. Tech., 31: 457.

297. Chung, H., Jaykus, L.-A., and Sobsey, M. D. (1996). Detection of human enteric viruses in oysters by in vivo and in vitro amplification of nucleic acids, Appl. Environ. Microbiol., 62: 3772.

298. Traore, O., Arnal, C., Mignotte, B., Maul, A., Laveran, H., Billaudel, S., and Schwartzbrod, L. (1998). Reverse transcriptase PCR detection of astrovirus, hepatitis A virus, and poliovirus in experimentally contaminated mussels: comparison of several extraction and concentration methods, Appl. Environ. Microbiol., 64: 3118.

299. Sobsey, M. D. (1987). Methods for recovering viruses from shellfish, seawater, and sediments, Methods for Recovering Viruses from the Environment (G. Berg, eds.), CRC Press, Boca Raton, FL, p.77.

300. Jiang, X., Estes, M. K., Metcalf, T. G., and Melnick, J. L. (1986). Detection of hepatitis A virus in seeded estuarine samples by hybridization with cDNA probes, Appl. Environ. Microbiol., 52: 711.

301. Dix, A. B., and Jaykus, L.-A. (1998). Virion concentration method for the detection of human enteric viruses in extracts of hard-shelled clams, J. Food Prot., 61: 458.

302. Crance, J. M., Apaire-Marchais, V., and Leveque, F., et al. (1995). Detection of hepatitis A virus in wild shellfish, Marine Pollution Bulletin, 30: 372.

303. Romalde, J. L., Estes, M. K., and Szucs, G., et al. (1994). In situ detection of hepatitis A virus in cell cultures and shellfish tissues, Appl. Environ. Microbiol., 60: 1921.

304. Arnal, C., Ferre-Aubineau, V., and Besse, B., et al. (1999). Comparison of seven RNA extraction methods on stool and shellfish samples prior to hepatitis A virus amplification, J. Virol. Meth., 77: 17.

305. Monceyron, C., and Grinde, B. (1994). Detection of hepatitis A virus in clinical and environmental samples by immunomagnetic separation and PCR, J. Virol. Meth., 46: 157.

306. Jothikumar, N., Cliver, D. O., and Mariam, T. (1998). Immunomagnetic capture PCR for rapid concentration and detection of hepatitis A virus from environmental samples, Appl. Environ. Microbiol., 64: 504.

307. Arnal, C., Ferre-Aubineau, V., Mignotte, B., Imbert-Marcille, B. M., and Billaudel, S. (1999). Quantification of hepatitis A virus in shellfish by competitive reverse transcription-PCR with coextraction of standard RNA, Appl. Environ. Microbiol., 65: 322.

308. Goswami, B. B., and Koch, W. H. (1994). Competitor template RNA for detection and quantification of hepatitis A virus by PCR, Biotechniques, 16: 114.

309. LeGuyader, F., Menard, D., and Dubois, et al. (1997). Use of an RT-PCR internal control to evaluate viral removal, Water Sci. Tech., 35: 461.

310. Franco, E., Toti, L., Fabrieli, R., Croci, L., De Medici, D., and Pana, A. (1990). Depuration Mytilus Galloprovincialis experimentally contaminated with hepatitis A virus, Int. J. Food Micro., 11: 321.

310a. Power, U. F., and Collins, J. K. (1990). Elimination of coliphages and Escherichia coli from mussels during depuration under varying conditions of temperature salinity and food availability, J. Food Prot., 53: 208.

311. Richards, G. (1985). Outbreaks of shellfish-associated enteric virus illness in the United States: requisite for development of viral guidelines, J. Food Prot., 48: 815.

312. Pina, S., Puig, M., Lucena, F., Jofre, J., and Girones, R. (1998). Viral pollution in the environment and in shellfish: human adenovirus detection by PCR as an index of human viruses, Appl. Environ. Microbiol., 64: 3376.

313. Jansen, R. W., Newbold, J. E., and Lemon, S. M. (1988). Complete nucleotide sequence of a cell culture–adapted variant of hepatitis A virus: comparison with wild-type virus with restricted capacity for in vitro replication, Virology, 163: 299.

314. New York State (1992). Chapter 529, Section 14-1.80, effective August 19, 1992.

315. Ming, YiD., and Cliver, D. O. (1995). Persistence of inoculated hepatitis A virus in mixed human and animal wastes, Appl. Environ. Microbiol., 61: 87.

316. Casteel, M. J., and Sobsey, M. D. (1998). Chlorine disinfection of strawberries to inactivate hepatitis A virus and other microbes. ASM Abstracts, P-102, p. 420.

317. Casteel, M. J., and Sobsey, M. D. (1999). Peroxyacetic acid disinfection of strawberries to inactivate hepatitis A virus (HAV) and other microbes, ASM Abstracts, P-58.

318. Neefe, J., Stokes, J., Baty, J., and Reinhold, J. (1945). Disinfection of water containing causative agent of infectious (epidemic) hepatitis, JAMA, 128: 1076.

319. Peterson, D. A., Hurley, T. R., Hoff, J. C., and Wolfe, L. G. (1983). Effect of chlorine treatment on infectivity of hepatitis A virus, Appl. Environ. Microbiol., 45: 223.

320. Grabow, W. O. K., Gaus-Muller, V., Prozesky, O. W., and Deinhardt, F. (1983). Inactivation of hepatitis A virus and indicator organisms in water by free chlorine residuals, Appl. Environ. Microbiol., 46: 619.

321. Hall, R. M., and Sobsey, M. D. (1993). Inactivation of hepatitis A virus and MS2 by ozone and ozone-hydrogen peroxide in buffered water, Water Sci. Tech., 27: 371.

322. Winokur, P. L., and Stapleton, J. T. (1992). Immunoglobulin prophylaxis for hepatitis A, Clin. Infect. Dis., 14: 580.

323. Pierce, P. F., Capello, M., and Bernard, K. W. (1990). Subclinical infection with hepatitis A in Peace Corps volunteers following immune globulin prophylaxis, Am. J. Trop. Med. Hyg., 42: 465.

324. Lange, W. R., and Frame, J. D. (1990). High incidence of viral hepatitis among American missionaries in Africa, Am. J. Trop. Med. Hyg., 43: 527.

325. Steffen, R. (1992). Risk of hepatitis A in travelers, Vaccine, 10: s69.

326. Flehmig, B., Heinricy, U., and Pfisterer, M. (1989). Immunogenicity of a killed hepatitis A vaccine in seronegative volunteers, Lancet, 1(8646): 1039.

327. Andre, F. E., Hepburn, A., and D'Hondt, E. D. (1990). Inactivated candidate vaccines for hepatitis A, Prog. Med. Virol., 37: 73.

328. Just, M., and Berger, R. (1992). Reactogenicity and immunogenicity of inactivated hepatitis A vaccines, Vaccine, 10: s110.

329. Wiedermann, G., Ambrosch, F., Kollaritsch, H., Hofmann, H., Kunz, C., D'Hondt, E., Delen, A., Andre, F. E., Safary, A., and Stephenne, J. (1990). Safety and immunogenicity of an inactivated hepatitis A candidate vaccine in healthy adult volunteers, Vaccine, 8: 581.

330. Riedemann, S., Reinhardt, G., and Frosner, G. G., et al. (1992). Placebo-controlled efficacy study of hepatitis A vaccine in Valdivia, Chile, Vaccine, 10: s152.

331. Provost, P. J., Bishop, R. P., Gerety, R. J., Hilleman, M. R., McAleer, W. J., Scolnick, E. M., and Stevens, C. E. (1986). New findings in live, attenuated hepatitis A vaccine development, J. Med. Virol., 20: 165.

332. Sjogren, M. H., Eckels, K. H., Binn, L. N., Burke, D. S., and Bancroft, W. H. (1987). Safety and immunogenicity of an inactivated hepatitis A vaccine in volunteers, J. Med. Virol., 21(Suppl. 74): 25A.

333. Ellerbeck, E., Lewis, J., and Midthun, K., et al. (1991). Safety and immunogenicity of an inactivated hepatitis A virus vaccine, Viral Hepatitis and Liver Disease (F. B. Hollinger, S. M. Lemon, and H. Margolis, eds.), Williams & Wilkins, Baltimore, p. 91.

334. Innis, B. L., Snitbhan, R., Kunasol, P., Laorakpongse, T., Poopatanakool, W., Kozik, C. A.,

Suntayakorn, S., Subnantapong, T., Safary, A., and Tang, D. B., et al. (1994). Protection against hepatitis A by an inactivated vaccine. JAMA 271: 1328.

335. Van Damme, P., Thoelen, S., De Groote, K., Safary, A., and Meheus, A. (1994). Inactivated hepatitis A vaccine: reactogenicity, immunogenicity, and long-term antibody persistence, J. Med. Virol., 44: 446.

336. Provost, P. J., Banker, F. S., Giesa, P. A., McAleer, W. J., Buynak, E. B., and Hilleman, M. R. (1982). Progress toward a live, attenuated human hepatitis A vaccine, Proc. Soc. Exp. Biol. Med, 170: 8.

337. Provost, P. J., Conti, P. A., Giesa, P. A., Banker, F. S., Buynak, E. B., McAleer, W. J., and Hilleman, M. R. (1983). Studies in chimpanzees of live, attenuated hepatitis A vaccine candidates, Proc. Soc. Exp. Biol. Med, 172: 357.

338. Mao, J. S., Dong, D. X., Zhang, H. Y., Chen, N. L., Zhang, X. Y., Huang, H. Y., Xie, R. Y., Zhou, T. J., Wan, Z. J., Wang, Y. Z., Hu, Z. H., Cao, Y. Y., Li, H. M., and Chu, C. M. (1989). Primary study of attenuated live hepatitis A vaccine (H2 strain) in humans, J. Infect. Dis., 159: 621.

339. Mao, J. S., Dong, D. X., Zhang, S. Y., Zhang, H. Y., Chen, N. L., Huang, H. Y., Xie, R. Y., Chai, C. A., Zhou, T. J., Wu, D. M., and Zhang, H. C. (1991). Further studies of attenuated live hepatitis A vaccine (strain H2) in humans, Viral Hepatitis and Liver Disease (F. B. Hollinger, S. M. Lemon, and H. S. Margolis, eds.), Williams & Wilkins, Baltimore, p. 110.

340. Midthun, K., Ellerbeck, E., Gershman, K., Calandra, G., Krah, D., McCaughtry, M., Nalin, D., and Provost, P. (1991). Safety and immunogenicity of a live attenuated hepatitis A virus vaccine in seronegative volunteers, J. Infect. Dis., 163: 735.

341. Johnston, J. M., Harmon, S. A., Binn, L. N., Richards, O. C., Ehrenfeld, E., and Summers, D. F. (1988). Antigenic and immunogenic properties of a hepatitis A virus capsid protein expressed in Escherichia coli, J. Infect. Dis., 157: 1203.

342. Karayiannis, P., O'Rourke, S., McGarvey, M. J., Luther, S., Waters, J., Goldin, R., and Thomas, H. C. (1991). A recombinant vaccinia virus expression hepatitis A virus structural polypeptides: characterization and demonstration of protective immunogenicity, J. Gen. Virol., 72: 2167.

343. Stapleton, J. T. (1995). Host immune response to hepatitis A virus, J. Infect. Dis., 171: S9.

344. Craig, A. S., Sockwell, D. C., Schaffner, W., Moore, W. L., Skinner, J. T., Williams, I. T., Shaw, F. E., Shapiro, C. N., and Bell, B. P. (1997). Use of hepatitis A vaccine in a community-wide outbreak of hepatitis A, Clin. Infect. Dis., 27: 531.

345. Wong, D. C., Purcell, R. H., Sreenivasan, M. A., Prasad, S. R., and Pavri, K. M. (1980). Epidemic and endemic hepatitis in India: evidence for non-A/non-B hepatitis virus etiology, Lancet, 2: 876.

346. Khuroo, M. S. (1980). Study of an epidemic of non-A, non-B hepatitis: possibility of another human hepatitis virus distinct from post-transfusion non-A, non-B type, Am. J. Med., 68: 818.

347. Rioche, M. K., Himmich, H., Cherkaoui, A., Mourid, A., Dobreuil, P., Zahroui, M. K., and Pillot, J. (1991). High incidence of sporadic non-A, non-B hepatitis in Morocco: epidemiologic study, Bull. Soc. Pathol. Exot. Filiales, 84: 117.

348. Clayson, E. T., Shrestha, M. P., Vaughn, D. W., Snitbhan, R., Shrestha, K. B., Longer, C. F., and Innis, B. L. (1997). Rates of hepatitis E virus infection and disease among adolescents and adults in Kathmandu, Nepal, J. Infect. Dis., 176: 763.

349. Favorov, M. O., Khudyakov, Y. E., Fields, H. A., Khudayakova, N. S., Padhue, N., Alter, M. J., Krawczynski, K., Mast, E., Polish, L., Shapiro, C., Yashina, T. L., Yarasheva, D. M., and Margolis, H. S. (1994). Enzyme immunoassay for the detection of antibody to hepatitis E virus based on synthetic peptides, J. Virol. Meth., 46: 237.

350. Arankalle, V. A., Tsarev, S. A., Chadha, M. S., Alling, D. W., Emerson, S. U., Banerjee, K., and Purcell, R. H. (1995). Age-specific prevalence of antibodies to hepatitis A and E viruses in Pune, India, 1982 and 1992, J. Infect. Dis., 171: 447.

351. Rab, M. A., Bile, M. K., Mubarik, M. M., Asghar, H., Sami, Z., Siddiqi, S., Dil, A. S., Barzgar, M. A., Chaudhry, M. A., and Burney, M. I. (1997). Water-borne hepatitis E virus epidemic in Islamabad, Pakistan: a common source outbreak traced to the malfunction of a modern water treatment plant, Am. J. Trop. Med. Hyg., 57: 1244.

352. Corwin, A. L., Khiem, HaB., Clayson, E. T., Sac, P. K., Nhung, V. T. T., Yen, V. T., Cuc, C. T. T., Vaughn, D., Merven, J., Richie, T. L., Putri, M. P., He, J., Grahma, R., Wignall, F. S., and Hyams, K. C. (1996). A waterborne outbreak of hepatitis E virus transmission in southwestern Vietnam, Am. J. Trop. Med. Hyg., 54: 559.

353. Sreenivasan, M. A., Banerjee, K., Pandya, P. G., Kotak, R. R., Pandya, P. M., Desai, N. J., and Vaghela, L. H. (1978). Epidemiological investigations of an outbreak of infectious hepatitis in Ahmedabad City during 1975–76, Indian J. Med. Res., 67: 197.

354. Sergeev, N. W., Paktoris, E. A., Ananev, W. A., Sinajko, G. A., Antinova, A. I., and Semenov, E. P. (1957). General characteristics of Botkin's disease occurring in Kirgiz Republic of USSR in 1955–1956, Soviet Healthcare Kirgizii, 5: 16.

355. Iarasheva, D. M., Favorov, M. O., Iashina, T. L., Shakgil'dian, I. V., Umarova, A. A., Sorokina, S. A., Kamardinov, K. K., and Mavashev, V. I. (1991). The etiological structure of acute viral hepatitis in Tadzhikistan in a period of decreased morbidity, Vopr. Virusol., 36: 454.

356. Tandon, B. N., Joshi, Y. K., Jain, S. K., Gandi, B. M., Mathiesen, L. R., and Tandon, H. D. (1982). An epidemic of non-A, non-B hepatitis in north India, Indian J. Med. Res., 75: 739.

357. Ray, R., Aggarwal, R., Salunke, P. N., Mehrotra, N. N., Talwar, G. P., and Naik, S. R. (1991). Hepatitis E virus genome in stools of hepatitis patients during large epidemic in north India, Lancet, 338: 783.

358. Khuroo, M. S., Duermeyer, W., Zargar, S. A., Ahanger, M. A., and Shah, M. A. (1983). Acute sporadic non-A, non-B hepatitis in India, Am. J. Epidemiol., 118: 360.

359. Hillis, A., Shrestha, S. M., and Saha, N. K. (1973). An epidemic of infectious hepatitis in the Kathmandu Valley, J. Nepal. Med. Assoc., 11: 145.

360. Shrestha, S. M., and Mala, D. S. (1957). J. Nepal. Med. Assoc., 13: 58.

361. De Cock, K. M., Bradley, D. W., Sandford, N. L., Govindarajan, S., Maynard, J. E., and Redeker, A. G. (1987). Epidemic non-A, non-B hepatitis in patients from Pakistan, Ann. Intern. Med., 106: 227.

362. Myint, H., Soe, M. M., Khin, T., Myint, T. M., and Tin, K. M. (1985). A clinical and epidemiological study of an epidemic of non-A, non-B hepatitis in Rangoon, Am. J. Trop. Med. Hyg., 34: 1183.

363. Ticehurst, J., Popkin, T. J., Bryan, J. P., Innis, B. L., Duncan, J. F., Ahmed, A., Iqbal, M., Malik, I., Kapikian, A. Z., Legters, L. J., and Purcell, R. H. (1992). Association of hepatitis E virus with an outbreak of hepatitis in Pakistan: serologic responses and pattern of virus excretion, J. Med. Virol., 36: 84.

364. Xia, X. (1991). An epidemiologic survey on a type E hepatitis (HE) outbreak, Chung Hua Liu Hsing Ping Hsueh Tsa Chih (China), 12: 257.

365. Zhuang, H., Cao, W.-Y., Liu, C.-B., and Wang, G.-M. (1991). Epidemiology of hepatitis E in China, Gastroenterol. Jpn., 26: 135.

366. Lok, A. S. F., Kwan, W.-K., Moeckli, R., Yarbrough, P. O., Chan, R. T., Reyes, G. R., Lai, C.-L., Chung, H.-T., and Lai, T. S. T. (1992). Seroepidemiological survey of hepatitis E in Hong Kong by recombinant-based enzyme immunoassays, Lancet, 340: 1205.

367. Bradley, D. W. and Balayan, M. S. (1988). Virus of enterically transmitted non-A, non-B hepatitis [Letter], Lancet, 2: 819.

368. Goldsmith, R., Yarbough, P. O., Reyes, G. R., Fry, K. E., Gabor, K. A., Kamel, M., Zakaria, S., Amer, S., and Gaffar, Y. (1992). Enzyme-linked immunosorbent assay for diagnosis of acute sporadic hepatitis E in Egyptian children, Lancet, 339: 328.

369. Sarthou, J. L., Budkowska, A., Sharma, M. D., Lhuillier, M., and Pillot, J. (1976). Character-

ization of an antigen–antibody system associated with epidemic non-A, non-B hepatitis in West Africa and experimental transmission of an infectious agent to primates, Ann. Inst. Pasteur/Virol., 137E: 225.

370. Centers for Disease Control (1987). Enterically transmitted non-A, non-B hepatitis–East Africa, MMWR, 36: 241.

371. Tsega, E., Krawczynski, K., Hansson, B.-G., Nordenfelt, E., Negusse, Y., Alemu, W., and Bahru, Y. (1991). Outbreak of acute hepatitis E virus infection among military personnel in northern Ethiopia, J. Med. Virol., 34: 232.

372. Diallo, A., Lazizi, Y., Le Guenno, B., and Pillot, J. (1991). Hepatitis-E-virus-associated antigen: improved detection in stools by protein Fv removal, Ann. Inst. Pasteur/Virol., 142: 449.

373. Velazquez, O., Stetler, H. C., Avila, C., Ornelas, G., Alvarez, C., Hadler, S. C., Bradley, D. W., and Sepulveda, J. (1990). Epidemic transmission of enterically transmitted non-A, non-B hepatitis in Mexico, 1986–1987, JAMA, 263: 3261.

374. Paul, K. Y., Schlauder, G. G., Carpenter, H. A., Murphy, P. J., Rosenblatt, J. E., Dawson, G. J., Mast, E. E., Krawczynski, K., and Balan, V. (1997). Acute hepatitis E by a new isolate acquired in the United States, Mayo Clin. Proc., 72: 1133.

375. Kane, M. A., Bradley, D. W., Shrestha, S. M., Maynard, J. E., Cook, E. H., Mishra, R. P., and Joshi, D. D. (1984). Epidemic non-A, non-B hepatitis in Nepal: recovery of a possible etiologic agent and transmission studies to marmosets, JAMA, 252: 3140.

376. Shakhgildian, I. V., Khukhlovich, P. A., Kuzin, S. N., Favorov, M. O., and Nedachin, A. E. (1986). Epidemiological characteristics of non-A, non-B hepatitis with a fecal-oral transmission mechanism, Vopr. Virusol., 31: 175.

377. Favorov, M. O., Khukhlovich, P. A., Zairov, G. K., and Listovskaia, E. K. (1986). Clinical-epidemiological characteristics and diagnosis of viral non-A, non-B hepatitis with fecal and oral mechanisms of transmission of the infection, Vopr. Virusol., 31: 65.

378. Bile, K., Isse, A., Mohamud, O., Allebeck, P., Nilsson, L., Norder, H., Mushahwar, I. K., and Magniums, L. (1994). Contrasting roles of rivers and wells as sources of drinking water on attack and fatality rates in hepatitis E epidemic in Somalia, Am. J. Trop. Med. Hyg., 51: 466.

379. Coursaget, P., Buisson, Y., Enogat, N., Bercion, R., Baudet, J.-M., Delmaire, P., Prigent, D., and Desrame, J. (1998). Outbreak of enterically-transmitted hepatitis due to hepatitis A and hepatitis E viruses, J. Hepatol. 28: 745.

380. Arankalle, V. A., Chadha, M. S., Mehendale, S. M., and Banerjee, K. (1988). Outbreak of enterically transmitted non-A, non-B hepatitis among school children, Lancet, 2: 1199.

381. Arankalle, V. A., Tsarev, S. A., Chadha, M. S., Alling, D. W., Emerson, S. U., Banerjee, K., and Purcell, R. H. (1995). Age-specific prevalence of antibodies to hepatitis A and E viruses in Pune, India, 1982 and 1992, J. Infect. Dis., 171: 447.

382. Bryan, J. P., Tsarev, S. A., Iqbal, M., Ticehurst, J., Emerson, S., Ahmed, A., Duncan, J., Rafiqui, A. R., Malik, I., Purcell, R. H., and Legters, L. J. (1994). Epidemic hepatitis E in Pakistan: patterns of serologic response and evidence that antibody to hepatitis E virus protects against disease, J. Infect. Dis., 170: 517.

383. Mast, E. E., Kuramoto, K., Favorov, M. O., Schening, V. R., Burkholder, B. T., Shapiro, C. N., and Holland, P. V. (1997). Prevalence of and risk factors for antibody to hepatitis E virus seroreactibity among blood donors in northern California, J. Infect. Dis., 176: 34.

384. Favorov, M. O., Kuzin, S. N., Iashina, T. L., Zairov, G. K., Gurov, A., V, Shavakhabov, ShS., Buriev, A. I., Zhantemirov, B. U., Shakhgildian, I., V., and Ketiladze, E. S. (1989). Characteristics of viral hepatitis non-A, non-B with a fecal- oral mechanism of transmission of the infection in southern Uzbekistan, Vopr. Virusol., 34: 436.

385. Bellabes, H., Benatallah, A., and Bourguermouh, A. (1984). Non-A, non-B epidemic viral hepatitis in Algeria: strong evidence for its water spread, Viral Hepatitis and Liver Disease (G. N. Vyas, J. L. Dienstag, and J. H. Hoffnagle, eds.), Grune and Stratton, Orlando, FL, p. 637.

387. Centers for Disease Control (1987). Enterically transmitted non-A, non-B hepatitis—Mexico, MMWR, 36: 597.

388. Melnick, J. L. (1957). A water-borne urban epidemic of hepatitis, Hepatitis Frontiers (F. W. Hartman, G. A. Logrippo, J. G. Matffer, et al., eds.), Little, Brown and Company, Boston, p. 211.

389. Khuroo, M. S., Teli, M. R., Skidmore, S., Sofi, M. A., and Khuroo, M. I. (1981). Incidence and severity of viral hepatitis in pregnancy, Am. J. Med., 70: 252.

390. Balayan, M. S., Andjaparidze, A. G., Savinskaya, S. S., Ketiladze, E. S., Braginsky, D. M., Savinov, A. P., and Poleschuk, V. F. (1983). Evidence for a virus in non-A, non-B hepatitis transmitted via the fecal-oral route, Intervirology, 20: 23.

391. Favorov, M., Fields, H., Musarov, A., Khudyakova, N., Yashina, T., Tesler, D., and Margolis, H. (1992). Clinical and epidemiological description and specific diagnosis of fulminant hepatitis E, J. Infect. Dis. (submitted).

392. Bradley, D. W., Krawczynski, K., Cook, E. H., Jr., McCaustland, K. A., Humphrey, C. D., Spelbring, J. E., Myint, H., and Maynard, J. E. (1987). Enterically transmitted non-A, non-B hepatitis: serial passage of disease in cynomolgus macaques and tamarins, and recovery of disease-associated 27 to 34 nm viruslike particles, Proc. Natl. Acad. Sci. USA, 84: 6277.

393. Tsarev, S. A., Tsareva, T. S., Emerson, S. U., Yarbough, P. O., Legters, L. J., Moskal, T., and Purcell, R. H. (1994). Infectivity titration of a prototype strain of hepatitis E virus in cynomologus monkeys, J. Med. Virol., 43: 135.

394. Bradley, D. W., Andjaparidze, A., Cook, E. H., Jr., McCaustland, K., Balayan, M., Stetler, H., Velazquez, O., Robertson, B., Humphrey, C., Kane, M., and Weisfuse, I. (1988). Aetiological agent of enterically transmitted non-A, non-B hepatitis, J. Gen. Virol., 69: 731.

395. Andjaparidze, A. G., Balayan, M. S., Savinov, A. P., Braginskiy, D. M., Poleschuk, V. F., and Zamyatina, N. A. (1986). Fecal-orally transmitted non-A, non-B hepatitis induced in monkeys, Vopr. Virusol., 1: 73.

396. Arankalle, V. A., Ticehurst, J., Sreenivasan, M. A., Kapikian, A. Z., Popper, H., Pavri, K. M., and Purcell, R. H. (1988). Aetiological association of a virus-like particle with enterically transmitted non-A, non-B hepatitis, Lancet, 1: 550.

397. Krawczynski, K., McCaustland, K., Mast, E., Yarbough, P. O., Purdy, M., Favorov, M. O., and Spellbring, J. (1996). Elements of pathogenesis of HEV infection in man and experimentally infected primates, Enterically-Transmitted Hepatitis Viruses (Y. Buisson, P. Coursaget, and M. Kane, eds.), La Simmarre, Tours, France, p.317.

398. Purdy, M. A., McCaustland, K. A., Krawczynski, K., Tam, A., Beach, M. J., Tassopoulos, N. C., Reyes, G. R., and Bradley, D. W. (1992). Expression of a hepatitis E virus (HEV)-trpE fusion protein containing epitopes recognized by antibodies in sera from human cases and experimentally infected primates, Arch. Virol., 123: 335.

399. Favorov, M. O., Fields, H. A., Purdy, M. A., Yashina, T. L., Aleksandrov, A. G., Alter, M. J., Yarasheva, D. M., Bradley, D. W., and Margolis, H. S. (1992). Serologic identification of hepatitis E virus infections in epidemic and endemic settings, J. Med. Virol., 36: 246.

400. Favorov, M. O., Khudyakov, Y. E., Mast, E. E., Yahsina, T. L., Shapiro, C. N., Khudyakova, N. S., Jue, D. L., Onischenko, G. G., Margolis, H. S., and Fields, H. A. (1996). IgM and IgG antibodies to hepatitis E virus (HEV) detected by an enzyme immunoassay based on an HEV-specific artificial recombinant mosaic protein, J. Med. Virol., 50: 50.

401. Reyes, G. R., Huang, C.-C., and Yarbough, P. O., et al. (1991). Hepatitis E virus: epitope mapping and detection of strain variation, Viral Hepatitis C, D, and E (T. Shikata, R. H. Purcell, and T. Uchida, eds.), Elsevier, Amsterdam, p.237.

402. Aggarwal, R. and Naik, S. R. (1994). Hepatitis E: intrafamilial transmission versus waterborne spread, J. Hepatol., 21: 718.

403. Meng, H., Yiang, X. C., and He, S. C. (1987). A foodborne outbreak of non-A, non-B hepatitis, Chin. J. Prev. Med., 21: 718.

404. Shi, G. R., Li, S. Q., and Qian, L. (1987). The epidemiological study on a foodborne outbreak of non-A, non-B hepatitis, J. Chin. Med. Univ., 16: 150.

405. Bai, F., Zhaorigetai, and Wu, B. R. (1987). A foodborne outbreak of non-A, non-B hepatitis, Neimenggu Med J., 7: 157.

406. Caredda, F., Antinori, S., Re, T., Pastecchia, C., Zavaglia, C., and Mauro, M. (1985). Clinical features of sporadic non-A, non-B hepatitis possibly associated with faecal-oral spread, Lancet, 2: 444.

407. Jothikumar, N., Aparna, K., Kamatchiammal, S., Paulmurugan, R., Saravanadeve, S., and Khanna, P. (1993). Detection of hepatitis E virus in raw and treated wastewater with the polymerase chain reaction, Appl. Environ. Microbiol., 59: 2558.

408. Pina, S., Jofre, J., Emerson, S. U., Purcell, R. H., and Girones, R. (1998). Characterization of a strain of infectious hepatitis E virus isolated from sewage in an area where hepatitis E is not endemic, Appl. Environ. Microbiol., 64: 4485.

409. Jothikumar, N., Khanna, P., Paulmurugan, R., Kamatchiammal, S., and Padmanabhan, P. (1995). A simple device for the concentration and detection of enterovirus, hepatitis E virus and rotavirus from water samples by reverse transcription-polymerase chain reaction, J. Virol. Meth., 55: 401.

410. Thomas, D. L., Yarbough, P. O., Vlahov, D., Tsarev, S. A., Nelson, K. E., Saah, A. J., and Purcell, R. H. (1997). Seroreactivity to hepatitis E virus in areas where the disease is not endemic, J. Clin. Microbiol., 35: 1244.

411. Hsieh, S. Y., Yang, P. Y., Ho, Y. P., Chu, C. M., and Liaw, Y. F. (1998). Identification of a novel strain of hepatitis E virus responsible for sporadic acute hepatitis in Taiwan, J. Med. Virol., 55: 300.

412. Bradley, D. W. (1990). Hepatitis non-A, non-B viruses become identified as hepatitis C and E viruses, Prog. Med. Virol., 37: 101.

413. Bradley, D. W. (1990). Enterically-transmitted non-A, non-B hepatitis, B. Med. Bull., 46: 42.

414. Uchida, T., Suzuki, K., Komatsu, K., Iida, F., Shikata, T., Rikihisa, T., Mizuno, K., Soe, S., Win, K. M., and Nakane, K. (1990). Occurence and character of a putative causative virus of enterically-transmitted non-A, non-B hepatitis in bile, Jpn. J. Exp. Med., 60: 23.

415. Reyes, G. R., Purdy, M. A., Kim, J. P., Luk, D.-C., Young, L. M., Tam, A. W., and Bradley, D. W. (1990). Isolation of a cDNA from the virus responsible for enterically transmitted non-A, non-B hepatitis, Science, 247: 1335.

416. Reyes, G. R., Purdy, M. A., Kim, J., Young, L. M., Luk, K.-C., and Bradley, D. W. (1990). Enterically transmitted NANB hepatitis—identification and characterization of a molecular clone, Viral Hepatitis and Hepatocellular Carcinoma (J. L. Sung and D. S. Chen, eds.), Excerpta Medica, Hong Kong, p.249.

417. Yarbough, P. O., Tam, A. W., Fry, K. E., Krawczynski, K., McCaustland, K. A., Bradley, D. W., and Reyes, G. R. (1991). Hepatitis E virus: identification of type-common epitopes, J. Virol., 65: 5790.

418. Yarbough, P. O., Tam, A. W., Krawczynski, K., Fry, K. E., McCaustland, K. A., Miller, A., Fernandez, J., Huang, C.-C., Bradley, D. W., and Reyes, G. R. (1991). Hepatitis E virus (HEV): identification of immunoreactive clones, Viral Hepatitis and Liver Disease (F. B. Hollinger, S. M. Lemon, and H. S. Margolis, eds.), Williams & Wilkins, Baltimore, p.524.

419. Reyes, G. R. and Kim, J. P. (1992). Sequence independent, single-primer amplification (SISPA) of complex DNA populations, Mol. Cell. Probes, 5: 473.

420. Meng, J., Pillot, J., Dai, X., Fields, H. A., and Khudyakov, Y. E. (1998). Neutralization of different geographic strains of the hepatitis E virus with anti-hepatitis E virus–positive serum samples obtain from different sources, Virology, 249: 316.

421. Tam, A. W., White, R., Reed, E., Short, M., Zhang, Y., Fuerst, T. R., and Lanford, R. E. (1996). In vitro propagation and production of hepatitis E virus from in vivo–infected primary macaque hepatocytes, Virology, 215: 1.

422. Tam, A. W., Smith, M. M., Guerra, M. E., Huang, C.-C., Bradley, D. W., Fry, K. E., and

Reyes, G. R. (1991). Hepatitis E virus (HEV): molecular cloning and sequencing of the full-length viral genome, Virology, 185: 120.

423. Zafrullah, M., Ozdener, M. H., Panda, S. K., and Jameel, S. (1997). The ORF3 protein of hepatitis E virus is a phosphoprotein that associates with the cytoskeleton, J. Virol., 71: 9045.

424. Tohya, Y., Taniguchi, Y., Utagawa, E., Takeda, N., Miyamura, K., Yamazaki, S., and Mikami, T. (1991). Sequence analysis of the 3′ end of feline calicivirus genome, Virology, 183: 810.

425. Cubitt, D., Bradley, D. W., Carter, M. J., Chiba, S., Estes, M. K., Saif, L. J., Schaffer, F. L., Smith, A. W., Studdert, M. J., and Thiel, H. J. (1995). Caliciviridae, Virus Taxonomy: Classification and Nomenclature of Viruses (F. A. Murphy, C. M. Fauquet, D. H. L. Dishop, S. A. Ghabrial, A. W. Jarvis, G. P. Martelli, M. A. Mayo, and M. D. Summers, eds.), in Arch Virol 10 (Suppl.): 359.

426. Tsarev, S., Emerson, S. U., Reyes, G. R., Tsareva, T. S., Legters, L. J., Malik, I. A., Iqbal, M., and Purcell, R. H. (1992). Characterization of a prototype strain of hepatitis E virus, Proc. Natl. Acad. Sci. USA, 89: 559.

427. Yin, S., Purcell, R. H., and Emerson, S. U. (1994). A new Chinese isolate of hepatitis E virus: comparison with strains recovered from different geographical regions, Virus Genes, 92: 23.

428. Gouvea, V., Snellings, N., Cohen, S. J., Warren, R. L., Myint, K. S. A., Shrestha, M. P., Vaughn, D. W., Hoke, C. H., and Innis, B. L. (1997). Hepatitis E virus in Nepal: similarities with the Burmese and Indian variants, Virus Res., 52: 87.

429. van Cuyck-Gandre, H., Zhang, H., Tsarev, S., Clements, N., Cohen, S., Caudill, J., Buisson, Y., Coursaget, P., Warren, R. L., and Longer, C. F. (1997). Characterization of hepatitis E virus (HEV) from Algeria and Chad by partial genome sequence, J. Med. Virol., 53: 340.

430. Chatterjee, R., Tsarev, S., Pillot, J., Coursaget, P., Emerson, S. U., and Purcell, R. H. (1997). African strains of hepatitis E virus that are distinct from Asian strains, J. Med. Virol., 53: 139

431. Gouvea, V., Cohen, S. J., Santos, N., Myint, K. S. A., Hoke, C., and Innis, B. L. (1997). Identification of hepatitis E virus in clinical specimens: amplification of hydroxyapatite-purified virus RNA and restriction endonuclease analysis, J. Virol. Meth., 69: 53.

432. Gouvea, V., Hoke, C. H., and Innis, B. L. (1998). Genotyping of hepatitis E virus in clinical specimens by restriction endonuclease analysis, J. Virol. Meth., 70: 71.

433. Aggarwal, R., McCaustland, K. A., Kilawari, J. B., Sinha, S. D., and Robertson, B. H. (1998). Genetic variability of hepatitis E virus within and between three epidemics in India, Virus Res., 59: 35.

434. Bradley, D. W., Krawczynski, K., Beach, M. J., and Purdy, M. K. A. (1991). Non-A, non-B hepatitis: toward the discovery of hepatitis C and E viruses, Semin. Liver Dis., 11: 128.

435. Cousaget, P., Buisson, Y., Depril, N., Cann, P., Chabaud, M., Molinie, C., and Roul, R. (1993). Mapping of linear B cell epitopes on open reading frames 2- and 3-encoded proteins of hepatitis E virus using synthetic peptides, FEMS Microbiol Lett., 109: 251.

436. Kaur, M., Hyams, K. C., Purdy, M. A., Krawczynski, K., Ching, W. M., Fry, K. E., Reyes, G. R., Bradley, D. W., and Carl, M. (1992). Human linear B-cell epitopes encoded by the hepatitis E virus include determinants in the RNA-dependent RNA polymerase, Proc. Natl. Acad. Sci. USA, 89: 3855.

437. Khudyakov, Y. E., Khudyakova, N. S., Fields, H. A., Jue, D., Starling, C., Favorov, M. O., Krawczynski, K., Polish, L., Mast, E., and Margolis, H. (1993). Epitope mapping in proteins of hepatitis E virus, Virology, 194: 89.

438. Khudyakov, Y. E., Favorov, M. O., Khudyakova, N. S., Cong, M., Holloway, B. P., Padhye, N., Lambert, S. B., Jue, D. L., and Fields, H. A. (1994). Artificial mosaic protein (MPr) containing antigenic epitopes of the hepatitis E virus, J. Virol., 68: 7067.

439. Dawson, G. J., Chau, K. H., Cabal, C. M., Yarbough, P. O., Reyes, G. R., and Mushahwar, I. K. (1992). Solid-phase enzyme-linked immunosorbent assay for hepatitis E virus IgG and IgM antibodies utilizing recombinant antigens and synthetic peptides, J. Virol. Meth., 38: 175.

440. Yarbough, P. O., Ram, A. W., Tam, K. E., Drawczynski, K., McCaustland, K. A., Bradley, D. W., and Reyes, G. R. (1991). Hepatitis E virus: identification of type common epitopes, J. Virol., 65: 5790.

441. Khuroo, M. S., Kamili, S., Dar, M. Y., Moecklii, R., and Jameel, S. (1993). Hepatitis E and long-term antibody status, Lancet, 341: 1355.

442. Favorov, M. O., Khudyakov, Y. E., Mast, E. E., Yashina, T. L., Shapiro, C. N., Khudyakova, N. S., Jue, D. L., Onischenko, G. G., Margolis, H. S., and Fields, H. A. (1996). IgM and IgG antibody to hepatitis E virus detected by an enzyme immunoassay based on HEV-specific artificial recombinant mosaic protein, J. Med. Virol., 50: 50.

443. Tsarev, S. A., Tsareva, T. S., and Emerson, S. U. (1993). ELISA for antibody to hepatitis E virus (HEV) based on complete open-reading frame-2 protein expressed in insect cells: identification of HEV infection in primates, J. Infect. Dis., 168: 369.

444. Thomas, D. L., Mahley, R. W., Badur, S., Paloglu, K. E., and Quinn, T. C. (1993). Epidemiology of hepatitis E virus infection in Turkey, Lancet, 341: 1561.

445. Arif, M., Qattan, I., Al-Faleh, F., and Ramia, S. (1994). Epidemiology of hepatitis E virus (HEV) infection in Saudi Arabia, Ann. Trop. Med. Parasitol., 88: 163.

446. Shapiro, C. N., and Margolis, H. S. (1993). Worldwide epidemiology of hepatitis A virus infection, J. Hepatol., 18(Suppl. 2): S11.

447. Mast, E. E., Alter, M. J., Holland, P. V., and Purcell, R. H. (1997). Evaluation of assays for antibody to hepatitis E virus by a serum panel, Hepatology, 27: 857.

448. Uchida, T., Suzuki, K., Iida, F., Shidata, T., Araki, M., Ichikawa, M., Rikihisa, T., Mizuno, K., Soe, S., and Win, K. (1991). Virulence of hepatitis E with serial passage to cynomolgus monkeys and identification of viremia, Viral Hepatitis and Liver Disease (F. B. Hollinger, S. M. Lemon, and H. S. Margolis, eds.), Williams & Wilkins, Baltimore, p.526.

449. McCaustland, K. A., Bi, S., Purdy, M. A., and Bradley, D. W. (1991). Application of two RNA extraction methods prior to amplification of hepatitis E virus nucleic acid by the polymerase chain reaction, J. Virol. Meth., 35: 331.

450. Bi, S. L., Purdy, M. A., McCaustland, K. A., Margolis, H. S., and Bradley, D. W. (1993). The sequence of hepatitis E virus isolated directly from a single source during an outbreak in China, Virus Res., 28: 233.

451. Cuyck-Gandre, H. V., Caudill, J. D., Zhang, H. Y., Longer, C. F., Molinie, C., Roue, R., Deloince, R., Coursaget, P., Mammouth, N. N., and Buisson, Y. (1996). Short report: polymerase chain reaction detection of hepatitis E virus in North African fecal samples, Am. J. Trop. Med. Hyg., 54: 134.

452. Aggarwal, R. and McCaustland, K. A. (1998). Hepatitis E virus RNA detection in serum and feces specimens with the use of microspin columns, J. Virol. Meth., 74: 209.

453. Nanda, S. K., Ansari, I. H., Acharya, S. K., Jameel, S., and Panda, S. K. (1995). Protracted viremia during acute sporadic hepatitis E virus infection, Gastroenterology., 108: 225.

454. Raimondo, G., Longo, G., Caredda, F., Saracco, G., and Rizzetto, M. (1986). Prolonged, polyphasic infection with hepatitis A, J. Infect. Dis., 153: 172.

455. Purdy, M., McCaustland, K., Krawczynski, K., Beach, M., Spelbring, J., Reyes, G., and Bradley, D. (1992). An expressed recombinant HEV protein that protects cynomologus macaques against challenge with wild-type hepatitis E virus. Proc. Second International Symposium on Persistent Virus Infections, Savannah, GA, p.41.

456. Tsarev, S. A., Tsareva, T. S., Emerson, S. U., Govindarajan, S., Shapiro, M., Gerin, J. L., and Purcell, R. H. (1994). Successful passive and active immunization of cynomolgus monkeys against hepatitis E, Proc. Natl. Acad. Sci. USA, 91: 10198.

457. Bradley, D. W., Beach, M. J., and Purdy, M. A. (1994). Molecular characterization of hepatitis C and E viruses, Arch. Virol., 7(s): 1.

458. CDC. (1996). Hepatitis Surveillance Report No. 56. U.S. department of Health and Human Services, Public Health Service, CDC, Atlanta, 6A.

459. Centers for Disease Control and Prevention (1998). Summary of Notifiable Diseases U.S., MMWR, 46: 54.

3

Norwalk Virus and the Small Round Viruses Causing Foodborne Gastroenteritis

Hazel Appleton
Central Public Health Laboratory, London, England

I. INTRODUCTION

Norwalk and Norwalk-like viruses are the most common cause of foodborne viral illness. Outbreaks of "food poisoning" or foodborne gastroenteritis over the years have been assumed to be caused by bacteria, whether or not such organisms are identified. In many incidents, no traditional food poisoning organism is found, and many notable incidents of unknown etiology have been recorded. It was the application of electron microscopy and the discovery of several gastroenteritis viruses in the 1970s that opened up the possibility of investigating apparent food poisoning outbreaks in which food poisoning bacteria were not detected. Viruses causing gastroenteritis usually are transmitted directly from person to person by the fecal-oral route, but it now is recognized clearly that on occasions some also may be transmitted through contaminated food and water. Of these, it is the Norwalk group of viruses that is the most important. Other gastroenteritis viruses rarely are involved.

Norwalk virus originated from an outbreak of nonbacterial gastroenteritis that occurred in a school in Norwalk, Ohio (1). It was suspected but never proven that the source of infection was a contaminated water supply. Illness spread by person-to-person contact within the community. Filtered stool suspension from one of the secondary cases was fed to healthy volunteers and induced illness (2). By means of immune electron microscopy (which already had been successful in revealing the identity of several other viruses), small virus particles were observed in the stools of volunteers during the period of illness (3). The virus could be propagated serially in volunteers and induced an antibody response. This was the first time a virus had been convincingly associated with gastroenteritis.

Other viruses morphologically similar to, but serologically distinct from, Norwalk virus have since been associated with gastroenteritis (4–6). In addition, other virus groups have been identified, also by electron microscopy, as causative agents of gastroenteritis; these groups include rotavirus, adenovirus types 40 and 41, and astrovirus. The Norwalk group of viruses cannot be cultured, and existing culture techniques for other gastroenteritis viruses are technically difficult and limited in application. The rapid development of molecular techniques during the 1990s has greatly facilitated studies and increased our understanding of the Norwalk group of viruses. Despite these advances, however, electron microscopy continues to be an essential tool in the investigation of all viruses causing gastroenteritis.

II. SMALL ROUND VIRUSES: CLASSIFICATION AND NOMENCLATURE

Various small round viruses in the size range of 20–40 nm may be found in feces; some of these have been associated with gastroenteritis. These viruses have a less striking appearance than rotaviruses and adenoviruses, which makes their detection and identification difficult. Furthermore, they are often excreted in smaller numbers than rotaviruses and adenoviruses. Electron microscopy is the method most widely used for detection. It is not a particularly sensitive method: a minimum of 10^6 particles/mL of specimen is required for detection. Shedding of viruses of the Norwalk group falls rapidly below detectable levels following onset of symptoms (7). The mechanical problems and small amounts of material available for study have made identification of the various small fecal viruses difficult and have restricted comparison of agents found in different outbreaks. Hence, there is considerable confusion in nomenclature in the literature.

However, a simple interim classification scheme based mainly on morphology has been used satisfactorily in the United Kingdom and elsewhere since 1982, and has proved a useful basis for reporting these viruses in the absence of a more formal classification (Table 1) (8). The small round fecal viruses are divided into five groups. Three of these groups have visible surface structure when viewed with the electron microscope: astrovirus, calicivirus, and small round structured virus (SRSV). Two groups have no discernible surface structure, i.e., enterovirus and parvovirus.

A. Structured Viruses

1. Astroviruses

Of the structured viruses, astroviruses form a novel group. They mainly infect the very young. A solid five- or six-point star can be seen on the surface of a small proportion of particles. The surface edge may appear smooth or slightly spiky (Fig. 1a).

2. Caliciviruses

Viruses with the classical appearance of a calicivirus have cuplike depressions visible on their surface, thus the name *calicivirus* from the Latin word *calix*, meaning ''cup'' (Fig.

Table 1 Small Round Viruses: Classification and Nomenclature

	Group	Characteristics	Examples
Viruses with surface structure	1. Small round structured virus (SRSV)	Size, 30–38 nm; RNA, amorphous surface, ragged outline	Norwalk,[a] Amulree, Hawaii, Montgomery County, Otofuke, Snow Mountain, Taunton
	2. Calicivirus	Size, 30–38 nm; RNA, surface hollows, ragged outline, ''Star of David'' configuration	Several human serotypes identified in United Kingdom and Japan
	3. Astrovirus	Size, 28–30 nm; RNA, 5–6 point surface star	8 human serotypes
Featureless viruses: Smooth entire outer edge and no discernible surface structure	1. Enterovirus (not normally associated with gastroenteritis)	Size, 25–30 nm; RNA, BD[b] 1.34 g/mL	Polioviruses, hepatitis A[c]
	2. Parvovirus	Size, 18–26 nm; DNA, BD 1.38–1.46 g/mL	Wollan, Ditchling, Parramatta, Cockle

[a] Norwalk is the prototype of the SRSV group. Norwalk and some other SRSVs now have been classified formally as caliciviruses.

[b] BD, buoyant density in cesium chloride

[c] Hepatitis A is morphologically identical to the enteroviruses but is now classified as a separate genus.

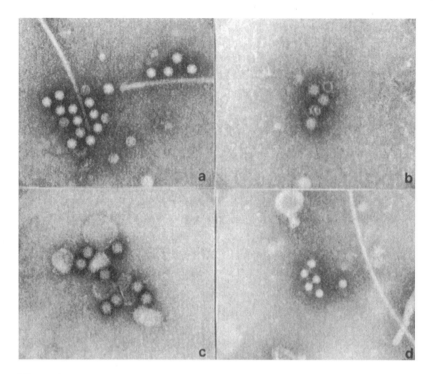

Figure 1 Electron micrographs of human enteric viruses, stained with phosphotungstic acid and magnified 200,000×: (a) astrovirus; (b) calicivirus; (c) small round structured viruses (SRSVs), identical in appearance to Norwalk virus; hence these viruses often are referred to as Norwalk-like or Norwalk group viruses; (d) human fecal parvovirus.

1b). Edges of the particles are spiky and in some orientations 10 projections can be counted clearly around the outer edge of the virus particle.

3. Small Round Structured Viruses

Small round structured viruses have an amorphous surface and ragged outline (Fig. 1c). Norwalk virus has the appearance of an SRSV and is the prototype of the group. Norwalk is an unusual virus in having only one major structural protein, which has a molecular weight of around 65 kDa. This led to the conclusion that SRSVs were very closely related to caliciviruses. However, SRSVs never seemed to show the classical surface structure of the calicivirus, but some well-characterized veterinary caliciviruses rarely show this classical appearance either. In 1992 the complete genome sequence of the Norwalk virus was described, resulting in its formal classification as a member of the Caliciviridae family by the International Committee on the Taxony of Viruses (9) (see Sec. III.F). Viruses with the appearance of SRSVs are variously known as Norwalk-like or Norwalk group viruses, SRSVs, or human caliciviruses.

B. Featureless Viruses

1. Enteroviruses

Enteroviruses measure 25–30 nm in diameter, have a smooth outer rim, and no discernible surface structure. Their buoyant density in cesium chloride is 1.34 g/mL. The classical culturable enteroviruses, which include polioviruses, coxsackieviruses, and echoviruses, normally are not associated with viral gastroenteritis; they rarely are seen by electron microscopy in stool samples, even when readily isolated in cell culture, and will not be considered further here.

Hepatitis A virus (HAV) is an RNA virus that is identical in appearance to the enteroviruses. It was originally classified with the enteroviruses as enterovirus 72. The genomic organization of HAV, however, differs from that of other enteroviruses, and it has more recently been classified as a member of the *Hepatovirus* genus, which is a separate genus in the Picornaviridae family.

2. Parvoviruses

Particles with the morphological and physical characteristics of parvovirus have been seen in association with some outbreaks, but their causative role in gastroenteritis is unclear. They measure 18–26 nm in diameter, and they also have a smooth outer rim and no discernible surface structure (Fig. 1d). Their buoyant density in cesium chloride is 1.38–1.46 g/mL, which distinguishes them from enteroviruses.

III. NORWALK VIRUS AND OTHER SMALL ROUND STRUCTURED VIRUSES

A. Clinical Features

Viral gastroenteritis usually is regarded as a mild, self-limiting illness lasting 24–48 hours. Symptoms commonly include malaise, abdominal pain, pyrexia, diarrhea, and/or vomiting. The incubation period is variable, ranging from about 12 to 60 hours, and is probably dose-dependent. Secondary transmission to close contacts is a characteristic feature of this group of viruses. Onset may be sudden and commence with projectile vomiting. This is a particular hazard where food is being prepared and laid out as virus can be disseminated over a wide area in aerosol droplets.

Although viral gastroenteritis is regarded as a trivial illness, people commonly feel debilitated for 2 or 3 weeks following infection. The economic implications of outbreaks of foodborne viral gastroenteritis, in terms of working days lost through illness and impaired performance, probably are grossly underrated. Caterers and food producers also sustain considerable financial losses following a foodborne outbreak.

B. Epidemiology

Viruses of the Norwalk group usually are transmitted from person to person by the fecal-oral route, but they also may be spread by food and water, causing common source outbreaks. This group of viruses is responsible for sporadic cases of gastroenteritis in the community and for outbreaks. These viruses in fact may be the most common cause of gastroenteritis in the developed world. Because of the technical difficulties of virus detection and the relatively mild nature of the illness produced, their incidence largely is underrecognized. To assess the role of the various intestinal pathogens, a large national community-based study of infectious intestinal disease took place in England in 1993–1996. Fecal specimens from patients and controls were examined for bacterial, viral, and parasitic organisms. Analysis of results indicates that SRSVs have a far greater role in gastrointestinal disease in the general community than reported figures have previously shown (10).

Norwalk group viruses can occur in vomit as well as feces, and hence may spread via aerosol droplets, although this is not true respiratory transmission. This airborne route probably contributes to the very rapid secondary spread observed in outbreaks. When projectile vomiting occurs, there is likely to be widespread dissemination of the virus over surfaces. These viruses appear to be infectious in very low doses. They are a particular hazard in such semiclosed communities as hospitals, schools, nurseries, cruise ships, and residential homes for the elderly; outbreaks in these settings occur commonly. Attack rates in outbreaks tend to be high, often exceeding 50% in the group at risk, and in some shellfish-associated outbreaks attack rates of 90% have been observed.

Some outbreaks have been very large and caused major disruption to the community. In one outbreak, involving a bakery product contaminated by an infected food handler, 3000 people were infected, and hospitals and schools in the area had to close (11). An outbreak in 1992, following a buffet meal at a training day for teachers from all of one county in the United Kingdom, caused major disruption in a large number of schools in the county.

Shellfish have been responsible for some very large community-wide outbreaks. In 1978, there were more than 2000 reported cases of viral gastroenteritis associated with the consumption of oysters in Australia (12). In Shanghai in 1988, there was an outbreak of hepatitis A associated with the consumption of clams involving at least 292,000 people (13). It was reported that in the month preceding this outbreak, there was also an outbreak of presumed viral gastroenteritis among the same population, and it was inferred that the grossly polluted clams were responsible for both outbreaks. In another outbreak in the United Kingdom, 70 senior medical and nursing staff of a children's hospital were ill after eating mussels at a staff party, and the hospital had to be closed to admissions for several days. Waterborne outbreaks in summer camps and youth hostels also occur commonly. It has been estimated that in the United States, Norwalk group viruses cause 40% of outbreaks in recreational camps, cruise ships, among communities and families, at schools and colleges, and in nursing homes (5). In the United Kingdom, since 1994 the number of outbreaks of SRSV infection reported annually has exceeded the number of outbreaks of *Salmonella* infection.

Infections by the Norwalk group of viruses occur throughout the year with no apparent seasonal peak. They infect any age group. In developed countries, acquisition of antibody to the Norwalk virus is gradual. In the United States, 20% of 5-year-old children have antibody, and levels rise to around 50% by 50 years of age (6). This is unlike rotavirus, to which most children have antibody by age 5. In more underdeveloped areas of the world, such as Bangladesh, Nepal, and Ecuador, acquisition of Norwalk antibody is similar to that of rotavirus antibody, and occurs at a young age. As such studies use radioimmunoassays (RIAs) and enzyme-linked immunoassays (ELISAs), it is likely that these antibody findings reflect exposure to any virus of the Norwalk group and not necessarily Norwalk virus itself (5). Unpublished studies in the United Kingdom indicate that illness due to Norwalk virus is relatively uncommon, whereas infection with other SRSVs is exceedingly common.

C. Importance as Foodborne Pathogens

Although it might be thought that any virus that infects via the gastrointestinal tract could be a foodborne pathogen, it is in fact Norwalk virus and the other SRSVs that are involved most frequently. In the United Kingdom, SRSVs have been implicated in more than 90%

of foodborne outbreaks of viral gastroenteritis where a virus was detected (14). Similarly, in the United States, Norwalk and Norwalk group viruses are the most frequently reported causes of foodborne viral illness.

D. Serological Relationships

Members of the Norwalk virus group have tended to be named after the geographic location of the incident in which they were identified. There have been few serological comparisons among the different agents because of the paucity of working material and the lack of any standard reference reagents. It is clear, however, that different serotypes exist (Table 2).

Initial comparisons were made by cross-challenge studies in volunteers, which demonstrated that Norwalk virus and Montgomery County virus were related, but the Hawaii virus was distinct (15). Most other comparisons have been made by immune electron microscopy (IEM). Hyperimmune animal sera have not been available, and studies have been done mainly with human acute and convalescent sera, which are not very satisfactory.

Radioimmunoassays and ELISAs have been developed for very few Norwalk group viruses. Epidemiological studies in the United States initially suggested that about 70% of outbreaks were due to Norwalk virus. However, it is now recognized that these tests were broadly cross-reactive to other serotypes, and the implication of Norwalk virus in such a high number of outbreaks was probably an overestimate of the importance of Norwalk virus itself, but not of the group as a whole (5). Some of the outbreaks investigated almost certainly were due to other serotypes.

Although Norwalk group viruses are detected frequently, few are compared antigenically. The extent of serological variation is unknown, and comparisons between viruses found in different countries are confusing. Different groups of workers have different systems of typing nomenclature, and shortage of reagents has made comparison difficult (16–18). It is possible that different serotypes are prevalent at different times and in different parts of the world. Of more than 20 SRSV outbreaks studied by immune electron microscopy in the United Kingdom in the 1980s, only 2 were due to Norwalk virus.

Table 2 Antigenic Relationship Between the Norwalk Group Viruses by Immune Electron Microscopy or Cross-Challenge Studies in Volunteers

Name		Relationship	Origin	Country
Norwalk	)		School	United States
Montgomery County	)	Related	Family	United States
Hawaii		Distinct	Family	United States
Otofuke	)	Related to one another but distinct from	Institute for mentally retarded	Japan
Sapporo	)	Norwalk and Hawaii	Children's home	
Taunton		Distinct from Norwalk	Hospital (foodborne)	United Kingdom
Amulree		Distinct from Norwalk and Hawaii	Hospital	United Kingdom
Snow Mountain		Distinct from Norwalk and Hawaii	Resort camp (waterborne)	United States
Belfast		Distinct from Norwalk	Army officers (shellfish)	United Kingdom

More recently, a source of antigen has been produced using the baculovirus expression system. Part of the genome of the Norwalk virus was incorporated into the genome of a baculovirus, an insect virus. When grown in suitable cells, large quantities of the Norwalk capsid protein are produced, identical in appearance and antigenic characteristics to empty capsids of naturally occurring Norwalk virus (19). This antigen can be concentrated and purified, and used in serological tests. It is highly immunogenic and has been used for the production of hyperimmune antiserum. Capsid antigens have now been produced for a few other SRSV strains, using either baculovirus or yeast cells (20,21).

Using these antigens, ELISA-based tests for detecting antibody have been developed, but care must be taken in interpreting results, as reactions can be broadly cross-reactive, as with the earlier ELISA and RIA tests. However, conversely, where hyperimmune sera to the capsid antigens have been made and used for detecting and identifying virus, the tests appear very type-specific. So far, expressed capsid antigens have only been made to a very small number of SRSV serotypes, but use of this type of antigen offers the possibility of developing simple tests for the Norwalk group of viruses. It is anticipated that such tests will become widely available for use in any microbiological laboratory.

E. Pathology

The pathogenic mechanisms of the Norwalk group viruses are poorly understood. Norwalk virus appears to infect mature enterocytes of the proximal small intestine (22,23). In biopsy specimens obtained during volunteer studies, the villi were stunted and the covering enterocytes were abnormal. The lamina propria was infiltrated by mononuclear inflammatory cells and neutrophils. Crypt cell hypertrophy was seen, particularly late in the disease process. These changes were accompanied by malabsorption of D-xylose, lactose, and fat for no more than 2 weeks from onset of illness. Gastric and rectal mucosa remained normal, and gastric secretion of hydrochloric acid, pepsin, and other intrinsic factors was unaltered. Gastric emptying, however, was delayed markedly. This abnormal gastric motor function probably explains the frequent nausea and vomiting that are characteristic of Norwalk viral illness (24).

Two antigenically distinct viruses (Newbury agents 1 and 2), which are similar morphologically to the Norwalk group viruses, have been isolated from calves (25). In calves, these cause anorexia, increased fecal output, change in fecal color, and malabsorption of xylose. In transmission studies with gnotobiotic calves, the pathogenic processes appear very similar to those of Norwalk virus. Such models seem to offer the opportunity of studying the pathogenic mechanisms of the Norwalk virus group in more detail.

Norwalk virus induces production of both local gut and serum antibody. The presence of serum antibody, however, does not necessarily confer protection against infection. This has been demonstrated in volunteer studies in which subjects have been infected repeatedly with Norwalk virus (26,27). In one study, 12 volunteers were given Norwalk virus and 6 developed symptoms. The same group was challenged again 27–42 months later. The six who developed illness on the first challenge became ill again, whereas those who remained well on the first challenge remained well on the second challenge also. There were significant rises in serum antibody levels following illness, even though these volunteers already had baseline antibody levels that suggested previous exposure to Norwalk virus. Those who were resistant to illness usually had low or absent serum antibody before and after exposure to the virus. In further studies, it was noted that titers of preexisting local gut antibody also were markedly greater in volunteers who subsequently developed illness than in those who remained well. This pattern of clinical immunity is unusual, and it is not known what factors determine the difference between the two groups. It appears that those who are able to generate antibody are more likely to develop illness. There does appear to be some degree of short-term clinical immunity, however. Of four volunteers challenged on a third occasion 6–14 weeks after their previous illness, only one developed symptoms again.

Whether a similar response to other Norwalk group viruses occurs is unknown. In view of the curious response to Norwalk virus, coupled with the existence of several serotypes within this group, it perhaps is not surprising that people experience repeated incidents of viral gastroenteritis during their lifetime.

F. Virology

Of all the SRSVs, Norwalk virus has been studied the most extensively. Despite extensive efforts in several centers, neither Norwalk virus nor any other SRSV has been propagated in cell culture or organ culture, and no animal model has been established. Chimpanzees inoculated with Norwalk virus did produce an antibody response, but they failed to develop any symptoms and there was no fecal shedding of virus although soluble antigen was detectable by RIA (28). Numerous volunteer studies, mainly with Norwalk virus and to a lesser extent with other strains, provided material for all of the early virological studies. Expression of recombinant viral capsids in baculovirus systems is now producing antigen for Norwalk virus and other SRSV strains (19–21).

When examined in the electron microscope without the addition of antibody, all of the Norwalk group viruses or SRSVs have the same appearance, comprising an amorphous surface and ragged outline (see Fig. 1c) (8). The suggestion of calici-like depressions occasionally may be seen. Immune electron microscopy techniques often have been used to aid detection, but care must be taken as a coating of antibody may mask the appearance of the virus particles and lead to erroneous identification.

The ragged outline of the particles and widespread use of IEM make estimation of size difficult, but most of these viruses seem to come within the range of 30–38 nm. Buoyant density in cesium chloride is 1.36–1.41 g/mL.

Radioimmunoassays and ELISAs have been developed, using virus extracted from human stool samples as antigen. Their use has been restricted to a very few laboratories. These assays have been used in investigating outbreaks, usually by looking for rises in serum antibody level rather than for the virus itself. As virus shedding rapidly falls below detectable levels, suitable stool specimens often are not available anyway. ELISAs using expressed recombinant viral capsid proteins as antigen have been used recently in some small epidemiological surveys (21,29,30).

Polymerase chain reaction (PCR) assays are now used in some laboratories for detection of virus, particularly for examining stool and vomitus samples from patients. Because of the genomic diversity within the SRSV group, there are problems in developing primers that will detect all strains (31–33). At present, the tests are insufficiently robust to be used routinely in all virology laboratories and are mainly confined to more specialized centers. Where it can be used, PCR offers a major advantage. Whereas SRSVs can often be detected by electron microscopy only for 1 or 2 days after onset of illness, PCR will detect virus for a longer period, sometimes up to a week. ELISAs have also been developed for detection of virus, using sera raised to the baculovirus-expressed recombinant antigens

(34,35). These tests seem highly specific and can be more easily used than PCR. They select for only one SRSV strain but are useful for epidemiological studies.

Polymerase chain reaction assays are being applied to the detection of SRSVs in shellfish, water, and other environmental samples (36–38). The primary extraction of viral RNA from such samples is complex and time consuming. Hence, such tests cannot be used on a routine basis at present.

Development of techniques other than electron microscopy is very welcome, since electron microscopy is labor-intensive and requires considerable skill on the part of the operator. However, other techniques select exclusively for one virus and often only for one serotype. Electron microscopy is the only catch-all technique with the potential to detect any virus group that might be present. Mixed infections with gastroenteritis viruses are not uncommon. More than one pathogen may be present, which is of particular relevance when considering a food such as shellfish or a water supply that may be grossly contaminated (4,12).

Volunteer studies have provided material for biochemical studies. Norwalk virus has a single structural protein with a molecular weight of 59 kDa and a soluble viral protein of 30 kDa (39). This protein structure is similar to that of the Caliciviridae, which led to the early suggestion that Norwalk virus is closely related to the caliciviruses. The serologically distinct Snow Mountain agent has a structural protein of 62 kDa (40). From Japan, it has been reported that an SRSV immunologically related to the Hawaii agent has a single structural protein of 63 kDa, and a human calicivirus from an outbreak in an orphanage also had a structural protein of 62 kDa (41). More recent cloning and sequencing studies have revealed that the genome of Norwalk virus is single-stranded RNA of positive sense with a poly-A (polyadenylated) tail at the 3′ end and a size of at least 7.5 kb (42,43).

The genome comprises three open reading frames (ORFs). ORF-1 encodes for nonstructural proteins and ORF-2 for the capsid protein. The function of ORF-3 is uncertain. This information has allowed Norwalk virus to be formally classified as a calicivirus (9). Other SRSV strains have now been sequenced, and these have the typical genomic arrangement of members of the Caliciviridae family.

G. Genomic Typing

Once the sequences of Norwalk virus and other SRSVs were known, reverse transcription polymerase chain reaction (RT-PCR) assays were developed by several groups of workers. The RT-PCR assays used primers directed at the RNA-dependent RNA polymerase region of ORF-1. Many different strains of SRSV have been identified. By cloning and sequencing the amplified PCR products, the strains have been divided into two genogroups. Genogroup 1 includes the prototype Norwalk virus. Genogroup 2 includes Hawaii virus, Southampton virus, and some morphologically typical caliciviruses. A third genogroup has also been identified. This third group includes morphologically typical caliciviruses, the prototype being the Japanese Sapporo strain. The genomic organization of the third group differs from that of genogroups 1 and 2, and from other caliciviruses, suggesting that it may be a separate genus within the Caliciviridae family (44).

The great diversity of strains within the SRSV group makes it difficult to design broadly reactive primers or sets of primers that will detect all SRSVs. As more sequence data becomes available, primers are constantly being improved, but there are still considerable numbers of SRSVs readily detected by electron microscopy that current PCR assays fail to detect.

H. Stability and Physical Characteristics

There is little precise information on the stability of the Norwalk group of viruses. Most information has been inferred from epidemiological observations in foodborne outbreaks and a small number of experiments with Norwalk virus in volunteers (45).

Viruses that infect via the gastrointestinal tract are acid-stable. Hence, they survive food processing and preservation conditions designed to produce the low pH that inhibits bacterial and fungal food spoilage organisms (e.g., pickling in vinegar and fermentation-type processes that produce such foods as yogurt). People have developed viral gastroenteritis after eating shellfish pickled in brine and vinegar. In one volunteer study, Norwalk virus was shown to retain its infectivity after exposure to pH 2.7 for 3 h at room temperature. It also was stable in 20% ether at 4°C for 18 h (45).

Like most viruses, the Norwalk group of viruses remains infectious after refrigeration and freezing. Frozen foods that have not received further cooking have been implicated in a number of incidents. It is ironic that those food storage conditions chosen to inhibit multiplication of bacterial and fungal organisms are ideal for the preservation of viruses. The Norwalk group viruses appear to be inactivated by normal cooking processes but are not always inactivated in shellfish given only minimal heat treatment. Norwalk virus heated to 60°C for 30 min remained infectious for volunteers. It is uncertain that it would be inactivated completely in many pasteurization processes.

The Norwalk group viruses appear to survive well on inanimate surfaces, as evidenced from the lingering outbreaks that have occurred in hospitals, in residential homes, and on cruise ships. Virus has been detected by PCR in environmental swabs from one hospital outbreak (46). Chlorine-based disinfectants are considered to be effective against this group of viruses. However, Norwalk virus is resistant to inactivation in the presence of 3.75–6.25 mg chlorine/L, equivalent to free residual chlorine of 0.5–1 mg/L and consistent with that present in a drinking water system (6). Norwalk virus is inactivated by 10 mg chlorine/L, which is the concentration used to treat a water supply after a contamination incident. It appears to be more resistant to chlorine than poliovirus or human rotavirus.

IV. FOODBORNE TRANSMISSION AND CONTROL

Foods may become contaminated with viruses in two main ways. They may be contaminated at the source in their growing or harvesting area, usually by coming into contact with polluted water; this is known as *primary contamination*. Shellfish are a particular problem and have been implicated in many outbreaks worldwide. Foods also may be contaminated during handling and preparation, often from infected food handlers; this is known as *secondary contamination*.

A. Shellfish

It is the bivalve mollusks, including oysters, mussels, clams, and cockles, that mainly are involved in transmitting viral illness (14,47–49). These shellfish grow in shallow coastal and estuarine waters that frequently are polluted with sewage. Scallops rarely have been associated with viral illness. They are free-living organisms that inhabit deeper offshore waters that are less subject to pollution.

The bivalves feed by filtering out particulate organic matter from the vast volumes of water passing over their gills, and this water can include potentially pathogenic microorganisms. Human viruses do not replicate in shellfish but can be concentrated in the mollusks. Thorough cooking destroys contaminating viruses, but it also causes shrinkage of the shellfish meat and renders it tough and unpalatable. Furthermore, such mollusks as oysters regularly are eaten raw by many people. However, mollusks are subject to some form of cleansing or heat treatment in an attempt to produce a microbiologically safe product before sale to the public. The efficiency of such procedures is monitored by bacteriological testing, but unfortunately absence of bacterial indicator organisms does not necessarily assure absence of viruses.

1. Heat Treatment

Some mollusks, such as cockles, are subject to a brief heat treatment by boiling or steaming. Outbreaks of viral illness have occurred from shellfish that have been heated sufficiently to kill bacterial contaminants, but obviously not enough to inactivate viruses (4,47,48).

Heat treatment studies on hepatitis A virus in the United Kingdom have led to recommendations that the internal temperature of shellfish meat be raised to 90°C and this temperature should be maintained for 1.5 min (50). These conditions are known to inactivate hepatitis A virus (and poliovirus) but still achieve a commercially acceptable product. The excellent surveillance system in the United Kingdom has shown that there have been no recorded incidents of viral illness—either hepatitis A or viral gastroenteritis—from shellfish so treated (51).

The Norwalk group of viruses cannot be cultured, and it has not been possible therefore to do similar heat inactivation experiments with these viruses in shellfish. Studies in the United Kingdom, using a feline calicivirus as a model for the Norwalk virus group, indicate that calicivirus is less heat-resistant than hepatitis A virus, which is consistent with the epidemiological observations (51). The same studies demonstrated a lack of correlation between culture and PCR results, for the detection of virus in the shellfish. This has important implications, if PCR assays for SRSVs are introduced for monitoring the safety for consumers of shellfish and other foods, or are used for designing treatment protocols for shellfish and production protocols for "ready-to-eat" foods (e.g., washed salads and vegetables, yogurts).

2. Relaying and Depuration

Some shellfish taken from polluted waters are treated by relaying or depuration. This particularly applies to oysters that may be eaten raw. Relaying involves moving shellfish from their original growing site to an area of cleaner water, where they are left for several weeks or months. The level of microbial contamination in the shellfish falls as microorganisms are washed out during the normal feeding process.

Depuration depends on the same principle, but the shellfish are placed in land-based tanks built for this purpose, usually for about 48 hr. *Escherichia coli*, the usual indicator of contamination, virtually is eliminated during this period, but viruses can remain for several weeks. There is no guarantee that shellfish that appear safe in bacteriological tests necessarily are free of viruses, and at present there is no satisfactory indicator system for viruses. The viruses that cause most problems, hepatitis A and the gastroenteritis viruses, particularly the Norwalk group, are difficult to detect in any type of specimen.

At present, there are no effective methods for detecting gastroenteritis viruses in shellfish on a routine basis. Viruses stick avidly to shellfish meat. Numerous extraction methods have been reported in the literature, but recovery rates in all are poor and none can be regarded as entirely satisfactory. It is probable also that different extraction methods are required for different types of virus (52). SRSVs and hepatitis A virus have been detected by PCR in shellfish samples, but these instances have been in special studies (36–38).

Because of the difficulties involved in the extraction and detection of viruses from shellfish, there has been considerable interest in the use of bacteriophages as indicators of viral contamination (53,54). Under depuration conditions, the elimination rate of bacteriophages appears to be similar to that of viruses. To exploit this further, it is necessary to choose a bacteriophage species that can be guaranteed always to be present.

The first commercial depuration plant was built in Conway in Wales in 1914 to reduce the number of typhoid cases arising from eating locally harvested mussels (55). This plant, which is still in operation, uses water disinfected with chlorine and is an effective system for treating mussels. However, oysters are very sensitive to chlorine; they close up and will not feed in the presence of chlorine. They may be cleansed in plants where the water is treated with ozone or ultraviolet irradiation.

Shellfish taken from grossly polluted water will not cleanse themselves adequately in depuration tanks. If shellfish are harvested from heavily polluted water, they should be relayed for several weeks before transfer to a depuration plant. It is apparent that different shellfish species require different conditions (e.g., temperature, salinity) to depurate optimally (56). This is a very complex subject and no satisfactory system has been devised for the removal of viruses as yet.

An attempt is being made within the European Union (EU) to classify the quality of shellfish-growing waters, and an EU directive on the treatments that must be applied to shellfish taken from different classes of water has been accepted by member states (57) (Table 3). The classification is based on the bacteriological quality of the water, determined by bacteriological testing of shellfish harvested from the waters, although it is recognized that viruses pose the main problem from bivalve mollusks. Virological testing of shellfish is not mandatory at this stage. If suitable virological methods become available, these could be incorporated at a later date.

3. Crustacea

Unlike the mollusks, most outbreaks of illness associated with crustacea are bacterial rather than viral, usually as a result of postcooking contamination. Crustacea may become

Table 3 Classification of Harvesting Areas for Bivalve Mollusks in the European Community

Water quality category	*E. coli*/100 g shellfish meat	Fecal coliforms/ 100 g shellfish meat	Conditions for shellfish production
A	<230	<300	May go for direct consumption
B	<4600	<6000	Must be depurated, heat-treated, or relayed to meet category A
C		<60,000	Must be relayed for long period (2 months) to meet categories A or B; also may be heat-treated by an approved method
D		>60,000	Harvesting prohibited

contaminated virally during handling by an infected food handler in the same way as any other food.

There has been concern in Europe and North America that warm-water shrimp and prawns, imported from tropical areas in which gastrointestinal illness is more prevalent, may pose a potential risk to consumers, particularly if local cooking procedures are unsatisfactory. Firm evidence that a real risk exists is lacking, although there are anecdotal reports of incidents of viral gastroenteritis from such sources.

B. Fruits and Vegetables

There is the potential for fruits, vegetables, and salad items to be contaminated with polluted water and untreated sewage at their source during irrigation and fertilization. Although several outbreaks of viral gastroenteritis have been attributed to salad items, contamination on these occasions usually is thought to have occurred at the time of preparation. Soft fruits, believed to have been contaminated at their source, have been implicated in the transmission of hepatitis A, and it must be assumed that incidents of viral gastroenteritis could occur in the same way.

C. Food Handlers

Viruses causing gastroenteritis are believed to be infectious in very low doses and thus are spread very easily from infected persons. It now is recognized that outbreaks arising from food contamination by infected food handlers are very common occurrences.

It is cold items such as sandwiches and salads, which require much handling during preparation, that are implicated most frequently. Without meticulous attention to personal hygiene and thorough and frequent hand washing, fecally contaminated fingers can contaminate foods and work surfaces. Viral gastroenteritis can be very sudden in onset and commence with projectile vomiting. Virus can be disseminated over a wide area in aerosol droplets, contaminating uncovered food and work surfaces. Gloves could reduce transfer of virus from fecally contaminated fingers to food but will not prevent transfer from contaminated work surfaces.

Persons with symptoms should of course be excluded from food handling. However, food handlers with very minimal symptoms have been implicated in the transmission of Norwalk group viruses. There is a little circumstantial evidence that Norwalk group viruses also may be excreted asymptomatically (58,59), but in the absence of more definite information, the recommendation that food handlers be allowed to return to work 48 hours after symptoms have ceased seems reasonable. That recommendation was based on the rapid decrease in virus shedding observed by electron microscopy and in practice appears to have worked satisfactorily. The period in which SRSVs can be detected in patients by PCR, however, is longer than that by electron microscopy. In some instances SRSVs can be detected for up to a week after onset of symptoms. It is not clear if persons shedding viruses detectable by PCR are still infectious once symptoms have ceased. Recommendations on how long to exclude people from work need to be kept under review.

Control of foodborne viral illness depends largely on strict attention to normal good hygienic practice in the kitchen and serving areas. Salad items, fruits, and raw vegetables should be washed thoroughly before use. Cross-contamination from uncooked shellfish also should be regarded as a potential hazard.

V. ASTROVIRUS

Astroviruses form a novel and morphologically distinct group of small viruses. They were observed first in 1975 in an outbreak of diarrhea in infants in a maternity unit, and later were named *astrovirus* for the solid, five- or six-point star visualized by electron microscopy on the surface of some particles (60,61).

Astroviruses normally are associated with gastroenteritis in young children, often under 1 year of age. Antibody acquisition is similar to that of rotavirus: in the United Kingdom, 70% of children have antibody by 3–4 years of age. Adults are infected infrequently although, as with rotavirus, outbreaks have been reported in the elderly.

Astroviruses have a worldwide distribution. They occur all year, although in temperate zones there is a winter/spring peak, similar to that of rotavirus. They mainly spread by the fecal-oral route but can be transmitted via food and water. Astroviruses have been observed in some adults with illness following the consumption of shellfish or contaminated water, but these incidents appear comparatively rare.

The incubation period is 3–4 days. Symptoms include diarrhea, fever, headache, malaise, nausea, and occasional vomiting. Diarrhea usually lasts 2–3 days, but may continue for 7–14 days and is accompanied by virus excretion. The virus infects mature enterocytes of the small intestine, causing a transient villous atrophy. Studies in lambs have shown that lost cells are replaced with immature cuboidal cells from the crypts, with histological recovery 5 days after infection.

Virus usually is detected by electron microscopy in stool specimens from infected persons. Astrovirus will undergo a nonproductive cycle of infection in primary human embryo kidney (HEK) cells and may be detected in the cytoplasm using fluorescent conjugated antibody. In the presence of trypsin, astrovirus undergoes productive replication in HEK with the release of free virus, which then may be adapted to grow in the continuous line of rhesus monkey kidney cells LLC-MK$_2$. A continuous cell line derived from human colonic carcinoma (CaCo-2) is susceptible to astrovirus infection, yielding higher virus titers than obtained in HEK or LLC-MK$_2$ (62), and is now more commonly used. Eight serotypes of human astrovirus have been recognized so far with type 1 being the predominant strain. Astrovirus occur in several animal species, but these strains are unrelated serologically to human strains.

Astroviruses are classified as a distinct family, the Astroviridae. The organization of the viral genome is unlike that of any other family of animal viruses. The genome is a positive sense, single-stranded RNA of approximately 7500 nucleotides. The complete genomes of types 1 and 5 have been sequenced. Like most enteric viruses, astroviruses are stable at pH 3. They are stable to lipid solvents and survive heating to 50°C for 30 min. At 60°C, the virus titer falls 3 $\log_{10}$ in 5 min and 6 $\log_{10}$ in 15 min.

VI. PARVOVIRUS

Other small viruses with the characteristics of parvoviruses have been observed in outbreaks of viral gastroenteritis, but their role as causative agents in humans remains uncertain (63–66). However, parvoviruses are proven and commercially important causes of gastroenteritis in some animal species.

These viruses measure 20–26 nm in diameter. They have a smooth outline and no discernible surface structure (Fig. 1d). Their buoyant density in cesium chloride is 1.38–

1.46 g/mL, which distinguishes them from enteroviruses. One particular virus (the cockle agent) from an outbreak in England was shown to contain single-stranded DNA, suggesting that this virus is indeed a parvovirus (67). At least three serotypes of human fecal parvoviruses have been reported by IEM (the Wollan/Ditchling group, the cockle agent, and the Parramatta agent) (4,65,66).

Parvovirus-like particles have been observed in schools and outbreaks of winter vomiting disease (e.g., Wollan agent, Ditchling agent, Parramatta agent) (63,64,66) and in a number of outbreaks associated with the consumption of shellfish (4,12,65). In 1977 in England, there was a large community-wide outbreak of gastroenteritis associated with consumption of cockles harvested from a particular coastal area (4,65). Around 800 cases were reported, but these were almost certainly only a small proportion of actual cases that occurred. Parvovirus was detected in almost all stool specimens examined from affected persons. This was in fact the first time that virus had been found in an outbreak of apparent food poisoning, and it led to the virological investigation of many subsequent "food poisoning" outbreaks.

Virus excretion may continue for several weeks after symptoms cease and low levels of virus also may be detected in some healthy people, making it difficult to establish a causative role. Parvovirus-like particles also have been detected occasionally in shellfish that have been implicated in illness and where similar virus has been found in the patients (4,12). In fecal specimens, parvovirus also may occur alongside such other established gastroenteritis viruses as rotavirus and Norwalk group viruses, but mixed infections among gastroenteritis viruses are not uncommon anyway. In some outbreaks parvovirus-like particles are excreted in vast numbers by a large proportion of ill persons. It has been suggested that such a virus may be part of the normal gut flora, and increased virus production occurs due to disturbance in gut physiology induced by other infecting organisms. This seems an unlikely explanation, as outbreaks in which most victims excrete these particles in large numbers are fairly uncommon. Furthermore, observations of these viruses have seemed to fluctuate over the years since 1971 when they were first described, suggesting there may be a cyclic pattern in their occurrence. The apparent infrequency of these viruses in association with illness has meant that little material has been available for study, a frustration familiar to that felt by those studying any of the small round gastroenteritis viruses.

VII. PROSPECTS

Continued surveillance and awareness are the most effective weapons in combating the foodborne spread of viruses. The particular problems associated with shellfish are encouraging efforts to establish optimum treatment procedures, either heat treatment or depuration, to render shellfish safe. It has been recommended by the World Health Organization (WHO) and by individual governments that in order to develop and evaluate more suitable treatment procedures there is a need for study of the actual viral pathogens of concern, i.e., those causing gastroenteritis and hepatitis (68,69). Sensitive methods that can be applied easily must be developed for detection and enumeration of these viruses. Only a culture system can ascertain residual infectivity unequivocally and at present only a model system can be used to represent the Norwalk group viruses. The anticipated development of rapid detection methods, which is arising from the ongoing work on the Norwalk virus genome, may give some reasonable indication of viral survival. Caution should be exercised in the interpretation of results, as detection of viral RNA by PCR may not necessarily correlate with infectivity.

Tests already have been developed for detection of Norwalk group viruses in stool specimens by PCR. More efficient methods for the extraction of nucleic acids need to be developed, so that such highly sensitive tests as PCR can be exploited further and more readily be applied to shellfish, other foods, water, and environmental specimens. The in vitro expression of viral antigens for Norwalk and Norwalk-like viruses is enabling the development of more specific detection tests for both antigen and antibody, which will lead to better understanding of the complex epidemiology of this group of viruses. Strain variation on a temporal and geographic basis is now being studied. Application of sensitive virus detection tests to foods, such as shellfish, and environmental samples is providing further epidemiological data.

For the present, continued epidemiological surveillance has indicated that heat treatment processes designed to inactivate hepatitis A also are preventing incidents of viral gastroenteritis following consumption of mollusks (48,51). Further studies to establish suitable heat treatment protocols for other types of shellfish are being pursued.

PCR assays are being used for detection of viruses in foods other than shellfish. Detection of SRSVs on ham and raspberries associated with outbreaks of illness has been reported. Extraction of viral RNA does not present as great a challenge as that encountered with virus avidly attached to shellfish meat. However, the number of virus particles present in these types of foods is very low as compared to specimens from patients, and available detection methods may not always be sufficiently sensitive. Viruses are concentrated in molluscan shellfish and hence are present in larger numbers than would be expected in other foods.

Recombinant capsid proteins can provide a relatively safe, noninfectious source of viral antigens. These antigens could be used for looking at food processing methods, including depuration conditions for shellfish. They could also be used, for instance, in assessing the efficacy of washing salads, vegetables, and fruits for the ready-to-eat market.

The recombinant capsid antigens, expressed by baculovirus and yeast cells, can be purified in large quantities. They are highly immunogenic and stable at low pH, which offers the potential for viral vaccines. Studies to express SRSV capsid antigens in transgenic plants, such as tobacco leaves and potato tubers, are underway. It has already been shown that antigens from these sources are immunogenic in mice, thus offering further potential for the development of oral vaccines (70).

Only meticulous attention to good food handling practices is likely to reduce the incidence of foodborne transmission from infected persons. Widespread education to prevent infected people—even those with the mildest symptoms—from handling food is essential. As techniques are developed to improve the detection of virus, which is now a realistic expectation, the true extent of this problem will be revealed, and it is likely to be very considerable. Sensitive detection tests will give more accurate estimates of the length of time virus excretion persists, and hence better informed advice can be formulated on such factors as how long food handlers should be excluded from work. Properly applied, these developments must lead to the reduced incidence of viral foodborne infections.

ABBREVIATIONS

CaCO-2   carcinoma of the colon
DNA   deoxyribonucleic acid
ELISA   enzyme-linked immunosorbent assay
HEK   human embryonic kidney

HAV   hepatitis A virus
IEM   immune electron microscopy
ORF   open reading frame
PCR   polymerase chain reaction
RNA   ribonucleic acid
RT-PCR   reverse transcriptase PCR
RIA   radioimmunoassay
SRSV   small round structured viruses

REFERENCES

1. Adler, J. L., and Zickl, R. (1969). Winter vomiting disease. J. Infect. Dis., 119:668–673.
2. Dolin, R., Blacklow, N. R., DuPont, H., Formal, S., Buscho, R. F., Kasel, J. A., Chames, R. P., Hornick, R., and Chanock, R. M. (1971). Transmission of acute infectious nonbacterial gastroenteritis to volunteers by oral administration of stool filtrates. J. Infect. Dis., 123:307–312.
3. Kapikian, A. Z., Wyatt, R. G., Dolin, R., Thornhill, T. S., Kalica, A. R., and Chanock, R. M. (1972). Visualization by immune electron microscopy of a 27-nm particle associated with acute infectious nonbacterial gastroenteritis J. Virol., 10:1075–1081.
4. Appleton, H. (1987). Small round viruses: classification and role in food-borne infections. In: Novel Diarrhoea Viruses (G. Bock and J. Whelan, eds.), Ciba Foundation Symposium 128, John Wiley and sons, Chichester, pp. 108–125.
5. Blacklow, N. R., and Greenberg, H. B. (1991). Viral gastroeneritis. N. Engl. J. Med., 325: 252–264.
6. Kapikian A. Z., Estes M. K., and Chanock RM. (1996). Norwalk group of viruses. In: Virology (B. N. Fields, D. M. Knipe, and P. M. Howley, eds.), Lipincott-Raven, Philadelphia pp. 783–810.
7. Thornhill, T. S., Kalica, A. R., Wyatt, R. G., Kapikian, A. Z., and Chanock, R. M. (1975). Pattern of shedding of the Norwalk particle in stools during experimentally induced gastroenteritis in volunteers as determined by immune electron microscopy. J. Infect. Dis., 132:28–34.
8. Caul, E. O., and Appleton, H. (1982). The electron microscopical and physical characteristics of small round human fecal viruses: an interim scheme for classification. J. Med. Virol., 9: 257–265.
9. Cubitt, D., Bradley, D., Carter, M., Chiba, S., Estes, M., Saif, L., Smith, A., Studdert, M., and Thiel, H.-J. (1995). Classification and nomenclature of viruses: Caliciviridae. Sixth Report of the International Committee on the Taxonomy of Viruses. Arch. Virol. Supplement 10, 359–363.
10. Wheeler J. G., Sethi D., Cowden J. M., Wall, P. G., Rodrigues L. C., Tompkins D. S., Hudson, M. J., and Roderick, P. J. on behalf of the Infectious Intestinal Disease Study Executive. (1999). Study of infectious intestinal disease in England: rates in the community, presenting to general practice, and reported to national surveillance. BMJ, 318:1046–1050.
11. Kuritsky, J. N., Osterholm, M. T., Greenberg, H. B., Korlath, J. A., Godes, J. R., Hedberg, C. W., Forfang, J. C., Kapikian, A. Z., McCullough, J. C., and White, K. E. (1984). Norwalk gastroenteritis: a community outbreak associated with bakery product consumption. Ann. Intern. Med., 100:519–521.
12. Murphy, A. M., Gorman, G. S., Christopher, P. J., Lopez, W. A., Davey, G. R., and Millsom, R. H. (1979). An Australia-wide outbreak of gastroenteritis from oysters caused by Norwalk virus. Med. J. Aust., 2:329–233.
13. Halliday, M. L., Kang, L.-Y., Zhou, T.-E., Hu, M. D., Pan, Q. C., Fu, T. Y., Huang, Y. S., and Hu, S. L. (1991). An epidemic of hepatitis A attributable to the ingestion of raw clams in Shanghai, China, J. Infect. Dis., 164:852–859.

14. Public Health Laboratory Service Working Party on Viral Gastroenteritis—Appleton, H., Caul, E. O., Clewley, J. P., Cruickshank, J. G., Cubitt, W. D., Kurtz, J. B., McSwiggan, D. A., Palmer, S. R., and Riordan, T. (1988). Foodborne viral gasteroenteritis: an overview (with a brief comment on hepatitis A). PHLS [Public Health Laboratory Service, London] Microbiol. Dig., 5:69–75.

15. Thornhill, T. S., Wyatt, R. G., Kalica, A. R., Dolin, R., Chanock, R. M., and Kapikian, A. Z. (1977). Detection by immune electron microscopy of 26- to 27-nm viruslike particles associated with two family outbreaks of gastroenteritis. J. Infect. Dis., 135:20–27.

16. Cubitt, W. D., Blacklow, N. R., Herrmann, J. E., Nowak, N. A., Nakata, S., and Chiba, S. (1987). Antigenic relationship between human caliciviruses and Norwalk virus. J. Infect. Dis., 156:806–814.

17. Lewis, D. C., Lightfoot, N. F., Pether, J. V. S. (1988). Solid-phase immune electron microscopy with human immunoglobulin M for serotyping of Norwalk-like viruses. J. Clin. Microbiol., 26:938–942.

18. Okada, S., Sekine, S., Ando, T., Hayashi, Y., Murao, M., Yabuuchi, K., Miki, T., and Ohashi, M. (1990). Antigenic characterization of small round structured viruses by immune electron microscopy. J. Clin. Microbiol., 28:1244–1248.

19. Jiang, X., Wang, M., Estes, M. K. (1992). Expression, self-assembly, and antigenicity of the Norwalk virus capsid protein. J. Virol., 66:6527–6532.

20. Jiang, X., Matson, D. O., Ruiz-Pallacias, G. M., Hu, J., Treanor, J., Pickering, L. K. (1995). Expression, self assembly, and antigenicity of a Snow Mountain–like calicivirus capsid protein. J. Clin. Microbiol., 33:1452–1455.

21. Green, K. Y., Kapikian, A. Z., Valdesuso, J., Sosnovtsev, S., Treanor, J. J., Lew, J. F. (1997). Expression and self-assembly of recombinant capsid protein from the antigenically distinct Hawaii human calicivirus. J. Clin. Microbiol., 35:1909–1914.

22. Bass, D. M., and Greenberg, H. B. (1991). Pathogenesis of viral gastroenteritis. In: Current Topics in Gastroenterology (M. Field, ed.), Elsevier, New York, pp. 139–157.

23. Schreiber, D. S., Blacklow, N. R., and Trier, J. S. (1973). The mucosal lesion of the small intestine in acute infectious nonbacterial gastroenteritis. N. Engl. J. Med., 288:1318–1323.

24. Meeroff, J. C., Schreiber, D. S., Trier, J. S., and Blacklow, N. R. (1980). Abnormal gastric motor function in viral gasteroenteritis. Ann. Intern. Med., 92:370–373.

25. Hall, G. A. (1987). Comparative pathology of infection by novel diarrhoea viruses. In: Novel Diarrhoea Viruses (G. Bock and J. Whelan, eds.), Ciba Foundation Symposium 128, John Wiley and Sons, Chichester, pp. 192–217.

26. Parrino, T. A., Schreiber, D. S., Trier, J. S., Kapikian, A. Z., and Blacklow, N. R. (1977). Clinical immunity in acute gasteroenteritis caused by Norwalk agent. N. Engl. J. Med., 297:86–89.

27. Blacklow, N. R., Herrmann, J. E., and Cubitt, W. D. (1987). Immunobiology of Norwalk virus. In: Novel Diarrhoea Viruses (G. Bock and J. Whelan, eds.), Ciba Foundation Symposium 128, John Wiley and Sons, Chichester, pp. 144–161.

28. Wyatt, R. G., Greenberg, H. B., Dalgard, D. W., Allen, W. P., Sly, D. L., Thornhill, T. S., Channock, R. M., and Kapikian, A. Z. (1978). Experimental infection of chimpanzees with the Norwalk agent of epidemic viral gastroenteritis. J. Med. Virol., 2:89–96.

29. Parker, S. M., Cubitt, D., Jiang, J. X., and Estes, M. (1993). Efficacy of a recombinant Norwalk virus protein enzyme immunoassay for the diagnosis of infections with Norwalk virus and other ''candidate'' caliciviruses. J. Med. Virol., 41:179–184.

30. Green, K. Y., Lew, J. F., Jiang, X., Kapikian, A. Z., and Estes, M. K. (1993). Comparison of the reactivities of baculovirus-expressed recombinant Norwalk virus antigen in serological assays and some epidemiological observations. J. Clin. Microbiol., 31:2185–2191.

31. Norcott, J. P., Green, J., Lewis, D., Estes, M. K., Barlow, K. L., and Brown, D. W. G. (1994). Genomic diversity of small round structured viruses in the United Kingdom. J. Med. Virol., 44:280–286.

32. Wang, J., Jiang, X., Maclore H. P., Desselberger, U., Gray, J., Ando, T., Seto, Y., Yamazaki, K., Oishi, I., and Estes, M. K. (1994). Sequence diversity of small, round-structured viruses in the Norwalk virus group. J. Virol., 68:5982–5990.

33. Green, J., Gallimore, C. I., Norcott, J. P., Lewis, D., and Brown, D. W. G. (1995). Broadly reactive reverse transcriptase polymerase chain reaction (RT-PCR) for the diagnosis of SRSV-associated gastroenteritis. J. Med. Virol., 47:392–398.

34. Herrmann, J. E., Blacklow, N. R., Matsui, S. M., Lewis, T. L., Estes, M. K., Ball. J. M., and Brinker, J. P. (1995). Monoclonal antibodies for detection of Norwalk virus antigen in stools. J. Clin. Microbiol., 33:2511–2513.

35. Hale, A. D., Lewis, D., Green, J., Xiang, X., and Brown, D. W. G. (1996). Evaluation of an antigen captire ELISA based on recombinant Mexico virus capsid antigen. Clin. Diag. Virol., 5:27–35.

36. Atmar, R. L., Neill, F. H., Ronalde, J. L., Guyader, F. Le, Woodley, C. M., Metcalf, T. G., and Estes, M. K. (1995). Detection of Norwalk virus and hepatitis A virus in shellfish tissues with the PCR. Appl. Environ. Microbiol., 61:3014–3018.

37. Lees, D. N., Henshilwood, K., Green, J., Gallimore, C. I., and Brown, D. W. G. (1995). Detection of small round structured viruses in shellfish by reverse transcription-PCR. Appl. Environ. Microbiol., 61:4418–4424.

38. Green, J., Henshilwood, K., Gallimore, C. I., Brown, D. W. G., and Lees, D. N. (1998). A nested reverse transcriptase PCR assay for detection of small round-structured viruses in environmentally contaminated molluscan shellfish. Appl. Environ. Microbiol., 64:858–863.

39. Greenberg, H. B., Valdesuso, J. R., Kalica, A. R., Wyatt, R. G., McAuliffe, V. J., Kapikian, A. Z., and Chanock, R. M. (1981). Proteins of Norwalk virus. J. Virol., 37:994–999.

40. Madare, H. P., Treanor, J. J., and Dolin, R. (1986). Characterization of the Snow Mountain agent of viral gastroenteritis. J. Virol., 58:487–492.

41. Tarashima, H., Chiba, S., Sakuma, Y., Kogasaka, R., Nakata, S., Minami, R., Horino, K., and Nakao, T. (1983). The polypeptide of a human calicivirus. Arch. Virol., 78:1–7.

42. Jiang, X., Graham, D. Y., Wang, K., and Estes, M. K. (1990). Norwalk virus genome cloning and characterization. Science, 250:1580–1583.

43. Matsui, S. M., Kim, J. P., Greenberg, H. B., Su, W., Sun, Q., Johnson, P. C., DuPont, H. L., Oshiro, L. S., and Reyes, G. R. (1991). The isolation and characterization of a Norwalk virus-specific cDNA. J. Clin. Invest., 87:1456–1461.

44. Lui, B. L., Clarke, I. N., Caul, E. O., and Lambden, P. R. (1995). Human enteric caliciviruses have a unique genome structure distinct from Norwalk-like viruses. Arch. Virol., 140:1345–1356.

45. Dolin, R., Blacklow, N. R., DuPont, H., Buscho, R. F., Wyatt, R. G., Kasel, J. A., Hornick, R., and Chanock, R. M. (1972). Biological properties of Norwalk agent of acute infectious nonbacterial gasteroenteritis. Proc. Soc. Exp. Biol. Med. 140:578–583.

46. Green, J., Wright, P. A., Gallimore, C. I., Mitchell, O., Morgan-Capner, P., Brown, D. W. G. (1998). The role of environmental contamination with small round structured viruses in a hospital outbreak investigated by reverse-transcriptase polymerase chain reaction assay. J. Hosp. Infect., 39:39–45.

47. Sockett, P. N., West, P. A., and Jacob, M. (1985). Shellfish and public health. PHLS [Public Health Laboratory Service-London] Microbiol. Dig., 2:29–35.

48. Appleton, H. (1990). Foodborne viruses. Lancet, 2:1362–1363.

49. Morse, D. L., Guzewich, J. J., Hanrahan, J. P., Stricof, R., Shayegani, M., Deibel, R., Grabau, J. C., Nowak, N. A., Herrmann, J. E., Cukor, G., and Blacklow, N. R. (1986). Widespread outbreaks of clam and oyster associated gasteroenteritis. Role of Norwalk virus. N. Engl. J. Med., 314:678–681.

50. Millard, J., Appleton, H., and Parry, J. V. (1987). Studies on heat inactivation of hepatitis A virus with special reference to shellfish. Epidemiol. Infect., 98:397–414.

51. Slomka, M. J., and Appleton, H. (1998). Feline calicivirus as a model system for heat inactivation studies of small round structured viruses in shellfish. Epidemiol. Infect. 121:401–407.

52. Yamashita, T., Sakae, K., Ishihara, Y., and Isomura, S. (1992). A two-year survey of the prevalence of enteric viral infections in children compared with contamination in locally-harvested oysters. Epidemiol. Infect., 108:155–163.

53. Power, U. F., and Collins, J. K. (1989). Differential depuration of poliovirus, Escherichia coli, and a coliphage by the common mussel, Mytilus edulis. Appl. Environ. Microbiol., 55:1386–1390.

54. Power, U. F., and Collins, J. K. (1990). Elimination of coliphages and Escherichia coli from mussels during depuration under varying conditions of temperature, salinity, and food availability. J. Food. Protect., 53:208–212.

55. Dodgson, R. W. (1928). Report on mussel purification. Fisheries Investigation Series II, no. 10, Her Majesty's Stationery Office, London.

56. Richards, G. P. (1988). Microbial purification of shellfish: a review of depuration and relaying. J. Food Protect., 51:218–251.

57. Council of the European Communities (1991). Council Directive of 15 July 1991 laying down the health conditions for the production and placing on the market of live bivalve molluscs (91/492/EEC). Official Journal of the European Communities, L268, 24 September 1991, pp. 1–14.

58. White, K. E., Osterholme, M. T., Mariotti, J. A. Korlath, J. A., Lawrence, D. H., Ristinen, T. L., and Greenberg, H. B. (1986). A foodborne outbreak of Norwalk virus gastroenteritis: evidence for post-recovery transmission. Am. J. Epidemiol., 124:120–126.

59. Iveson, A. M., Gill, M., Bartlett, C. L. R., Cubitt, W. D., and McSwiggan, D. A. (1987). Two outbreaks of foodborne gastroenteritis caused by a small round structured virus: evidence of prolonged infectivity in a food handler, Lancet, 2:556–558.

60. Kurtz, J. B. (1994). Astroviruses. In: Viral Infections of the Gastrointestinal Tract, 2nd ed. (A. Z. Kapikian, ed.), Marcel Dekker, New York, pp. 569–580.

61. Willcocks, M. M., Carter, M. J., and Madeley, C. R. (1992). Astroviruses. Rev. Med. Virol., 2:97–106.

62. Willcocks, M. M., Carver, M. J., Laidler, F. R., and Madelay, C. R. (1990). Growth and characterization of human faecal astroviruses in a continuous cell line. Arch. Virol., 113:73–82.

63. Paver, W. K., Caul, E. O., Ashley, C. R., and Clarke, S. K. R. (1973). A small virus in human faeces. Lancet, 1:237–240.

64. Appleton, H., Buckley, M., Thom, B. T., Cotton, J. L., and Henderson, S. (1977). Virus-like particles in winter vomiting disease. Lancet, 1:409–411.

65. Appleton, H., and Pereira, M. S. (1977). A possible virus aetiology in outbreaks of food poisoning from cockles. Lancet, 1:780–781.

66. Christopher, P. J., Grohmann, G. S., Millsom, R. H., and Murphy, A. M. (1978). Parvovirus gastroenteritis. A new entity for Australia. Med. J. Aust., 1:121–124.

67. Turton, J., Appleton, H., and Clewley, J. P. (1990). Similarities in nucleotide sequences between serum and faecal human parvovirus DNA. Epidemiol. Infect., 105:197–201.

68. World Health Organization (1990). WHO Consultation of Public Health Aspects of Seafoodborne Zoonotic Diseases. WHO.CDS/VPH/190.86.

69. Committee on the Microbiological Safety of Food, Chairman Sir Mark Richmond (1991). The Microbiological Safety of Food. Part II, Report of the Committee on the Microbiological Safety of Food. Her Majesty's Stationery Office, London.

70. Ball J. M., Estes, M. K., Hardy, M. E., Conner, M. E., Opekun, A. R., Graham, D. Y. (1996). Recombinant Norwalk virus-like particles as an oral vaccine. Arch. Virol. (Suppl.), 12:243–249.

4

Rotavirus

Syed A. Sattar, V. Susan Springthorpe, and Jason A. Tetro
University of Ottawa, Ottawa, Ontario, Canada

I. INTRODUCTION

Since their discovery in 1973 (1), rotaviruses have frequently been identified as the most common diarrheal disease requiring treatment or hospitalization in children under 5 years of age (2–4). Worldwide, rotaviruses account for approximately 9% of the 1.5 billion diarrheal episodes (2) and 20–40% of the 4.6 million diarrhea-associated deaths in children under 5 years of age (5). The overall picture, however, is quite different between the developed and developing parts of the world. In the developed world, 70% of all children under the age of 5 years are expected to develop rotavirus-based gastroenteritis, with only

1 in 8 requiring further care from a physician: death occurs only rarely (6–8). In the developing world, the prevalence of rotaviruses in the etiology of gastroenteritis is between 20% and 46% of all cases, but the absolute number of cases recorded, as well as the death toll, which stands at around 10% of the infections in children under 5 years of age (9), ensures an immense socioeconomic impact of this disease. Development of efficient rotavirus vaccines is therefore a priority. While early attempts had limited success, more recent approaches offer some promise for application in the developing world (10). However, widespread vaccine trials will be necessary before the true efficacy of these candidate vaccines can be evaluated.

Rotaviruses are highly infectious. In general, they are considered to cause only minor infections in adults. Many infections in adults result from infection transmission from an initial case of infantile gastroenteritis (11), but child involvement is not always necessary (12,13). The true extent of rotavirus diarrhea in adults has yet to be determined. Some studies also link transmission of rotaviruses to juveniles to asymptomatic infection in adults (14). Given the differences in severity of the disease and the numbers of asymptomatic infections among both adults and children, overt cases of rotaviral disease may represent only the tip of the iceberg.

Even in the developed world there is a significant economic impact from rotavirus disease. In the United States, rotaviruses cause about 3.5 million infections every year (15). These infections result in approximately 49,000–55,000 hospitalizations and less than 40 deaths annually in children under 5 years of age (3,6). The economic costs of rotaviral diarrhea in the United States are evaluated at $564 million dollars per year without including the financial burden due to parental work time lost, inconvenience, and fatalities (16). In Canada, during 1980–1988, rotaviruses were the most frequently identified of all viral agents from laboratory-diagnosed cases of infections in children under 4 years of age (17). In England and Wales, there were an estimated 17,810 cases of rotaviral-related hospitalizations during the 1993–1994 fiscal year. Economic losses from adult rotaviral infections are undocumented because symptoms may be mild or transient. However, losses due to rotavirus infections in domestic animals of economic importance are considerable (18).

Ever since the etiological significance of rotaviruses has been realized, and the understanding of their biology and genetics has increased, numerous reviews on rotaviruses have been written. The earlier reviews focused largely on the clinical and epidemiological aspects of rotaviral infections in humans and animals (4,19–27). These references have been retained here for reader information. More recent reviews tend to address more specific aspects of rotavirus biology such as genetics (28), pathogenicity and animal models (29–32), molecular epidemiology (33,34), serotyping (35), and vaccine development (10,36). This chapter will focus primarily on those aspects of rotaviruses that relate directly to their capacity to spread through food and drinking water. Where relevant, reference will be made to other reviews. In addition, a searchable rotavirus web site (http://www. rotavirus.com) is now available for the latest information on these important pathogens.

II. BASIC CHARACTERISTICS AND SEROLOGICAL GROUPING

A. Morphology and Physicochemical Properties

Rotaviruses, established as a genus of the family Reoviridae, are named for their wheel-like structure where the double icosahedral shell of protein subunits forms a ''rim'' around the nucleic acid core. Their genome consists of 11 segments of double-stranded RNA (37). In fixed and negatively stained preparations, complete rotavirus particles are 70–75 nm in diameter. When unfixed and unstained preparations are examined by cryoelectron microscopy, these viruses show a particle diameter of 80 nm with 60 surface spikes (38,39).

Heating at 50°C for at least 30 min was found necessary to reduce rotavirus infectivity by 99%; heating at 50°C in the presence of magnesium chloride enhanced the rate of virus inactivation, whereas the addition of magnesium sulfate to the virus suspension before heating had a stabilizing effect (40). Chelating agents such as EDTA disrupt the outer shell of rotaviruses, thus rendering them noninfectious (40). Calcium chloride and other chaotropic agents can degrade the inner capsid and release viral cores (41). Rotavirus infectivity is rapidly lost at pH values below 3 (42) and above 10 (43). Proteolytic enzymes such as trypsin and pancreatin enhance rotavirus infectivity in vitro and their presence is often necessary for virus isolation and quantitation in cell culture (44,45). The potential role of proteolytic enzymes during an infection in vivo is not well defined.

B. Rotavirus Diversity and Serological Grouping

The high rate of divergence of RNA virus genomes (46) has led to an unknown but wide variety of different human rotaviruses (47), which presumably continue to evolve. Coinfection with more than one of these can, because of the segmented nature of the rotavirus genome, lead to new reassortants (20). This occurs in an efficient manner early during experimental coinfections and is not believed to be under immune pressure (48). While at first glance such an event may be considered rare in nature, reports of multiple strains of human rotaviruses circulating concurrently or sequentially within a geographically limited area during an outbreak are quite common (49–51). In fact, rotaviruses can be regarded as "a heterogeneous population of reassortants and related variants" (33). This leads to a number of potential issues with regard to rotavirus infections: (a) the possibility of repeated infections with different rotavirus strains, (b) potential difficulties in protection from live virus vaccines, and (c) potential difficulties in diagnostic tests due to nonrecognition by immunological reagents.

Rotaviruses are classified into seven major groups (A–G) based on genomic and serological properties (52). All seven groups of rotavirus infect animals but only those in groups A–C are known to infect humans (19). Group A rotaviruses are the most common, and antibodies to this group antigen have been detected in nearly 100% of children by the age of 5 years in serological surveys conducted in many parts of the world (23). In contrast, antibodies to group C rotaviruses have been detected in less than 50% of the general population (53). Group B rotaviruses have mainly been detected in large epidemics of gastroenteritis in adults in China. Statistical data on antibodies in the general population, however, have yet to be compiled.

Rotaviruses are further classified based on antigenic relationships defined by the capsid proteins. The inner capsid consists of trimers of the virus protein VP6, which is highly immunogenic and specifies the group and subgroup antigens. The outer capsid contains the major glycoprotein VP7 and a minor protease-sensitive protein, VP4. These segregate independently and define the 11 G and 7 P serotypes, respectively. When this is coupled with the other grouping and electropherotyping schemes, the identification of individual rotaviral strains can be overwhelming to the nonexpert. A brief summary of this topic is available (47); for a fuller discussion of this issue, please refer to Theil (48), Kapikian and Chanock (23), Coulson (36), and Holmes (34).

III. ROTAVIRAL GASTROENTERITIS

A. Clinical Manifestations

Ingestion of viable rotavirus particles can infect humans (54,55), and oral challenge studies in humans (55) show the minimal infective dose (MID) to be as little as one cell culture infective unit. Such a low MID is not surprising in view of the ease with which rotavirus infections can spread. Furthermore, spread of rotavirus in nature may be favored by the fecal shedding of large clumps of virus or membrane-associated complexes (56), which are particularly resistant to inactivation by drying (57) or chemical disinfectants (58,59). In many institutional outbreaks the pattern of spread of rotaviral gastroenteritis is highly reminiscent of airborne transmission (60). Rotaviruses survive well in air, and limited studies in animals have shown that entry of rotaviruses through the nose can lead to gastro-enteritis (61).

The incubation period is 1–3 days and infection in the nonimmune host usually starts with a sudden onset of acute diarrhea often accompanied by vomiting, fever, and respiratory illness (15). Virus shedding in the feces occurs for 5–7 days (62). In severe cases, rapid dehydration leads to electrolyte imbalance, and unless rehydration therapy is instituted, renal shutdown and death can ensue (63–65). The nutritional state of the host determines the severity of the disease. Diarrhea and malnutrition are interlinked and have been shown to be synergistic causes of death in young children. Malnutrition lowers the resistance to rotavirus infection, and the diarrhea resulting from infection can, in turn, lead to reduced food intake and malabsorption (66). When both vomiting and diarrhea are seen, water loss may be extremely rapid. Recovery is generally complete except for temporary intolerance to lactose in some children (15). Reinfections with rotaviruses occur frequently (48,67), but the severity of the disease may be somewhat reduced in subsequent attacks.

In at least three cases of rotaviral infection, hemorrhagic shock and encephalopathy (HSE) has resulted (68). Rotavirus has also been identified in the cerebral spinal fluid of eight Japanese children with gastroenteritis. Although the mechanism by which rotavirus breached the intestine to colonize the cerebral spinal fluid remains unclear, the symptoms of those infected are quite well understood, i.e., convulsions accompanying acute gastroen-teritis (69).

Although overt cases of diarrheal disease comprise the majority of infections in children under the age of 5, 16–30% are subclinical (4). Asymptomatic infection can, however, lead to excretion of rotaviruses and potentiate secondary infections. In adults, group A rotaviruses generally cause mild or asymptomatic infections with the same poten-tial for transmission through the fecal–oral route.

Outbreaks due to non–group A rotaviruses also purport severe diarrhea and vomiting and some degree of dehydration (70) in both children and adults. In particular, group C rotavirus has caused two mass outbreaks in which the rate of infection among adults was higher than that of children (67).

The primary site of rotavirus infection is the epithelium lining the small intestinal villi. The proximal part of the small intestine is affected first with the subsequent involve-ment of the jejunum and ileum (22). Virus-induced destruction of villous enterocytes leads to villous atrophy and this loss of the absorptive surface results in dehydration. The sever-ity of illness does not correspond to the amount of virus excreted.

Apart from young children, the elderly (11,71,72) or those suffering from other clinical conditions (73) are at a higher risk of acquiring rotaviral infections; chronic infections have been reported in patients with immunodeficiency diseases (74,75). Outbreaks of rotaviral infections in transplant patients can be particularly serious (76).

The pathogenicity of rotaviruses has been widely studied in relation to serotype. Although most severe infections in young children have been caused by serotypes G1–4 in recent years, with G1 predominating (19), rotavirus pathogenicity is complex. It depends on multiple viral and host factors that are of different importance in different host systems, and no single pathogenicity factor has been determined (29,30).

B. Laboratory Diagnosis

Clinical manifestations of rotaviral gastroenteritis are not distinct enough to permit a differential diagnosis and testing of clinical specimens is generally required for confirmation of the etiology. Several simple and reliable techniques are now available for a rapid detection and identification of at least groups A and B rotaviruses (77,78), including culture and subsequent confirmatory analysis (79). Group C rotaviruses are, however, noncultivatable and require direct testing on fecal matter for diagnosis. Methods to detect group C rotaviruses include enzyme-linked immunosorbent assay (ELISA) (80), latex agglutination and passive hemagglutination (81). Reverse transcription polymerase chain reaction (RT-PCR) has also been described as a simple and reliable method for the detection of rotaviruses (82).

Stools collected at the peak of the diarrheal phase are the specimens of choice for direct electron microscopy; the presence of large numbers of morphologically distinct virus particles makes this technique very rapid and highly reliable. Stool samples can also be used for antigen detection, RNA gel electrophoresis, or virus isolation. When stool samples are unavailable, soiled diapers could be used (77). Virus isolation can be attempted from rectal swabs. Even though rotavirus particles as well as rotavirus antigens have been detected in vomitus (83), the usefulness of such a clinical sample in laboratory diagnosis of this disease remains to be established. Paired sera are necessary for a proper serological diagnosis. Kapikian and Chanock (84) have presented a detailed comparison of the strengths and weaknesses of the laboratory diagnostic techniques available up to that time. More recently developed techniques such as PCR and nucleic acid probes have also been successfully applied to the detection of rotaviruses in stool samples (33,85).

Table 1 presents the recommendations of the U.S. Centers for Disease Control (CDC) for the collection, storage, and transportation of stool and serum specimens from suspected cases of acute viral gastroenteritis (78). Specific forms are available from the CDC for reporting water-and foodborne outbreaks (86).

C. Treatment

No specific treatment for rotaviral gastroenteritis is available at the present time. As nonspecific treatment, oral or parenteral rehydration is highly effective and safe (87–89), and guidelines for the administration of this therapy are available (90–92). When oral rehydration solutions are not available commercially, they can be readily prepared from common food items (93). The increased use of oral rehydration therapy in developing countries has already saved the lives of millions of children (87).

Several trials are currently described which point to the use of lactobacilli as a treatment in acute rotaviral gastroenteritis (94). The use of a simple culture of *Lactobacil-*

Table 1 U.S. CDC's Instructions for Collection, Storage, and Transportation of Stool and Serum Specimens for Laboratory Diagnosis of Suspected Cases of Viral Gastroenteritis[a]

	Stool specimens	Serum specimens
When to collect	Within 48–72 h after onset of illness.	Acute phase serum within 1 week and convalescent serum within 6 weeks of onset of illness.
How much to collect	As much stool sample from each of 10 persons as possible (at least 10 cm³ each person); samples from 10 controls may also be submitted.	If possible, obtain paired sera from the same 10 persons from whom stools were obtained; samples from 10 well persons may be submitted as controls.
Method of collection	Place fresh stool specimens (liquid preferable), unmixed with urine, in clean dry containers (e.g., urine specimen cups).	Collect blood (15 mL from adults and 3 mL from children) in tubes without anticoagulants. Desirable to separate serum and send for analysis.
Storage of specimen after collection	Immediately refrigerate at 4° C. *DO NOT FREEZE* if electron microscopy is anticipated.	Refrigerate or freeze sera until shipped. Refrigerate, but *DO NOT FREEZE* blood specimens.
Transportation	Keep refrigerated. Place bagged and sealed specimens on ice or with frozen refrigerant packs in an insulated box. Send by overnight mail. *DO NOT FREEZE.*	Ship sera refrigerated (follow instructions for stool specimens) or froozen on dry ice. *DO NOT FREEZE* blood samples.

[a] Label each specimen container with a waterproof marker. Put samples in sealed, waterproof containers (e.g., plastic bags). Batch collection and send by overnight mail, scheduled to arrive at destination on a weekday during business hours.
Source: Modified from Ref. 78.

lus casei subsp. *casei* strain GG, which is regularly found in yogurt, has been shown to promote an increased immune response against rotavirus and commence stabilization of the gut mucosal barrier (95). As the mechanisms behind this are not yet clear, the use of yogurt to treat rotaviral gastroenteritis has not been approved.

IV. EPIDEMIOLOGY OF ROTAVIRAL GASTROENTERITIS

The capacity of rotaviruses to cause outbreaks depends on a complex interaction of factors such as the immune status of the individual, nonspecific host resistance, underlying or concurrent infections, the predominating type(s) of rotavirus in circulation, nutritional status of the population, the level of public health and hygiene, climatic factors, and social and cultural peculiarities of the community. Outbreaks of rotaviral gastroenteritis occur in the general community as well as in many types of institutional settings. Rotaviruses have been shown to be responsible for 6–24% of diarrheal disease in children in community-based studies (37). Rotaviral infections can readily spread within families as well and the index case is usually a young child (96–98). Adults can also be sources of infection (99); caregivers (100,101) and food handlers may be particularly important in this regard.

Outbreaks of rotaviral infection are common in infants and young children in hospitals (102,103), day care centers (103,104), and schools (67,105). In two cases of group C rotaviral outbreaks in Japanese schools, teachers also presented symptoms of infection (68).

Rotaviruses have a wide host range and are only partially restricted to specific hosts. Although human–human contact through the fecal–oral route causes the majority of secondary cases of rotaviral gastroenteritis, the potential for interspecies infection has now been demonstrated. Under experimental conditions, rotaviruses isolated from one animal species have been shown to infect other types of animals. This indicates that rotaviruses can cross species boundaries (105) and suggests that certain cases of human rotavirus infection may be zoonotic in origin or the result of reassortants between animal and human strains (33). Certain isolates of human rotaviruses have been found to be closely related to those from cats and dogs (106,107). Gouvea et al. (33) reported that the causative agent of an outbreak in Brazilian children was a rotavirus normally associated with disease in horses and piglets.

The epidemiology of rotaviruses is further complicated by the existence of multiple serotypes and repeated infections. Studies in animals have shown that levels of rotavirus antibody in the small intestine correlate more closely with protection than do levels of serum antibody (108). In humans also, the presence of rotavirus neutralizing antibodies in the serum did not confer immunity to reinfection (55). Since intestinal immunity is only of limited duration, rotavirus reinfections are likely to be common. Sonza and Holmes (109) showed a peak of fecal antibody to rotaviruses 2–4 weeks after infection, which declined to undetectable levels within 2 months. Limited studies in animals show that cellular immunity may also protect against rotaviruses, but its role in humans remains to be elucidated (37). The lower rates of clinical rotavirus infections in neonates are believed to be due to antibodies acquired from the mother through transplacental transfer.

V. SEASONALITY OF ROTAVIRAL GASTROENTERITIS

In temperate regions of the world, outbreaks of rotaviral gastroenteritis generally occur in the colder and drier periods (25,61,110). In Australia, Japan, the United Kingdom, South Africa, and Canada, there is a ''winter peak'' in which epidemic-like spikes of rotaviral-based hospitalizations are observed (4,8,111). This is not due to the absence of rotavirus infections because infectious rotavirus particles, albeit at reduced levels (112), are detected in sewage and surface waters in the nonoutbreak seasons. Based on studies on children in Nigeria (113), there seems to be an inverse relationship between the relative humidity of the region and outbreaks of rotavirus. As if to confirm this observation, the annual outbreak peaks of hospitalization due to rotaviral gastroenteritis in the United States appear first in the southwestern region during the month of November and move toward the northeastern region as the winter progresses (114).

The apparent lack of seasonality of rotavirus infections in some countries may be due to the fact that only a small proportion of their population has access to safe drinking water (115) and food. It is therefore quite likely that rotavirus spread occurs through these vehicles on a year-round basis. This is supported by the reports of outbreaks of gastroenteritis due to group B rotaviruses in China (70). In addition, there can be marked seasonal variations in water quality; in the dry season, the demand for water is higher and at the same time increased evaporation of surface water reduces the dilution of discharged waste. In the rainy season, land runoff can increase the level of contamination of both surface and groundwater sources. On the other hand, in developed countries, proper systems for the treatment and disposal of sewage and the treatment, distribution, and storage of potable water considerably reduce the possibility of rotavirus spread through water. The recorded waterborne outbreaks of rotaviral gastroenteritis in such areas occur due to occasional problems with water quality.

VI. VEHICULAR SPREAD OF ROTAVIRAL INFECTIONS

In spite of the remarkable advances in the past 30 years in our understanding of the etiology of acute nonbacterial gastroenteritis, no causative agent(s) can be incriminated in many cases of this disease. Recent data from the United States show that in nearly 50% (116) and 62% (117) of the water- and foodborne outbreaks of gastroenteritis, respectively, no etiological agent could be identified. Cliver (118) has suggested that, in spite of improvements in surveillance systems, foodborne outbreaks in the United States are underreported by a factor of 10 or more; in the developing world, as many as 99% of foodborne illnesses are unreported (119). An analysis of foodborne diseases in Canada during 1975–1984 also showed that no etiology could be determined in more than 75% of the incidents and 50% of cases (120). More than 53% of waterborne outbreaks in Sweden during 1975–1984 also remained unassociated with an etiological agent (121).

Figure 1 shows the complex interrelationship between various vehicles that may play a role in the spread of rotaviral gastroenteritis. The relative importance of these vehicles in the transmission of the infection remains undetermined in spite of the frequent

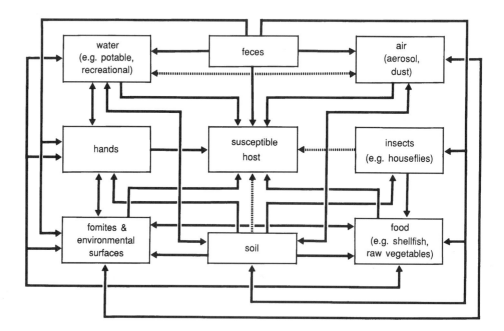

Figure 1   Interrelationship of vehicles that may play a role in the transmission of rotaviral gastroenteritis. (From Ref. 61.)

occurrence of community and institutional outbreaks. This is particularly true with regard to our understanding of rotavirus spread through food.

Appleton et al. (122) suggested this to be due to the insufficient investigation of foodborne outbreaks of gastroenteritis rather than the inability of rotaviruses to spread by food. Although more published information is available on the demonstrated and potential role of drinking water as a vehicle for rotaviruses, there are major gaps in our knowledge in this area as well. In general, waterborne pathogens are also capable of spreading through food and rotaviruses may not be an exception in this regard.

The general public is also somewhat responsible for the lack of information on foodborne rotaviral infection. In a recent study of practice consultations in the United Kingdom (123), only 1 in every 26 actual incidents of foodborne illness was reported. Of those who did report, 18% were positive for rotavirus, the predominant etiological agent in winter. Extrapolation of this data would suggest that infections by rotavirus are far more common than previously thought.

Although the relative importance of food and water as a rotavirus vehicle in the developed world is still being determined, there is no doubt of their importance in the developing world. Yet, even in these settings, the true extent of rotavirus spread through food and drinking water would be very difficult to determine. In a recent review of 34 studies (124), consumption of contaminated food, water, or ice cubes is believed to have been responsible for rotaviral gastroenteritis in 0–36% of American adults traveling to Latin America, Africa, and Asia.

A priori, for a pathogen such as rotavirus to be spread by animate or inanimate vehicles, it must be capable of survival on or in the vehicle for sufficient time to make contact with the next susceptible host. The following is a summary of what is known about rotavirus spread through food and drinking water and how other vehicles such as fomites, environmental surfaces, and human hands may contribute to such spread.

A. Foodborne Spread

1. Sources of Rotaviral Contamination

Rotaviral contamination of various types of food can occur through (a) the use of fecal wastes (night soil/biosolids) for fertilization and/or sewage effluents for irrigation of crops; (b) washing of vegetables that are consumed raw in contaminated water; (c) preparation or handling of food by persons with rotavirus-contaminated hands; (d) placement of food in contaminated vessels or on contaminated environmental surfaces; and (e) cultivation of shell-fish in sewage-polluted waters.

2. Rotavirus Recovery from Food

Techniques are now available for recovering infectious rotavirus particles from raw vegetables (125), mussels, oysters, and blue crab (126–130). Rotaviruses have been found in market lettuce (131) at concentrations of 10,000 virions per gram. The likely source of contamination was unsafe water, which was used to wash the vegetables. Infectious rotaviruses have also been detected in mussels collected from the northern Adriatic (132) and the Mediterranean off the coast of Barcelona (133).

Badawy et al. (134) have shown that simian rotavirus SA-11 could survive on raw vegetables for 25–30 days at 4°C and 5–25 days at 20°C. Virus survival of this duration allows for international translocation of rotavirus-contaminated foodstuffs. It is anticipated, however, that such contamination of imported food items would be sporadic and would most likely give rise to a few initial cases of infection; secondary spread of the virus through other means may mask the role of food as a vehicle.

Even though shellfish, harvested from sewage-polluted waters and eaten raw, may be the most likely food to be linked directly to rotaviral gastroenteritis, no such outbreaks have as yet been reported. It should, however, be noted here that shellfish is rarely consumed raw by very young children, and adults infected by rotavirus-contaminated shellfish may go unnoticed because of the absence of any clinical symptoms.

3. Foodborne Outbreaks

Hara et al. (135) reported outbreaks of acute gastroenteritis in primary schools and other institutions from several districts of Japan. The explosive nature of these outbreaks resembled mass food poisoning and school lunches were suspected as the vehicles. Matsumoto et al. (67) have described a similar outbreak, due to a group C rotavirus, which occurred simultaneously in seven elementary schools in Fukui City, Japan, and affected more than 3000 individuals. Since one particular preparation center supplied lunch to all of the affected schools, the outbreak is believed to have spread by food; no specific food item could be incriminated through a questionnaire nor could rotavirus be detected in the samples of food and water tested.

Table 2 summarizes the information on confirmed and suspected foodborne outbreaks of rotaviral gastroenteritis reported in New York State since 1985 (136–140). The 11 such outbreaks reported during this period affected a total of 460 individuals. Eight (61.5%) of the outbreaks were associated with food service establishments. The dates of onset of the outbreaks show no particular seasonal pattern. The reasons for the paucity of similar reports from other parts of the United States are not known at this stage. This may relate more to the differences in the degree of surveillance and less to a total absence of such outbreaks elsewhere.

4. Rotavirus Elimination During Food Preparation

Very little published information is available on the removal and inactivation of rotaviruses during food preparation. Wood and Adams (141) have shown that lactic acid fermentation of weaning foods was not sufficient to inactivate rotavirus SA-11 unless this was combined with heat treatment generally used to gelatinize starch. Panon et al. (142) found a bovine rotavirus in milk to be quite stable during the process for making soft cheeses and suggest that such cheeses made from virus-contaminated milk may carry infectious rotaviruses. Normal cooking temperatures and times should be adequate for inactivating rotaviruses, but we have not been able to find any specific studies in this regard. This may be particularly important for short durations of microwave cooking and exposure to ionizing radiation, which is often used for preservation of foodstuffs.

5. Antirotaviral Activity in Food

Foods that contain antirotaviral antibodies (143,144), lactic acid bacteria (94), lactoferrin, interferon-γ and interleukin-1 (145) have shown the potential to interfere with the course of rotavirus infection in vitro and, in some cases, in vivo. Panon et al. (142) have demonstrated antirotaviral activity in raw milk, while Yolken et al. (76) have reported the presence of antirotaviral antibodies in pasteurized bovine milk. These antibodies, presumably derived from natural rotavirus infections of cattle, were able to prevent cell cultures and experimental animals against infections by several types of human rotaviruses (76); however, methods used to convert bovine milk into infant food resulted in significant losses

Table 2 Foodborne Outbreaks of Rotaviral Gastroenteritis in New York State as Reported by the New York State Department of Health[a]

Year and location	Outbreak	Date of onset	No. of cases	Vehicle(s)	Place of prep. of contam. food
1985					
Erie	Suspected[b]	Dec. 01	8	Cold foods	Food service establishment
Erie	Confirmed	Dec. 06	114	Cold foods	Food service establishment
Ontario	Confirmed	Nov. 10	35	Unknown	Unknown
Suffolk	Confirmed	Oct. 03	114	Water/ice, Greek salad	Food service establishment
1986					
Warren	Suspected	May 04	41	Unknown	Food service establishment
Yates	Suspected	Aug. 01	75	Unknown	Food service establishment
1988					
Columbia	Suspected	Nov. 21	5	Unknown	Food service establishment
Jefferson	Confirmed	Dec. 06	18	Shepherd's pie	Food service establishment
1989					
Onandaba	Confirmed	Mar. 20	6	Unknown	Unknown
Tompkins	Confirmed	April 09	3	Strawberry shortcake	Home
1990					
Erie	Suspected	June 05	41	Hamburger, brownies	Camp
1990–1995					
TBA	Confirmed	Mar. 20	6	Unknown	Unknown
	Confirmed	April 09	3	Strawberry shortcake	Home
	Suspected	June 05	41	Hamburger, brownies	Camp

[a] Table based on data from New York Department of Health (April 1987; April 1988; Dec. 1989; Dec. 1990; Feb. 1992).
[b] When epidemiological evidence implicates the virus but no laboratory confirmation is available.

in this activity. Bovine colostrum containing antibodies to human rotavirus has been administered in the passive immunization of children (146) and piglets (147). Human maternal breast milk has shown equivocal results in nonimmunized mothers (148); mothers immunized in the postpartum period are able to boost the immune system of their breast-fed children (149). The IgY antibody found in chicken eggs has also been shown to interfere with rotaviral infection (150). The method by which IgY prevents infection has yet to be determined.

B. Spread Through Drinking Water

1. Rotavirus Contamination

Rotaviruses can enter into drinking water through (a) the seepage of wastes into groundwaters, (b) improper treatment of water by private or public supply systems, (c) contamination of treated water in the distribution system, (d) contamination of water during collection and storage.

2. Survival of Rotaviruses in Water

Simian rotavirus SA-11 could survive in experimentally contaminated freshwater for several days at 20°C (151–153). Studies with a human rotavirus (154) showed that in river water, before and after conventional drinking water treatment, it could survive for several weeks at 20°C or 4°C; virus survival was even longer when samples of raw river water were filtered through a 0.45-μm diameter membrane filter prior to virus inoculation.

Disposal of domestic wastes by land spreading is being used increasingly. Many of the group A rotaviruses can be found in sewage and other wastes (155). Rotaviruses can survive in anaerobically stored liquid and semiliquid animal wastes for up to 6 months (156). The lack of adsorption of rotaviruses to soil particles (157,158) could lead to rotavirus contamination of groundwater (159,160). Rotaviruses also survive well in groundwaters (161), which are often used for drinking without treatment.

The reasons for the prolonged stability of rotaviruses in water are not clearly understood. Viruses that do not adsorb to suspended solids tend to be less stable in the environment than those that adsorb to solid matter (162). Rotaviruses are often excreted as clumps in association with cellular debris (163) or are antibody-coated, providing a biologically stable shield against the environment. These two factors may provide some insight into the relatively strong stability of the virus in varying environments.

3. Waterborne Outbreaks

Rotavirus spread is believed to have occurred through water in a number of outbreaks (164–70) and information on these has been summarized before (61,170,171). In addition to these, similar outbreaks have been reported from East Germany (172), Israel (173), the former USSR (174), Sweden (121), and Brazil (175).

On a global basis, water may spread rotaviral gastroenteritis more commonly than is suggested by published reports. As is the case with food, one or a few index cases may initiate an outbreak but the water origin may be masked by secondary spread by other routes. The true extent of waterborne spread of this disease in developing countries remains undetermined in spite of rampant sewage pollution and the common occurrence of acute diarrhea. In such regions, the relative importance of water as a vehicle is particularly difficult to assess because rotavirus transmission may occur simultaneously through several vehicles (Fig. 1). In Kenya, an increase in rainfall has been directly linked to an increase in gastroenteritis among children (176); rotavirus was the main etiological agent. A similar trend is seen in Costa Rica, where a rise in endemic diarrhea correlates to the winter season (177). In China, the initial vehicle for rotaviral gastroenteritis of adults was sewage-polluted water followed by person-to-person spread of the disease (178).

It has been suggested (23) that group A rotaviruses may not spread through water, as the large waterborne outbreaks in China have been shown to be due to non–group A rotaviruses (179). However, it is not known if non–group A rotaviruses were also responsible for other such outbreaks (61). It should be noted here that the published data on rotaviruses in water comes from studies with group A rotaviruses and it is difficult to determine how much of it will be applicable to non–group A rotaviruses.

4. Rotavirus Detection in Water

Traditional techniques for the concentration and detection of rotaviruses from drinking waters have already been reviewed (61,180,181).

Dahling et al. (182) evaluated several commercial ELISA and latex agglutination test kits for the detection of rotaviruses in concentrates of seeded tap water samples. Whereas many of these kits were found to be quite sensitive, economical, and simple to use, their application in the routine monitoring of water and food samples is limited. They detect viral antigens (or viral nucleic acid) and not infectious viral particles. Strappe (183) compared several methods of second step concentration to recover infectious rotaviruses from seeded samples of tap water and found hydroextraction to give the highest rate of recovery.

Reverse transcription-polymerase chain reaction has been adapted to detect rotaviruses and other enteric viruses in water (184,185). This relatively inexpensive and rapid test has only recently become available and evaluations of RT-PCR tests have yet to be published.

A study of viral pollution of water in Mexico (160) was the first to report the detection of infectious rotaviruses in samples of treated drinking water. Thirteen of the 31 samples (42%) tested were rotavirus-positive, and the level of contamination in them ranged from 16 to 2500 immunofluorescent focus-forming units/20 L. Many of the virus-positive samples met the conventional bacteriological standards.

Hepatitis A virus nucleic acid as well as infectious entero- and rotaviruses were detected in samples of drinking water collected at a Spanish military camp experiencing an outbreak of hepatitis A (133). All of the five potable water concentrates tested were positive for rotavirus and the level of rotavirus contamination in them ranged from 20 to 85 immunofluorescent focus-forming units/1000 L. The water samples had a total chlorine residual of 0.2 mg/L and they did not reveal the presence of any fecal coliforms.

Leon and Gerba (186) used a 2500-bp clone of segment 4 of a human rotavirus as a radiolabeled probe to detect rotaviral contamination in samples of treated and untreated potable waters. Six percent (1/16) of tap water samples and 16% (5/32) of groundwater samples were found to be positive.

During a survey (1987–1991) of raw waters in several locations in Croatia (132; personal communication), 3% (4/134) of the samples were found to be positive for rotaviruses; a 4-year (1986–1989) survey of drinking water from a middle-sized town showed enteric viruses in 3.7% (5/135) of the samples and one of these was positive for rotaviruses.

5. Rotavirus Elimination During Treatment of Water for Drinking

Infectious rotavirus particles have been recovered from a variety of natural waters, including those used for drinking (159,160,187,188). The efficiency of rotavirus removal and inactivation by various stages of water treatment has been studied (188). When compared to enteroviruses, rotavirus SA-11 adsorbed poorly to aluminum hydroxide (alum) and a variety of soil types (161). Rotaviruses have been shown to resist a variety of chemical germicides, including chlorine, in laboratory carrier and suspension tests (58,59,189). Raphael et al. (190) have been shown to resist inactivation by chlorine residuals in samples of conventionally treated drinking water collected either as effluent at a treatment plant or out of a tap. This implies that, even if rotaviruses are completely removed and inactivated during conventional water treatment, posttreatment contamination of water supplies with rotaviruses may lead to disease transmission (191). Failure to chlorinate drinking water has also been blamed for waterborne outbreaks of rotaviral gastroenteritis (70,

Table 3 Factors that Might Contribute to Waterborne Spread of Rotaviral Gastroenteritis

Factors	Reference(s)
(1) Conventional methods of sewage treatment have been found to be less efficient in removing rotaviruses when compared to enteroviruses.	161
(2) Discharge of effluents from conventional sewage treatment plants results in rotaviral pollution of receiving waters and soils.	151, 154, 160
(3) Rotaviruses can retain their infectivity for several days in the water environment.	152, 154, 160
(4) Infectious rotaviruses can be recovered from polluted waters.	151, 154, 159, 160
(5) The ability of rotaviruses to survive on land and their poor retention by soil particles can contaminate groundwaters in areas practicing spray irrigation or land disposal of municipal wastes.	151, 158
(6) Conventional methods of potable water treatment may be less efficient in removing rotaviruses when compared to enteroviruses. Furthermore, levels of chlorine used for the terminal disinfection of drinking water have been found to be ineffective against rotaviruses.	153, 154, 160
(7) In many communities, particularly those in the developing world, contamination of treated potable water with feces can occur in the distribution system or during storage.	127, 160, 170
(8) Rotaviruses can retain their infectivity in treated potable waters for several days.	171, 189
(9) Infectious rotaviruses have been detected in treated drinking waters from certain communities.	127
(10) Ingestion of relatively small numbers of infectious rotavirus particles has been shown to result in infection.	

169). Major factors associated with waterborne spread of rotavirus are summarized in Table 3.

Panangala et al. (192) showed that *N*-halamine biocidal polymers have shown up to 6 log reductions of rotavirus, based on flow rate, as a last-stage filtration step in handheld water processing. The levels of leaching of impurities was also within governmental regulations.

The U.S. Environmental Protection Agency (EPA) considers one infection/10,000 per year as an acceptable health risk of infections from potable waters (193). Reliable field data for the level of risk of acquiring an infection from drinking rotavirus-contaminated water are not available. However, results from feeding experiments in humans have been used to develop models to determine the potential health risks assuming a consumption of 2 L of drinking water per person per day. For an acceptable risk for rotaviral infections the water to be consumed should not contain more than 0.3 infective unit of virus per 100 liters per day (194,195). Microbial risk assessment models for drinking water may have been reviewed (196) with the implication that models for rotavirus may have overestimated the risk based on assumptions of virus monodispersion. Whether this is true or not remains to be established. Certainly, rotaviruses are almost certainly shed mainly in clumps or aggregates (56), and may remain so in the field.

C. Fomites and Environmental Surfaces

Contamination of food and drinking water with enteric pathogens can occur either directly by contact with feces or indirectly from virus-contaminated fomites and environmental surfaces. Indeed, contaminated equipment has been among the major contributing factors in foodborne disease outbreaks in the United States (117). In actual practice, the recovery of pathogens from incriminated fomites and environmental surfaces is hardly ever attempted. To a large degree, this is due to the general lack of availability of simple and efficient methods for the microbiological examination of such samples. The available indirect evidence shows that fomites and environmental surfaces possess a strong potential for the contamination of food and drinking water.

Studies in our laboratory (57) and elsewhere (104,197,198) have demonstrated that rotaviruses can remain viable on inanimate surfaces for several days when dried from a fecal suspension. Rotavirus survival on inanimate surfaces is also influenced by relative humidity (RH) and air temperature. At low or medium RH, these viruses could remain viable on plastics, glass, and stainless steel for more than 10 days at room temperature. Raising the RH to 80% resulted in a more rapid and irreversible loss in virus infectivity.

In our studies, the survival of the Wa strain of human rotavirus on porous materials was more variable than on nonporous surfaces (57). The reasons for this are not clear at this stage, but the apparent loss in infectivity could be due to either the irreversible adsorption of virus particles to the fibers, a poor recovery of virions from porous materials, or the inactivation of the virus by biocides and other chemicals present in the materials tested.

Infectious rotavirus particles have been recovered from a variety of surfaces (including refrigerator handles) and objects in a day care center (104). Chemical disinfection of rotavirus-contaminated nonporous inanimate surfaces may be particularly difficult due to the relatively high resistance of these viruses to many commonly used products (58,59). An in-depth study of viral contamination in a pediatric ward in Saudi Arabia showed that such objects as toys, furniture, sinks, toilet handles, and charts were found to be contaminated with rotavirus, supposedly several hours or days after use. In two cases, viral contamination was responsible for an outbreak of gastroenteritis. Although cleaning practices were performed regularly, only general disinfectants were used, against which rotaviruses are highly resistant (189). The study also suggests that these data can be extrapolated into other areas of high child traffic, such as child care centers.

D. Hands

Rotaviral antigens (199), as well as infectious rotaviral particles (104,200), have been recovered from human hands. According to the CDC (117), poor personal hygiene of the food handler was the second most important factor responsible for foodborne outbreaks; during 1983–1987, it was reported in nearly 38% of the outbreaks due to infectious agents. To further emphasize this point, the act of proper hand washing has been shown to significantly diminish the number of diarrheal episodes as well as a complete prevention of secondary cases (201).

A study has shown that when volunteers were asked to touch a rotavirus-contaminated surface with a finger and then put the finger in the mouth, 63% (5/8) of them became infected (198).

The Wa strain of human rotavirus, when suspended in feces, was found to survive well on the fingerpads of volunteers (202). More than 4 h after experimental contamination, nearly 7% of the deposited infective units were still detectable on the skin surface. This virus was also found to survive better on human hands when compared to rhinovirus 14, parainfluenzavirus 3, *Escherichia coli*, and *Staphylococcus aureus* (203). Ansari et al. (204) have shown that certain commonly used antiseptics are less effective in the elimination of rotaviruses from hands when compared to *E. coli*.

Rotavirus survival on the hands of six volunteers was similar (202), but the rate of loss in virus infectivity on hands was higher than that seen on nonporous inanimate surfaces (57). The higher body temperature, moisture on the skin, the enzymatic activity of the normal skin microflora, and specific or nonspecific virus inhibitors in sweat may account for the difference.

Infectious rotavirus in dried fecal suspensions on hands can be readily transferred to inanimate surfaces or other hands upon contact (202); likewise, touching or handling of rotavirus-contaminated objects can transfer infectious virus to hands. This suggests that animate and inanimate surfaces may play a complementary, and possibly a synergistic, role in the transmission of rotaviruses. In view of this, measures for the prevention and control of rotavirus spread through food should include the proper disinfection of hands as well as environmental surfaces.

VII. PREVENTION AND CONTROL

In the past 5 years, a rhesus reassortant rotavirus vaccine (RRV) has been developed for children. The vaccine is specific for four serotypes within group A (G1–G4), the four main causes of human rotavirus infection (205). The RRV has demonstrated its effectiveness in developing protection against rotaviral challenges (206–208) and has shown no significant side effects. Recently, the Vaccine Research and Development Unit of WHO recommended that this vaccine be used in a large-scale trial in Latin America and Venezuela, with small trials to be conducted in Asia and Africa. The hope of the Unit is that outbreaks due to group A rotaviruses may be significantly reduced if not eliminated in these areas. Although evidence exists for the development of vaccine-independent rotaviral reassortants (33,209,210), there has been no concrete evidence of such rotaviruses in the environment.

In the United States, the issue of vaccination programs has been hotly debated based on cost effectiveness. As previously mentioned, up to 70% of children will contract a rotaviral infection but will almost certainly recover. A recent study by Tucker et al. (4) suggested that even at a cost of $20 per vaccination, the United States would save approximately $296 million dollars to society with a net loss of over $100 million to the health care system. Conversely, both areas would save money if the cost of the vaccine was reduced to $9 per person. To date, no concrete figure has been given on the cost of RRV.

Other efforts to develop effective means of controlling and preventing outbreaks of rotaviral diarrhea have met with very limited success. This may be due to the fact that there are major gaps in our knowledge of how rotaviruses spread in the general community and in institutional settings and why outbreaks of rotaviral gastroenteritis are predominantly seasonal in nature. Furthermore, considerable work on a global scale is required to determine the number and relative significance of non–group A rotaviruses, the temporal and spatial variations in serotypes belonging to group A (47,211). In the absence of specific chemotherapy and suitable vaccines, answers to these and related questions will be necessary before rational means of controlling these infections could be instituted.

VIII. CONCLUDING REMARKS AND DIRECTIONS FOR THE FUTURE

Since the discovery of human rotaviruses nearly 30 years ago, considerable progress has been made in our understanding of their nature and the role they play in disease production in humans. We have also made rapid strides in the development and application of relatively simple and specific tests for the laboratory diagnosis of rotaviral infections. Although no specific and safe antirotaviral drugs are as yet available, nonspecific treatment strategies such as oral rehydration have shown considerable promise in the treatment of severe dehydration brought on by rotaviral diarrhea. A number of experimental vaccines have been developed for the immunization of children against rotaviral gastroenteritis (208). Even the newer vaccines now being considered may require additional modifications as a more complete understanding of the antigenic spectrum of rotaviruses becomes available (211). Strategies for antiviral vaccines have recently been reviewed (10) and one tetravalent vaccine (RotaShield) has now received approval from the U.S. Food and Drug Administration for human use. Recently, concerns have been raised on the safety of such vaccines (211).

Infection and resulting outbreaks, however, are still a grim reality both in the developed and the developing worlds and there are few answers to their continuing presence. Considering the trend of the ''winter peak'' in the developed world, does the cold weather reduce the rotavirus-eliminating efficiency of drinking water treatment systems, thereby leading to sporadic foci of infections? Or does the increase in international air travel between endemic areas and temperate zones play a role in this observed seasonality? The possible acquisition of an apparent or inapparent case of rotavirus infection by adults traveling abroad (124) makes introduction from this source a likely event.

On an individual level, rotaviruses survive quite well on human hands (203) and are relatively resistant to many commonly used hygienic handwash agents (193), yet simple handwashing has been shown to reduce the surge of secondary cases of infection (202). The fecal–oral route is obviously suspect, but what about the role of fomites and environmental surfaces used for food preparation and storage? Handwashing is rendered ineffective if contamination occurs afterward. The findings of Abad et al. (189) suggest that proper decontamination of environmental surfaces may interrupt the spread of rotaviral infections. These two simple factors in combination may significantly reduce the rate of foodborne infections and rotaviral outbreaks due to food transfer.

In developing countries, general improvements in water quality, public health, and hygiene should lead to a reduction in cases of rotaviral gastroenteritis, and some studies have shown this trend (9,212). However, outbreaks continue to arise and burden already strained economies. Even if vaccination programs are possible, there may not be enough money to run them efficiently. Other, more cost-efficient methods, such as the cessation of plant irrigation with sewage-contaminated waters, must be legislated to decrease the spread of rotaviruses.

The frequency of outbreaks of gastroenteritis due to group B and C rotaviruses is rising. Do these viruses survive in food or water as well as, or better than, group A rotaviruses? It is also not clear as to why the major waterborne outbreaks of rotaviral gastroenteritis in adults due to group B rotaviruses are so far limited to China. As group A rotaviruses of animal origin are increasingly being found to infect humans, what is the potential for groups D–G to cause gastroenteritis in humans? The entire picture of rotaviral infection, vehicular spread, and, indeed, vaccination will have to be reexamined should this occur. Until then, the progress made should aid in the reduction of rotavirus prevalence and perhaps increase the safety and health of the world's children.

ABBREVIATIONS

CDC   Centers for Disease Control and Prevention (Atlanta, GA)
DNA   deoxyribonucleic acid
EDTA   Ethylenediaminetetraacetic acid
ELISA   enzyme-linked immunosorbent assay
PCR   polymerase chain reaction
RT-PCR reverse transcriptase PCR
RNA   ribonucleic acid
PFU   plaque-forming units
WHO   World Health Organization

REFERENCES

1. G Davidson, RF Bishop, RRW Townley, IH Holmes, BJ Ruck. Importance of a new virus in acute sporadic enteritis in children. Lancet 1:242–245, 1975.
2. M Kuzuya, R Fujii, M Hamano, J Nakamura, M Yamada, S Nii, T Mori. Molecular analysis of outer capsid glycoprotein VP7 genes from two isolates of human group C rotavirus with different genome electropherotypes. J Clin Microbiol 34:3185–3189, 1996.
3. JL Liddle, MA Burgess, GL Gilbert, RM Hanson, PB McIntyre, RF Bishop, MJ Ferson. Rotavirus gastroenteritis: impact on young children, their families and the health care system. Med J Austr 167:304–307, 1996.
4. AW Tucker, AC Haddix, JS Bresee, RC Holman, UD Parashar, RI Glass. Cost-effectiveness analysis of a rotavirus immunization program for the united states. JAMA 279:1371–1376, 1998.
5. PS Mead, L Slutsker, V Dietz, LF McCaig, JS Bresee, C Shapiro, PM Griffin, RV Tauxe. Food-related illness and death in the United States. Emerg Infect Dis 5:607–625, 1999.
6. P Desikan, JD Daniel, CN Kamalarathnam, MM Mathan. Molecular epidemiology of nosocomial rotavirus infection. J Diarrh Dis Res 14(1):12–15, 1996.
7. RI Glass, PE Kilgore, RC Holman, S Jin, JC Smith, PA Woods, MJ Clarke, MS Ho, JR Gentsch. The epidemiology of rotavirus diarrhea in the United States: surveillance and estimates of disease burden. J Infect Dis 174, Suppl 11, 1996.
8. MJ Ferson. Hospitalisations for rotavirus gastroenteritis among children under five years of age in New South Wales. Med J Austr 164:273–276, 1996.
9. MJ Ryan, M Ramsay, D Brown, NJ Gay, CP Farrington, PG Wall. Hospital admissions attributable to rotavirus infection in England and Wales. J Infect Dis 174, Suppl 8:S12–S18, 1996.
10. UD Parashar, JS Bresee, JR Gentsch, RI Glass. Rotavirus. Emerg Infect Dis 4:561–570, 1998.
11. DB Hrdy. Epidemiology of rotavirus infection in adults, Rev Infect Dis 9:461, 1987.
12. CD Brandt, HW Kim, WJ, Rodriguez, JO Arrobio, BC Jeffries, EP Stallings, C Lewis, AJ Miles, RM Chanock, AZ Kapikian, RH Parrott. Pediatric viral gastroenteritis during eight years of study. J Clin Microbiol 18:71, 1983.

13. OH Meurman, MJ Laine. Rotavirus epidemic in adults. N Engl J Med 296:1298–1299, 1977.

14. MD Holdaway. Management of gastroenteritis in early childhood. Drugs 14:383–389, 1977.

15. CW LeBaron, NP Furutan, J Lew, JR Allen, V Gouvea, C Moe, S Monroe. Viral agents of gastroenteritis: public health importance and outbreak management. MMWR 39:1, 1990.

16. RI Glass, PF Pinsky, N Young-Okoh, WM Sappenfidd, JW Buehler, N Gunter, L Anderson. Diarrheal deaths in American children: are they preventable? JAMA 260:3281, 1988.

17. JM Weber, CA Parker. Laboratory diagnosed human viral infections in Canada, 1980–1988: trends and clinico-epidemiological characteristics. Diag Microbiol Infect Dis 14:225, 1991.

18. LJ Saif. Introduction. In: LJ Saif and KW Theil, eds, Viral Diarrheas of Man and Animals. CRC Press, Boca Raton, FL, p 3, 1990.

19. RF Bishop. Natural history of human rotavirus infection. Arch Virol Suppl 12, 1:19–128, 1996.

20. RF Ramig, RL Ward. Genomic segment reassortment in rotaviruses and other reoviridae, Adv Virus Res 39:163, 1991.

21. AR Bellamy, GW Both. Molecular biology of rotaviruses. Adv Virus Res 38:1–43, 1990.

22. LJ Saif. Comparative aspects of enteric viral infections. In: LJ Saif and KW Theil, eds. Viral Diarrheas of Man and Animals. CRC Press, Boca Raton, FL, p 9, 1990.

23. AZ Kapikian, RM Chanock. Rotaviruses, in Fields. Virology, Raven Press, New York, p 1353, 1990.

24. MK Estes, J Cohen. Rotavirus gene structure and function, Microbiol Rev 53:410, 1989.

25. SA Sattar, VS Springthorpe. Rotavirus gastroenteritis. Int J Sci Technol 1:26, 1988.

26. MR Christensen. Human viral gastroenteritis. Clin Microbiol Rev 2:51, 1988.

27. MK Estes, DY Graham, DH Dimitrov. The molecular epidemiology of rotavirus gastroenteritis. Prog Med Virol 29:1–22, 1984.

28. RF Ramig. Genetics of the rotaviruses. Annu Rev Microbiol 51:225–255, 1997.

29. B Burke, U Desselburger. Rotavirus pathogenicity. Virology 218:299–305, 1996.

30. U Desselburger. Viral factors determining rotavirus pathogenicity. Arch Virol Suppl 13:131–139, 1997.

31. MA Franco, N Feng, HB Greenberg. Molecular determinants of immunity and pathogenicity of rotavirus infection in the mouse model. J Infect Dis 174 Suppl 1:S47–50, 1996.

32. LJ Saif, LA Ward, L Yuan, BI Rosen, TL To. The gnotobiotic piglet as a model for studies of disease pathogenesis and immunity to rotaviruses. Arch Virol Suppl 12:153–161, 1996.

33. V Gouvea, M Brantly. Is rotavirus a population of reassortants? Trends Microbiol 3:159–162, 1995.

34. IH Holmes, Development of rotavirus molecular epidemiology: electropherotyping. Arch Virol Suppl 12:87–91, 1996.

35. BS Coulson. VP4 and VP7 typing using monoclonal antibodies. Arch Virol Suppl 12:113–118, 1996.

36. RI Glass, R Compas, D Lang. Part of Fifth Rotavirus Vaccine Workshop: current issues and future developments. J Infect Dis 124: Suppl 1 S1–S126, 1996.

37. NR Blacklow, HB Greenberg. Viral gastroenteritis, N Engl J Med 325:252, 1991.

38. BVV Prasad, JW Burns, E Marietta, MK Estes, W Chiu. Localization of the VP4 neutralization sites in rotavirus by three dimensional cryoelectron microscopy. Nature 343:476, 1990.

39. M Yeager, KA Dryden, NH Olson, HB Greenberg, TS Baker. Three dimensional structure of rhesus rotavirus by cryoelectron microscopy and image reconstruction. J Cell Biol 110: 2133, 1990.

40. MK Estes, DY Graham, EM Smith, CP Gerba. Rotavirus stability and inactivation. J Gen Virol 43:403, 1979.

41. P Bican, J Cohen, A Charpilienne, R Scherrer. Purification and characterization of bovine rotavirus cores. J Virol 43:1113, 1982.

42. C Weiss, HF Clark. Rapid inactivation of rotaviruses by exposure to acid buffer or acid gastric juice. J Gen Virol 66:2725, 1985.

43. EL Palmer, ML Martin, FA Murphy. Morphology and stability of infantile gastroenteritis virus: comparison with reovirus and blue tongue virus. J Gen Virol 35:403, 1977.

44. MK Estes, DY Graham, BB Mason. Proteolytic enhancement of rotavirus infectivity: molecular mechanisms. J Virol 39:879, 1981.

45. S Ramia, SA Sattar. The role of trypsin in plaque formation by simian rotavirus SA-11. Can J Comp Med 44:433–439, 1980.

46. EG Strauss, JH Strauss, AJ Levine. Virus evolution. In: BN Fields, DM Knipe, and PM Hawley, eds. Fields Virology, 3rd ed. Raven Press, New York, p. 167, 1996.

47. JR Gentsch, PA Woods, M Ramachandran, BK Das, JP Leite, A Alfieri, R Kumar, MK Bhan, RI Glass. Review of G and P typing results from a global collection of rotavirus strains: implications for vaccine development. J Infect Dis 175 Suppl 1:S30–36.

48. KW Theil. Group A rotaviruses, In: LJ Saif and KW Theil, eds. Viral Diarrheas of Man and Animals. CRC Press, Boca Raton, FL, p 36, 1990.

49. Y Chiba, C Miyazaki, Y Makino, LN Mutanda, A Kibue, EO Licenga, PM Tukei. Rotavirus infection of young children in two districts of Kenya from 1982–1983 as analyzed by electrophoresis of genomic RNA. J Clin Microbiol 19:579–582, 1984.

50. MH Lourenco, JC Nicolas, J Cohen, R Scherrer, F Bircout. Study of human rotavirus genome by electrophoresis: attempts of classification among strains isolated in France. Ann Virol 122:161, 1981.

51. SM Rodger, RF Bishop, C Birch, B McClean, IH Holmes. Molecular epidemiology of human rotavirus in Melbourne, Australia, from 1973–1979, as determined by electrophoresis of genome ribonucleic acid. J Clin Microbiol 13:272, 1981.

52. H Tsunemitsu, B Jiang, LJ Saif. Sequence comparison of the VP7 gene encoding the outer capsid glycoprotein among animal and human group C rotaviruses. Arch Virol 141:705–713, 1996.

53. VL James, PR Lambden, CO Caul, IN Clarke. Seroepidemiology of human group C rotavirus in the U.K. J Med Virol 52:86–91, 1997.

54. AZ Kapikian, RG Wyatt, MM Levine, RW Yolken, D VanKirk, R Dolin, HB Greenberg, RM Chanock. Oral administration of human rotavirus to volunteers: induction of illness and correlates of resistance. J Infect Dis 147:95, 1983.

55. RL Ward, DI Bernstein, EC Young, JR Sherwood, DR Knowlton, GM Schiff. Human rotavirus studies in volunteers: determination of infectious dose and serological response to infection. J Infect Dis 154:871, 1986.

56. FP Williams. Membrane-associated viral complex observed in 5 stools and cell culture. Appl Environ Microbiol S0:S23, 1985.

57. SA Sattar, N Lloyd-Evans, VS Springthorpe RC Nair. Institutional outbreaks of rotavirus diarrhoea: potential role of fomites and environmental surfaces as vehicles for virus transmission. J Hyg 96:277, 1986.

58. N Lloyd-Evans, VS Springthorpe, SA Sattar. Chemical disinfection of human rotavirus-contaminated inanimate surfaces. J Hyg 97:163, 1986.

59. VS Springthorpe, JL Grenier, N Lloyd-Evans, SA Sattar. Chemical disinfection of human rotaviruses: efficacy of commercially available products in suspension tests. J Hyg 97:163–173, 1986.

60. SA Sattar, MK Ijaz. Spread of viral infections by aerosols. CRC Crit Rev Environ Contr 17: 89, 1987.

61. SA Ansari, VS Springthorpe, SA Sattar. Survival and vehicular spread of human rotaviruses: possible relation to seasonality of outbreaks. Rev Infect Dis 13:448, 1991.

62. TH Flewett. Clinical features of rotavirus infections. In: DAJ Tyrrell and AZ Kapikian, eds. Virus Infections of the Gastrointestinal Tract. Marcel Dekker, New York, p 12S, 1988.

63. JAK Carlson, PJ Middleton, MT Szymanski, J Huber, M Petric. Fatal gastroenteritis: an analysis of 21 cases. Am J Dis Child 132:477, 1978.

64. JG McCormack. Clinical features of rotavirus gastroenteritis. J Infect Dis 4:167–174, 1982.

65. EO Caul, CR Ashley, JM Darville, JC Bridger. Group C rotavirus associated with fatal enteritis in a family outbreak. J Med Virol 30:201, 1990.

66. DA Sack, M Rhoads, A Molla, AM Molla, MA Wahed. Carbohydrate malabsorption in infants with rotavirus diarrhea, Am J Clin Nutr 36:1112, 1982.

67. K Matsumoto, M Hatano, K Kobayashi, A Hasegawa, S Yamazaki, S Nakata, Y Chiba, Y Kimura. An outbreak of gastroenteritis associated with acute rotaviral infection in children. J Infect Dis 160:611, 1989.

68. L Holmberg. Haemorrhagic shock and encephalopathy syndrome. Acta Paediatrica 85:535–536, 1996.

69. H Komori, M Wada, M Eto, H Oki, K Aida, T Fujimoto. Benign convulsions with mild gastroenteritis: a report of 10 recent cases detailing clinical varieties. Brain Dev 17:334–337, 1995.

70. T Hung. Rotavirus and adult diarrhea. Adv Virus Res 35:193, 1988.

71. DC Lewis, NF Lightfoot, WD Cubitt. Outbreaks of astrovirus type 1 and rotavirus gastroenteritis in a geriatric in-patient population. J Hosp Infect 14:9–14, 1989.

72. BM Totterdell, JE Banatvala, IL Chrystie, G Ball, WD Cubitt. Systemic lymphoproliferative responses to rotavirus. J Med Virol 25:37–44, 1988.

73. H Holzel, DW Cubitt, DA McSwiggan, PJ Sanderson, J Church. An outbreak of rotavirus infection among adults in a cardiology ward. J Infect 2:33, 1980.

74. FT Saulsbury, JA Winkelstein, RH Yolken, RH. Chronic rotavirus infection in immunodeficiency, J Paediatr 97:61, 1980.

75. AL Cunningham, GS Grohman, J Harkness, C Law, D Marriot, B Tindall, DA Cooper. Gastrointestinal viral infections in homosexual men who were symptomatic and seropositive for human immunodeficiency virus. J Infect Dis 158:386, 1988.

76. RH Yolken, GA Losonsky, S Vonderfecht, F Laister, SB Wee. Antibody to human rotavirus in cow's milk, N Engl J Med 312:605, 1985.

77. JE Herrmann, NR Blacklow. Gastroenteritis viruses. In: NJ Schmidt and RW Emmons, eds. Diagnostic Procedures for Viral, Rickettsial and Chlamydial Infections. American Public Health Association, Washington, DC, p. 925, 1989.

78. JF Lew, CW LeBaron, RI Glass, T Tok, PM Griffin, JG Wells, DD Juranek, SP Wahlquist. Recommendations for collection of laboratory specimens associated with gastroenteritis, MMWR 39(RR-14):1, 1990.

79. G Donelli, F Superti. The rotavirus genus. Comparative. Immunol Microbiol Infect Dis 17(3-4):305–320, 1994.

80. R Fuji, M Kuzuya, M Hamano, M Yamada, S Yamazaki. Detection of human group C rotaviruses by an enzyme-linked immunosorbent assay using monoclonal antibodies. J Clin Microbiol 30:1307, 1992.

81. M Kuzuya, R Fujii, M Hamano, T Nagabayashi, H Tsunemitsu, M Yamada, S Nii, T Mori. Rapid detection of human group C rotaviruses by reverse passive hemagglutination and latex agglutination tests using monoclonal antibodies. J Clin Microbiol 31:1308–1311, 1993.

82. J Buesa, J Colomina, J Raga, A Villanueva, J Prat. Evaluation of reverse transcription and polymerase chain reaction RT/PCR for the detection of rotaviruses: applications of the assay. Res Virol 147:353–361, 1996.

83. H Cotterill, A Curry, T Riordan. Rotavirus in vomit. J Infect 16:206, 1988.

84. AZ Kapikian, RM Chanock. Viral gastroenteritis. In: AS Evans, ed. Viral Infections of Humans: Epidemiology and Control, 4th ed. Plenum Press, New York, p 293, 1989.

85. JJ Eiden, J Wilde, F Firoozmand, R Yolken. Detection of animal and human group B rotaviruses in fecal specimens by polymerase chain reaction, J Clin Microbiol 29:539, 1991.

86. BL Herwaldt, GF Craun, SL Stokes, DD Juranek. Waterborne-disease outbreaks, 1989–1990, MMWR 40(SS-3):1, 1991.

87. JG Banwell. Worldwide impact of oral rehydration therapy. Clin Therapeut 12A:29, 1990.

88. KA Grisanti, DM Jaffe. Dehydration syndromes. Oral rehydration and fluid replacement. Emergen. Med Clin North Am 9:565, 1991.

89. RJ Kallen. The management of diarrheal dehydration in infants using parenteral fluids. Pediatr Clin North Am 37:256, 1990.

90. AK Leung, WL Robson. Acute diarrhea in children. What to do and what not to do, Postgrad Med 86:167, 1989.

91. World Health Organization. A Manual for the Treatment of Diarrhoea, Document No. WHO/ CDD/SER/80.2, WHO, Geneva, 1990.

92. DA Sack. Use of oral rehydration therapy in acute watery diarrhea. A practical guide. Drugs 41:566, 1991.

93. A Mackenzie, G Barnes. Oral rehydration in infantile diarrhea in the developing world. Drugs 36(S-4):48, 1988.

94. H Majamaa, E Isolauri, M Saxelin, T Vesikari. Lactic acid bacteria in the treatment of acute rotavirus gastroenteritis. J Pediatr Gastroenterol Nutr 20(3):333–338, 1995.

95. S Salminen, E Isolauri, E. Salminen. Clinical uses of probiotics for stabilizing the gut mucosal barrier: successful strains and future challenges. Antoine van Leewwuwenhoek 70:347–358, 1996.

96. K Grimwood, GD Abbot, DM, Fergusson, LC Jennings, JM Allan. Spread of rotavirus within families: a community based study. Br Med J 287:575, 1983.

97. HW Kim, CD Brandt, AZ Kapikian, RG Wyatt, JO Arrobio, WJ Rodriguez, RM Chanock, RH Parrott. Human reovirus-like agent infection. Occurrence in adult contacts of pediatric patients with gastroenteritis. JAMA 238:404, 1977.

98. WJ Rodriguez, HW Kim, CD Brandt, RH Yolken, M Richard, JO Arrobio, RH Schwartz, AZ Kapikian, RM Chanock, RH Parrott. Common exposure outbreak of gastroenteritis due to type 2 rotavirus with high secondary attack rate within families. J Infect Dis 140:353, 1979.

99. J Tietzova, A Petrovicova, P Pazdiora. Characterization of human rotaviruses isolated from symptomatic and asymptomatic infections. Acta Virologica 39(4):211–214, 1995.

100. BL Barron-Romero, J Barreda-Gonzalez, R DovalUgalde, L Zermeno Eguia Liz, M Huerta-Pena. Asymptomatic rotavirus infection in day care centers. J Clin Microbiol 22:116, 1985.

101. K Hjelt, PC Graduballe, L Henriksen, PA Krasilnikoff. Rotavirus infections among the staff of a general pediatric department. Acta Paediatr Scand 74:617, 1985.

102. A Di Matteo, A Sarasini, MS Scotta, M Parea, G Licardi, G Gerna. Nosocomial outbreak of infant rotavirus diarrhea due to the appearance of a new serotype 4 strain. J Med Virol 27:100, 1989.

103. EL Ford-Jones. The special problems of nosocomial infection in rotavirus the pediatric patient. In: RP Wenzel, ed. Prevention and Control of Nosocomial Infections. Williams and Wilkins, Baltimore, p 494, 1987.

104. BH Keswick, LK Pickering, HL DuPont, WE Woodward. Survival and detection of rotaviruses on environmental surfaces in daycare centers. Appl Environ Microbiol 46:813, 1983.

105. DWG Brown, L Campbell, DS Tomkins, MH Hambling. School outbreaks of gastroenteritis due to atypical rotavirus. Lancet 2:737, 1989.

106. O Nakagomi, A Ohshima, Y Aboudy. Molecular identification by RNA-RNA hybridization of a human rotavirus that is closely related to rotaviruses of feline and canine origin. J Clin Microbiol 28(1):198, 1990.

107. O Nakagomi, M Mochizuki, Y Aboudy, I Shif, I Silberstein, T Nakagomi. Hemagglutination by a human rotavirus isolate as evidence for transmission of animal rotaviruses to humans. J Clin Microbiol 30:1011, 1992.

108. PA Offit, HF Clark. Protection against rotavirus-induced gastroenteritis in a murine model by passively acquired gastrointestinal but not circulating antibodies, J Virol 54:58, 1985.

109. S Sonza, IH Holmes. Coproantibody response to rotavirus infection. Med J Aust 2:496, 1980.

110. SM Cook, RI Glass, CW LeBaron, MS Ho. Global seasonality of rotavirus infections. Bull WHO 68:171, 1990.

111. P Shears, A Wright. Community-acquired infections among children in an urban environment: a 2-year prospective study in Liverpool, U.K. J Infect 30(2):173–177, 199.

112. TW Hejkal, EM Smith, CP Gerba. Seasonal occurrence of rotavirus in sewage. Appl Environ Microbiol 47:588, 1984.

113. MO Paul, EA Erinle. Influence of humidity on rotavirus prevalence among Nigerian infants and young children with gastroenteritis. J Clin Microbiol 15:212–215, 1982.

114. MS Ho, RI Glass, PF Pinsky, N Young-Okoh, WM Sappenfield, JW Buehler, N Gunter, LJ Anderson. Diarrheal deaths in American children: are they preventable? JAMA 260: 3281, 1988.

115. M Chaudhuri, SA Sattar. Domestic water treatment in developing countries. In: GA McFeters, ed. Drinking Water Microbiology. Springer-Verlag, New York, p 168, 1990.

116. WC Levine, WT Stephenson, GF Craun. Waterborne disease outbreaks. MMWR 39 (SS-I): I, 1990.

117. NH Bean, PM Griffm, JS Goulding, CB Ivey. Foodborne disease outbreaks, 5-year summary, 1983–1987. MMWR 39 (SS-1):15, 1990.

118. DO Cliver. Foodborne disease in the United States, 1946–1986. Int J Food Microbiol 4:269, 1987.

119. IE Haffejee. The epidemiology of rotavirus infections: a global perspective J Pediatr Gastroenterol 20:275–286, 1995.

120. ECD Todd. Foodborne disease in Canada—a 10-year summary from 1975 to 1984. J Food Protect S5:123, 1992.

121. Y Andersson, TA Stenstrom, Waterborne outbreaks in Sweden—causes and etiology. Water Sci Technol 19:575, 1975.

122. H Appleton, EO Caul, JP Clewley, JG Cruickshank, WD Cubitt, JB Kurtz, DA McSwiggan, SA Palmer, T Riordan. Foodborne viral gastroenteritis: an overview. Public Health Laboratory Service Microbiol Dig 5:69, 1988.

123. VL James, PR Lambden, EO Caul, SJ Cooke, IN Clarke. Seroepidemiology of human group C rotavirus in the UK. J Med Virol 52:86–91, 1997.

124. RE Black. Epidemiology of travelers' diarrhea and relative importance of various pathogens. Rev Infect Dis 12:S73, 1990.

125. AS Badawy, CP Gerba, LM Kelly. Development of a method for recovery of rotavirus from the surface of vegetables. J Food Protect 48:261, 1985.

126. C Beril, S Boher, L Schwartzbrod. Detoxification by Sephadex LH20 of seafood concentrates for rotavirus assay. Water Sci Technol 24:433, 1991.

127. S Boher, C Beril, D Terver, L Schwartzbrod. Comparison of two methods for the recovery of rotavirus from mussels and oysters. Water Sci Technol 24:423, 1991.

128. KM Seidel, SM Goyal, VC Rao, JL Melnick. Concentration of rotavirus and enteroviruses from blue crabs Callinectes sapidus. Appl Environ Microbiol 46:1293, 1983.

129. JI Speirs, RD Pontefract, J Gaewig. Methods for recovering poliovirus and rotavirus from oysters. Appl Environ Microbiol 53:2666–2670, 1987.

130. Y Zhou MK Estes, X Jiang, TG Metcalf. Concentration and detection of hepatitis A virus and rotavirus from shellfish by hybridization tests. Appl Environ Microbiol 57:2963, 1991.

131. F Hernandez, R Monge, C Jimenez, L Taylor. Rotavirus and hepatitis A virus in market lettuce Latuca sativa in Costa Rica. Int J Food Microbiol 37:221–223, 1997.

132. D Cecuk, M Grce. July 1991. Personal communication.

133. A Bosch, F Lucena, JM Diez, R Gajardo, M Blasi, J Jofre. Waterborne viruses associated with hepatitis outbreak. J Am Water Works Assoc 83:80, 1991.

134. AS Badawy, CP Gerba, LM Kelly. Survival of rotavirus SA-II on vegetables. Food Microbiol 2:199–205, 1985.

135. M Hara, J Mukoyama, T Tsuruhara, Y Ashiwara, Y Saito, I Tagaya. Acute gastroenteritis among schoolchildren associated with reovirus-like agent. Am J Epidemiol 107:161, 1978.

136. New York State Department of Health. April 1987. A Review of Foodborne Disease Outbreaks in New York State—1985. Bureau of Community Sanitation and Food Protection, Albany, p 34.

137. New York State Department of Health. April 1988. A Review of Foodborne Disease Outbreaks in New York State—1986. Bureau of Community Sanitation and Food Protection, Albany, p 36.

138. New York State Department of Health. December 1989. A Review of Foodborne Disease Outbreaks in New York State—1988. Bureau of Community Sanitation and Food Protection, Albany, p 46.

139. New York State Department of Health. December 1990. A Review of Foodborne Disease Outbreaks in New York State—1989. Bureau of Community Sanitation and Food Protection, Albany, p 49.

140. New York State Department of Health. February 1992. A Review of Foodborne Disease Outbreaks in New York State—1990. Bureau of Community Sanitation and Food Protection, Albany, p 56.

141. GW Wood, MR Adams. Effects of acidification, bacterial fermentation temperature on the survival of rotavirus in a model weaning food. J Food Protect 55:52, 1992.

142. G Panon, S Tache, C Labie. Respective stability of rotavirus and coronavirus in bovine milk. Le Lait 68:49, 1988.

143. G Panon, S Tache, C Labie. Antiviral substances in raw bovine milk active against bovine rotavirus and coronavirus. J Food Protect S0:862, 1987.

144. S Ylitalo, M Uhari, S Rasi, J, Pudas, J Leppaluoto. Rotaviral antibodies in the treatment of acute rotaviral gastroenteritis. Acta Paediatrica 87:264–267, 1998.

145. DM Bass. Interferon gamma and interleukin 1, but not interferon alfa, inhibit rotavirus entry into human intestinal cell lines. Gastroenterology 113:81–89, 1997.

146. GP Davidson, PBD Whyte, E Daniels, K Franklin, H Nunan, PI McCloud, AG Moore, DJ Moore. Passive immunization of children with bovine colostrum containing antibodies to human rotavirus. Lancet 2:709, 1989.

147. JP Schaller, LJ Saif, CT Cordle, E Candler, TR Winship, KL Smith. Prevention of human rotavirus-induced diarrhea in gnotobiotic piglets using bovine antibody. J Infect Dis 165: 623, 1992.

148. RI Glass, BJ Still, RG Wyatt, Y Hoshino, H Banu, AZ Kapikian. Observations questioning a protective role for breast-feeding in severe rotavirus diarrhea. Acta Paediatrica Scand 75: 713, 1986.

149. LK Pickering, AL Morrow, I Herrera, M O'Ryan, MK Estes, SE Guilliams, L Jackson, S Carter-Campbell, DO Matson. Effect of maternal rotavirus immunization on milk and serum antibody titers. J Infect Dis 172:723–728, 1995.

150. T Ebina, K Tsukada. Protease inhibitors prevent the development of human rotavirus–induced diarrhea in suckling mice. Microbiol Immunol 35:583–588, 1991.

151. CJ Hurst, CP Gerba. Stability of simian rotavirus in fresh and estuarine waters. Appl Environ Microbiol 39:1, 1980.

152. OC Pancorbo, BG Evanshen, WF Campbell, SL Lambert, SK Curtis, TW Woolley. Infectivity and antigenicity reduction rates of human rotavirus strain Wa in fresh waters. Appl Environ Microbiol 53:1803, 1987.

153. SA Sattar, RA Raphael, VS Springthorpe. Rotavirus survival in conventionally treated drinking water. Can J Microbiol 30:653, 1984.

154. RA Raphael, SA Sattar, VS Springthorpe. Long-term survival of human rotavirus in raw and treated river water. Can J Microbiol 31:124, 1985.

155. R Gajardo, N Bouchriti, RM Pinto, A Bosch. Genotyping of rotaviruses isolated from sewage. Appl Environ Microbiol 61:3460–3462, 1995.

156. F Pesaro, I Sorg, A Metzler. In situ inactivation of animal viruses and a coliphage in nonaerated liquid and semiliquid animal wastes. Appl Environ Microbiol 61:92–97, 1995.

157. CJ Hurst, CP Gerba, I Cech. Effects of environmental variables and soil characteristics on virus survival in soil. Appl Environ Microbiol 40:1067, 1980.

158. VC Rao, TG Metcalf, JL Melnick. Human viruses in sediments, sludges soils. Bull WHO 64:1, 1986.

159. BH Keswick, CP Gerba. Viruses in ground water. Environ Sci Technol 14:1290, 1980.

160. TR Deetz, E Smith, SM Goyal, CP Gerba, JJ Vollet, L Tsai, HL Dupont, BH Keswick. Occurrence of rota-and enteroviruses in drinking and environmental waters in a developing nation. Water Res 18:567, 1984.

161. SM Goyal, CP Gerba. Comparative adsorption of human enteroviruses, simian rotavirus and selected bacteriophages to soils. Appl Environ Microbiol 38:241, 1979.

162. VC Rao, TG Metcalf, JL Melnick. Human viruses in sediments, sludges, and soils. Bull WHO 64:1, 1986.

163. FP Williams. Membrane-associated viral complexes observed in stools and cell culture, Appl Environ Microbiol 50:523, 1985.

164. RS Hopkins, B Gaspard, FP Williams, RJ Karlin, G Cukor, NR Blacklow. A community water-borne gastroenteritis outbreak: evidence for rotavirus as the agent, Am J Publ Health 74:263, 1984.

165. AC Linhares, FE Pinheiro, RB Freitas, YB Gabbay, JA Shirley, GM Beards. An outbreak of rotavirus diarrhea among a non-immune, isolated South American Indian community. Am J Epidemiol 113:703, 1981.

166. E Lycke, J Blomberg, G Berg, A Eriksson, L Madsen. Epidemic acute diarrhea in adults associated with infantile gastroenteritis. Lancet 2:1056, 1978.

167. DM Morens, RM Zweighaft, TM Vernon, GW Gary, JJ Eslien, BT Wood, RC Holman, R Dolin. A waterborne outbreak of gastroenteritis with secondary person-to-person spread association with a viral agent. Lancet 1:964, 1979.

168. A Murphy, G Grohman, M Sexton. Viral gastroenteritis on Norfolk Island. In: JS Mackenzie, ed. Viral Diseases in South-East Asia and the Western Pacific. Academic Press, New York, p. 437, 1982.

169. F Sutmoller, RS Azerado, MD Lacerda, OM Barth, HG Perdra, E Hoffer, H Schatzmayer. An outbreak of gastroenteritis caused by both rotavirus and Shigella sonnet in a private school in Rio de Janeiro. J Hyg 88:285, 1982.

170. JB Rose, CP Gerba. A review of viruses in drinking water. Curr Pract Environ Eng 2:119, 1986.

171. CP Gerba, JB Rose, SN Singh. Waterborne gastroenteritis and viral hepatitis. CRC Crit Rev Environ Contr 15:213, 1985.

172. R Walter, HJ Dobberkau, J Durkop. A virological study of the health hazards associated with the indirect reuse of water. In: M Butler, AR Medlin, and R Morris, eds. Viruses and Disinfection of Water and Wastewater. University of Surrey, Guildford, United Kingdom, p 144, 1982.

173. TH Tulchinsky, I Levine, TA Swartz. Waterborne enteric disease outbreaks in Israel, Monogr Virol 15:61, 1984.

174. BA Zamotin, LT Libiysinen, FL, Bortnik, EP Chernitskaya, ZI, Emina, NF Rossikhin, VI Veselov, SA Kharyutkin, VA Egerev, VA Nikulin, ZD, Dorozhkine, KP Kostyukova, GV Dorozhkin, VG Shironina, VS Chernyi, EI Agapov, VI Vlasor. Waterborne group infection of rotavirus etiology. Microbiol Epidemiol Immunol 11:99, 1981.

175. MJ Cox, VLA James, RS Azevedo, E Massad, GF Medley. Infection with group C rotavirus in a suburban community in Brazil. Trop Med Int Health 3:891–895, 1998.

176. S Saidi, Y Iijima, WK Sang, AK Mwangudza, JO Oundo, K Taga, M Aihara, K Nagayama, H Yamamoto, PG Waiyaki, T Honda. Epidemiological study on infectious diarrheal diseases in children in a coastal rural area of Kenya. Microbiol Immunol 41:773–778, 1987.

177. P González, A Sánches, P Rivera, C Jiménez, F Hernández, F. Rotavirus and coronavirus outbreak—etiology of annual diarrhea in Costa Rican children. Revista de Biologia Tropical 45:989–991, 1997.

178. T Hung, G Chen, C Wang, H Yao, Z Fang, T Chao, Z Chou, W Ye, X Chang, S Den, X Liang, W Chang. Water-borne outbreak of rotavirus in adults in China caused by a novel rotavirus. Lancet 1.1139, 1984.

179. G Chen, T Hung, JC Bridger, MA McCrae. Chinese adult rotavirus is a group B rotavirus. Lancet, 2:1123, 1985.

180. S Ramia, SA Sattar. Concentration of seeded simian rotavirus SA-I from potable waters by using talc-Celite layers and hydroextraction. Appl Environ Microbiol 39:493, 1980.

181. SM Goyal, CP Gerba. Concentration of viruses from water by membrane filters. In: CP Gerba and SM Goyal, eds. Methods in Environmental Virology. Marcel Dekker, New York, p 59, 1982.

182. DR Dahling, BA Wright, FP Williams Jr. Evaluation of rotavirus test kits for possible use with environmental samples. In: JR Hall and GD Glysson, eds. Monitoring Water in the 1990s: Meeting New Challenges. Am Soc Test Mater, Philadelphia, 1991.

183. P Strappe. Rotavirus detection—a problem that needs concentration. Water Sci Technol 24: 221, 1991.

184. N Jothikumar, P Khanna, R Paulmurugan, S Kamatchiammal, P Padmanabhan. A simple device for the concentration and detection of enterovirus, hepatitis E virus and rotavirus from water samples by reverse transcription–polymerase chain reaction. J Virol Meth 55:401–415, 1995.

185. B Grinde, TO Jonassen, H Ushijima. Sensitive detection of group A rotaviruses by immuno-magnetic separation and reverse transcription–polymerase chain reaction. J Virol Meth 55: 327–338, 1995.

186. RD Leon, CP Gerba. Detraction of rotaviruses in water samples by gene probes. Water Sci Technol 24:281, 1991.

187. Hurst, CJ. Presence of enteric viruses in freshwater and their removal by the conventional drinking water treatment process. Bull WHO 69:113, 1991.

188. P Payment, R Armon. Virus removal by drinking water treatment processes. CRC Crit Rev Environ Contr 19:15, 1989.

189. FX Abad, RM Pinto, JM Diez, A Bosch. Disinfection of human enteric viruses on fomites. FEMS Microbiol Lett 156:107–111, 1997.

190. RA Raphael, SA Sattar VS Springthorpe. Lack of human rotavirus inactivation by residual chlorine in municipal drinking water systems. Rev Int des Sci de l'Eau, 3:67, 1987.

191. RW Ryder, CA Oquist, H Greenberg, DN Taylor, F Orskov, I Orskov, AZ Kapikian, RB Sack. Traveller's diarrhea in Panamanian tourists in Mexico. J Infect Dis 144:442, 1981.

192. VS Panangala, L Liu, G Sun, SD Worley, A Mitra. J Virol Meth 66:263–268, 1997.

193. U.S. Environmental Protection Agency. National primary drinking water regulations: filtration; disinfection; turbidity, Giardia lamblia, viruses, Legionella heterotrophic bacteria. Final rule. Fed Reg 54:27486, 1989.

194. JB Rose, CP Gerba. Use of risk assessment for development of microbial standards. Water Sci Technol 24:29, 1991.

195. S Regli, JB Rose, CN Haas, CP Gerba. Modeling the risk from Giardia and viruses in drinking water. J Am Water Works Assoc 83:76, 1991.

196. P Gale. Developments in microbiological risk assessment models for drinking water—a short review. J Appl Bacteriol 81:403–410, 1996.

197. K Moe, JA Shirley. The effects of relative humidity and temperature on the survival of human rotavirus in faeces. Arch Virol 72:179, 1982.

198. RL Ward, DI Bernstein, DR Knowlton, JR Sherwood, EC Young, TM Cusack, JR Rubino, GM Schiff. Prevention of surface-to-human transmission of rotavirus by treatment with disinfectant spray. J Clin Microbiol 29:1991–1996, 1991.

199. AR Samadi, MI Haq, QS Ahmed. Detection of rotavirus in handwashings of attendants of children with diarrhoea, Br Med J 286:188, 1983.
200. LK Pickering, AV Bartlett, WE Woodward. Acute infectious diarrhea among children in daycare: epidemiology and control. Rev Infect Dis 8:539, 1986.
201. NS Shahid, WB Greenough II, WB II, AR Samadi, MI Huq, R Rahman. Hand washing with soap reduces diarrhoea and spread of bacterial pathogens in a Bangladesh village. J Diarrh Dis 14:85–89, 1996.
202. SA Ansari, SA Sattar, VS Springthorpe, GA Wells, W Tostowaryk. Rotavirus survival on human hands and transfer of infectious virus to animate and non-porous inanimate surfaces. J Clin Microbiol 26:1513, 1988.
203. SA Ansari, VS Springthorpe, SA Sattar, S Rivard. Potential role of hands in the spread of respiratory viral infections: studies with human parainfluenzavirus 3 and rhinovirus 14. J Clin Microbiol 29:2115, 1991.
204. SA Ansari, SA Sattar, VS Springthorpe, G Wells, W Tostowaryk. In vivo protocol for testing efficacy of hand-washing agents against viruses and bacteria: experiments with rotavirus and Escherichia coli. Appl Environ Microbiol 55:3113, 1989.
205. RH Foster, AJ Wagstaff. tetravalent human-rhesus reassortant rotavirus vaccine—a review of its immunogenicity, tolerability and protective efficacy against paediatric rotavirus gastro-enteritis. Biodrugs 9:155–178, 1998.
206. Anonymous. Vaccine research and development. Rotavirus vaccines for developing countries. Wkly Epidemiol Rec 72:35–40, 1997.
207. RL Ward, DR Knowlton, ET Zito, BL Davidson, R Rappaport, ME Mack. Serologic correlates of immunity in a tetravalent reassortant rotavirus vaccine trial. US Rotavirus Vaccine Efficacy Group. J Infect Dis 176:570–577, 1997.
208. AZ Kapikian, Y Hoshino, RM Chanock, I Perez-Schael. Efficacy of a quadrivalent rhesus rotavirus-based human rotavirus vaccine aimed at preventing severe rotavirus diarrhea in infants and young children. J Infect Dis 174(Suppl 72), 1996.
209. AD Steele, MC van Niekerk, MJ Mphahlele. Geographic distribution of human rotavirus VP4 genotypes and VP7 serotypes in five South African regions. J Clin Microbiol 33:1516–1519, 1995.
210. Y Suzuki, T Gojobori, O Nakagomi. Intragenic recombinations in rotaviruses. FEBS Lett 427:183–187, 1998.
211. Centers for Disease Control and Prevention. Withdrawal of rotavirus vaccine recommendation. MMWR 48:1007, 1999.
212. S Aijaz, K Gowda, HV Jagannath, RR Reddy, PP Maiya, RL Ward, HB Greenberg, M Raju, A Babu, CD Rao. Epidemiology of symptomatic human rotaviruses in Bangalore and Mysore, India, from 1988 to 1994 as determined by electropherotype, subgroup and serotype analysis. Arch Virol 141:715–726, 1996.

5

Other Foodborne Viruses

Syed A. Sattar and Jason A. Tetro

University of Ottawa, Ottawa, Ontario, Canada

I. INTRODUCTION

The global impact of foodborne disease is immense. In the United States alone, the yearly estimate of health care costs and lost revenue due to illness is close to U.S. $50 billion (1). Worldwide the tally is many times greater although actual numbers are hard to establish as individual cases of foodborne disease, which may account for as much as 99% of all cases in some parts of the developing world (2), are either unreported or are attributed to water (3). Viruses represent an important cause of foodborne disease in humans and the major groups of foodborne viruses have been dealt with in other chapters in this book. This chapter will discuss the known or suspected role of other types of viruses in causing foodborne disease in humans; it will also summarize the available information on the foodborne spread of prions.

Theoretically speaking, any virus that is excreted in feces in sufficiently large numbers and manages to survive outside the body of the host long enough to contaminate food has the potential to spread through this route. Contamination of food with other types of foodborne viruses generally occurs exogenously and this is primarily due to direct or indirect contact with virus-containing fecal material (3). However, endogenous contamination of meat and milk can occur with agents such as certain types of tick-borne encephalitis viruses, foot-and-mouth disease virus, immunodeficiency viruses, cytomegalovirus, and the proteinaceous prions.

In addition to this, several animal viruses have demonstrated an ability to cross the species barrier and may become involved in human disease. These agents, which include Bornavirus, influenzavirus H5:N1, pestiviruses, picobirnaviruses, and toroviruses have the potential for foodborne spread and should be placed under close surveillance for their mechanisms of cross-species spread.

II. VIRUSES THAT INFREQUENTLY SPREAD THROUGH FOODS

A. Tick-borne Encephalitis

Tick-borne encephalitis (TBE) is one of the few well-documented examples of a nonenteric foodborne virus (4). The etiological agent of this disease is an acid-resistant filovirus (5). The spread of TBE in nature is predominantly through the bite of ticks of the genus *Ixodes*, which feed primarily on animals and, in rare cases, on humans (6). Infectious virus may be shed in the milk of infected animals (7,8) and subsequent consumption of contaminated raw milk may lead to infection in humans (9,10).

Upon ingestion of the virus, infection starts in the alimentary tract but spreads over a few weeks to the perivascular areas through the cerebrospinal fluid (8). Inflammation ensues resulting in fever, severe headache, nausea, vomiting, and photophobia (11). As the virus spreads, continual neurological damage ensues leading to paresis, paralysis, and sensory loss. Mortality occurs in approximately 20% of infected individuals, usually a week after the onset of clinical symptoms (5).

Outbreaks of TBE are generally localized to areas of central Europe and Russia where the ticks *I. ricinus* and *I. persulcatus* are respectively distributed but the virus is considered endemic to the majority of Eurasia (12). Documented cases of TBE have been found in Austria (13), Czech Republic (10), Germany (14), Russia (12), and Sweden (15). Recently, a case of TBE was reported from Japan (16). In all cases but one (6), ingestion of TBE-contaminated raw milk has been the cause of infection. Proper pasteurization of milk is the simplest and best means of eliminating the threat of foodborne spread of this disease.

B. Enteric Viruses

Among enteric viruses, enteroviruses are worldwide agents of disease in humans as well as many species of warm-blooded animals, though no evidence exists of cross-species virus transfer. Currently, the 67 members of this group that have been isolated from humans include three types of polioviruses (causative agents of poliomyelitis), 24 coxsackieviruses type A, 6 coxsackieviruses type B, 30 echoviruses, and 4 more agents (enterovirus 68–71) with no specific subgrouping (17). In addition to poliomyelitis, enteroviruses can cause a variety of clinical conditions in humans affecting nearly every organ in the body.

The lining of the gut is the primary site of replication of most enteroviruses and as a result they are excreted in large numbers in the feces of infected individuals. In general, these viruses are relatively resistant to heat, pH, and many chemicals and are also quite stable in the environment (18,19). They are found regularly in fecally polluted waters, sewage, raw and digested sludges (20), and have also been isolated from edibles such as vegetables (21,22) and seafood (23,24). Most cases of enterovirus infection remain asymptomatic while shedding infectious virus.

In spite of their common occurrence, fecal shedding in large numbers, and relative stability in the environment (22), there are surprisingly few published reports of foodborne spread of enteroviral infections. This may be explained perhaps by their widespread distribution, low minimal infective dose (25), propensity to cause a large proportion of subclinical infections, ability to induce long-lasting immunity, and potential for spread through other vehicles such as water and air. The fact that most cases of foodborne infections are not recorded and reported outbreaks not sufficiently investigated (24) is also undoubtedly responsible for our current ignorance with regard to the true contribution of enteroviruses in terms of foodborne spread of infections. A summary of recorded outbreaks of foodborne enteric viral infections follows.

1. Polioviruses

A raw milk–associated outbreak of poliomyelitis in 1914 is believed to be the first recorded foodborne outbreak of a viral disease (26). Consumption of raw milk was implicated in 10 more such outbreaks in the United States and United Kingdom by 1949 (27). Starting in the early 1950s, milkborne outbreaks of poliomyelitis became a rare event, at least in the industrialized parts of the world. This could to a large extent be ascribed to the rapid and nearly universal acceptance of the practice of milk pasteurization around that time. Sustained and widespread immunization against poliomyelitis over the past few decades has drastically reduced its numbers and therefore the threat of spread of polioviruses through food or any other means has now virtually disappeared.

2. Echoviruses and Coxsackieviruses

While we have knowledge of a total of no fewer than 60 different serotypes of echo- and coxsackieviruses for more than three decades, there are only a handful of foodborne outbreaks where these viruses have been implicated. Even in these reports, the evidence for the role of these enteroviruses is at best circumstantial.

Both recorded foodborne outbreaks due to an echovirus occurred in the United States. The first of these was in 1976 in Pennsylvania where 80 cases of aseptic meningitis were recorded among those who ate coleslaw served at a picnic; echovirus type 4 was identified as the causative agent but the actual source of the virus could not be determined (28). The second outbreak occurred in the State of New York in 1988 and led to 161 cases of an unspecified illness. The causative agent proved to be echovirus type 5 but the specific type of food and the actual source of the virus remained undetermined (29).

The sole report of a foodborne outbreak due to the coxsackievirus B subgroup comes from the former Soviet Union (30). A day care center was the site of the outbreak and coxsackievirus types B1, B3, and B5 were reported to have been isolated from certain items of food, surfaces in food preparation areas, as well as the infected individuals.

3. Coronaviruses/Toroviruses

Coronaviruses are a major cause of disease in many types of animals of economic importance. Transmissible gastroenteritis (TGE) of swine is a case in point (31). In humans, the major health impact of coronaviruses is as agents of acute infections of the respiratory tract (32); their role as etiological agents of gastroenteritis in humans is still unclear.

However, a foodborne outbreak of acute gastroenteritis at a sports club in Scotland was ascribed to a coronavirus (33), although the actual type of virus involved, the nature of the incriminated food, and the source of the virus could not be ascertained.

More recently, enveloped viruses very similar in size and appearance to coronaviruses have been detected in the stools of animals and humans (34) with diarrhea. These agents have been named "toroviruses" because closer electron microscopic examination shows them to be doughnut-like in shape. In animals, ingestion of fecally excreted torovirus particles has been shown to result in gastroenteritis (35). It is not know if this is true for humans as well.

4. Adenoviruses

Of the nearly 47 known adenoviruses of humans, types 40 and 41 are now regarded as significant causes of acute gastroenteritis in children under 24 months of age (36). Their prevalence in immunocompromised individuals is also on the rise (37). In immunocompromised individuals, adenoviruses in general can cause symptomatic infection; viral shedding can last for months and result in severe morbidity and mortality (38,39).

Primary infection by the fastidious adenoviruses through food has been suggested due to their presence in mussels (23,24) and vegetable matter (21), but evidence for a firm association is lacking (40,41).

III. PRIONS AND THEIR FOODBORNE SPREAD

While slow-progressing degenerative neurological disorders, known collectively as transmissible spongiform encephalopathies (TSEs), have been known to exist in humans and animals for many decades (42), their etiological agents, the proteinaceous prions (43), have only recently been identified. Formerly, these agents were referred to as "slow viruses" because of the unusually long incubation period for the diseases caused by them. Prions and their role in the spread of foodborne disease are being discussed here mainly because of their historical link with viruses.

The primary target of prions is the central nervous system, and the slow degeneration of neurons caused by them may take several months to a few decades to result in noticeable clinical symptoms. The clinical phase of the disease commences with a change in behavior, dementia, motor movement disorders, spasticity, rigidity, and tremors. Death usually ensues within 8–12 months of the onset of symptoms. A case of prion disease always proves fatal.

In animals, prions cause many types of diseases but of note are scrapie in sheep and a spongiform encephalopathy in cattle. In humans, prions have been linked to Creutzfeldt-Jakob disease (CJD) and a cannibalism-related disease known as kuru, the first documented example of a prion-caused infection (42,44). Recently, a variant Creutzfeldt-Jakob disease (vCJD) has also been identified as a prion-caused infection (42) in humans. Unlike classical CJD, vCJD has a shorter incubation period and tends to occur in younger individuals (45).

In 1986, an outbreak of bovine spongiform encephalopathy (BSE) was identified in the United Kingdom (46). As of March 31, 1996, there had been approximately 160,000 confirmed cases of BSE-infected cattle (46). The illness, also termed "mad cow disease," was subsequently shown to be transmissible to humans through exposure to or ingestion of material from affected animals. Kuru, which was linked to ritualistic ingestion of human body parts, became the base for a warning concerning the ingestion of meat from prion-infected animals (42).

Initially, there was no proof that the bovine prion could jump the species barrier and cause a CJD-like illness in humans. However, in 1994 (47) the proof appeared with the discovery of vCJD, which after molecular analysis proved to be the same agent that causes BSE (48). How BSE jumped the species barrier is still unknown but experimental studies using mice and nonhuman primates have demonstrated the relationship between consumption of meat from BSE-infected animals and the development of spongiform encephalopathy (49). As of November 30, 1998, 35 cases of the new variant CJD had been detected or confirmed (50).

The now documented link between consumption of meat and infection has led to an interest in the mass media (51) and a general call to avoid the consumption of animal brains. In December 1997, the British government passed a law banning the sale of beef on the bone as ingestion of bone marrow has also been implicated in the development of vCJD (52). More recently, prions from squirrels have also been isolated as agents of disease in humans (53).

IV. VIRUS TRANSMISSION THROUGH HUMAN BREAST MILK

Systemic infections can lead to viral shedding in saliva, semen, urine, feces, and breast milk (54). Human breast milk is a high-risk route of postnatal transmission of human immunodeficiency viruses (HIV), cytomegalovirus (CMV), and human T-lymphotrophic virus (HTLV-I), of which CMV infection is the most prevalent. CMV has been found in almost three-quarters of the human population under the age of 10 years (55) but is mainly a subclinical condition. Usually, CMV-infected immunocompromised individuals (56,57) develop clinical symptoms, which include hepatitis, occlusions in the GI tract, and respiratory complications.

Although incidence rates are much lower, the consequences of infection by HIV and HTLV are much more serious, i.e., immunodeficiency and T-cell leukemia, respectively (58). The prevalence of HIV in the breast milk of infected mothers can be as high as 80% but the actual postnatal transmission rate of HIV is about 26%, far less than the rate of perinatal vertical transmission (59). Rates of transmission of HTLV-I are comparable at around 20% (60).

Infants who feed on the breast milk from infected mothers are the only ones at risk of contracting HIV or HTLV-I. Unfortunately, many areas of the developing world do not have alternate options for infected mothers (59). Even in the developed world, infected mothers may still choose to breast-feed (59). In these cases, breast milk should be collected and heated to at least 56°C to ensure inactivation of these highly labile viruses (54) prior to consumption.

V. ANIMAL VIRUSES WITH POTENTIAL FOR FOODBORNE SPREAD

A. Foot-and-Mouth Disease Virus

Foot-and-mouth disease (FMDV) is caused by a picornavirus. The primary hosts of the disease are cloven-hoofed animals, including goats, cattle, pigs, and sheep. A few cases of human infection by this virus have been recorded (61). The usual portal of entry for the virus is the respiratory tract, where it infects a wide range of tissues (62) and then becomes systemic. Infection of the mammary tissue results in virus discharge in milk (63). FMDV also survives well in aerosolized milk (64), thus posing a risk of inhalation of infectious airborne particles during handling/consumption of raw milk from FMDV-infected animals. However, no documented cases of such spread have thus far been reported. The same can be said for the handling of sausage meat, which has also been shown to harbor FMDV (65).

Currently, the risk, however small, of foodborne spread of FMDV exists only in countries where the virus still causes infections in animals. Any animal products that might contain infectious FMDV cannot be imported into virus-free regions (27).

B. Influenzaviruses

Recently, an uncommon strain of avian influenzavirus (H5:N1) caused 18 cases of infection in humans in the Hong Kong region with 8 fatalities (66). In at least 6 of the cases there were signs of GI distress (67). The virus isolates from human cases demonstrated an almost perfect homology with the avian counterpart indicating an avian species, more specifically chickens, as the most likely source of the virus (68,69). In most of the 18 cases in Hong Kong, interactions with live chickens had occurred either directly or indirectly in markets where selling live chickens as food is commonplace (66). Spread of the virus to humans is thought to have been mediated by aerosols rather than through consumption of contaminated chicken meat (66).

VI. SUMMARY

As noted by Cliver (27), ''Viruses causing human illness by transmission via foods emanate principally from the human intestines.'' While other chapters in this book have covered the major classes of foodborne viruses of human origin, this chapter has summarized the information on prions and other viral agents that either infrequently spread through foods or show a potential to cause foodborne outbreaks. The inclusion of these examples here is also to encourage research to better elucidate the role of various vehicles in the spread of such infectious agents. This point is well illustrated by a recent report on the foodborne spread of Lassa virus (70); hunting of peridomestic rodents and the consumption of their meat has been found to be a risk factor in the spread of the virus to humans.

The general scope of foodborne disease transmission is changing as we alter our eating habits and preferences and as international trade in foodstuffs increases even further. The profound yet somewhat delayed impact on human health that our modern practices of rearing and feeding domesticated animals can have is dramatically illustrated by the recent BSE fiasco in the United Kingdom. Consumption of contaminated meat and milk, or foodstuffs contaminated by unsafe water, human or animal fecal matter, and, to some extent droplets, is still the only route by which foodborne transmission can occur. But the number of novel viruses suspected in foodborne transmission of disease is rising. Therefore, research must continue to observe and analyze mechanisms of infection and their relation to food. Evidence does support the theory that enteric animal diseases can cross the species barrier and cause similar symptoms in humans. If the now famous outbreak of influenza in Hong Kong has not ensured the validity of this theory, the prion epidemic has all but solidified the world's acceptance of the hypothesis.

The coming years will prove to be interesting and exciting in foodborne disease research. Currently, there is a haze of inconsistent scientific observations, generalized and sometimes incorrect statements made by the media, and a lack of epidemiological information. With increased focus on mechanisms of spread for these viruses, that fog will lift and a clearer perspective on foodborne disease in general will appear.

ABBREVIATIONS

AIDS	acquired immunodeficiency syndrome
BSE	bovine spongiform encephalopathy
CMV	cytomegalovirus
CJD	Creutzfeldt-Jakob disease
FMDV	Foot-and-mouth disease virus
HIV	Human immunodeficiency virus
HTLV-I	Human-T-cell lymphotrophic virus I
vCJD	variant CJD
PCR	polymerase chain reaction
TBE	tick-borne encephalitis
TGE	transmissible gastroenteritis
TSE	transmissible spongiform encephalopathies

REFERENCES

1. GJ Jackson. Principles and costs in the regulation of microbially contaminated foods. Southeast Asian J Trop Med Public Health 22S:382–383, 1991.
2. Y Motarjemi, FK Kaferstein. Global estimation of foodborne diseases. Wld Health Stat Quart 50(1–2):5–11, 1997.
3. TM Lüthi. Food and waterborne viral gastroenteritis: a review of agents and their epidemiology. Mitt Gebete Lebensm Hyg 88:119–150, 1997.
4. Anonymous. Tick borne encephalitis. Wkly Epidemiol Rec 70:120–122, 1995.
5. CJ Peters, A Sanchez, PE Rollin, TG Ksiazek, FA Murphy. Filoviridae: Marburg and Ebola viruses. In: BN Fields, DM Knipe, and PM Howley, eds. Fields Virology, 3rd ed. Raven Press, New York, pp 1161–1176, 1996.
6. PK Bjerre. Virus encephalitis after tick bite. Ugeskrift for Laeger 156:7065, 1994.
7. LA Vereta, VZ Skorobrekha, SP Nikolaeva, VI Aleksandrov, VI Tolstonogova, TA Zakharycheva, AP Red'ko, MI Lev, NA Savel'eva. O peredache virusa kleshchevogo entsefalita s korov'im molokom. Meditsinskaia Parazitologiia I Parazitarnye Bolezni 24:54–56, 1994.
8. W Sixl, D Stunzner, H Withalm, M Kock. Rare transmission mode of FSME (tick-borne encephalitis) by goat's milk. Geographia Medica Suppl 2:11–14, 1989.
9. Anonymous Outbreak of tick-borne encephalitis presumeably, milk-borne. Wkly Epidemiol Rec 69:140–141, 1994.
10. M Gresikova, M Kaluzova. Biology of tick-borne encephalitis virus. Acta Virologica 41:115–124, 1997.
11. RE Shope, JM Meegan. Arboviruses. In: A. Evans ed. Viral Infections of Humans: Epidemiology and Control, 4th ed. Plenum, New York, pp 154–176, 1997.

12. AL Botvinkin, OV Mel'nikov, FA Danchinova, LB Badueva, NA Makarchick. Raspredelenie infitsirovannykh virusom kleshchevogo entsefalita kleshchei vdol' lineinogo uchetnogo marshruta. Meditsinskaia Parazitologiia I Parazitarnye Bolezni 3:24–28, 1996.

13. M Labuda, D Stunzner, O Kozuch, W Sizl, E Kocianoca, R Schaffler, V Vyrostekova, Tickborne encephalitis virus activity in Styria, Austria. Acta Virologica 37:187–190, 1993.

14. R Kaiser. Tick-borne encephalitis in southern Germany. Lancet 345:463, 1995.

15. B Niklasson, S Vene. Vector-borne diseases in Sweden—a short review. Arch Virol Suppl 11:49–55, 1996.

16. I Takashima, K Morita, M Chiba, D Hayasaka, T Sato, C Takezawa, A Igarashi, H Kariwa, K Yoshimatsu, J Arikawa, N Hashimoto. A case of tick-borne encephalitis in Japan and isolation of the virus. J Clin Microbiol 35:1943–1947, 1997.

17. JL Melnick. Enteroviruses: polioviruses, coxsackieviruses, echoviruses and newer enteroviruses. In: BN Fields, DM Knipe, and PM Howley, eds. Fields Virology, 3rd ed. Raven Press, New York, pp 655–714, 1996.

18. MC Mahl, C Sadler. Virus survival on inanimate surfaces. Can J Microbiol 21(6):819–823 1975.

19. ML McGeady, JS Siak, RL Crowell. Survival of coxsackievirus B3 under diverse environmental conditions. Appl Environ Microbiol 37(5):972–977, 1979.

20. LG Irving, FA Smith. One-year survey of enteroviruses, adenoviruses and reoviruses isolated from effluent at an activated-sludge purification plant. Appl Environ Microbiol 41:51–59, 1981.

21. BK Ward, CM Chenoweth, LG Irving. Recovery of viruses from vegetable matter. Appl Environ Microbiol 44:1389–1394, 1982.

22. JT Tierney, R Sullivan, EP Larkin. Persistence of poliovirus 1 in soil and on vegetables grown in soil previously flooded with inoculated sewage sludge or effluent. Appl Environ Microbiol 33:109–113, 1977.

23. FX Abad, RM Pinto, C Villena, R Gajardo, A Bosch. Viruses in mussels: public health implications and depuration. J Food Prot 60:677–681, 1997.

24. T Yamashita, K Sakae, Y Ishihara, S Isomura. A 2-year survey of the prevalence of enteric viral infections in children compared with contamination in locally-harvested oysters. Epidemiol Infect 108:155–163, 1992.

25. CJ Hurst, RM Clark, SE Regli. Estimating the risk from ingestion of microbially contaminated water. In: CS Hurst, ed. Modeling Disease Transmission and Its Prevention by Disinfection. Cambridge University Press, Cambridge, UK, p 109, 1996.

26. G Jubb. The third outbreak of epidemic poliomyelitis at West Kirby. Lancet 1:67, 1915.

27. DO Cliver. Other foodborne viral diseases. In: DO Cliver, ed. Foodborne Disease Handbook, 2nd ed. Marcel Dekker, New York, pp 137–144, 1994.

28. Centers for Disease Control. Aseptic Meningitis Surveillance, U.S. Department of Health, Education and Welfare, Atlanta, p 11, 1979.

29. New York Department of Health. A Review of Foodborne Disease Outbreaks in New York State, 1988. Bureau of Community Sanitation and Food Protection, Albany, NY, 1989.

30. AM Opsherovich, GS Chasovnikova. Opyt vydeleniya enterovirusov iz obyektov vneshnei sredy. In: AM Opsherovich, ed. Materialy Problemnoi Komissii Akad, Akad. Med. Nauk SSSR, Moscow, pp 90–92, 1967.

31. ST Yanga, IA Gardner, HS Hurd, KA Eernisse, P Willeberg. Management and demographic factors associated with seropositivity to transmissible gastroenteritis virus in US swine herds, 1989–1990. Prev Vet Med 24(3):213–228, 1995.

32. AS Monto. Coronaviruses. In: AS Evans, ed. Viral Infections of Humans: Epidemiology and Control, 4th ed. Plenum, New York, pp 221–223, 1997.

33. Anonymous Outbreak of diarrhoeal illness associated with coronavirus. Commun Dis Environ Hlth Scotland Wkly Rep 25, 1, 1991.

34. G Beards, C Hall, J Green, TH Flewett, F Lamouliatte, P du Pasquier. An enveloped virus

in stools of children and adults with gastroenteritis that resembles the Breda virus of calves. Lancet 1:1050–1052, 1984.

35. GN Woode. Breda and Breda-like viruses: diagnosis, pathology and epidemiology. Ciba Found Symp 128:175–191, 1987.

36. CD Brandt, HW Kim, WJ Rodriguez, JO Arrobio, BC Jeffries, EP Stallings, C Lewis, AJ Miles, MK Gardner, RH Parrott. Adenoviruses and pediatric gastroenteritis. J Infect Dis 151(3):437–443, 1985.

37. EN Janoff, JM Orenstein, JF Manischewitz, PD Smith. Adenovirus colitis in the acquired immunodeficiency syndrome. Gastroenterology 100:976–979, 1991.

38. I Uhnoo, E Olding-Stenkvist, A Kreuger. Clinical features of acute gastroenteritis associated with rotavirus, enteric adenoviruses, and bacteria. Arch Dis Child 61:732–738, 1986.

39. KP Schofield, DJ Morris, AS Bailey, JC de Jong, G Corbitt. Gastroenteritis due to adenovirus type 41 in an adult with chronic lymphocytic leukemia. Clin Infect Dis 19:311–312, 1994.

40. J Noel, A Mansoor, U Thaker, J Herrmann, D Perron-Henry, WD Cubitt. Identification of adenoviruses in faeces from patients with diarrhoea at the Hospitals for Sick Children, London, 1989–1992. J Med Virol 43:84–90, 1994.

41. M Brown. Laboratory identification of adenoviruses associated with gastroenteritis in Canada from 1983 to 1986. J Clin Microbiol 28:1525–1529, 1990.

42. NR Cashman. A prion primer. Can Med Assoc J 157, 1381–1385, 1997.

43. SB Pruisner. Molecular biology and pathogenesis of prion diseases. Trends Biochem Sci 21: 482–487, 1996.

44. SB Pruisner. Genetic and infectious prion diseases Arch Neurol 50, 1129–1153, 1993.

45. C Weissmann, A Aguzzi. Bovine spongiform encephalopathy and early onset variant Creutzfeldt-Jakob disease. Curr Opin Neurobiol 7:695–700, 1997.

46. SN Cousens, M Zeidler, TF Esmonde, R de Silva, JW Wilesmith, PG Smith, RG Will. Sporadic Creutzfeldt-Jakob disease in the United Kingdom: analysis of epidemiological surveillance data for 1970–96. Br Med J 315:389–395, 1997.

47. RG Will, JW Ironside, M Zeidler, SN Cousens, K Estibeiro, A Alperovitch, S Poser, M Pocchiari, A Hofman, PG Smith. A new variant of Creutzfeldt-Jakob disease in the UK. Lancet 347: 921–925, 1996.

48. ME Bruce, RG Will, JW Ironside, I McConnell, D Drummond, A Suttie, L McCardle. Transmissions to mice indicate that ''new variant'' CJD is caused by the BSE agent. Nature 390: 448, 1997.

49. RM Ridley, HF Baker. Oral transmission of BSE to primates. Lancet, 348, 1174.

50. Monthly Creutzfeldt-Jakob Disease Statistics. United Kingdom Department of Health. Jan. 4, 1999. http://www.doh.gov.uk/cjd/jan99.htm, 1996.

51. J Walsh. A fatal beef crisis. Time Int 147(14), http://pathfinder.com/time/international/1996/960401/coverbeef.html, 1996.

52. J Geary. Hitting a nerve: new links to mad cow disease lead to a potential ban of beef on the bone and a debate on how to assess consumer risk. Time Europe, 150(24) http://pathfinder.com/time/magazine/int/971215/europe.hitting_a_ner.html, 1997.

53. JR Berger, E Waisman, B Weisman. Creutzfeldt-Jakob disease and eating squirrel brains. Lancet 350(9078):642, 1997.

54. J Golding. Unnatural constituents of breast milk—medication, lifestyle, pollutants, viruses. Early Hum Dev 49 Suppl:S29–S43, 1997.

55. MD Yow, NH White, LH Taber, AL Frank, WC Gruber, RA May, HJ Norton. Acquisition of cytomegalovirus infection from birth to 10 years: a longitudinal serologic study. J Pediatr 110:37–42, 1987.

56. ND Francis, AW Boylston, AHG Roberts, JM Parkin, AJ Pinching. Cytomegalovirus infection in gastrointestinal tracts of patients infected with HIV-1 or AIDS. J Clin Pathol 42:1055–1064, 1989.

57. K Numazaki. Human cytomegalovirus infection of breast milk. FEMS Immunol Med Microbiol 18:91–98, 1997.

58. P Lewis, R Nduati, JK Kreiss, GC John, BA Richardson, D Mbori-Ngacha, J Ndinya-Achola, J Overbaugh. Cell-free human immunodeficiency virus type 1 in breast milk. J Infect Dis 177:34–39, 1998.

59. P van de Perre. Postnatal transmission of human immunodeficiency virus type 1: the breast feeding dilemma. Am J Obstet Gynecol 173:483–487, 1995.

60. S Hino, S Katamine, T Miyamoto, H Doi, Y Tsuji, T Yamabe, JE Kaplan, DL Rudolph, RB Lal. Association between maternal antibodies to the external envelope glycoprotein and vertical transmission of human T-lymphotrophic virus type I. Maternal anti-env antibodies correlate with protection in non-breast-fed children. J Clin Invest 95:2920–2925, 1995.

61. Z Jebavy. Slintavka a kulhavka u lidi. Ceskoslovenska Stomatologie 76:200–203, 1976.

62. JL Melnick. Enteroviruses: Polioviruses, coxsackieviruses, echoviruses and newer enterovirus. In: BN Fields, DM Knipe, and PM Howley, eds. Fields Virology, 3rd ed. Raven Press, New York, pp 655–714, 1996.

63. F Fenner, BR McAuslan, CA Mims, J Sambrook, DO White. Pathogenesis: the spread of viruses through the body. In: F Fenner, BR McAuslan, CA Mims, J Sambrook, DC White, eds. The Biology of Animal Viruses, 2nd ed. Academic Press, New York, p 390, 1974.

64. RF Sellers, AL Donaldson, KAJ Herniman. Inhalation, persistence and dispersal of foot-and-mouth disease virus by man. J Hyg 68:565, 1970.

65. GF Panina, A Civardi, I Massirio, F Scatozza, P Baldini, F Palmia. Survival of foot-and-mouth disease virus in sausage meat products (Italian salami). Int J Food Microbiol 8:141–148, 1989.

66. Anonymous Cohort on avian flu. Http://www.info.gov.hk/dh/new/bulletin/13-11-98.htm, 1998.

67. Anonymous. Information on influenza A H5N1 (''bird flu'' or ''avian flu'') for tourists coming to Hong Kong. Http://www.info.gov.hk/dh/new/20-01-98.htm, 1998.

68. KY Yuen, PKS Chan, M Peiris, DNC Tsang, TL Que, KF Shortridge, PT Cheung, WK To, ETF Ho, R Sung, AFB Cheung. Clinical features and rapid viral diagnosis of human disease associated with avian influenza A H5N1 virus. Lancet 351:467–471, 1998.

69. EC Claas, AD Osterhaus, R Wanbeek, JC dejong, GF Rimmelzwann, DA Senne, S Krauss, KF Shortridge, RG Webster. Human influenza virus A H5N1 virus related to a highly pathogenic avian influenza virus. Lancet 351:472–477, 1998.

70. J Ter Meulen, I Lukashevich, K Sidibe, A Inapogui, M Marx, A Dorlemann, ML Yansane, K Koulemou, J Chang-Claude, H Schmitz. Hunting of peridomestic rodents and consumption of their meat as possible risk factors for rodent-to-human transmission of Lassa virus in the Republic of Guinea. J Trop Med Hyg 55(6):661–666, 1996.

6

Detection of Human Enteric Viruses in Foods

Lee-Ann Jaykus
North Carolina State University, Raleigh, North Carolina

I. INTRODUCTION

Human enteric viruses are now recognized as significant causes of foodborne illness (1–3), recently ranked as fifth and sixth among identified causes of foodborne disease in the United States (4). Additional data indicate that foodborne disease outbreaks of unconfirmed etiology frequently meet some of the clinical criteria for outbreaks of acute viral gastroenteritis (1–3). The apparent failure to confirm a viral etiology in such outbreaks has been due largely to the lack of available tests and the reluctance of public health officials to use epidemiological criteria in the classification of foodborne viral disease (1,2,5,6). The unavailability of food specimens and the failure to report outbreaks of mild gastrointestinal (GI) disease has also contributed to reporting difficulties. All of these factors have resulted in a drastic underestimate of the true scope and significance of food-borne viral infection (3).

Human enteric viruses replicate in the intestines of infected human hosts, are excreted in the feces, and are almost always detectable in domestic sewage effluents (7). They are transmitted by the fecal–oral route, and as such, the ultimate source of contamination for foods is through contact with human fecal pollution. This contamination may occur directly, through poor personal hygiene practices of infected food handlers, or indirectly, via contact with fecally contaminated waters or soils. Since the viruses must survive the pH variations and enzymes present in the human GI tract, they are regarded as highly environmentally stable, allowing virtually any food to serve as a vehicle for transmission and enabling them to withstand a wide variety of food storage and processing conditions. While human enteric viruses are unable to replicate in contaminated foods and are generally present in low numbers, their infectious doses are presumed to be low and therefore any level of contamination may pose a public health threat.

The most common types of foodborne viral disease are infections with hepatitis A virus (HAV; infectious hepatitis) and acute viral gastroenteritis (8,9). The severity of hepatitis A viral infection, the lack of immunity in the younger age group, and the clear epidemiological link between the consumption of contaminated foods and hepatitis A infection (10–12) have made this an important agent of concern in foodborne transmission. Related to HAV are the human enteroviruses which cause a wide range of clinical syndromes, many of which are remote to the GI tract. These viruses are occasionally associated with foodborne disease (8,13) and are the most commonly isolated agents in surveys regarding naturally occurring viral contamination in foods, particularly shellfish (14,15). The prototype virus in this family is poliovirus, which continues to be a problem in developing countries (8). As the human enteroviruses are readily detectable in laboratory systems, they continue to serve as an excellent model for virus detection methods and as a potential indicator of human fecal contamination of water and food.

While the enteroviruses and HAV have been extensively studied over the years, the agents of acute nonbacterial gastroenteritis are less well characterized. Although acute viral gastroenteritis is mild and self-limiting, infectious doses are presumed to be low and the absence of effective and available diagnostic techniques has complicated investigation of the clinical and economic impact of this disease. Furthermore, the incidence of acute viral gastroenteritis in the United States appears to be increasing and many outbreaks are unreported (3,16). Recent evidence indicates only temporary immunity to these viruses, suggesting that exposed individuals may continue to be susceptible throughout life (17). The prototype virus in this group is the Norwalk agent. Other related viruses, classified as small round-structured viruses (SRSVs), have also been implicated in many foodborne disease outbreaks (3,18,19). These viruses have been listed among the pathogens that are newly recognized as predominantly foodborne in the United States over the last 20 years (20).

Over the past 8 years, the genomes of several SRSVs of human origin have been cloned and sequenced, including the Norwalk agent (21,22) and the Southhampton agent (23). The genome size, buoyant density, and genomic organization of these viruses is similar to that of other viruses in the family Caliciviridae (24), prompting investigators to classify these Norwalk-like SRSVs as belonging to this virus family (22,23). None of these human enteric viruses have been cultivated in vitro and there exists no practical animal model for their propagation. Additionally, recent molecular evidence suggests considerable sequence diversity among human SRSVs with distinct antigenicity (24–26), suggesting additional complexities for the development of detection methods.

Other human enteric viruses have also been associated with outbreaks of foodborne disease. Rotaviruses, which cause acute and often severe gastroenteritis in infants and children, and are implicated in many waterborne outbreaks worldwide, have occasionally been associated with foodborne transmission (8). A group of small round viruses (SRVs) have been reported as the cause of several outbreaks of shellfish-associated gastroenteritis (27). These agents are poorly characterized and most likely represent more than one virus type, including the human caliciviruses and perhaps members of the genus *Parvovirus* (28,29). Astroviruses have been detected in the feces of children with gastroenteritis (30) and at least one outbreak has been associated with the consumption of oysters (31). Enteral non-A, non-B hepatitis has been transmitted predominantly by sewage-contaminated waters and person-to-person contact, and although foodborne transmission has not been documented, it might possibly occur. A human calicivirus designated hepatitis E virus (HEV) is the etiological agent of this disease (32,33), which is endemic in Asia, Africa, and Latin America, but not in the United States (8).

II. EPIDEMIOLOGY AS A MEANS OF DETECTION

From an epidemiological standpoint, both HAV and the human SRSVs constitute the most important groups of enteric viruses transmitted by the foodborne route. Consumption of contaminated molluscan shellfish is a frequent source of infection and acute viral gastroenteritis is considered the most common shellfish-borne disease (11). Contamination of shellfish-growing waters with human sewage provides the necessary source of human enteric viruses that are accumulated and concentrated by the shellfish during the feeding process. Outbreaks have been linked to shellfish harvested from approved waters (12), grossly fecally contaminated waters (34,35), malfunctioning boat sewage disposal systems (36,37), and ill shellfish harvesters who routinely discharged their sewage overboard (38). Unfortunately, there is poor correlation between the levels of bacteriological indicators on which the sanitary quality of shellfish is based, and the presence of enteric viruses in shellfish meats or growing waters (39). Furthermore, viral persistence has been found in shellfish after heat processing, γ irradiation, and controlled purification (40).

While shellfish have been important regional vehicles in the transmission of foodborne viral disease, poor personal hygiene of infected food handlers is another key source of infection. For instance, foodborne disease surveillance data from Minnesota from 1984 to 1991 indicated that viral gastroenteritis was the most common foodborne illness, accounting for 39/100 (39%) of the reported outbreaks, predominantly linked to infected food handlers (3). Key outbreaks with food handler involvement have been associated with breads, baked goods, and salads (41–43). The possibility of distributor level contamination for chopped or shredded products has also been suggested (3). The persistence of human enteric viruses in handled foods makes prevention of disease transmission after postprocessing contamination even more difficult (40).

Fecally contaminated water and/or soil in food production and processing may also provide a source of viral contamination for foods. Vegetables and fruits may become contaminated through wastewater and sludge used for irrigation (44) or, alternatively, through the use of nonpotable water during washing (45). Outbreaks associated with contaminated fruits and vegetables have been reported (46,47), including a highly publicized 1997 outbreak of hepatitis A linked to the consumption of contaminated frozen strawberries (48). Since human enteric viruses may survive for long periods on the surface of these products and may not be eliminated by washing alone (49), fruits and vegetables can serve indirectly in the transmission of waterborne infections. Another related and epidemiologically important source of foodborne viral disease has been the consumption of contaminated commercial ice (50). A summary of several key foodborne viral disease outbreaks is provided in Table 1.

Most of the epidemiological information about viral foodborne disease comes from outbreak investigations conducted by state and local health departments and the surveillance programs directed by the U.S. Centers for Disease Control and Prevention (CDC). However, since many of these diseases are not reportable and surveillance is based on voluntary reporting by state health departments, the magnitude of this disease problem is underestimated. This is exacerbated by a reluctance to use epidemiological criteria in the classification of foodborne viral disease and the failure to report mild outbreaks of gastrointestinal illness. Furthermore, since investigation generally occurs after the fact, important information and samples may have been destroyed, consumed, or lost to inaccurate recall. Since many of these outbreaks are small and confined, and state and local health departments have limited resources, many outbreaks may pass without significant epidemiological investigation (51).

However, the fact remains that prior to the introduction of molecular biological techniques, most detection of human enteric viruses associated the foods was achieved by investigation of isolated outbreaks. Recent applications of molecular epidemiology have aided significantly in the investigation of viral foodborne disease. In a 1993 outbreak, investigators were able to use molecular epidemiological approaches to determine the presence of a mixed SRSV infection in clusters of patients experiencing viral gastroenteritis attributable to the consumption of contaminated raw/steamed oysters (52). More recently, the Parkville virus, associated with a foodborne outbreak attributable to an infected food handler, was characterized and found to be genetically distinct from the Norwalklike SRSVs and more related to the Sapporo virus clade (53). These findings further illustrate the genetic diversity of the human SRSV group and demonstrate the power of molecular epidemiological approaches in differentiating and determining the emerging role of these viruses in food and waterborne disease outbreaks (41,52–55).

III. DETECTION OF HUMAN ENTERIC VIRUSES

A. Detection of Human Enteric Viruses in Clinical Specimens

Suspicion of illness caused by human enteric viruses can be established epidemiologically by incubation period and illness duration analysis, the demonstration of classic viral gastroenteritis symptoms, and the absence of bacterial and/or protozoal pathogens in stool samples (5,6). Beyond this, laboratory confirmation of human enteric viral infection has been based on a rise in specific antibody to the virus or, alternatively, the demonstration of virus particles, antigen, or nucleic acid in clinical samples, usually stool specimens. Immune electron microscopy (IEM), radioimmunoassay (RIA), and enzyme immunoassay (EIA) have all been applied to facilitate detection of these agents in clinical samples (56). Historically, the utility of these assays has been limited by poor detection limits

Table 1 Recent Outbreaks of Foodborne Viral Disease

Agent	Location	Date	No. cases	Food	Confirmation[a]	Ref.
Norwalk	U.S.	1987	191	Commercial ice	Clinical only	50
Norwalk	U.S.	1988	1440	Celery/chicken salad	Clinical only	45
HAV	China	1988	300,000 (4% total population)	Raw clams	Yes—IEM, hybridization, cell culture, expt. infection	34, 35
HAV	U.S.	1988	61	Raw oysters	Yes—Antigen capture-RT-PCR	10
HAV	U.S.	1988	202	Lettuce	Clinical only	47
HAV	U.K.	1989	50	Bread	Clinical only	45
HAV	U.S.	1990	28	Strawberries	Clinical only	46
SRSV	U.S.	1993	25 clusters	Raw/steamed oysters	Possible-RT-PCR, slight difference in amplicon sequence	38, 100, 112 134, 135
HAV	U.S.	1994	46	Baked goods	Clinical only	42
SRSV	U.S.	1994	188	Salad	Clinical only	41
SRSV	U.S.	1994	34 clusters	Steamed/roasted oysters	Clinical only	136
SRSV	U.S.	1994	46 cases	Various—food handler	Clinical only	53
SRSV	U.S.	1997	60 clusters	Oysters	Clinical only	36
HAV	U.S.	1997		Strawberries	In progress	48

[a] Confirmation of the virus, viral antigen, or viral nucleic acid in food specimens.

($\geq 10^4 - 10^5$ particles/mL) and the inability to cultivate the human SRSVs in vitro, which has limited the supply of viral antigen available for developing reagents (3). This has been exacerbated by the fact that the Norwalk agent is only one of several SRSVs that cause outbreaks with similar clinical and epidemiological features (3). Molecular biological techniques such as nucleic acid amplification have improved detection capabilities and reagent availability. For instance, the development of ELISA methods based on baculovirus-expressed Norwalk capsid antigen may facilitate commercialization of methods (57,58). Currently, there are some excellent methods available for the detection of human enteric viruses in clinical specimens and these have recently been reviewed by Kapikian et al. (56) (SRSVs), Hollinger and Ticehurst (59) (HAV), and Melnick (60) (human enteroviruses).

B. Detection of Human Enteric Viruses in Foods

Unfortunately, development of effective methods to detect human enteric viruses in foods has lagged behind the clinical capabilities. Like bacterial pathogens in food and water, viruses are frequently present in small numbers. However, unlike traditional food microbiological techniques, which have relied on cultural enrichment and selective plating to increase bacterial cell numbers and differentiate pathogens in the presence of background microflora, human enteric viruses require live mammalian cells in order to replicate. For this reason, standard methods for the detection of enteric bacteria in foods cannot be used, and detection of viruses requires instead an initial concentration step, often from large volumes of food or water, followed by subsequent detection, traditionally by mammalian cell culture infectivity assay or, more recently, using immunological or molecular techniques (51). In general, virus detection from foods relies on the sequential steps of (a) sampling; (b) virus concentration (removal of inhibitors); (c) detection; and (d) confirmation.

1. Sampling

The major issues associated with sampling include adequate representation and sample size. An optimal sample size of 25–100 g has been recommended (9). This sample size should be small enough to manipulate in the lab yet provide sufficient amount for adequate concentration of small numbers of viruses. Dividing complex food products, such as sandwiches, into component parts for separate assay may also facilitate detection. For instance, some investigators have limited virus assays from shellfish to dissected digestive tissues in an effort to improve virus recoveries and minimize sample inhibition (61,62). Adequate sample representation is also key, and since contamination may be intermittent and/or unevenly distributed in a given lot of food, the testing of multiple samples from the same item of food is recommended. Although stock cultures of human enteric viruses are stable for long periods of time at −80°C, little is known about the stability of these agents in naturally contaminated foods (C. L. Moe, personal communication). Ideally, food samples should be refrigerated soon after collection and processed for virus concentration immediately upon receipt by the testing facility.

2. Virus Concentration

Due to its historical association with viral foodborne disease, most of the developmental work in virus concentration from foods has been done using shellfish as the model food commodity. Few if any methods have been applied to other food commodities. Two general schemes for the concentration of human enteric viruses from shellfish have been reported and these are designated extraction–concentration methods and adsorption–elution–concentration methods. The purpose of these methods is to provide a high recovery of infectious virus along with reduction of sample volume and removal of cytotoxic or interfering compounds that may later compromise detection. All of the sample manipulations take advantage of the tendency of most enteric viruses to behave as proteins in solution and their exquisite ability to remain infectious at extremes of pH or in the presence of organic solvents such as chloroform or Freon. Extraction–concentration methods use a combination of techniques such as ultracentrifugation, ultrafiltration, polyelectrolyte flocculation, and acid precipitation to separate and concentrate viruses from shellfish into an aqueous phase of low volume.

Adsorption–elution–concentration methods seek to initially adsorb viruses to food particulates by alternation of pH and ionic conditions. Elution techniques are then applied to dissociate the virions from sample precipitates. Concentration of the virus may then be achieved through centrifugation, polyethylene glycol (PEG) and/or acid precipitation. Additional methods to aid in the removal of bulk food materials may be applied such as filtration and solvent extraction. Organic solvent extraction is particularly useful in the removal of lipids and other nonpolar food compounds without substantial reduction in virus levels. These methods have been reviewed extensively by Jaykus et al. (40). The success of adsorption–elution–precipitation methodology for recovering viruses from naturally contaminated shellfish has made this the method of choice in recent years (40,63). Virus yields after concentration have been reported to range from 10% to 90% (64,65). A representative virus concentration scheme is shown in Fig. 1.

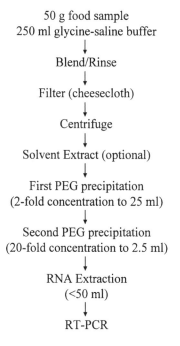

Figure 1 Representative virus concentration and purification method for foods.

3. Virus Detection

a. Mammalian Cell Culture Infectivity Assay. Assay of human enteric viruses from concentrates of shellfish has historically been based on the infectivity of the viruses for susceptible laboratory hosts. The major classes of host systems for detecting enteric virus infectivity are mammalian cell cultures of either primate or suckling mouse origin (66). Choice of the host system depends primarily on the viruses to be detected. Currently, there is no single universal host system available to enable replication and corresponding detection of a majority of the numerous human enteric viruses that may contaminate foods.

Primary and secondary human embryonic kidney and monkey kidney cell cultures are the preferred host systems for an array of culturable human enteric viruses. While few laboratories have the facilities and financial resources to work with cultures of primary cells, several convenient and alternative secondary cell lines are routinely used. Buffalo green monkey kidney (BGMK) cells, a continuous cell line from African green monkey kidney (AGMK) cells, are sensitive to many enteroviruses and are often used by laboratories that evaluate human enterovirus recovery and detection in environmental samples, particularly when using poliovirus type 1 (PV1) as a model (65,67–69). MA-104, a continuous rhesus monkey kidney–derived cell line, is sensitive to rotaviruses as well as many enteroviruses and thus provides complementary selectivity in virus isolation (70,71). The RD cell line is a continuous line of human rhabdomyosarcoma cells that have been shown to support the replication of certain group A coxsackieviruses as well as a wide variety of other enteroviruses (72). HeLa (human cervical carcinoma), HEp-2 (human carcinoma of the larynx), LLC-MK2 (rhesus monkey kidney), MFK (fetal rhesus monkey kidney), BSC-1 (monkey kidney), and Vero (AGMK) cells are also permissive to many human enteroviruses (66). Numerous studies have compared the relative sensitivities of these cell lines in the isolation of enteric viruses from environmental samples, often with conflicting results (72,73). Investigators agree that the use of at least two different host systems will facilitate detection of a wide spectrum of human enteroviruses from shellfish and environmental samples. For HAV, fetal rhesus monkey kidney–derived (FRhK-4) cells are used to detect the HM-175 lab-adapted strain which, unlike wild-type HAV, can induce cytopathic effects (CPEs) in these permissive cells (74).

There are two basic procedures for detection and quantification of enteric viruses by mammalian cell culture infectivity: quantal methods (TCID$_{50}$, tissue culture infective dose, or MPN) and enumerative methods (plaque assay). Quantal assays score replicate cultures inoculated with samples or sample dilutions as negative or positive for CPEs over a period of 1–4 weeks or until obvious cell degeneration has occurred. The virus titer is quantified statistically using an MPN approach (66). A potential problem with quantal assays is the potential for the presence of two or more viruses in the same inoculum; if differential rates of growth occur, the viruses that replicate more slowly may fail to be detected. In this case, either limiting dilution analysis is required to permit serological identification of distinct viruses.

Plaque assays may be more precise than quantal assays because relatively large numbers of individual infectious units can be counted directly as localized areas of infection (zones of clearing, known as plaques) under a semisolid phase (agar) overlaying a host cell monolayer, facilitating direct enumeration of viruses (66). However, some investigators believe that the quantal assay may be better suited for detecting extremely low levels of viral contamination and for the detection of slow-growing and nonplaquing viruses. Regardless, both techniques have been applied to the detection of human enteric viruses in naturally contaminated environmental and shellfish samples (15,75,76).

A major problem associated with quantitative assay of viruses in shellfish concentrates by mammalian cell culture infectivity is cytotoxicity. Sample toxicity is usually the result of residual food components, which may cause the destruction of the entire cell monolayer, or, alternatively, localized areas of necrosis that are indistinguishable from virus plaques (77,78). The fact that no universal host system is available for the detection of all or most of the human enteric viruses is an additional limitation. For instance, many of the epidemiologically important viruses do not replicate (Norwalk, human caliciviruses, and astroviruses) or replicate poorly (wild-type hepatitis A virus) in currently available host systems. These methods are further limited because levels of viral contamination in foods can be low and mammalian cell culture–based detection assays are slow and expensive.

b.   Immunological Methods.   Given the technological limitations of mammalian cell culture infectivity assays, the antigenic specificity of the viral coat protein has been used to develop alternative detection strategies based on RIA or EIA. Early work focused on using immunological techniques to visualize the presence of noncytopathic viruses in cell culture. For instance, Lemon et al. (79) developed a radioimmunofocus assay (RIFA) for the quantitation of HAV in cell cultures, and this has been applied to the detection of the virus in shellfish. Later modifications sought to apply direct EIA methods for the detection of viruses in shellfish without the need for concurrent amplification of the agents in cell culture (80,81).

RIA, EIA, ELISA, and IEM have all been adapted for the detection of SRSVs in clinical (fecal) specimens with a high degree of success. Both enzyme immunoassays and solid phase IEM have been used to detect Norwalk virus in fecal specimens associated with a foodborne outbreaks (3,82) as well as to demonstrate serological diversity of the Norwalk-like virus group (83–85). While these methods are effective for clinical specimens, the general unavailability of reagents and their poor detection limits ($\geq 10^4$ infectious virus particles) restricts their practical application for the detection of viral contamination in foods.

c.   Nucleic Acid Hybridization Methods.   Gene probes and other nucleic acid hybridization techniques have also been applied for the detection of human enteric viruses, with most applications reported for clinical or environmental samples. Detection limits for hybridization to genomic viral RNA have been reported at 2.5×10^5 physical particles of rotavirus SA-11 (86), 10^4 physical particles of HAV (87), and $5 \times 10^2 – 1 \times 10^3$ PFU of rotavirus SA-11 and coxsackievirus B3, respectively (76,88). Margolin et al. (89) were able to detect $1–10$ PFU of poliovirus using probes of high specific activity. In general, RNA probes are five- to eightfold more sensitive than DNA probes (76,90). Investigators have used gene probes to detect human enteroviruses (89,91), HAV (90,92), and rotaviruses (92) in environmental waters and shellfish.

d.   Nucleic Acid Amplification Methods.   In vitro enzymatic nucleic acid amplification methods such as the polymerase chain reaction (PCR) offer an opportunity to enrich a single specific nucleic acid sequence up to a million fold and hence provide a sensitive and specific detection method with a theoretical detection limit of one virus unit. In essence, these nucleic acid amplification methods have the potential of replacing standard cultural enrichment methods with faster nucleic acid enrichment. This has distinct advantages for the detection of human enteric viruses, where cultural enrichment approaches may not be possible. The method is readily adaptable to the detection of RNA viruses by preceding the PCR with a rapid (1-h) reverse transcription step; hence the designation RT-PCR. Over the last 10 years, several RT-PCR approaches have been reported for the detection of human enteric viruses in foods, predominantly shellfish. Development of these methods has been complicated by low levels of viral contamination in foods, the need to assay samples of large size, and the presence of residual food components that might interfere with enzymatic nucleic acid amplification reactions. To address these issues, four alternative approaches have been reported to simultaneously concentrate viruses or viral nucleic acids, reduce sample volumes, and decrease the level of interfering compounds. The most frequently applied approach involves isolating and purifying total nucleic acids (RNA) from the food sample prior to RT-PCR (61,62,69,93–101). A second approach combines capture of the virus with specific antibody followed by nucleic acid amplification using RT-PCR (10,102–104). In the third approach, the intact virus particles are concentrated and purified from the complex food matrix resulting in sample volume reduction and removal of inhibitors, followed by subsequent heat release of viral nucleic acid from the virion capsid and RT-PCR (67,75,105) (Table 2). In a fourth method, investigators have preceded RT-PCR with cell culture propagation of viruses, providing an approach resembling cultural enrichment for the detection of foodborne bacterial pathogens (75,76). These methods are summarized in Table 2.

Nucleic acid extraction-RT-PCR. The most popular approach to facilitate the detection of viruses in environmental and shellfish samples involves direct RNA extraction followed by RT-PCR. In a representative method (61,62,69), viruses in oyster homogenates are purified and concentrated fourfold by solvent extraction and PEG precipitation (106). RNA is then extracted directly from these concentrates by sodium dodecyl sulfate–proteinase K digestion, phenol-chloroform extraction, and ethanol precipitation. Further removal of inhibitors has been accomplished by treatment of crude nucleic acid extracts with the cationic detergent cetyltrimethylammonim bromide (CTAB). For example, in one method, 9 mL shellfish homogenates was extracted with one-half the volume of chloroform butanol, flocculated with Cat-Floc T (Calgon Corp., Pittsburg, PA), and viruses in the aqueous phase precipitated with 6.5 mL PEG-2000 (24% wt/vol) and 1.2 M NaCl. PEG pellets were suspended in 3 mL water and the virus digested with proteinase K at 56°C for 30 min, extracted twice with an equal volume of phenol/chloroform/water, and the aqueous phase precipitated in ethanol. The resulting pellet was suspended in water and adjusted to a final concentration of 1.4% CTAB and 0.11 M NaCl. After pelleting again by centrifugation, the nucleic acids were precipitated a second time in ethanol/sodium acetate and resuspended in 100 mL of water; 20 mL of this final concentrate was used for each RT-PCR reaction. This method has been further refined by using dissected shellfish tissues (stomach and digestive diverticula) in place of whole shellfish and by the addition of an internal RNA standard to simultaneously evaluate the presence of RT-PCR inhibitors (62). The method has been tested collaboratively using shellfish tissues artificially contaminated with the Norwalk virus (61), and reportedly has been used to detect virus in oysters implicated in a viral gastroenteritis outbreak (100).

Other documented methods also use PEG precipitation and Freon extraction to initially purify viruses from up to 50 g shellfish (97,98) followed by extraction of total nucleic acids by guanidinium isothiocyanate and glass powder. Essentially, in these studies, a reaction mix of glass powder matrix and guanidinium isothiocyanate was used to extract

Table 2 RT-PCR Detection Methods for Human Enteric Viruses in Foods

Pathogen	Sample	Detection limit	Field samples	Ref.
Nucleic acid extraction/RT-PCR				
HAV	Clams	2000 particles/g	No	94
Poliovirus	Oysters	38 PFU/20 g	No	69
HAV	Clams/oysters	100 PFU/1.5 g	No	69
Norwalk/SRSV	Clams/oysters	5–10 PCRU[a]/1.5 g	Yes (SRSV)	62, 100
Norwalk	Various	20–200 PCRU/10 g (nested)	No	95
Poliovirus	Shellfish	20 MPNCU/g (nested)	Yes (EV)	101
HAV	Shellfish	20 TCID $_{50}$/g (nested)	No	101
Rotavirus	Shellfish	100 FF/g (nested)	No	101
Poliovirus	Oysters/mussels	10 PFU/5 g	No	98
SRSV	Oysters/mussels	Not specified	Yes (SRSV)	93
Poliovirus, HAV	Clams	100 PFU/50 g	No	93
Norwalk	Clams	1000 PCRU/50 g	No	93
Poliovirus	Shellfish	3–30 TCID$_{50}$/g (nested)	No	96
SRSV, rotavirus, HAV	Shellfish	Not specified (nested)	Yes (RV)	96
Antibody capture/RT-PCR				
HAV	Clams/oysters	Not specified	No	103
HAV	Oysters	Not specified	Yes (HAV)	10
HAV	Oysters	8 PFU/g	No	102
HAV	Oysters	10 PFU/20 g	No	104
Virion concentration/RT-PCR				
Poliovirus, HAV	Oysters	10 PFU/50 g	Yes (EV, HAV)	67, 75
Norwalk	Oysters	4500 PCRU/50 g	No	67
Poliovirus, HAV	Clams	1000 PFU/50 g	No	105
Norwalk	Clams	100 PCRU/50 g	No	105

[a] PCRU, PCR-amplifiable unit.

total nucleic acids from shellfish tissue weight equivalents of 1–8 g. The guanidinium isothiocyanate served to lyse the samples and protect the RNA from enzymatic degradation. RNA effectively bound to the glass powder and could be washed with ethanol and acetone prior to elution in Tris buffer and a final precipitation in ethanol. The entire final RNA precipitate was used in RT-PCR reactions. These investigators reported significant removal of shellfish-related inhibitors using this method, with sample tolerances exceeding 6 g equivalent weight. While exact detection limits were not described, they were able to detect SRSVs in 5 of 31 shellfish samples harvested from both pristine and fecally impacted sites and also from shellfish associated with four separate foodborne disease outbreaks (97). In many cases PCR products were not visible by agarose gel electrophoresis alone but could be visualized by subsequent Southern hybridization. DNA sequence analysis of RT-PCR products was carried out to assure that detection was not the result of cross-contamination and to investigate sequence diversity among the amplicons obtained from naturally contaminated samples.

Gouvea et al. (95) were the first to apply PCR-based detection methods to other food products. Ten-gram samples of representative foods (shellfish, orange juice, milk, cole slaw, melon, and lettuce) were artificially contaminated with Norwalk virus inoculum, blended or washed, clarified by centrifugation, and then the pellet was deproteinized by sequential extractions using guanidinium isothiocyanate. The pooled, clarified supernatants (shellfish, cole slaw, and melon) were then extracted with Freon. RNA in the clarified homogenates was isolated as previously reported (107). Briefly, RNA was bound by the addition of 100 mL of hydroxyapatite, collected by centrifugation, washed with 10 mM potassium phosphate (pH 6.8), and finally eluted from the hydroxyapatite by sequential washes with 150 and 200 mM potassium phosphate, respectively. The RNA was then precipitated using CTAB and NaCl (final concentration of 0.8% CTAB, 0.5% NaCl, 0.05% EDTA) at 56°C for 30 min. The RNA-CTAB salt was collected by centrifugation, resuspended in water, and precipitated a second time in ethanol. The final RNA pellet was resuspended in 40 mL of Rnasin-protected water, with 4 mL used in each RT-PCR reaction. These investigators could not detect Norwalk virus after a single PCR amplification but were able to detect it using a "nested" or double-PCR approach, reporting detection limits of 20–200 genomic copies of Norwalk virus per 10 g food sample. They confirmed the identity of the RT-PCR amplicons by restriction endonuclease digestion analysis.

Most recently, Leggitt (99) used a method employing homogenization, filtration through cheesecloth, Freon extraction (hamburger only), and two sequential PEG precipitations to concentrate poliovirus, HAV, and Norwalk virus from 50-g samples of artificially contaminated hamburger and lettuce. The sequential PEG precipitations were performed at increasing PEG concentration (6% and 12% PEG, respectively), and resulted in a 10- to 20-fold sample volume reduction from 50 g to approximately 2.5 mL (Fig. 1). This resuspended PEG precipitate could be assayed for virus recovery by mammalian cell culture infectivity assay, with reported recoveries of up to 50% for poliovirus and 4% for HAV. The total RNA in the entire 2.5-mL sample concentrate could be extracted using a combined guanidinium thiocyanate/phenol-chloroform approach, resulting in a 25-mL sample, or an additional 100-fold sample volume reduction. Viral RNA was consistently detected by RT-PCR at initial inoculum levels of $\geq 10^2$ PFU/50-g food sample for poliovirus and $\geq 10^3$ PFU/50-g food sample for HAV. Norwalk virus was detected at inoculum levels of $\geq 10^3$ RT-PCR-amplifiable units (PCRU)/50-g food sample. All RT-PCR amplicons were confirmed by subsequent internal oligoprobe hybridization. A representative agarose gel and subsequent Southern hybridization can be seen in Fig. 2.

Although this approach results in the detection of relatively pure RNA, disadvantages include multiple sample manipulation steps, which may result in incomplete recovery and/or potential degradation of RNA during the extraction procedure, and difficulty in making direct correlations with infectivity due to loss of virion integrity. Direct assay of the viruses from the concentrated sample prior to RNA extraction can circumvent this problem, although this is rarely done.

Antibody-capture RT-PCR. An alternative to RNA extraction would involve direct isolation of viruses from environmental and food samples. Jansen et al. (108) detected HAV in stool specimens using an antigen capture RT-PCR (AC-PCR) approach, and Desenclos et al. (10) implicated HAV in oyster outbreak specimens by immunocapture of the virus, heat release of the viral RNA, and subsequent RT-PCR detection. The most comprehensive study of this approach for the detection of viruses in foods has been reported by Deng et al. (103) and Lopez-Sabater et al. (104). These investigators initially purified HAV from 20-g samples of artificially contaminated oysters and clams by elution in 100 mL buffer, addition of 2 mL of a 1% solution of Cat-Floc T, followed by sequential filtration through a Whatman GF/F filter and a 0.2-mm membrane filter. Alternatively, viruses in the first filtrate were concentrated 10-fold (to 10 mL) by ultrafiltration through

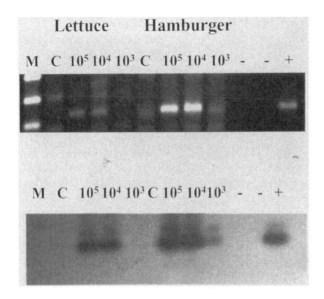

Figure 2 Representative RT-PCR detection of Norwalk virus from artificially contaminated lettuce and hamburger. Fifty-gram lettuce or hamburger samples were inoculated with 10^3–10^5 PFU PCR-amplifiable units; processed for virus concentration by the sequential steps of homogenization, filtration, solvent extraction, PEG precipitation, and RNA extraction; and subjected to RT-PCR for virus detection. Top panel is agarose gel electrophoresis for virus detection; bottom panel is Southern hybridization of RT-PCR amplicons for virus confirmation. The food commodity and initial level of Norwalk virus inoculum is given above each gel lane. C, uninoculated food control; (−), negative controls for RNA extraction and RT-PCR; (+), positive Norwalk virus control for RT-PCR.

a PM 30 membrane, followed by filtration through a 0.2-mm membrane filter. The viruses in the filtrate were assayed for recovery and detection by mammalian cell culture infectivity assay and AC-PCR. In the AC-PCR method, 0.5 mL polypropylene microfuge tubes were coated with diluted (1:1000) anti-HAV IgG as described by Jansen et al. (110). Filtrate (80–100 mL) was then added to each tube and the tube incubated overnight at 4°C. The tubes were washed six times with a Tris-KCl-Mg$_2$Cl buffer and then used directly in the RT-PCR reactions after heat release of the viral nucleic acid. In a later study, paramagnetic beads were coated with biotinylated human anti-HAV IgG. For HAV capture, 1 mL of oyster extract was mixed with the antibody-coated beads for 2 h at room temperature. After incubation, the beads were trapped against the side of the microfuge tube using a magnetic particle separator and the supernatant fluid removed by aspiration. The beads were washed six times as above, the viral nucleic acids released by heating at 95°C for 5 min, and the supernatant fluid transferred for reverse transcription, discarding the magnetic beads in the process. By using magnetic immunocapture, viral particles were concentrated 500-fold with removal of RT-PCR inhibitory compounds. Detection limits of 10 PFU HAV in 20 g of oyster meat were reported (104).

This virus concentration/detection strategy has advantages in that a positive reaction demonstrates the presence of antigen-associated HAV RNA, which may be more highly associated with infectious virus than detection of free viral nucleic acid (103). Additionally, there are fewer sample manipulations, making the method potentially simpler than other approaches. A disadvantage may be the specificity of the antibody, since only a single antigenic type of virus can be detected. However, other investigators have coated paramagnetic beads with a pooled source of IgG in the form of human serum immunoglobulin and successfully detected the human enteroviruses, HAV, and the Norwalk virus in both seeded and naturally contaminated surface water samples (109).

Virion concentration-RT-PCR. De Leon et al. (110) demonstrated that direct RT-PCR amplification of the Norwalk virus in 10–20% stool samples was possible by initial clarification by fluorocarbon extraction and centrifugation, followed by further purification using spin column chromatography with Sephadex G-200. Viruses were recovered in the excluded volume, which averaged 70–80 mL per column; 10 mL was used in each subsequent RT-PCR reaction after heating to release and denature the viral nucleic acid. A direct virion isolation method has been reported for shellfish that seeks to further concentrate virions from oyster and clam samples that have previously been extracted for virus concentration using adsorption–elution–precipitation methods (65). This method resulted in extracts of small volume with recovery of infectious viruses and removal of enzymatic inhibitors (67,75,105). In this method, RT-PCR inhibitors were initially removed by two consecutive extractions using equal volumes of Freon followed by virus precipitation with 4–12% PEG-6000 and 0.3 M NaCl. The precipitated viruses were recovered by centrifugation and resuspended in one-seventh the original sample volume using a Tris-Tween buffer at pH 9.0. Viruses in the PEG eluants were precipitated for 15 min by the addition of an equal volume of Pro-Cipitate (LigoChem, Fairfield, NJ). The solid phase was recovered by centrifugation and the adsorbed viruses eluted using 4 mL of Tris-Tween buffer followed by centrifugation to remove the excess Pro-Cipitate. Viruses in the eluant were further concentrated by a secondary PEG precipitation step using 10% PEG and 0.3 M NaCl, reconstituted in buffer to a final volume of approximately 200 mL, with 10-mL aliquots used in RT-PCR reactions. The final sample concentrate represented an overall 1000-fold sample volume reduction with recovery of infectious viruses. Using this approach, viruses in processed samples could be readily detected by both RT-PCR and cell culture infectivity. Virus recoveries by cell culture infectivity were 25–35% for PV1 and 5–10% for HAV for both oysters and hard-shelled clams and RT-PCR detection limits were between 10^1 and 10^3 PFU/50 g shellfish extract. The method was readily adaptable to the concentration and subsequent detection of the Norwalk virus using RT-PCR (67,105). This method has also been applied for the detection of human enteric viruses in naturally contaminated oyster samples, where 5 of 31 samples were positive for human enterovirus sequences by this direct RT-PCR approach (75). Later methodological adaptations used the proprietary agent Viraffinity in place of Pro-Cipitate (93).

The virion concentration method has advantages in that significant sample volume reductions were achieved with recovery of infectious viruses, enabling direct comparison of detection by nucleic acid amplification and cell culture infectivity assay. Furthermore, previous studies have reported good correlation between detection by viral infectivity and RT-PCR using this approach (67). Disadvantages include substantial virus loss during the extraction steps, which may reduce overall detection limits.

Cell culture infectivity and RT-PCR. Shieh et al. (76) combined traditional cell culture and molecular techniques by amplifying viruses from environmental water concentrates in mammalian cell culture prior to RNA extraction of cell monolayers and probing with single-stranded RNA probes. Later, Chung et al. (75) examined 31 oyster specimens harvested from fecally impacted waters and found 12 of 31 samples positive for human enteroviruses by CPE and 6 of 31 positive by RT-PCR of RNA extracts from CPE-negative cultures. In this study, HAV was also detected in RNA extracts of two CPE-positive samples by RT-PCR and subsequent internal oligoprobe hybridization.

Mammalian cell culture amplification followed by RT-PCR accomplishes the dual purpose of increasing the number of copies of target nucleic acid and incorporating an infectivity assay. Potential disadvantages to such an approach include the lack of susceptible cell lines for the epidemiologically important human enteric viruses, cytotoxicity associated with food extracts, extended testing times, and high cost.

4. Virus Confirmation and Identification

In mammalian cell culture infectivity assay, inoculated cultures are incubated and periodically examined for CPE or plaques for 1–3 weeks or until the cells begin to deteriorate with age. Material from CPE-positive cultures or plaques can be confirmed by further passage into fresh cell cultures. Material from CPE-negative cultures may be further passaged in a process known as "blind passage," which occasionally yields additional virus isolates (66). Confirmed virus isolates are typically identified using conventional serological techniques (56,59,60).

While PCR will amplify DNA molecules several thousand fold, the presence of specifically amplified product must be identified, much as presumptively positive results obtained in ELISA assays require cultural confirmation. This is of particular importance in food systems, where nonspecific nucleic acid amplification may be a problem (67). While gel electrophoresis is commonly employed to visualize the amplified product, it is not specific. Alternative PCR confirmation methods have included internal oligonucleotide probe hybridization (61,62,67,69,75,99,105), 'nested' PCR (95,96), restriction endonuclease digestion (96), and RT-PCR amplicon sequencing (97,100).

5. Methodological Barriers

Although prototype nucleic acid amplification methods for the detection of human enteric viruses in foods have been reported, there are still several barriers to overcome before these methods can be applicable to routine testing. In order to obtain sample representation and detection sensitivity adequate for the low infectious doses of most human enteric viruses, large samples of the incriminated foods need to be processed. While some of the virus detection methods reported in the literature begin with large (50-g) sample sizes, most reported methods involve extraction from very small volumes of sample aliquots, which limits the sensitivity of the assay procedure from the very beginning. The approaches then applied to concentrate and purify the viruses from the homogenate not only need to be reasonably efficient but also need to produce a concentrate that is low in both volume and enzymatic inhibitors (63). To complicate matters further, the prototype extraction and detection procedures may well differ for different virus strains or different food commodities. For instance, shellfish have extremely high levels of enzymatic inhibitors, and similar RT-PCR inhibition has been noted for extracts from fruit and vegetable products (99). Complex carbohydrates (particularly glycogen) are considered a major cause of shellfish-related PCR inhibition, although other food-related molecules have been found to be inhibitory as well (111). Most of these food-related inhibitors have not been classified and many have remained recalcitrant to almost all removal processes applied.

IV. FUTURE DIRECTIONS

All of the methodological approaches described above have been applied to artificially contaminated shellfish species (Table 2). Unfortunately, only a few methods have been applied to naturally contaminated shellfish (10,62,75,96,97,100,101) and on only three occasions have the methods successfully detected viruses in food samples epidemiologically linked to disease (10,97,100). Only two studies have attempted detection in food products other than shellfish (95,99). Some key future research directions include the following: (a) improvements to virus concentration methods; (b) the relationship between detected signal and infectious virus; (c) alternative nucleic acid amplification methods; (d) alternative confirmation methods; (e) SRSV detection; and (f) molecular epidemiology.

A. Improvements to Virus Concentration Methods

Despite enormous strides in the ability to detect viral agents in foods, there remains a need to further refine extraction and detection methods to enable the detection of the low levels of virus (10^1–10^2 infectious units/50-g sample) anticipated in naturally contaminated food samples. This will require serious consideration of sample size when designing virus extraction methods for foods. Ideally, and when possible, virus recoveries should be assessed in parallel by cell culture infectivity assay at various steps during the extraction process. Key extraction parameters need to be optimized, such as elution conditions for HAV, replacement of Freon with more environmentally acceptable alternatives, and removal of residual RT-PCR inhibitors. Nonspecific amplification or reduced amplification efficiency at low copy numbers of template nucleic acid have been reported in food systems (67), presumably due to the presence of residual food components or, alternatively, food-derived DNA/RNA (67). Modifications of protocols to reduce nonspecific amplification in food systems may include single-tube amplification strategies, alternative recombinant *Taq* DNA polymerases, pretreatment of food sample concentrates with DNAase and/or treatment of post-RT cDNA with ribonuclease H, and addition of RT-PCR enhancement agents (67,112).

B. Relationship Between Detection by RT-PCR and Infectious Virus

Recent studies have addressed the relationship between detection of viral nucleic acid by RT-PCR and the presence of infectious virus in environmental samples (113,114). Both studies indicated a relatively short life of viral RNA in natural waters, noting that detection of virus by RT-PCR in environmental samples is indicative of either recent fecal pollution or the presence of well-protected virus particles. Similar studies need to be undertaken for foods, particularly in relationship to potential food processing techniques such as thermal inactivation or γ irradiation.

C. Alternative Nucleic Acid Amplification Methods

Most RT-PCR assays developed for the detection of viruses in foods or clinical specimens target individual viruses. The multiplex approach, which incorporates specific primers for two or more targets in a single reaction tube, enables the detection of more than one target nucleic acid in a single test. Multiplex RT-PCR offers several advantages over monoplex methods, including reduced labor and reagent costs and potentially faster detection (115). A multiplex RT-PCR format has been reported for the simultaneous detection of polioviruses, HAV, and rotaviruses (116), and a model system for the detection of the human enteroviruses, HAV, and Norwalk virus is in development (117). In most cases, the detection limit for multiplex assays drops 10- to 100-fold when compared to monoplex assays.

However, successful detection of viruses using multiplex methods has been reported in naturally contaminated clinical (118,119) and environmental specimens (116).

While semiquantitative PCR approaches for the detection of human enteric viruses have been reported (62), these methods need to be further refined. Alternative nucleic acid amplification methods are also on the horizon. Of particular interest may be nucleic acid sequence–based amplification (NASBA) (120). This method uses three enzymes (reverse transcriptase, RNase H, and T7 RNA polymerase) and two specific oligonucleotide primers (one that contains the bacteriophage T7 promoter sequence) to amplify target RNA or DNA sequences. The method has advantages over PCR because it is more rapid, can be made quantitative, and the amplification is carried out at a single temperature, obviating the need for specialized thermocyling equipment. NASBA methods are currently marketed in kit format, making this method readily available to clinical labs without the need for expensive equipment or extensive training. While one could perhaps expect similar problems with food-related inhibition using NASBA, this may be an important alternative nucleic acid amplification method in coming years. Furthermore, since it is isothermal, NASBA may be more readily adaptable to biosensor formats.

D. Alternative Confirmation Methods

While PCR confirmation methods such as internal oligonucleotide probe hybridization, nested PCR, restriction endonuclease digestion, and RT-PCR amplicon sequencing have been applied, these methods can take up to 48 h to complete and may be subject to cross-contamination. Within the last few years, techniques based on linking ELISA approaches to DNA hybridization have resulted in the potential for real-time PCR confirmation methods. In these methods, a solid phase capture is linked to immunomolecular detection using chromogenic, chemiluminescent, or fluorescent substrates (121). More recently, fluorescent DNA probes such as molecular beacons (122) and the Taq-Man (PE Applied Biosystems, Foster City, CA) assay (123) have been reported for the real-time measurement of hybridization kinetics. Efforts must be placed on reducing overall PCR confirmation time by bypassing gel electrophoresis, specifically in the investigation of the use of biotinylated/fluorescent dye–labeled probes and alternative enzyme-enhanced detection strategies. The ultimate goal would be to achieve endpoint detection within 30 min, thereby considerably streamlining confirmation methods. An alternative confirmation method may be the ligase chain reaction (LCR) (124). This method is capable of differentiating amplicons based on single-base-pair differences and has been used for the differentiation of genetically related virus strains (125) and *Salmonella* serovars (126).

E. SRSV Detection

Recent advances in molecular virology have enabled investigators to delineate the phylogenetic relationship between the various SRSVs (52–55,127–131). Previously, a series of SRSVs originating from the United Kingdom were classified by IEM and solid phase IEM into four distinct antigenic groups designated UK1, UK2, UK3, and UK4. More recently, sequence information on many of these strains has become available (21–25,52,54,127,128,131,132). Well over 40 of these strains have now been classified into two genogroups (G1 and G2) by sequence analysis of a segment of the RNA polymerase gene, with Norwalk virus (NV) now designated is the prototype for genogroup I and Snow Mountain, Hawaii, and Mexico viruses as the prototypes for genogroup II. Further

Table 3 RT-PCR Primers and Probes for SRSV/Norwalk-like Viruses

Sequence/sense	Location	Size (bp)	Viruses	Ref.
Primers				
5′-GCACCATCTGAGATGGATGT-3′	4673–4878	206	Norwalk (SMA, TV)	26, 137
5′-GTTGACACAATCTCATCATC-3′				
Probe				
5′-GGCCTGCCATCTGGATTGCC-3′				
Primers				
5′-TGTCACGGATCTCATCATCACC-3′ (−)	4856–4876 G1, G2	G2-123	G2-UK1, UK3, UK4, SMA	54
5′-TGGAATTCCATCGCCCACTGG-3′ (+)	4754–4773 G2			
5′-TGTCACGGATCTCATCATCACC-3′ (−)	4856–4876 G1, G2	G1-123	G1-UK2, Norwalk	
5′-GTGAAACAGCATAAAATCACTGG-3′ (+)	4754–4773 G1			
5′-GTGAAACAGTATAAACCACTGG-3′ (+)	4754–4773 G1			
5′-GTGAAACAGTATAAACCATTGG-3′ (+)	4754–4773 G1			
Probes				
5′-ACATCAGGAGAGTGCCCACT-3′ (−)	4804–4823 P1-A		UK2	
5′-ACATCAGGTGATAAGCCAGT-3′ (−)				
5′-ACATCGGGTGATAGGCCTGT-3′ (−)				
5′-ACATCTGGTGAGAGACCTGA-3′ (−)	4804–4823 P1-B		2/7 UK1	
5′-ATGTCGGGGCCTAGTCCTGT-3′ (−)	4804–4823 P2-A		4/7 UK1	
5′-ATGTCAGGGGACAGGTTTGT-3′ (−)	4804–4823 P2-B		1/7 UK1, UK3, UK4	
Primers				
5′-AC(A/T/G)AT(C/T)TCATCATCACCATA-3′ (−)	4865–4884	Various	UK1, UK2, UK3, UK4	133
Probes				
5′-CTCTGTGCACTTTCTGAAGT-3′ (+)	4796–4815		NV, SMA, 2/2 UK3, 1/2 UK4	
5′-ACCTTGTGTGCCATGTCTGA-3′ (+)	4793–4812		2/3 UK2	
5′-CTATGTGCACTGTCAGAAGT-3′ (+)	4796–4815		1/2 UK1, 1/2 UK2	
5′-CTGCAGGCATACGAAAGCCA-3′ (+)	4796–4815		Sapporo	
Primers				
5′-GAATTCCATCGCCCACTGGCT-3′ (+)	4756–4776	113	UK1, UK2, UK3 UK4	138
5′-ATCTCATCATCATCACCATA-3′ (−)				
Primers				
5′-CAGCCCACTAGTGTCATGTG-3′ (+)	57–76	237	Sapporo/Parkville	53
5′-TCACCATAAGTGTGAACAGTCTC-3′ (−)	271–293			

subdivisions have been made based on internal oligoprobe hybridization of RT-PCR product (52,54). A third genogroup, with Sapporo virus as the prototype, is antigenically and genetically distinct and more closely related to animal caliciviruses than to other known human caliciviruses (53,128,130).

While numerous RT-PCR formats have been described (Table 3) and sequence information on a variety of strains continues to be collected, all primer pairs described to date have failed to detect a proportion of the SRSV strains tested (133). Clearly, development of universal detection methods is limited until more complete information about this diverse group of human enteric viruses is available. Excellent researchers are in the process of gathering additional sequence information on which improvement in primer design can be made. Once these are available, both endemic and epidemic strains of SRSVs can be characterized, providing further information about the role of this virus group in foodborne disease outbreaks.

F. Molecular Epidemiology

Questions remain regarding the predominant types of SRSVs associated with foodborne disease outbreaks in the United States and abroad. Using panels of specific primers, investigators should be able to amplify domains of the polymerase region of SRSV fecal extracts associated with foodborne disease outbreaks. By sequencing these RT-PCR amplicons and comparing homologies, phylogenetic relationships between strains can be established (52–55,127–131). As new primers become available, perhaps even corresponding to different regions of the SRSV genome, further strain differentiation may be possible. With the availability of recombinant SRSV antigen (57,58), acute and convalescent sera from foodborne viral disease outbreaks can also be tested to investigate the immunogenic relationships between virus groups. This information will be crucial in evaluating the importance of key virus strains, in assessing the relationships between these viruses, and in evaluating serodiagnosis as a tool in foodborne disease outbreak investigation. The acquisition of additional sequence information will also enable the improved detection of these viruses in both clinical and food specimens.

Recent strides in molecular biology have facilitated the development of inexpensive and widely available virus detection reagents (56). This will increase the number of facilities that are capable of performing diagnostic testing, which will ultimately facilitate epidemiological investigation. By linking detection and identification in both clinical and food specimens, a more comprehensive approach for the investigation of outbreaks of foodborne viral disease may be undertaken.

In cases where both clinical and food specimens are available, the ability to detect viruses in both types of samples will help in the establishment of strong causal associations. Using nucleic acid sequencing techniques, RT-PCR amplicons from both food and clinical samples may be sequenced and compared in an effort to link specific food articles to disease using a molecular epidemiological approach. This may be particularly effective in tracking the transmission of these viruses within and between communities by linking clusters of cases and identifying sources and reservoirs of infection.

V. CONCLUSIONS

Research is needed to develop and refine the prototype virus detection protocols into collaboratively tested methods that can be used routinely and expeditiously to evaluate the microbiological safety of foods (63). For this to be achieved, several developmental efforts will be necessary, most specifically: (a) simple, rapid, and cost-effective methods to extract viruses from foods; (b) simple and reliable methods for the removal of enzyme inhibitors; (c) methods that are applicable to multiple food products and not restricted to a single virus; (d) quantitative approaches for evaluating relative levels of viral contamination; and (e) technology transfer. Furthermore, additional experimentation is needed to establish the relationship between detection using molecular amplification techniques and the presence of infective virus particles in foods. With sufficient developmental effort, effective methods for the rapid detection of human enteric viruses in foods at naturally occurring levels of contamination may be available in the near future. When taken together, this work will significantly increase our understanding of these agents and facilitate the implementation of effective strategies to prevent the transmission of viruses through foods. Ultimately, this will provide the information necessary to improve the safety of food products, protect public health, and minimize financial losses due to viral contamination of foods.

ACKNOWLEDGMENTS

This manuscript is Paper No. FSR 98-24 in the Journal Series of the Department of Food Science, North Carolina State University, Raleigh, NC 27695-7624. The use of trade names in this publication does not imply endorsement by the North Carolina Agricultural Research Service or criticism of similar products not mentioned. Thanks to the many students and technicians who have contributed to the development of methods to detect human enteric viruses in foods, most specifically Alissa Dix, Paris Leggitt, Soraya Rosenfield, Danielle Robins Datz, and Nancy Stanley.

ABBREVIATIONS

AGMK   African green monkey kidney
BGMK   Buffalo green monkey kidney
CDC   Centers for Disease Control and Prevention (Atlanta, GA)
CPE   cytopathic effect
CTAB   cetyltrimethylammonium bromide
DNA   deoxyribonucleic acid
EIA   enzyme immunoassay
HAV   hepatitis A virus
HEV   hepatitis E virus
IEM   immune electron microscopy
LCR   ligase chain reaction
MPN   most probable number
NASBA   nucleic acid–based sequence amplification
PCR   polymerase chain reaction
RNA   ribonucleic acid
PFU   plaque-forming unit
RT-PCR   reverse transcriptase PCR
RIA   radioimmunoassay

RIFA   radioimmunofocus assay
SRSV   small round structured virus
$TCID_{50}$ tissue culture infectious dose for 50%

REFERENCES

1. NH Bean, JS Goulding, MT Daniels, FJ Angulo. Surveillance for foodborne disease outbreaks: United States, 1988-1992. J Food Protect 60:1265–1286, 1997.
2. NH Bean, PM Griffin. Foodborne disease outbreaks in the United States, 1973-1987: pathogens, vehicles, and trends. J Food Protect 53:804–817, 1990.
3. CW Hedberg, MT Osterholm. Outbreaks of food-borne and waterborne viral gastroenteritis. Clin Microbiol Rev 6:199–210, 1993.
4. Council for Agricultural Science and Technology (CAST). Foodborne Pathogens: Risks and Consequences. Washington, DC: Library of Congress, 1994.
5. JE Kaplan, R Feldman, DS Campbell, C Lookabaugh, W Gary. The frequency of a Norwalk-like pattern of illness in outbreaks of acute gastroenteritis. Am J Pub Health 72:1329–1332, 1982.
6. JE Kaplan, GW Gary, RC Baron, N Singh, LB Schonberger, R Feldman, HB Greenberg. Epidemiology of Norwalk gastroenteritis and the role of Norwalk virus in outbreaks of acute nonbacterial gastroenteritis. Ann Intern Med 96:756–761, 1982.
7. JS Slade, BJ Ford. Discharge in the environment of viruses in wastewater, sludges, and aerosols. In: G. Berg, ed. Viral Pollution of the Environment. Boca Raton, FL: CRC Press, 1983, pp 3–18.
8. DO Cliver. Virus transmission via food. Food Technol 51:71–78, 1997.
9. DO Cliver, RD Ellender, GS Fout, PA Shields, MD Sobsey. Foodborne Viruses. In: C Vanderzant and DF Splittstoesser, eds. Compendium of Methods for the Microbiological Examination of Foods. Washington, DC: American Public Health Association, 1992, pp 763–787.
10. JCA Desenclos, KC Klontz, MH Wilder, OV Nainan, HS Margolis, RA Gunn. A multistate outbreak of hepatitis A caused by the consumption of raw oysters. Am J Publ Health 81:1268–1272, 1991.
11. Centers for Disease Control. Food-borne surveillance data for all pathogens in fish/shellfish for years 1973–1987. Atlanta, GA: US Public Health Service, US Department of Health and Human Services, 1989.
12. GP Richards. Outbreaks of shellfish-associated enteric virus illness in the United States: requisite for development of viral guidelines. J Food Protect 48:815–823, 1985.
13. Centers for Disease Control. Foodborne and waterborne disease outbreaks annual summary 1976. HEW Publication No. CDC 78-8185. Washington, DC: US Department of Health, Education and Welfare, 1978.
14. CP Gerba, SM Goyal. Detection and occurrence of enteric viruses in shellfish: a review. J Food Protect 41:743–754, 1978.
15. DA Wait, CR Hackney, RJ Carrick, G Lovelace, MD Sobsey. Enteric bacterial and viral pathogens and indicator bacteria in hard shell clams. J Food Protect 46:493–496, 1983.
16. LS Rosenblum, IR Mirkin, DT Allen, S Safford, SC Hadler. A multifocal outbreak of hepatitis A traced to commercially distributed lettuce. Am J Publ Health 80:1075–1079, 1990.
17. DY Graham, X Jiang, T Tanaka, AR Opekun, HP Madore, MK Estes. Norwalk virus infection of volunteers: new insights based on improved assays. J Infect Dis 170:34–43, 1994.
18. PHLS Virology Committee, Viral Gastroenteritis Sub-Committee. Outbreaks of gastroenteritis associated with SRSVs. PHLS Microbiol Dig 10:2–8, 1993.
19. RL Frankhauser, JS Noel, SS Monroe, T Ando, RI Glass. Molecular epidemiology of small round structured viruses in outbreaks of gastroenteritis in the United States. Abstracts of the Annual Meeting of the American Society for Microbiology, Atlanta, 1998.

20. RV Tauxe. Emerging foodborne diseases: an evolving public health challenge. Emerg Infect Dis 3:425–434, 1997.

21. X Jiang, M Wang, K Wang, MK Estes. Sequence and genomic organization of Norwalk virus. Virology 195:51–61, 1993.

22. X Jiang, DY Graham, K Wang, MK Estes. Norwalk virus genome cloning and characterization. Science 250:1580–1583, 1990.

23. PR Lambden, EO Caul, CR Ashley, IN Clarke. Sequence and genome organization of a human small round-structured (Norwalk-like) virus. Science 259:516–519, 1993.

24. J Wang, X Jiang, HP Madore, J Gray, U Desselberger, T Ando, Y Seto, I Oishi, JF Lew, KY Green, MK Estes. Sequence diversity of small, round-structured viruses in the Norwalk virus group. J Virol 68:5982–5990, 1994.

25. J Green, CI Gallimore, JP Norcott, D Lewis, DWG Brown. Broadly reactive reverse transcriptase polymerase chain reaction for the diagnosis of SRSV-associated gastroenteritis. J Med Virol 47:392–398, 1995.

26. CL Moe, J Gentsch, T Ando, G Grohmann, SS Monroe, X Jiang, J Wang, MK Estes, Y Seto, C Humphrey, S Stine, RI Glass. Application of PCR to detect Norwalk virus in fecal specimens from outbreaks of gastroenteritis. J Clin Microbiol 32:642–648, 1994.

27. H Appleton, MS Pereira. A possible virus aetiology in outbreaks of food-poisoning from cockles. Lancet 2:780–781, 1977.

28. H Appleton. Small round viruses: classification and role in food-borne infections. In: G Bock and J Whelan, eds. Novel Diarrhoea Viruses. Chichester, UK: John Wiley and Sons, 1987, pp 108–125.

29. EO Caul. Discussion. Small round viruses: classification and role in food-borne infections. In: G Bock and J Whelan, eds. Novel Diarrhoea Viruses. Chichester, UK: John Wiley and Sons, 1987, pp 120–125.

30. H Appleton, PG Higgins. Viruses and gastroenteritis in infants. Lancet 1:1297, 1975.

31. EO Caul. Discussion. Astroviruses: human and animal. In: G Bock and J Whelan, eds. Novel Diarrhoea Viruses. Chichester, UK: John Wiley and Sons, 1987, pp 102–107.

32. GR Reyes, MA Purdy, JP Kim. Isolation of a cDNA from the virus responsible for enterically transmitted non-A, non-B hepatitis. Science 247:1335–1339, 1990.

33. DW Bradley, A Andjaparidze, EH Cook. Etiologic agent of enterically transmitted non-A, non-B hepatitis. J Gen Virol 69:731–738, 1988.

34. ZY Xu, ZH Li, JX Wang, ZP Xiao, DH Dong. Ecology and prevention of a shellfish-associated hepatitis A epidemic in Shanghai, China. Vaccine Suppl 1:S67–S68, 1992.

35. ML Halliday, LY Kanf, TK Zhou, MD Hu, QC Pan, TY Fu, YS Huang, SL Hu. A nepidemic of hepatitis A attributable to the ingestion of raw clams in Shanghai, China. J Infect Dis 164:852–859, 1991.

36. Centers for Disease Control and Prevention. Viral gastroenteritis associated with eating oysters—Louisiana, December 1996–January 1997. MMWR 46:1109–1112, 1997.

37. Centers for Disease Control and Prevention. Viral gastroenteritis associated with consumption of raw oysters–Florida, 1993. MMWR 43:446–449, 1994.

38. MA Kohn, TA Farley, T Ando, M Curtis, SA Wilson, Q Jin, SS Monroe, RC Baron, LM McFarland, RI Glass. An outbreak of Norwalk virus gastroenteritis associated with eating raw oysters: Implications for maintaining safe oyster beds. JAMA 273:466–471, 1995.

39. CP Gerba. Viral disease transmission by seafoods. Food Technol 42:99–103, 1988.

40. LA Jaykus, MT Hemard, MD Sobsey. Human enteric pathogenic viruses. In: CR Hackney and MD Pierson, eds. Environmental Indicators and Shellfish Safety. New York: Chapman and Hall, 1994, pp 92–153.

41. PE Kilgore, ED Belay, DM Hamlin, JS Noel, CD Humphrey, HE Gary Jr, T Ando, SS Monroe, PE Kludt, DS Rosenthal, J Freeman, RI Glass. A university outbreak of gastroenteritis due to a small round-structured virus: application of molecular diagnostics to identify the etiological agent and patterns of transmission. J Infect Dis 173:787–793, 1996.

42. AC Weltman, NM Bennett, JH Misage, JJ Campana, LS Fine, AS Doniger, GJ Balzano, DA Ackman, GS Birkhead. Multi-county outbreak of hepatitis A, New York state. Am J Epidemiol 139:S80, 1994.

43. ARE Warburton, TG Wreghitt, A Pamling, R Buttery, KN Ward, KR Perry, JV Parry. Hepatitis A outbreak involving bread. Epidemiol Infect 106:199–202, 1991.

44. EP Larkin. Viruses in wastewater sludges and in effluents used for irrigation. Environ Int 7:29, 1982.

45. RD Warner, RD, RW Carr, FK McClesky, PC Johnson, LMG Elmer, VE Davidson. A large nontypical outbreak of Norwalk virus: gastroenteritis associated with exposing celery to nonpotable water and with Citrobacter freundii. Arch Intern Med 151:2419–2424, 1991.

46. MT Niu, LB Polish, BH Robertson, BK Khanna, BA Woodruff, CN Shapiro, MA Miller, DJ Smith, JK Gedrose, MJ Alter, HS Margolis. Multistate outbreak of hepatitis A associated with frozen strawberries. J Infect Dis 166:518–524, 1992.

47. LS Rosenblum, IR Mirkin, DT Allen, S Safford, SC Hadler. A multifocal outbreak of hepatitis A traced to commercially distributed lettuce. Am J Pub Health 80:1075–1087, 1990.

48. Centers for Disease Control and Prevention. Hepatitis A associated with consumption of frozen strawberries—Michigan, March, 1997. MMWR 46:288–295, 1998.

49. AS Bradawy, CP Gerba, LM Kelley. Survival of rotavirus SA-11 on vegetables. Food Microbiol 2:199, 1985.

50. RO Cannon, JR Poliner, RB Hirschhorn, DC Rodeheaver, PS Silverman, EA Brown, GH Talbot, SE Stine, SS Monroe, DT Dennis, RI Glass. A multistate outbreak of Norwalk virus gastroenteritis associated with consumption of commercial ice. J Infect Dis 164:860–863, 1991.

51. L Jaykus. Epidemiology and detection as options for the control of viral and parasitic foodborne disease. Emerg Infect Dis 3:529–539, 1997.

52. T Ando, Q Jin, JR Gentsch, SS Monroe, JS Noel, SF Dowell, HG Cicirello, MA Kohn, RI Glass. Epidemiologic applications of novel molecular methods to detect and differentiate small round structured viruses (Norwalk-like viruses). J Med Virol 47:145–152, 1995.

53. JS Noel, BL Liu, CD Humphrey, EM Rodrigues, PR Lambden, IN Clarke, DM Dawyer, T Ando, RI Glass, SS Monroe. Parkville virus: a novel genetic variant of human calicivirus in the Sapporo virus clade, associated with an outbreak of gastroenteritis in adults. J Med Virol 52:173–178, 1997.

54. T Ando, SS Monroe, JR Gentsch. Detection and differentiation of antigenetically distinct small round-structured viruses (Norwalk-like viruses) by reverse transcription-PCR and Southern hybridization. J Clin Microbiol 33:64–71, 1995.

55. J Lew, AZ Kapikian, J Valdesuso, KY Green. Molecular characterization of Hawaii virus and other Norwalk-like viruses: evidence for genetic polymorphism among human caliciviruses. J Infect Dis 170:535–542, 1994.

56. AJ Kapikian, MK Estes, RM Chanock. Norwalk group of viruses. In: BN Fields, DM Knipe, PM Howley, eds. Fields Virology. Philadelphia: Lippencott-Raven, 1996, pp 783–810.

57. JJ Gray, X Jiang, P Morgan-Capner, U Desselberger, MK Estes. The prevalence of antibody to Norwalk virus in England. Detection by ELISA with baculovirus-expressed Norwalk capsid antigen. J Clin Microbiol 31:1022–1025, 1993.

58. JJ Treanor, X Jiang, HP Madore, MK Estes. Subclass specific serum antibody response to recombinant Norwalk capsid antigen (rNV) in adults infected with Norwalk, Snow Mountain, or Hawaii viruses. J Clin Microbiol 31:1630–1634, 1993.

59. FB Hollinger, JR Ticehurst. Hepatitis A virus. In: BN Fields, DM Knipe, and PM Howley, eds. Fields Virology. Philadelphia: Lippincott-Raven, 1996, pp 735–782.

60. JL Melnick. Enteroviruses: Polioviruses, coxsackieviruses, echoviruses, and newer enteroviruses. In: BN Fields, DM Knipe, and PM Howley, eds. Fields Virology. Philadelphia: Lippincott-Raven, 1996, pp 655–712.

61. RL Atmar, FH Neill, CM Woodley, R Manger, GS Fout, W Burkhardt, L Leja, ER McGov-

ern, F LeGuyader, TG Metcalk, MK Estes. Collaborative evaluation of a method for the detection of Norwalk virus in shellfish tissues by PCR. Appl Environ Microbiol 62:254–258, 1996.

62. RL Atmar, HF Neill, JL Romalde, F LeGuyader, CM Woodley, TG Metcalf, MK Estes. Detection of Norwalk virus and hepatitis A virus in shellfish tissues with the PCR. Appl Environ Microbiol 61:3014–3018, 1995.

63. R DeLeon, L Jaykus. Detection of bacteria and viruses in shellfish. In: CJ Hurst, GR Knudsen, MJ McInerney, LD Stenzenbach, MV Walter eds. Manual of Environmental Microbiology. Washington, DC: American Society for Microbiology, 1996.

64. MD Sobsey. Methods for recovering viruses from shellfish, seawater, and sediments. In: G Berg, ed. Methods for Recovering Viruses from the Environment. Boca Raton, FL: CRC Press, 1987.

65. MD Sobsey, RJ Carrick, HR Jensen. Improved methods for detecting enteric viruses in oysters. Appl Environ Microbiol 36:212–128, 1978.

66. N Guttman-Bass. Cell cultures and other host systems for detecting and quantifying viruses in the environment. In: G Berg, ed. Methods for Recovering Viruses from the Environment. Boca Raton: CRC Press, 1987 pp 195–228.

67. L Jaykus, R DeLeon, MD Sobsey. A virion concentration method for detection of human enteric viruses in oysters by PCR and oligoprobe hybridization. Appl Environ Microbiol 62:2074–2080, 1996.

68. L Jaykus, R DeLeon, M Sobsey. 1995. Development of a rapid molecular method for the detection of human enteric viruses in oysters. J Food Protect 58:1357–1362, 1995.

69. RL Armar, TG Metcalf, HF Neill, MK Estes. Detection of enteric viruses in oysters by using the polymerase chain reaction. Appl Environ Microbiol 59:631–635, 1993.

70. EM Smith, CP Gerba. Laboratory methods for the growth and detection of animal viruses. In: CP Gerba and SM Goyal, eds. Methods in Environmental Virology. New York: Marcel Dekker, 1982.

71. SE Oglesbee, DA Wait, AF Meinhold. MA-104: A continuous cell line for the isolation of human enteric viruses from environmental samples. Abstracts of the Annual Meeting of the American Society for Microbiology 81:219, 1981.

72. NJ Schmidt, HH Ho, JL Riggs, RH Lennette. 1978. Comparative sensitivity of various cell culture systems for isolation of viruses from wastewater and fecal samples. Appl Environ Microbiol 36:480–486, 1978.

73. DN Ridinger, RS Spendlove, BB Barnett, DB George, JC Roth. Evaluation of cell lines and immunofluorescence and plaque assay procedures for quantifying reoviruses in sewage. Appl Environ Microbiol 43:740–746, 1982.

74. T Cromeans, MD Sobsey, HA Fields. Development of a plaque assay for a cytopathic, rapidly replicating isolate of hepatitis A. J Med Virol 22:45–56, 1987.

75. H Chung, L Jaykus, MD Sobsey. Improved detection of human enteric viruses in field oyster specimens by in vivo and in vitro amplification of nucleic acids. Appl Environ Microbiol 62:3772–3778, 1996.

76. YSC Shieh, RS Baric, MD Sobsey, J Ticehurst., TA Miele, R DeLeon, R Walter. Detection of hepatitis A virus and other enteroviruses in water by ssRNA probes. J Virol Meth 31:119–136, 1991.

77. KF Fannin, SH Abid, JJ Bertucci, JM Reed, SC Vanna, C Lue-Hing. Significance of reporting infectious viral or plaque forming unit concentrations from environmental samples. Abstracts of the Annual Meeting of the American Society for Microbiology 78:200, 1978.

78. LYC Leong, SJ Barret, RR Trussell. False positives in testing of secondary sewage for enteric viruses. Abstracts of the Annual Meeting of the American Society for Microbiology 78:200, 1978.

79. SM Lemon, LN Binn, RH Marchwicki. Radioimmunofocus assay for quantitation of hepatitis A virus in cell cultures. J Clin Microbiol 17:834–839, 1983.

80. M Deng, DO Cliver. A broad-spectrum enzyme-linked immunosorbant assay for the detection of human enteric viruses. J Virol Meth 8:87–98, 1984.

81. PC Loh, MA Dow, RS Fujioka. Use of the nitrocellulose-enzyme immunosorbent assay for rapid, sensitive, and quantitative detection of human enteroviruses. J Virol Meth 12:225–234, 1985.

82. EM Heun, RL Vogt, PJ Hudson, SS Parren, GW Gary. Risk factors for secondary transmission in households after a common-source outbreak of Norwalk gastroentertitis. Am J Epidemiol 126:1181–1186, 1987.

83. D Lewis, T Ando, CK Humphrey, SS Monroe, RI Glass. Use of solid phase immune electron microscopy for classification of Norwalk-like viruses into six antigenic groups from 10 outbreaks of gastroenteritis in the United States. J Clin Microbiol 33:501–504, 1995.

84. DC Lewis. Three serotypes of Norwalk-like virus demonstrated by solid-phase immune electron microscopy. J Med Virol 30:77–81, 1990.

85. SP Parker, WD Cubitt, X Jiang, MK Estes. Efficacy of a recombinant Norwalk virus protein enzyme immunoassay for the diagnosis of infections with Norwalk virus and other human "candidate" caliciviruses. J Med Virol 41:179–184, 1993.

86. DJ Dimitrov, DY Graham, MK Estes. Detection of rotavirus by nucleic acid hybridization with cloned DNA of simian rotavirus SA-11 genes. J Infect Dis 152:293–300, 1985.

87. X Jiang, MK Estes, TG Metcalf, JL Melnick. Detection of hepatitis A virus in seeded estuarine samples by hybridization with cDNA probes. Appl Environ Microbiol 52:711–717, 1986.

88. R DeLeon. Use of gene probes and an amplification method for the detection of rotaviruses in water. PhD dissertation, University of Arizona, Tuscon, 1989.

89. AB Margolin, MJ Hewlett, CP Gerba. Use of a cDNA dot-blot hybridization technique for detection of enteroviruses in water. Proceedings of the Water Quality Technology Conference, Denver, CO, 1986, pp 87–95.

90. X Jiang, MK Estes, TG Metcalf. Detection of hepatitis A virus by hybridization with single-stranded RNA probes. Appl Environ Microbiol 53:2487–2495, 1987.

91. AB Margolin, KJ Richardson, R DeLeon, CP Gerba. Application of gene probes to the detection of enteroviruses in water. In: RA Larson, ed. Biohazards of Drinking Water Treatment. Chelsea, MI: Lewis Publishers, 1989.

92. Y Zhou, MK Estes, X Jiang, TG Metcalf. Concentration and detection of hepatitis A virus and rotavirus from shellfish by hybridization tests. Appl Environ Microbiol 57:2963–2968, 1991.

93. AB Dix. Development of methods to extract human enteric viruses from hard-shelled clams for detection by reverse transcriptase–polymerase chain reaction (RT-PCR) and oligoprobe hybridization (OP). MS thesis, North Carolina State University, Raleigh, 1997.

94. BB Goswami, WH Koch, TA Cebula. Detection of hepatitis A in *Mercenaria mercenaria* by coupled reverse transcription and polymerase chain reaction. Appl Environ Microbiol 59:2765–2770, 1993.

95. V Gouvea, N Santos, M Carmo-Timenetsky, MK Estes. Identification of Norwalk virus in artificially seeded shellfish and selected foods. J Virol Meth 48:177–187, 1994.

96. D Hafliger, M Gilen, J Luthy, PH Hubner. Seminested RT-PCR systems for small round structured viruses and detection of enteric viruses in seafood. Int J Food Microbiol 37:27–36, 1997.

97. DN Lees, K Henshilwood, J Green, CI Gallimore, DWG Brown. Detection of small round structured viruses in shellfish by reverse transcription–PCR. Appl Environ Microbiol 61:4418–4424, 1995.

98. DN Lees, K Henshilwood, WJ Dore. Development of a method for detection of enteroviruses in shellfish by PCR with poliovirus as a model. Appl Environ Microbiol 60:2999–3005, 1994.

99. PR Leggitt. Development of methods to purify and concentrate human enteric viruses from

representative food commodities for detection by reverse transcription–polymerase chain reaction (RT-PCR) and oligoprobe hybridization (OP). MS thesis, North Carolina State University, Raleigh, 1998.

100. F LeGuyader, FH O'Neill, MK Estes, SS Monroe, T Ando, RL. Detection and analysis of a small round-structured virus strain in oysters implicated in an outbreak of acute gastroenteritis. Appl Environ Microbiol 62:4268–4272, 1996.

101. F LeGuyader, E Dubois, D Menard, M Pommepuy. Detection of hepatitis A virus, rotavirus, and enterovirus in naturally contaminated shellfish and sediment by reverse transcription–seminested PCR. Appl Environ Microbiol 60:3665–3671, 1994.

102. T Cromeans, OV Nainan, HS Margolis. Detection of hepatitis A virus RNA in oyster meat. Appl Environ Microbiol 63:2460–2463, 1997.

103. MY Deng, SP Day, DO Cliver. Detection of hepatitis A virus in environmental samples by antigen-capture PCR. Appl Environ Microbiol 60:1927–1933, 1994.

104. EI Lopez-Sabater, MY Deng, DO Cliver. Magnetic immunoseparation PCR assay (MIPA) for detection of hepatitis A virus (HAV) in American oyster (Crassostrea virginica). Lett Appl Microbiol 24:101–104, 1997.

105. AB Dix, L Jaykus. A virion concentration method for the detection of human enteric viruses in extracts of hard-shelled clams. J Food Protect 61:458–465, 1998.

106. GD Lewis, TG Metcalf. Polyethylene glycol precipitation for recovery of pathogenic viruses, including hepatitis A virus and human rotavirus, from oyster, water, and sediment samples. Appl Environ Microbiol 54:1983–1987, 1988.

107. N Santos, V Gouvea. Improved method to recover viral RNA from fecal specimens for rotavirus detection. J Virol Meth 46:11–21, 1994.

108. RW Jansen, G Siegl, SM Lemon. Molecular epidemiology of human hepatitis A virus defined by antigen-capture polymerase chain reaction. Proc Natl Acad Sci USA 87:2867–2871,1990.

109. KJ Schwab, R DeLeon, MD Sobsey. Immunoaffinity concentration and purification of waterborne enteric viruses for detection by reverse transcriptase PCR. Appl Environ Microbiol 62:2086–2094, 1996.

110. R DeLeon, SM Matsui, RS Baric, JE Herrmann, NR Blacklow, HB Greenburg, MD Sobsey. Detection Norwalk virus in stool specimens by reverse transcriptase–polymerase chain reaction and non-radioactive oligoprobes. J Clin Microbiol 30:3151–3157, 1992.

111. L Rossen, P Norskov, K Holmstrom, OF Rasmussen. Inhibition of PCR by components of food samples, microbial diagnostic assays and DNA-extraction solutions. Int J Food Microbiol 17:37–45, 1992.

112. SF Dowell, C Goves, KB Kirkland, HC Cicirello, T Ando, Q Jin, JR Gentsch, SS Monroe, CD Humphrey, C Slemp, DM Dwyer, RA Meriwether, RI Glass. A multistate outbreak of oyster-associated gastroenteritis: implications for interstate tracing of contaminated shellfish. J Infect Dis 172:1497–1503, 1995.

113. SS Limsewat, SS Ohgaki. Fate of liberated viral RNA in wastewater determined by PCR. Appl Environ Microbiol 63:2932–2933, 1997.

114. YL Tsai, B Tran, CJ Palmer. Analysis of viral RNA persistence in seawater by reverse transciptase–PCR. Appl Environ Microbiol 61:363–366, 1995.

115. JS Chamberlain, JR Chamberlain. Optimization of multiplex PCRs. In: KB Mullis, F Ferre, RA Gibbs, eds. The Polymerase Chain Reaction. Boston: Birkhauser, 1994, pp 38–46.

116. YL Tsai, B Tran, LR Sangermano, CJ Palmer. Detection of poliovirus, hepatitis A virus, and rotavirus from sewage and ocean water by triplex reverse transcriptase PCR. Appl Environ Microbiol 60:2400–2407, 1994.

117. SI Rosenfield. Development of reverse transcription polymerase chain reaction (RT-PCR) approaches to the detection of bacterial and viral foodborne disease agents. MS thesis, North Carolina State University, Raleigh, 1998.

118. R Jackson, DJ Morris, RJ Cooper, AS Bailey, PE Klapper, GM Cleator, AB Tullo. Multiplex polymerase chain reaction for adenovirus and herpes simplex virus in eye swabs. J Virol Meth 56:41–48, 1996.

119. LM McElhinney, RJ Cooper, DJ Morris. Multiplex polymerase chain reaction for human herpesvirus-6, human cytomegalovirus, and human B-globin DNA. J Virol Meth 53:223–233, 1995.

120. R Sooknanan, B van Gemen, LT Malek. Nucleic acid sequence based amplification (NASBA). In: DL Wiedbrauk and DH Farkas, eds. Molecular Methods for Virus Detection. Orlando, FL: Academic Press, 1995, pp 261–284.

121. L Andreoletti, D Hober, SS Belaich, PE Lobart, A Dewilde, P Wattre. Rapid detection of enteroviruses in clinical specimens using PCR and microwell capture hybridization assay. J Virol Meth 62:1–10, 1996.

122. SS Tyagi, FR Kramer. Molecular beacons: probes that fluoresce upon hybridization. Nature Biotechnol 14:303–308, 1996.

123. PK Witham, CT Yamashiro, KJ Livak, CA Batt. A PCR-based assay for the detection of Escherichia coli shiga-like toxin genes in ground beef. Appl Environ Microbiol 62:1347–1353, 1996.

124. F Barany. The ligase chain reaction in a PCR world. PCR Meth Applic 1:5–16, 1991.

125. M Pfeffer, H Meyer, M Weidmann. A ligase chain reaction targeting two adjacent nucleotides allows the differentiation of cowpox virus from other Orthopoxvirus species. J Virol Meth 49:353–360, 1994.

126. GG Stone, RD Oberst, MP Hays, ss McVey, MM Chengappa. Combined PCR-oligonucleotide ligation assay for rapid detection of Salmonella serovars. J Clin Microbiol 33:2888–2893, 1995.

127. WD Cubitt, K Jiang, J Wang, MK Estes. Sequence similarity of human caliciviruses and small round structured viruses. J Med Virol 43:252–258, 1994.

128. X Jiang, WD Cubitt, T Berke, W Zhong, X Dai, S Nakata, LK Pickering, DO Matson. Sapporo-like human caliciviruses are genetically and antigenically diverse. Arch Virol 142:1813–1827, 1997.

129. BL Liu, IN Clarke, EO Caul, PR Lambden. Human enteric caliciviruses have a unique genome structure and are distinct from the Norwalk-like viruses. Arch Virol 140:1345–1356, 1995.

130. DO Matson, WM Zhong, SS Nakata, K Numata, X Jiang, LK Pickering, SS Chiba, MK Estes. Molecular characterization of a human calicivirus with sequence relationships closer to animal caliciviruses. J Med Virol 45:215–222, 1995.

131. WD Cubitt, K Jiang, MK Estes. Sequence similarity of human caliciviruses and small round structured viruses. J Med Virol 43:252–258, 1994.

132. AS Kahn, CL Moe, RI Glass, SS. Monroe, MK Estes, LE Chapman, X Jiang, C Humphrey, E Pon, JK Islander, LB Schonberger. Norwalk virus-associated gastroenteritis traced to ice consumption aboard a cruise ship in Hawaii: comparison and application of molecular-based assays. J Clin Microbiol 32:318–322, 1994.

133. F LeGuyader, MK Estes, ME Hardy, FH Neill, J Green, DWG Brown. Evaluation of a degenerate primer for the PCR detection of human caliciviruses. Arch Virol 141:2225–2235, 1996.

134. Centers for Disease Control and Prevention. Viral gastroenteritis associated with consumption of raw oysters—Florida, 1993. MMWR 43:446–449, 1994.

135. Centers for Disease Control and Prevention. State outbreak of viral gastroenteritis related to consumption of oysters—Louisiana, Maryland, Mississippi, and North Carolina, 1993. MMWR 42:945–948, 1993.

136. Centers for Disease Control and Prevention. Multistate outbreak of viral gastroenteritis associated with consumption of oysters—Apalachicola Bay, Florida, December 1994–January 1995. MMWR 44:37–39, 1995.

137. X Jiang, J Wang, MK Estes. Characterization of SRSVs using RT-PCR and a new antigen ELISA. Arch Virol 140:363–374, 1995.

138. J Green, CI Gallimore, JP Norcott, D Lewis, DWG Brown. Broadly reactive reverse transcriptase polymerase chain reaction for the diagnosis of SRSV-associated gastroenteritis. J Med Virol 47:392–398, 1995.

7

Medical Management of Foodborne Viral Gastroenteritis and Hepatitis

Suzanne M. Matsui and Ramsey C. Cheung
Stanford University School of Medicine, Stanford, and VA Palo Alto Health Care System, Palo Alto, California

I. INTRODUCTION

Food and water are the means through which numerous viruses causing gastroenteritis and hepatitis are transmitted. Norwalk virus, a human calicivirus and major cause of epidemic gastroenteritis, and hepatitis A virus (HAV), a picornavirus and well-recognized cause of "infectious" hepatitis, are the most common of these agents. More recently, other viruses, including other strains of human caliciviruses, astroviruses, and rotaviruses, and a unique, non-A, non-B, enterically transmitted hepatitis virus (hepatitis E virus or HEV), have been identified in outbreaks of food- and waterborne illness.

Between 1988 and 1992, viruses accounted for 4% of foodborne disease outbreaks and 6% of the cases among those outbreaks reported to the U.S. Centers for Disease Control and Prevention (CDC) in which a bacterial, parasitic, chemical, or viral cause could be established (1). These figures do not include outbreaks that occurred on cruise ships, those in which the food was eaten outside of the United States, or if the route of transmission between the contaminated food and the infected individual was deemed indirect. Of the 59% of reported outbreaks in which the etiology was not determined, it was estimated that one-third or more were likely caused by viruses that eluded detection (2).

Primary contamination occurs when such foods as bivalve mollusks are harvested from waters contaminated with sewage (3). Oysters, clams, cockles, and mussels efficiently concentrate viral and bacterial pathogens from the water in which they filter-feed (4). Depuration of shellfish in rapidly circulating, UV–irradiated, salt water for 48–72 h effectively diminishes bacterial contamination but does not ensure that the shellfish are also equally free of viruses (5). Though viruses do not appear to replicate in oysters and other molluscan shellfish, mollusks concentrate large volumes of water, so that eating a small amount of contaminated shellfish may cause disease. Oysters, in particular, are frequently consumed raw or undercooked and have been implicated as the source of outbreaks of gastroenteritis and/or hepatitis (6). Food may also be contaminated during preparation or storage (7). Prevention of such secondary contamination requires adherence to good hygienic practices, proper attention to handwashing, disinfecting contaminated surfaces, and separation of potentially contaminated food from uncontaminated food.

Changes in dietary practices and preferences also play a role in foodborne illness. As an example, greater awareness of the cardiovascular health benefits of a diet rich in fresh fruits and vegetables has led to greater demand for domestic and imported produce and, consequently, exposure to a wider array of enteric pathogens (8). The popular trend of dining outside the home more frequently also increases exposure to enteric pathogens that may be transmitted through food handled and prepared in commercial kitchens.

This chapter focuses on the management of gastrointestinal (GI) and hepatic illnesses transmitted through food and water. We consider both public health issues as well as the treatment of affected individuals.

II. VIRAL GASTROENTERITIS

Acute GI illness is a commonly encountered problem throughout the world. Each year, an estimated 3–5 billion cases of diarrheal illness occur globally, with a large proportion of cases affecting children under the age of 5 in developing countries (9,10). Severe illness and a significant mortality rate are associated with acute gastroenteritis in these countries. In developed countries, where acute diarrhea is also a frequently encountered symptom, morbidity due to acute gastroenteritis is significant when measured in days lost from work or school (11). It is estimated that viral agents account for 30–40% of infectious diarrhea in the United States (12). The diagnosis of a viral etiology is often one of exclusion since detection techniques—culture, serology, microscopy, and molecular probes—have not been extensively developed for many enteric viruses (see Chapter 6). Foodborne viral pathogens of the human gastrointestinal tract include Norwalk and related small round structured viruses (SRSVs, or Norwalk-like viruses) that share similar morphology, protein structure, and nucleic acid composition, as well as other human caliciviruses, astroviruses, and rotaviruses (see Chapters 4 and 5).

A. Norwalk Virus, Norwalk-like Viruses, and Caliciviruses

The molecular characterization of Norwalk virus, combined with detailed ultrastructural analysis and previously obtained biochemical data, has allowed its classification in the

family Caliciviridae (13–15). Although the classically recognized human calicivirus particles have a more distinctive surface morphology than Norwalk virus and related small
round structured viruses (SRSVs) by direct electron microscopy, they share the features
of a polyadenylated, plus-stranded RNA genome enclosed in a capsid composed of a single
structural protein of approximately 60 kDa. When compared by cryoelectron microscopy,
all of these caliciviruses have the characteristic structural feature of arch-like capsomeres
(16,17). The immunological relationships between Norwalk virus, SRSVs (e.g., Snow
Mountain Agent, Hawaii Agent, and Otofuke Agent), and caliciviruses appear to be quite
intricate when evaluated with human infection sera (18,19). In more recent analyses that
complement these initial immunological investigations, the genetic relationships between
Norwalk virus and other SRSVs have been studied and indicate that these viruses fall into
one of two major genogroups (20). Norwalk virus is the prototype of genogroup I, which
also includes Southampton and Desert Shield viruses. Genogroup II has as its prototype
the Snow Mountain virus and includes Mexico (MX), Hawaii, and Toronto viruses. Human
caliciviruses with classic morphology belong to a separate genogroup. Recombinant virus-
like particles (VLPs) of Norwalk virus and several other SRSVs that have recently been
produced are an abundant source of reagents of uniform quality that can be used to evaluate
serum antibody response (21).

Norwalk virus has been shown by radioimmunoassay (RIA), enzyme immunoassay
(EIA), and immune electron microscopy (IEM) to be a major cause of acute nonbacterial
gastroenteritis in the United States (20). Large outbreaks of food- and waterborne illness
have been attributed to Norwalk and morphologically similar viruses. Shellfish contaminated with Norwalk and related viruses have most frequently been the source of large
foodborne outbreaks in the United States and abroad (6,22,23). Ill food handlers may also
transmit infection to consumers by preparing salads and other foods that require much
handling and no further cooking (7,24–27). Infected individuals may shed infectious virus
in feces for 48 h or more after resolution of acute GI symptoms (25,26). Finally, contaminated drinking and recreational water have been implicated in the transmission of Norwalk
virus infection (28–35). Transmission by fomites and aerosols has been suggested but is
not well established (36).

Norwalk virus appears to be relatively resistant to chlorine treatment of water and
may require levels of 10 mg/L to be inactivated (37). It has been reported that Norwalk
virus continues to be infectious after exposure to chlorine concentrations of 6.25 mg/L
for 30 min. Of note, the peak chlorine concentration in water in many municipalities is
generally less than 5 mg/L. It is not surprising, then, that outbreaks of gastroenteritis due
to Norwalk virus have occurred in a variety of settings including schools, nursing homes,
hospitals, cruise ships, hotels, catered events, recreational camps, lakes, and pools (7,24–
36,38,39). Norwalk virus has also been found in association with a syndrome resembling
travellers' diarrhea in U.S. troops in Operation Desert Shield (40).

Illness due to Norwalk virus infection may run the full spectrum from vomiting
only, to diarrhea only, to a combination of vomiting and diarrhea with other constitutional
symptoms (41). Transient abnormalities in carbohydrate and fat absorption, as well as
gastric emptying time, have been observed in volunteers administered a bacteria-free,
safety-tested inoculum. Severe dehydration and deaths due to dehydration and electrolyte
imbalance have been reported (42). Curiously, infection does not appear to confer long-
term immunity, although short-term immunity can be demonstrated (43). Furthermore,
human volunteers who developed illness after the initial inoculum was administered were
equally susceptible when challenged with the same inoculum 2–4 years later. The immu-

nological response to SRSVs and the determinants of protection are areas that require more detailed analysis.

The first reports of Norwalk virus genetic sequence (13,14) laid the foundation for sequence-specific detection of Norwalk and related viruses. A large effort has been devoted to determining the optimal oligonucleotide primer sequences that can be used to amplify a wide range of Norwalk-like viruses by the reverse transcriptase polymerase chain reaction (RT-PCR; 20,21). A second diagnostic tool that has been developed recently is the recombinant virus-like particle (VLP) for several SRSVs (21,44–47). These VLPs can be applied to detect serum antibody responses, or can be used to generate antiserum that can then be used in assays to detect SRSV antigens, in an EIA format. The EIAs designed to detect SRSV antigen are extremely specific for either Norwalk virus or MX virus, and are unable to detect human caliciviruses with classic morphology. These recombinant reagents are not commercially available at this time. Direct electron microscopy can be used to detect viral particles in infected feces, if shed in sufficient concentration (more than 10^6 particles per mL). For the Norwalk outbreak, IEM was used successfully to aggregate the small number of viral particles in the specimen and to make the definitive connection between the virus and the outbreak (48). IEM, not available in most diagnostic laboratories, is a labor-intensive procedure that requires an experienced microscopist and the appropriate immunological reagents.

Treatment of individual patients consists of such supportive measures as oral rehydration therapy (49), depending on the severity of symptoms. In one study, bismuth subsalicylate was shown to decrease the amount of abdominal cramping but did not have an effect on the duration of diarrhea (50). From a public health standpoint, action must be directed at adequate chlorination of drinking water, establishment of valid standards for evaluating shellfish safety, identification of infected food-service workers, and avoidance of secondary contamination by maintaining a clean and hygienic food preparation environment. Public awareness of the risks of eating raw or undercooked shellfish, as well as proper handling and preparation of shellfish, is essential. For example, in order to inactivate viruses that may be present in such shellfish as clams, the internal temperature must reach 100°C and requires 4–6 minutes of steaming. The popular practice, however, is to steam clams just until the shells open, a process that can occur in 1 min (51). As a counterpoint, two recent reports (52,53) suggest that cooking oysters (by grilling, stewing, frying, or steaming) may not adequately inactivate contaminating SRSVs. In these epidemiological studies, those who ate only ''thoroughly cooked'' oysters and those who ate raw oysters were equally likely to develop gastroenteritis.

Human caliciviruses with classic morphology are a cause of gastroenteritis in young children, among whom transmission is believed to be person-to-person (42). Outbreaks of gastroenteritis due to calicivirus-contaminated shellfish, cold foods prepared by an infected food handler, and drinking water have been reported (54,55).

B. Astroviruses

Astroviruses are 27–34 nm in diameter. They are nonenveloped viruses with a characteristic, starlike morphology by electron microscopy (56). In endemic infections, astroviruses may account for up to 8.6% of diarrhea in young children (57,58). Outbreaks of gastroenteritis due to astrovirus have occurred in nursing homes, hospitals, schools, and among groups of adults exposed to contaminated water or shellfish (59–62). Large outbreaks, involving thousands of school-age children and adults in Japan, have been attributed to astrovirus (63,64). Whether epidemic or endemic, gastroenteritis due to astrovirus is generally milder than illness caused by group A rotavirus, with symptoms lasting 2–4 days in immunocompetent individuals (36,57). Although astrovirus infections are usually self-limiting, chronic diarrhea and shedding of astrovirus has been noted in certain immunocompromised individuals (65–67). Supportive treatment to prevent dehydration is indicated. The role of immunoprophylaxis has not been evaluated.

C. Rotaviruses

Group A rotaviruses are the single most important cause of severe, dehydrating diarrhea in young children throughout the world (68) (see also Chapter 4). Adults, particularly parents of infected children, and elderly, debilitated patients, also may be at increased risk for developing rotaviral gastroenteritis (69). Food- and waterborne outbreaks are rarely associated with group A rotavirus (70,71).

Although rotaviruses appear to be more sensitive than Norwalk virus to chlorine treatment of water, complete disinfection may be difficult to achieve for a variety of reasons. First, infected individuals shed large quantities of virus in stool, on the order of 10^{10} particles per gram feces. Second, rotaviruses may remain infective in water for several days. Finally, at room temperature, viruses may be infective for up to 7 months (71). It should also be noted that rotavirus infections cause significant disease in both developing and developed countries, without regard to the level of hygiene, suggesting that other lines of defense (e.g., vaccination) are needed to control the development of this illness.

Oral rehydration therapy has been responsible for stemming the morbidity rate due to infectious diarrhea throughout the world (72). These orally administered isoosmotic solutions, composed of sodium chloride, potassium chloride, trisodium citrate, glucose, and water, take advantage of the cotransport of sodium and glucose into intestinal villus cells, which effects the passive absorption of water and other ions (73). Although these solutions do not decrease the duration of the diarrheal illness, patients with mild to moderate dehydration improve clinically as fluids are replaced and electrolyte balance is restored. Patients with severe dehydration ($\geq$10% fluid deficit) initially require intravenous fluid resuscitation.

Early feedings and maintenance of adequate caloric intake are important adjuncts to oral rehydration solutions. It is recommended that breast-fed infants continue to nurse ad libitum, whereas bottle-fed infants should be given formulas that are reduced in, or free of, lactose. Older children should continue to eat a regular diet with emphasis on starches, cereals, yogurt, fruits, and vegetables. Soft drinks, gelatin desserts, and foods high in simple sugars may provide too large an osmotic load and are best avoided during the diarrheal illness (74).

An effective, live, orally administered, tetravalent rotavirus vaccine (RotaShield® by Wyeth Lederle Vaccines) was licensed by the FDA in 1998 for vaccination of infants in the United States (75). However, post-marketing reports to the Vaccine Adverse Event Reporting System and studies performed in California and Minnesota health care systems (76) suggested a strong association between the development of intussusception and vaccination with RotaShield. After consideration of these and other scientific data, the Advisory Committee on Immunization Practices *no longer recommends* vaccination with Rota-Shield, and the manufacturer has voluntarily withdrawn the vaccine from the market (77). Other vaccine formats including the use of rotavirus reassortants that express several different neutralizing epitopes, polyvalent vaccines, and human strains that cause asymptomatic infection (78), as well as strategies based on recombinant VLPs (79) and DNA vaccination (80), are under study.

Human group B rotavirus (adult diarrhea rotavirus, ADRV) has been responsible for several epidemics of gastroenteritis that have involved millions of people in developing areas of China (81). Unlike group A rotaviruses, however, group B rotaviruses tend to infect a larger proportion of adults. ADRV is thought to be transmitted by contaminated drinking water. The illness is self-limiting and characterized by a severe, large-volume, watery (''cholera-like'') diarrhea, vomiting, and dehydration (82). Recently, outbreaks due to human group C rotavirus also have been reported (83). Infections with group C rotavirus are most commonly found in children aged 4–7 years and is characterized by abdominal pain, nausea, vomiting, and fever, with diarrhea being less prevalent (28%) (82,84). The ability to detect infection due to group B and C rotaviruses is restricted to a number of primary research laboratories.

III. VIRAL HEPATITIS

There are currently six well-characterized hepatitis viruses: A, B, C, D, E, and G. Hepatitis viruses B, C, D, and G are transmitted parenterally. Only HAV and HEV are enterically transmitted foodborne pathogens that have been found to cause outbreaks of hepatitis throughout the world (86–91).

A. Hepatitis A Virus

Illness due to HAV, a 27-nm, nonenveloped picornavirus, becomes apparent 15–50 days (average incubation period of 28 days) after ingestion of contaminated food or water. The incubation period may be inversely related to the viral titer of the HAV inoculum. It is often difficult to identify the specific contaminated food or exposure. The symptomatic phase of HAV infection typically presents as abrupt onset of anorexia, nausea, fever, malaise, abdominal pain, dark urine, and jaundice. Diarrhea occurs in about half of all infected children but is uncommon in adults. Infection with HAV is more likely to be symptomatic among older children and adults (88). In developing countries where sanitation and hygiene are poor, children are exposed to HAV at an early age and frequently develop asymptomatic infection. Treatment of patients who develop symptomatic disease is generally conservative, supportive, and includes bed rest as well as maintenance of adequate food and fluid intake. On average, the jaundiced patient will recover fully in approximately 6 weeks, with normalization of serum bilirubin and ALT (alanine aminotransferase). Since infectivity and HAV shedding in stool peak during the 2 weeks preceding the symptomatic phase of infection, it is neither necessary to isolate infected patients nor useful to examine the stool for HAV antigen. HAV rarely (<1% of cases) causes fulminant hepatitis, occurring almost exclusively in individuals with chronic liver disease (92) or those who are over the age of 50 when infected (88).

Diagnosis of HAV infection is based on the detection of IgM antibody to HAV, which is present in more than 99% at the time of presentation. This antibody peaks within the first month following exposure and declines to undetectable levels within 12 (usually 6) months (88). Subsequently, IgG antibody to HAV persists and confers lifelong immunity to HAV infection. Passive immunization with immune globulin does not influence the result of the commercially available assay that measures anti-HAV antibodies in serum. Chronic HAV infection does not develop.

Outbreaks associated with food that has been contaminated before reaching the food-service establishment are on the rise in recent years (92). Foodborne causes of HAV infection include drinks made with contaminated ice or water, unpeeled fruits and vegetables contaminated by polluted water, and undercooked or raw shellfish harvested from polluted water (87). Shellfish appear to be responsible for the majority of epidemics (86,87). Soft fruits, such as berries (consumed raw or processed in ice cream and other products), and fresh vegetables have been implicated occasionally in outbreaks (93–95). Shellfish can concentrate up to 100 times the original level of HAV from the surrounding water, and HAV can survive for about 7 days in contaminated mussels (96). Unfortunately, the common practice of steaming shellfish until they are just opened does not kill HAV, and cooking for a longer time is necessary (87). As is the case for Norwalk virus, boiling shellfish for 4–6 minutes is needed to inactivate HAV and prevent its transmission (51,97). Contaminated drinking water has also been associated with outbreaks of HAV infection (98). Recently, a waterborne outbreak of enterically transmitted hepatitis in East Africa was found to be due to both HAV and HEV (99).

Efforts to control the incidence of HAV illness are aimed at prevention. Traditionally, this is achieved by passive transfer of antibodies with administration of immune globulin. Immune globulin, 0.02 mL/kg, is also recommended as postexposure prophylaxis (as soon as possible after exposure, but within 2 weeks) for close personal (household and sexual) contacts of index cases, in day care centers where an index case has been identified, and for common source exposure (92). Two highly efficient vaccines consisting of inactivated whole-virus particles and empty capsids are licensed in the United States: Havrix (Smith Kline Beecham) and Vaqta (Merck). Both are given as a single dose, followed by a booster dose 6–12 months later (in adults, Havrix, 1440 ELISA units, or Vaqta, 50 units). Clinical trials on inactivated HAV vaccine to date have involved well over 50,000 vaccines in over 27 countries. Vaccinees developed anti-HAV IgG (and occasionally low titer of IgM) at much lower levels than those acquired after natural infection. After the first dose, 94–100% of adult recipients developed neutralizing antibodies, but up to 45% lacked neutralizing antibodies, in 14 days (92,100). The vaccine is expected to be protective for at least 10 years.

Only 2%–3% of hepatitis A cases in the United States are associated with foodborne outbreaks (92,101), but such outbreaks can be very costly. For example, the total cost from a foodborne outbreak in Denver, Colorado in 1992 was estimated to be over $800,000 (102). In this outbreak, a food handler at a large Denver catering facility was diagnosed as having acute HAV infection, and subsequently 43 cases of hepatitis A associated with the catering facility were reported to the Colorado Department of Public Health and Environment. Over half of the total cost was for the 16,293 immune globulin injections administered ($450,397). Serological studies and physicians' fees ($133,218), as well as health department personnel time ($105,699), also contributed significantly to the total cost. By contrast, the direct medical cost for the infected patients totaled only $46,064.

The food, through which HAV is transmitted in these types of outbreaks, is usually contaminated by an HAV-infected food handler during preparation (101). Ironically, food handlers pose a higher risk in countries with low rates of HAV infection (such as the United States). In areas of the world where HAV is endemic, most food handlers acquired immunity during childhood and are not susceptible to HAV infection. The incidence of HAV infection among food handlers in the United States is no different from that of the general population (92). Immunization of food handlers with the highly effective HAV vaccine could prevent common source outbreaks. The most recent guidelines from the Advisory Committee on Immunization Practices (92) recommend vaccination of food handlers only if state and local health authorities or private employers determine that such vaccination is cost-effective. If a food handler is diagnosed with HAV infection, passive immunization with immune globulin of other food handlers at the same location is recommended. Passive immunization of patrons is recommended only if (a) the food handler both directly handled uncooked (or previously cooked) foods and had diarrhea or poor hygiene during the time when he or she was likely to be infectious, and (b) patrons at risk can be identified and treated within 2 weeks of exposure (92). It is important to follow these guidelines rigorously. As the Denver outbreak illustrates, indiscriminate use of immune globulin prophylaxis can be costly: over 16,000 people received immune globulin, although only 5000 people were potentially at risk (102).

Travel to areas where HAV is endemic is associated with increased risk of infection. The risk of acquiring HAV infection among unprotected Swiss travelers to developing countries is 3 cases per 1000 persons per month of stay. For those such as backpackers, who may live and eat under poor hygienic conditions while traveling, the risk is even higher (20 cases per 1000 persons per month (103). HAV vaccination or immune globulin prior to departure is recommended for all susceptible persons traveling to, or working in, countries that have high or intermediate HAV endemicity (92,104). Immune globulin has traditionally been used to prevent disease in international travelers to endemic areas (0.02 mL/kg, if traveling for 1–2 months; 0.06 mL/kg every 5 months, if traveling for more than 3–5 months). Prescreening for immunity in certain populations and age groups before vaccination may be more cost-effective. According to one study, HAV vaccination is most cost-effective for travelers who expect to travel three or more times in 10 years (with an average stay of 3–4 weeks), or for trips longer than 6 months (105). A clinical trial among German travelers showed an overall anti-HAV seroconversion rate of 96.5%, 2 weeks after primary vaccination. Coadministration of other vaccines for travelers has no influence on the response to HAV vaccine (106). Administration of the first dose of HAV vaccine is recommended at least 4 weeks before travel. For those who are traveling to a high-risk area within 4 weeks of the initial dose of vaccine, immune globulin (0.02 mL/kg) should also be given at a different site (92).

In the absence of active or passive immunization, travelers could reduce the risk of foodborne HAV infection by taking simple precautions. The infectivity of purified virus is totally destroyed by heating to 100°C for 5 min, but only partially reduced by heating to 60°C for 1 h (107). HAV is also inactivated by heating food to >185°F (85°C) for 1 min or disinfecting surfaces with sodium hypochlorite (household beach) at a dilution of 1:100 (108). The standard recommendation from the Office of Medical Services, U.S. Department of State, is to boil water for 3 min. Most cases of hepatitis A could be prevented if travelers avoid eating uncooked (or undercooked) shellfish, uncooked vegetables, and unpeeled fruits. Travelers should also avoid drinks containing ice, or those made with unpurified water, and swimming in contaminated water.

The cost of one dose of immune globulin is less expensive than a dose of vaccine (~U.S. $60) (average wholesale price, *Red Book*, 1999). Thus, immune globulin is probably most cost-effective and sufficient for preventing illness in close contacts of index cases or travelers on short-term visits to endemic areas. Vaccination is recommended for travelers destined to be in endemic countries for extended periods (diplomatic corps, military units), those with chronic liver disease, persons with occupational risk of exposure, and children in communities where rates of hepatitis A are at least twice the national average (92).

B. Hepatitis E Virus

Hepatitis E virus is a recently identified enterically transmitted agent that causes hepatitis (109). It is unrelated to the other hepatitis viruses. Outbreaks due to HEV in fecally contaminated water have been documented in Southeast Asia, India, China, Central Asia, the Middle East, Central America, and Africa. Sporadic cases also occur in these endemic areas. HEV infection has also been found in a small number of travelers returning to the United States from endemic areas (110).

Acute hepatitis develops after an incubation period of 4–6 weeks (range of 2–9 weeks), and cholestasis is more common in HEV infection than other viral hepatitis infections. HEV is excreted in stool from 4 days before to 6 days after the onset of clinical symptoms. Like HAV infection, the illness due to HEV is self-limiting and resolves within 3 weeks (range 1–6 weeks), without development of chronic infection or a carrier state. Symptomatic infection primarily occurs in young to middle-aged adults (15–40 years old). Lower rates of disease are found in younger age groups. Secondary spread among household contacts is low (0.7–2.2%), compared to infection with HAV ($>50\%$) (91) or Norwalk virus (10–30%) (28,30). A high, unexplained mortality rate (15–25%) is found among infected women who are pregnant (91,111). Medical management consists of conservative supportive measures. Antibody to HEV of the IgM class is detectable only during the acute phase, but antibody of the IgG class persists (89,112) and may be protective in an outbreak setting (89). The role of passive immunization in preventing illness has not been established.

Hepatitis E virus is an icosahedral nonenveloped virus, 32 nm in diameter, with a single-stranded, plus-sense polyadenylated RNA genome, 7.5 kb in length (113). Four distinct isolates of HEV have now been completely sequenced including the Burma prototype strain, and isolates from Mexico, Pakistan, and China. Partial sequence of several Asian and African strains are also available. The Mexican strain is more divergent than the two subgroups of Asian strains. The overall nucleic acid homology between the Mexican strain and the Burmese and Pakistani strains is 76% and 77%, respectively. Serological assays are available for detecting HEV antibody using synthetic peptides or recombinantly expressed proteins consisting of the immunodominant epitopes of the putative structural regions of HEV (ORFs 2 and 3). HEV isolates from different geographic areas have been shown to contain at least one major cross-reactive epitope. Most assays also utilize antigenic domains from the two geographically distinct HEV strains (isolates from Burma and Mexico). Although these assays have not been standardized, most can detect anti-HEV antibodies in more than 95% of patients with acute hepatitis during an outbreak of HEV. Evaluation of 12 anti-HEV assays by a panel of well-characterized sera showed variable sensitivity (17–100%) for known HEV-positive sera. However, when sera from U.S. blood donors without a history of hepatitis (i.e., normal sera) were tested by the 12 anti-HEV assays and compared in a pairwise fashion, there was poor concordance among reactive sera, ranging from 0% to 89% (114). These data indicate that until the basis for the high rate of seroreactivity among volunteer blood donors in the United States (21%) is elucidated, anti-HEV serology in asymptomatic individuals from nonendemic areas should be interpreted with caution (115).

Hepatitis E virus is transmitted by the fecal–oral route. At least three outbreaks in China have been postulated to be caused by food or utensils contaminated by HEV-infected patients during the incubation period. One of the outbreaks had an overall attack rate of 19.8% (90). Fecally contaminated drinking water has been the most commonly implicated vehicle of transmission in most outbreaks. Several outbreaks occurred in a seasonal pattern after heavy rain (116). Modern water treatment appears to reduce the risk of HEV infection in endemic areas. One common source waterborne HEV epidemic in Pakistan was caused by malfunction at a modern water treatment plant (117). Epidemiological evidence suggests that boiling water may inactivate HEV (118), but the efficacy of chlorination of water in inactivating HEV is unknown.

Foreign travel to HEV endemic areas is a risk factor for developing HEV seroreactivity (119) and acute HEV hepatitis (110,120). From 1989 to 1992, acute HEV infection was documented among six persons in the United States who had returned from international travel (110). Molecular analysis of HEV isolates also confirmed travel to HEV endemic areas to be the major source of acute HEV infection in Taiwan (120). To reduce the risk of infection, travelers to HEV endemic areas should eat cooked food or fruits that can be peeled, drink only boiled or disinfected water, and practice good personal hygiene. The role of immune globulin is unknown. Immune globulin from donors in HEV-endemic countries was ineffective in outbreak settings (90,121). However, the presence of preexisting anti-HEV IgG seemed to prevent HEV infection among contacts of patients with acute HEV hepatitis in an outbreak (89). In addition, rhesus monkeys infected with one Indian isolate of HEV were immune when subsequently challenged with other Indian HEV isolates (122). Passive immunization with IgG from a well-characterized patient, who recovered from HEV infection, did not protect rhesus monkeys from HEV infection (123). This is in contrast to a recombinant vaccine from the ORF-2 region of the Pakistani strain that protected rhesus monkeys from challenge with homologous or heterologous virus (124). Thus, vaccination with a recombinant HEV antigen may be effective in preventing HEV infection in the future.

IV. CONCLUSION

In this chapter, we briefly reviewed the common foodborne pathogens that cause viral gastroenteritis (Norwalk virus and SRSVs, astrovirus, and rotavirus) and hepatitis (HAV and HEV). Infection is caused by consumption of water or food that is contaminated by these agents. Contaminated shellfish has been responsible for large foodborne outbreaks of viral gastroenteritis and hepatitis, with substantial medical and economic impact. Treatment of these maladies is mostly aimed at relief of symptoms and supportive care. Prevention is perhaps the most effective "therapy." For individuals in situations where risk of infection is high, maintaining good personal hygiene, being aware of food safety issues, and practicing such simple measures as boiling water for at least 3 min before consumption can be protective. Since contaminated shellfish is a common source of infection, proper handling and preparation of shellfish is necessary.

Natural infection by rotavirus and HAV confers long-term immunity to reinfection. Preexisting antibody to HEV may be protective in an outbreak setting. By contrast, Norwalk virus does not appear to confer long-term immunity. Effective vaccines currently exist for HAV and rotavirus, but only the former is approved by the FDA. Animal and epidemiological studies suggest that vaccination against HEV may be effective in disease prevention.

Public health measures, including maintenance of safe water supplies, timely inspection of restaurants or other commercial food preparation enterprises, and education of food service workers as well as the public, are also extremely important. The national initiative to improve the U.S. food supply should stimulate more research in the area and yield improved strategies with which to screen food for pathogens and to investigate outbreaks that do occur.

ABBREVIATIONS

ADRV	adult diarrhea rotavirus
CDC	Centers for Disease Control and Prevention (Atlanta, GA)
EIA	enzyme immunoassay
ELISA	enzyme-linked immunosorbent assay
FDA	Food and Drug Administration
GI	gastrointestinal
HAV	hepatitis A virus
HEV	hepatitis E virus
IEM	immune electron microscopy
IgG	immunoglobulin G
ORF	open reading frame
RIA	radioimmunoassay
RNA	ribonucleic acid
RT-PCR	reverse transcriptase PCR
SRSV	small round structured viruses
UV	ultraviolet
VLP	virus-like particle

REFERENCES

1. NH Bean, JS Goulding, C Lao, FJ Angulo. Surveillance for foodborne-disease outbreaks—United States, 1988–1992. MMWR 45(SS-5):1–66, 1996.
2. CG Helmick, PM Griffin, DG Addiss, RV Tauxe, DD Juranek. Infectious diarrheas. In: JE Everhart, ed. Digestive Diseases in the United States: Epidemiology and Impact. Washington DC: US Department of Health and Human Services, Public Health Service, National Institutes of Health, National Institute of Diabetes and Digestive and Kidney Diseases, 1994, pp 85–123.
3. H Appleton. Foodborne viruses. Lancet 336:1362–1364, 1990.
4. PW Chang, OC Liu, LT Miller, SM Li. Multiplication of human enteric viruses in Northern quahogs. Proc Soc Exp Biol Med 136:1380–1384, 1971.
5. GS Grohmann, AM Murphy, PJ Christopher, E Auty, HB Greenberg. Norwalk virus gastroenteritis in volunteers consuming depurated oysters. Austr J Exp Biol Med Sci 59:219–228, 1981.
6. DL Morse, JJ Guzewich, JP Hanrahan, R Stricof, M Shayegani, R Deibel, JC Grabau, NA Nowak, JE Herrmann. Widespread outbreaks of clamand oyster-associated gastroenteritis: role of Norwalk virus. N Engl J Med 314:678–681, 1986.
7. MR Griffin, JJ Surowiec, DI McCloskey, B Capuano, B Pierzynski, M Quinn, R Wojnarski, WE Parkin, H Greenberg, GW Gary. Foodborne Norwalk virus. Am J Epidemiol 115:178–184, 1982.
8. CW Hedberg, KL MacDonald, MT Osterholm. Changing epidemiology of food-borne disease: a Minnesota perspective. Clin Infect Dis 18:671–82, 1994.

9. JD Snyder, MH Merson. The magnitude of the global problem of acute diarrhoeal disease: a review of active surveillance data. Bull WHO 60:605–613, 1982.

10. C Bern, J Martines, I de Zoysa, RI Glass. The magnitude of the global problem of diarrhoeal disease: a ten-year update. Bull WHO 70:705–714, 1992.

11. National Center for Health Statistics, U.S. Department of Health, Education, and Welfare. Current Estimates. From the Health Interview Survey. Series 10, No. 85. United States— 1972, DHEW publication (HRA) 74-1512. U.S. Government Printing Office, Washington, DC, 1973.

12. HL DuPont, LK Pickering. Relative importance of enteropathogens in acute endemic diarrhea and food-borne diarrheal illness. In: HL DuPont, LK Pickering, eds. Infections of the Gastrointestinal Tract: Microbiology, Pathophysiology, and Clinical Features. New York: Plenum Press, 1980, pp 195–213.

13. X Jiang, DY Graham, K Wang, MK Estes. Norwalk virus genome cloning and characterization. Science 250:1580–1583, 1990.

14. SM Matsui, JP Kim, HB Greenberg, W Su, Q Sun, PC Johnson, HL DuPont, LS Oshiro, GR Reyes. The isolation and characterization of a Norwalk virus-specific cDNA. J Clin Invest 87:1456–1461, 1991.

15. HB Greenberg, JR Valdesuso, AR Kalica, RG Wyatt, VJ McAuliffe, AZ Kapikian, RM Chanock. Proteins of Norwalk virus. J Virol 37:994–999, 1981.

16. BVV Prasad, R Rothnagel, X Jiang, MK Estes. Three-dimensional structure of baculovirus-expressed Norwalk virus capsids. J Virol 68:5117–5125, 1994.

17. BV Prasad, DO Matson, AW Smith. Three-dimensional structure of calicivirus. J Mol Biol 240:256–264, 1994.

18. WD Cubitt, NR Blacklow, JE Herrmann, NA Nowak, S Nakata, S Chiba. Antigenic relationships between human caliciviruses and Norwalk virus. J Infect Dis 156:806–814, 1987.

19. S Okada, S Sekine, T Ando, Y Hayashi, M Murao, K Yabuuchi, T Miki, M Ohashi. Antigenic characterization of small round-structured viruses by immune electron microscopy. J Clin Microbiol 28:1244–1248, 1990.

20. AZ Kapikian, MK Estes, RM Chanock. Norwalk group of viruses. In: BN Fields, DM Knipe, PM Howley, et al., eds. Fields Virology. Philadelphia: Lippincott-Raven Publishers, 1996, pp 783–810.

21. X Jiang, DO Matson, WD Cubitt, MK Estes. Genetic and antigenic diversity of human caliciviruses (HuCVs) using RT-PCR and new EIAs. Arch Virol (Supplementum) 12:251–262, 1996.

22. BI Truman, HP Madore, MA Menegus, JL Nitzkin, R Dolin. Snow Mountain agent gastroenteritis from clams. Am J Epidemiol 126:516–525, 1987.

23. AM Murphy, GS Grohmann, PJ Christopher, WA Lopez, GR Davey, RH Millsom. An Australia-wide outbreak of gastroenteritis from oysters caused by Norwalk virus. Med J Austr 2:329–333, 1979.

24. JN Kuritsky, MT Osterholm, HB Greenberg, JA Korlath, JR Godes, CW Hedberg, JC Forfang, AZ Kapikian, JC McCullough, KE White. Norwalk gastroenteritis: a community outbreak associated with bakery product consumption. Ann Intern Med 100:519–521, 1984.

25. KE White, MT Osterholm, JA Mariotti, JA Korlath, DH Lawrence, TL Ristinen, HB Greenberg. A foodborne outbreak of Norwalk virus gastroenteritis: evidence for post-recovery transmission. Am J Epidemiol 124:120–126, 1986.

26. JA Reid, EO Caul, DG White, SR Palmer. Role of infected food handler in hotel outbreak of Norwalk-like viral gastroenteritis: implications for control. Lancet 2:321–323, 1988.

27. L Sekla, W Stackiw, S Dzogan, D Sargeant. Foodborne gastroenteritis due to Norwalk virus in a Winnipeg hotel. Can Med Assoc J 140:1461–1464, 1989.

28. JW Taylor, GW Gary Jr, HB Greenberg. Norwalk-related gastroenteritis due to contaminated drinking water. Am J Epidemiol 114:584–592, 1981.

29. RA Goodman, JW Buehler, HB Greenberg, TW McKinley, JD Smith. Norwalk gastroenteri-

tis associated with a water system in a rural Georgia community. Arch Environ Health 37: 358–360, 1982.

30. RC Baron, FD Murphy, HB Greenberg, CE Davis, DJ Bregman, GW Gary, JM Hughes, LB Schonberger. Norwalk gastrointestinal illness: an outbreak associated with swimming in a recreational lake and secondary person-to-person transmission. Am J Epidemiol 115:163–172, 1982.

31. R Wilson, LJ Anderson, RC Holman, GW Gary, HB Greenberg. Waterborne gastroenteritis due to the Norwalk agent: clinical and epidemiologic investigation. Am J Publ Health, 72: 72–74, 1982.

32. JS Koopman, EA Eckert, HB Greenberg, BC Strohm, RE Isaacson, AS Monto. Norwalk virus enteric illness acquired by swimming exposure. Am J Epidemiol 115:173–177, 1982.

33. JE Kaplan, RA Goodman, LB Schonberger, EC Lippy, GW Gary. Gastroenteritis due to Norwalk virus: an outbreak associated with a municipal water system. J Infect Dis 146:190–197, 1982.

34. HW Lawson, MM Braun, RI Glass, SE Stine, SS Monroe, HK Atrash, LE Lee, SJ Englender. Waterborne outbreak of Norwalk virus gastroenteritis at a southwest US resort: role of geological formations in contamination of well water. Lancet 337:1200–1204, 1991.

35. RO Cannon, JR Poliner, RB Hirschhorn, DC Rodeheaver, PR Silverman, EA Brown, GH Talbot, SE Stine, SS Monroe, OT Dennis, RI Glass. A multistate outbreak of Norwalk virus gastroenteritis associated with consumption of commercial ice. J Infect Dis 164:860–863, 1991.

36. LA Sawyer, JJ Murphy, JE Kaplan, PF Pinsky, D Chacon, S Walmsley, LB Schonberger, A Phillips, K Forward, C Goldman, J Brunton, RA Fralick, AO Carter, GW Gary Jr, RI Glass, DE Low. 25- to 30-nm virus particle associated with a hospital outbreak of acute gastroenteritis with evidence for airborne transmission. Am J Epidemiol 127:1261–1271, 1988.

37. BH Keswick, TK Satterwhite, PC Johnson, HL DuPont, SL Secor, JA Bitsura, GW Gary, JC Hoff. Inactivation of Norwalk virus in drinking water by chlorine. Appl Environ Microbiol 50:261–264, 1985.

38. JL Adler, R Zickl. Winter vomiting disease. J Infect Dis 119:668–673, 1969.

39. RA Gunn, WA Terranova, HB Greenberg, J Yashuk, GW Gary, JG Wells, PR Taylor, RA Feldman. Norwalk virus gastroenteritis aboard a cruise ship: an outbreak on five consecutive cruises. Am J Epidemiol 112:820–827, 1980.

40. KC Hyams, AL Bourgeois, BR Merrell, P Rozmajzl, J Escamilla, SA Thornton, GM Wasserman, A Burke, P Echeverria, KY Green, AZ Kapikian, JN Woody. Diarrheal disease during Operation Desert Shield. N Engl J Med 325:1423–1428, 1991.

41. NR Blacklow, R Dolin, DS Fedson, H DuPont, RS Northrup, RB Hornick, RM Chanock. Acute infectious nonbacterial gastroenteritis: etiology and pathogenesis. Ann Intern Med 76: 993–1008, 1972.

42. CW LeBaron, NP Furutan, JF Lew, JR Allen, V Gouvea, C Moe, SS Monroe. Viral agents of gastroenteritis: public health importance and outbreak management. MMWR 39(RR-5): 1–24, 1990.

43. TA Parrino, DS Schreiber, JS Trier, AZ Kapikian, NR Blacklow. Clinical immunity in acute gastroenteritis caused by Norwalk agent. N Engl J Med 297:86–89, 1977.

44. X Jiang, M Wang, DY Graham, MK Estes. Expression, self-assembly, and antigenicity of the Norwalk virus capsid protein. J Virol 66:6527–6532, 1992.

45. X Jiang, DO Matson, GM Ruiz-Palacios, J Hu, J Treanor, LK Pickering. Expression, self-assembly, and antigenicity of a Snow Mountain agent-like calicivirus capsid protein. J Clin Microbiol 33:1452–1455, 1995.

46. JP Leite, T Ando, JS Noel, B Jiang, CD Humphrey, JF Lew, KY Green, RI Glass, SS Monroe. Characterization of Toronto virus capsid protein expressed in baculovirus. Arch Virol 141: 865–875, 1996.

47. KY Green, AZ Kapikian, J Valdesuso, S Sosnovtsev, JJ Treanor, JF Lew. Expression and self-assembly of recombinant capsid protein from the antigenically distinct Hawaii human calicivirus. J Clin Microbiol 35:1909–1914, 1997.

48. AZ Kapikian, RG Wyatt, R Dolin, TS Thornhill, AR Kalica, RM Chanock. Visualization by immune electron microscopy of a 27nm particle associated with acute infectious non-bacterial gastroenteritis. J Virol 10:1075–1081, 1972.

49. RC Baron, HB Greenberg, G Cukor, NR Blacklow. Serologic responses among teenagers following natural exposure to Norwalk virus. J Infect Dis 150:531–534, 1984.

50. MC Steinhoff, RG Douglas Jr, HB Greenberg, DR Callahan. Bismuth subsalicylate therapy of viral gastroenteritis. Gastroenterology 78:1495–1499, 1980.

51. RS Koff, HS Sear. Internal temperature of steamed clams. N Engl J Med 276:737–739, 1967.

52. KB Kirkland, RA Meriwether, JK Leiss, WR Mac Kenzie. Steaming oysters does not prevent Norwalk-like gastroenteritis. Pub Health Rep 111:527–530, 1996.

53. S McDonnell, KB Kirkland, WG Hlady, C Aristeguieta, RS Hopkins, SS Monroe, RI Glass. Failure of cooking to prevent shellfish-associated viral gastroenteritis. Arch Intern Med 157:111–116, 1997.

54. ON Gill, WD Cubitt, DA McSwiggan, BM Watney, CLR Bartlett. Epidemic of gastroenteritis caused by oysters contaminated with small round structured viruses. Br Med J 287:1532–1534, 1983.

55. WD Cubitt. Human, small round structured viruses, caliciviruses and astroviruses. Baillière's Clin Gastroenterol 4:643–656, 1990.

56. H Appleton. Small round viruses: Classification and role in food-borne infections. In: G Bock. J Whelan, eds. Novel Diarrhoea Viruses. Chichester: John Wiley and Sons, 1987, pp 108–125.

57. JE Herrmann, DN Taylor, P Echeverria, NR Blacklow. Astroviruses as a cause of gastroenteritis in children. N Engl J Med 324:1757–1760, 1991.

58. JR Cruz, AV Bartlett, JE Herrmann, P Caceres, NR Blacklow, F Cano. Astrovirus-associated diarrhea among Guatemalan ambulatory rural children. J Clin Microbiol 30:1140–1144, 1992.

59. J Kurtz, WD Cubitt. Astroviruses and caliciviruses. In: MJG Farthing, GT Keusch, eds. Enteric Infection: Mechanisms, Manifestations, and Management. New York: Raven Press, 1989, pp 205–215.

60. LS Oshiro, CE Haley, RR Roberto, JL Riggs, M Croughan, H Greenberg, A Kapikian. A 27 nm virus isolated during an outbreak of acute infectious non-bacterial gastroenteritis in a convalescent home: a possible new serotype. J Infect Dis 143:791–795, 1981.

61. T Konno, H Suzuki, N Ishida, R Chiba, K Mochizuki, A Tsunoda. Astrovirus-associated epidemic gastroenteritis in Japan. J Med Virol 9:11–17, 1982.

62. JF Lew, CL Moe, SS Monroe, JL Allen, BM Harrison, BD Forrester, SE Stine, PA Woods, JC Hierholzer, JE Herrmann. Astrovirus and adenovirus associated with diarrhea in children in day care settings. J Infect Dis 164:673–678, 1991.

63. I Oishi, K Yamazaki, T Kimoto, Y Minekawa, E Utagawa, S Yamazaki, S Inouye, GS Grohmann, SS Monroe, SE Stine, C Carcamo, T Ando, RI Glass. A large outbreak of acute gastroenteritis associated with astrovirus among students and teachers in Osaka, Japan. J Infect Dis 170:439–443, 1994.

64. ET Utagawa, S Nishizawa, S Sekine, Y Hayashi, Y Ishihara, I Oishi, A Iwasaki, I Yamashita, K Miyamura, S Yamazaki, S Inouye, RI Glass. Astrovirus as a cause of gastroenteritis in Japan. J Clin Microbiol 32:1841–1845, 1994.

65. AD Phillips, SJ Rice, JA Walker-Smith. Astrovirus within human small intestinal mucosa. Gut 23:A923, 1982.

66. JB Kurtz, TW Lee. Astroviruses: human and animal. In: G Bock, J Whelan, eds. Novel Diarrhoea Viruses. Chichester: John Wiley and Sons, 1987, pp 92–107.

67. M Björkholm, F Celsing, G Runarsson, J Waldenström. Successful intravenous immunoglob-

ulin therapy for severe and persistent astrovirus gastroenteritis after fludarabine treatment in a patient with Waldenström's macroglobulinemia. Int J Hematol 62:117–120, 1995.

68. SM Cook, RI Glass, CW LeBaron, M-S Ho. Global seasonality of rotavirus infections. Bull WHO 68:171–177, 1990.

69. DB Hrdy. Epidemiology of rotaviral infection in adults. Rev Infect Dis 9:461–469, 1987.

70. RS Hopkins, GB Gaspard, FP Williams, Jr, RJ Karlin, G Cukor, NR Blacklow. A community waterborne gastroenteritis outbreak: evidence for rotavirus as the agent. Am J Pub Health 74:263–265, 1984.

71. S Ramia. Transmission of viral infections by the water route: implications for developing countries. Rev Infect Dis 7:180–188, 1985.

72. DA Sack, AMAK Chowdhury, A Eusof, MA Ali, MH Merson, S Islam, RE Black, KH Brown. Oral hydration in rotavirus diarrhoea: a double blind comparison of sucrose with glucose electrolyte solution. Lancet 2:280–283, 1978.

73. N Hirschhorn, WB Greenough III. Progress in oral rehydration therapy. Sci Am 264:50–56, 1991.

74. C Duggan, M Santosham, RI Glass. The management of acute diarrhea in children: oral rehydration, maintenance, and nutritional therapy. MMWR 41(RR-16):1–20, 1992.

75. CDC. Rotavirus vaccine for the prevention of rotavirus gastroenteritis among children— recommendations of the Advisory Committee on Immunization Practices. MMWR 1999; 48(no. RR-2).

76. CDC. Intussusception among recipients of rotavirus vaccine—United States, 1998–1999. MMWR 1999;48:577–581.

77. CDC. ACIP Recommendation: U.S. Rotavirus Vaccine. National Immunization Program Website, October 22, 1999.

78. NR Blacklow, HB Greenberg. Viral gastroenteritis. N Engl J Med 325:252–264, 1991.

79. ME Conner, CD Zarley, B Hu, S Parsons, D Drabinski, S Greiner, R Smith, B Jiang, B Corsaro, V Barniak, et al. Virus-like particles as a rotavirus subunit vaccine. J Infect Dis 174 (suppl):S88–S92, 1996.

80. JE Herrmann, SC Chen, EF Fynan, JC Santoro, HB Greenberg, HL Robinson. DNA vaccines against rotavirus infections. Arch Virol (Suppl)12:207–215, 1996.

81. ME Penaranda, MS Ho, ZY Fang, H Dong, XS Bai, SC Duan, WW Ye, MK Estes, P Echeverria, T Hung. Seroepidemiology of adult diarrhea rotavirus in China, 1977 to 1987. J Clin Microbiol 27:2180–2183, 1989.

82. ER Mackow. Group B and C Rotaviruses. In: MJ Blaser, PD Smith, JI Ravdin, HB Greenberg, RL Guerrant. Infections of the Gastrointestinal Tract. New York: Raven Press, 1995, pp 983–1008.

83. C-H VonBonsdorf, L Svensson. Human serogroup C rotavirus in Finland. Scand J Infect Dis 20:475–478, 1988.

84. DW Brown, L Campbell, DS Tomkins, MH Hambling. School outbreak of gastroenteritis due to atypical rotavirus. Lancet 2:737–738, 1989.

85. K Matsumoto, M Hatano, K Kobayashi, A Hasegawa, S Yamazaki, S Nakata, S Chiba, Y Kimura. An outbreak of gastroenteritis associated with acute rotaviral infection in school-children. J Infect Dis 160:611–615, 1989.

86. BL Portnoy, PA Mackowiak, CT Caraway, JA Walker, TW McKinley, CA Klein Jr. Oyster-associated hepatitis: failure of shellfish certification programs to prevent outbreaks. JAMA 233:1065–1068, 1975.

87. ML Halliday, LY Kang, TK Zhou, MD Hu, QC Pan, TY Fu, YS Huang, SL Hu. An epidemic of hepatitis A attributable to raw clams in Shanghai, China. J Infect Dis 164:852–859, 1991.

88. SM Lemon. Type A viral hepatitis: epidemiology, diagnosis, and prevention. Clin Chem 43: 1494–1499, 1997.

89. JP Bryan, SA Tsarev, M Iqbal, J Ticehurst, S Emerson, A Ahmed, J Duncan, AR Rafiqui,

IA Malik, RH Purcell, LJ Legters. Epidemic hepatitis E in Pakistan: Patterns of serological response and evidence that antibody to hepatitis E virus protets against disease. J Infect Dis 170:517–521, 1994.

90. H Zhuang. Hepatitis E and strategies for its control. In: Y-M Wen, Z-Y Xu, JL Melnick, eds. Viral Hepatitis in China: Problems and Control Strategies. Monographs in Virology. Basel: Karger, 1992, Vol 19, pp 126–139.

91. EE Mast, K Krawczynski. Hepatitis E: an overview. Annu Rev Med 47:257–266, 1996.

92. Prevention of hepatitis A through active or passive immunization: recommendations of the Advisory Committee on Immunization Practices. MMWR 48 (RR-12):1–38, 1999.

93. TMS Reid, HG Robinson. Frozen raspberries and hepatitis A. Epidemiol Infect 98:109–112, 1987.

94. MT Niu, LB Polish, BH Robertson, BK Khanna, BA Woodruff, CN Shapiro, MA Miller, JD Smith, JK Gedrose, MJ Alter, et al. Multistate outbreak of hepatitis A associated with frozen strawberries. J Infect Dis 166:518–524, 1992.

95. LS Rosenblum, IR Mirkin, DT Allen, S Safford, SC Hadler. A multifocal outbreak of hepatitis A traced to commercially distributed lettuce. Am J Publ Health 80:1075–1059, 1990.

96. R Enriquez, GG Frosner, V Hochstein-Mintzel, S Riedemann, G Reinhart. Accumulation and persistence of hepatitis A virus in mussels. J Med Virol 37:174–179, 1992.

97. G Giusti, GB Gaeta. Doctors in the kitchen: experiments with cooking bivalve mollusks. N Engl J Med 304:1371–1372, 1981.

98. AC Moore, BL Herwaldt, GF Craum, RL Calderon, AK Highsmith, DD Juranek. Survillance for waterborne disease outbreaks—United States, 1991–1992. MMWR CDC Surveill Sum 42:1–22, 1993.

99. P Coursaget, Y Buisson, N Enogat, R Bercion, JM Baudet, P. Delmaire, D Prigent, J Desrame. Outbreak of enterically-transmitted hepatitis due to hepatitis A and hepatitis E viruses. J Hepatol 28:745–750, 1998.

100. Hepatitis A vaccine. Med Lett 37:51–52, 1995.

101. Centers for Disease Control and Prevention. Foodborne hepatitis A—Missouri, Wisconsin, and Alaska, 1990–1992. MMWR 42:526–529, 1993.

102. CB Dalton, A Haddix, RE Hoffman, EE Mast. The cost of a food-borne outbreak of hepatitis A in Denver, Colo. Arch Intern Med 156; 1013–1016, 1996.

103. R Steffen. Hepatitis A in travelers: the European experience. J Infect Dis 171 (Suppl 1): S24–28, 1995.

104. Advice for travellers. Med Let 40:47–50, 1998.

105. BCA Langer, GG Frosner. Relative importance of the enterically transmitted human hepatitis virus type A and E as a cause of foreign travel associated hepatitis. Arch Virol (Suppl)11: 171–179, 1996.

106. U Bienzle, HL Bock, R Clemens, J Kruppenbacher. Immunogenicity, safety and interference of SB Biologicals' inactivated hepatitis A vaccine with other simultaneously: experience of multicenter trial in Germany (Abstr. 039). Scientific Program and Abstracts of the International Symposium on Viral Hepatitis and Liver Disease (8th Triennial Congress), Tokyo, 1993.

107. PJ Provost, BS Wolanski, IVJ Muller, OL Ittensohn, WJ McAleer, MR Hilleman. Physical, chemical, and morphological dimensions of human hepatitis A virus strain CR 326 (38578). Proc Soc Exp Biol Med 148:532–539, 1975.

108. MS Favero, WW Bond. Disinfection and sterilization. In: AJ Zuckerman, HC Thomas, eds. Viral Hepatitis: Scientific Basis and Clinical Management. New York: Churchill Livingstone, 1993, pp 565–575.

109. DW Bradley. Enterically-transmitted, non-A, non-B hepatitis. Br Med Bull 46:442–461, 1990.

110. Centers for Disease Control and Prevention. Hepatitis E among U.S. travelers, 1989–1992. MMWR 42:1–4, 1993.

111. ID Gust, RH Purcell. Report of a workshop: Waterborne non-A, non-B hepatitis. J Infect Dis 156:630–635, 1987.

112. A Koshy, S Grover, KC Hyams, MA Shabrawy, A Pacsa, B al-Nakib, SA Zaidi, AA al-Anezi, S al-Mufti, J Burans, M Carl, AL Richards. Short-term IgM and IgG antibody responses to hepatitis E virus infection. Scand J Infect Dis 28:439–441, 1996.

113. DW Bradley. Hepatitis E virus: a brief review of the biology, molecular virology, and immunology of a novel virus. J Hepatol 22(Suppl 1):140–145, 1995.

114. EE Mast, MJ Alter, PV Holland, Purcell RH for the hepatitis E virus antibody serum panel evaluation group. Evaluation of assays for antibody to hepatitis E virus by a serum panel. Hepatology 27:857–861, 1998.

115. DL Thomas, PO Yarbough, D Vlaov, SA Tsarev, KE Nelson, AJ Saah, RH Purcell. Seroreactivity to hepatitis E virus in areas where the disease is not endemic. J Clin Microbiol 35:1244–1247, 1997.

116. K Bile, A Isse, O Mohamud, P Allebeck, L Nilsson, H Norder, IK Mushahwar, LO Magnius. Contrasting roles of rivers and wells as sources of drinking water on attack and fatality rates in a hepatitis E epidemic in Somalia. Am J Trop Med Hyg 51:466–474, 1994.

117. MA Rab, MK Bile, MM Mubarik, H Asghar, Z Sami, S Siddiqi, AS Dil, MA Barzgar, MA Chaudhry, MI Burney. Water-borne hepatitis E epidemic in Islamabad, Pakistan: a common source outbreak traced to the malfunction of a modern water treatment plant. Am J Trop Med Hyg 57:151–157, 1997.

118. A Corwin A, K Jarot, I Lubis, K Nasution, S Suparmawo, A Sumardiati, S Widodo, G Orndorff, Y Choi, et al. Two years investigation of epidemic hepatitis E virus transmission in West Kalimantan (Borneo), Indonesia. Trans Roy Soc Trop Med Hyg 89:262–265, 1995.

119. EE Mast, IK Kuramoto, MO Favorov, VR Schoening, BT Burkholder, CN Shapiro, PV Holland. Prevalence and risk factors for antibody to hepatitis E virus seroreactivity among blood donors in northern California. J Infect Dis 176:34–40, 1997.

120. JC Wu, IJ Sheen, TY Chiang, WY Sheng, YJ Wang, CY Chan, SD Lee. The impact of traveling to endemic areas on the spread of hepatitis E virus infection: epidemiological and molecular analysis. Hepatology 27:1415–1420, 1998.

121. MS Khuroo, MY Dar. Hepatitis E: evidence for person-to-person transmission and inability of low dose immune serum glogulin from an Indian source to prevent it. Ind J Gastroenterol 11:113–116, 1992.

122. VA Arankalle, MS Chadha, LP Chobe, R Nair, K Banerjee. Cross-challenge studies in rhesus monkeys employing different Indian isolates of hepatitis E virus. J Med Virol 46:358–363, 1995.

123. A Chauhan, JB Dilawari, R Sharma, M Mukesh, SR Saroa. Role of long-persisting human hepatitis E virus antibodies in protection. Vaccine 16:755–756, 1998.

124. SA Tsarev, TS Tsareva, SU Emerson, S Govindarajan, M Shapiro, JL Gerin, RH Purcell. Recombinant vaccine against hepatitis E: dose response and protection against heterologous challenge. Vaccine 15:1834–1838, 1997.

8

Epidemiology of Foodborne Viral Infections

Thomas M. Lüthi
Official Food Control Authority of the Canton of Solothurn,
Solothurn, Switzerland

I. INTRODUCTION

Diarrheal diseases affect millions worldwide and have the greatest impact among children, especially in developing countries. However, they are also of public health significance in developed countries and are associated with considerable morbidity and a substantial number of hospitalizations among children and the elderly (1). In these countries, the estimated median percentage of diarrheal episodes in children associated with specific agents was given by Bern et al. (1) as approximately 30% caused by rotaviruses, 15% by bacteria, and another 15% by other viruses. However, in about 40% of the episodes of diarrhea an agent could not be recognized.

Until 25 years ago, most causes of acute nonbacterial gastroenteritis were unknown, even though in 1929 (2) Zahorsky, an American pediatrician, had described a syndrome named "winter vomiting disease." The first viral agent was described in 1972 (3) as the etiological agent of the syndrome following an outbreak in Norwalk, Ohio (4).

Foodborne or waterborne disease is defined by the World Health Organization (WHO) as "a disease of an infectious or toxic nature caused by, or thought to be caused by, the consumption of food or water" (5). It is an important burden of communicable diseases in developed countries. The Centers for Disease Control and Prevention (CDC) in Atlanta, Georgia estimate that contaminated food causes 6.6 million cases of acute illness and 9000 associated deaths annually in the United States alone.

In an overview of the environmental health situation in Europe by WHO's European Centre for Environment and Health, it has been estimated that each year 130 million Europeans (15% of the total population of the WHO European Region) are affected by episodes of foodborne disease (6). A community study in The Netherlands reported an incidence of more than 700 episodes of gastrointestinal illness per 1000 person-years. For the whole country, with its 15 million inhabitants, 10 million episodes were estimated to occur each year (7).

Nevertheless, death from foodborne infections is relatively uncommon. Those most at risk are usually infants, the very old, or those debilitated by preexisting conditions such as AIDS (8). Although episodes of diarrhea usually are acute in nature, studies suggest that about 1% of affected people may develop chronic diarrhea and long-term sequelae (8).

Emerging infectious diseases also include foodborne viral illnesses when they are defined as infections that are either new or preexisting but are rapidly increasing by means of incidence or geographic range and for which a particular route of transmission is newly recognized and agents are now known because of advances in detection (9). Unlike many other communicable diseases, foodborne illnesses are largely preventable (10).

The three main groups of foodborne and waterborne viral diseases are (a) acute viral gastroenteritis (e.g., small round-structured viruses, astroviruses); (b) hepatitis (e.g., hepatitis A and E); and (c) others such as tickborne encephalitis and poliomyelitis.

Transmission by food or water has been documented mainly for hepatitis A virus (HAV), astroviruses, caliciviruses, rotaviruses, and the group defined as small round-structured viruses (SRSVs), also known as Norwalk-like viruses and now believed to belong to the calicivirus group. However, as the laboratory diagnosis of viral gastroenteritis improves, other etiological agents may be implicated as causes of foodborne viral disease (11). Table 1 gives an overview of the most important food- and waterborne viruses (see also Chapters 3, 4, and 5).

Table 1 Summary of the Most Important Food- and Waterborne Viruses

Virus	Family	Food-/ waterborne?	Foodborne/ waterborne outbreaks?
Astrovirus	Astroviridae	Yes	Yes
Adenovirus	Adenoviridae	No?	No
Calicivirus	Caliciviridae	Yes	Yes
Coronavirus	Coronaviridae	No	No
Enterovirus: poliovirus	Picornaviridae	Yes	Yes
Enterovirus: coxsackievirus	Picornaviridae	Yes	Yes
Enterovirus: echovirus	Picornaviridae	Yes	Yes
Hepatitis A	Picornaviridae	Yes	Yes
Hepatitis E	Caliciviridae	Yes	Yes
Parvovirus	Parvoviridae	?	?
Rotavirus	Reoviridae	Yes	Yes
Small round-structured viruses (SRSVs), e.g., Norwalk virus	Caliciviridae	Yes	Yes
Small round featureless virus (SRV)	Parvoviridae ?	?	?
Tickborne encephalitis virus	Flaviviridae	Yes	No

Norwalk, hepatitis A, and other viruses were the fifth, sixth, and tenth leading causes of foodborne illness, respectively, in the United States during 1983 through 1987 (12).

A. Comparison to Bacterial Infections

Compared to other foodborne diseases, those caused by viruses are less severe and seldom fatal. This might be why the problem of viral contamination of food has been neglected. Yet, because many foodborne viral diseases are not recognized either as foodborne or as caused by viruses, the actual number of cases must be assumed to be significantly higher than that reported (13).

In contrast to many foodborne bacterial pathogens, viruses cannot multiply in food or water but can remain infectious for prolonged periods. Where testing could be conducted, the infectious dose for enteric viruses was found to be quite low.

Analysis of food samples for viruses remains difficult. Methods to detect foodborne viruses in different foods and water are far from routine and are expensive, cumbersome, time consuming, and require skilled personnel. Viruses in food and water are generally present only in small numbers and elaborate sample concentration steps are often necessary for their detection while keeping them viable. Many important foodborne viral pathogens cannot be grown in laboratory animals or cell culture systems. Therefore, molecular biological approaches (see Chapter 6) are currently regarded as the methods of choice for the analysis of food samples incriminated in outbreaks of infections. Even when viruses do not grow in food or water, their epidemiological potential is high because secondary cases may occur after infection with a common food source.

Foodborne bacterial zoonotic infections are well known (e.g., *Salmonella* spp., *Campylobacter* spp.). However, foodborne viral zoonoses are rare.

As stated above, many different groups of viruses can cause foodborne infections. Even within a given taxonomic group of foodborne viruses there are considerable differences among the different types involved. These differences make surveillance for foodborne viruses even more difficult.

Epidemiological investigations of foodborne outbreaks of suspected viral etiology are also hampered by the unavailability of defined reference strains.

B. Surveillance

For reasons enumerated above, epidemiological investigations and surveillance of foodborne viral diseases cannot be conducted on a routine basis in most countries. Only a few of the 46 countries of the European Region reported surveillance data on foodborne viruses to the FAO/WHO Collaborating Centre for Research and Training in Food Hygiene and Zoonosis in Berlin, Germany, for the period 1990–1992. It has been estimated that no more than 10% of the cases of all foodborne illness are reported to official agencies (6). This underscores the fact that cases of foodborne viral infection are grossly underestimated (9).

In many countries sporadic cases of viral gastroenteritis are not notifiable and health authorities are informed of outbreak situations only.

II. PREVALENCE

A. Human Astroviruses

Astroviruses are distributed worldwide and cause infections throughout the year but with peaks during winter and spring months. Gastroenteritis caused by human astroviruses (HastVs) is most common among 1- to 3-year-old infants. With the improvements in the detection method, the appreciation of the role of astrovirus-associated gastroenteritis has changed. In early surveys, which were based on electron microscopy, astroviruses were rarely recognized as a cause of gastroenteritis and were found in less than 1% of children with diarrhea (14).

Currently, seven serotypes of these viruses are known. Testing of sera collected from staff and patients at two children's hospitals in London, England with recombinant virus antigen showed that over 50% of the tested population were infected by serotype 1 between the ages of 5 and 12 months. These numbers rose to 90% by 5 years. Serotype 6, on the other hand, was relatively uncommon in all age groups (15). Similar findings were reported from other parts of the world. Other serotypes are reported to vary in frequency between 1% and 7%.

Astroviruses are the most common cause of diarrhea in adults immunocompromised with HIV (16).

B. Human Calicivirus

Surveys of children with gastroenteritis in different countries have been performed and the prevalence of children excreting human calicivirus (HuCV) was 1.3% (Norway), 1.8% (UK), 1.2% (Japan), 0.5% (Australia), 0.3% (USA), and 1.5% (China) (17). Based on immunological prevalence studies of pooled immunoglobulin and serum samples from different areas of the world, most people appear to be infected by the age of 12 years. The peak of acquisition is found between 6 months and 2 years. A seasonality is not yet known(18).

C. Norwalk and Norwalk-like Viruses

Norwalk and Norwalk-like viruses (SRSVs), have been associated with epidemic viral gastroenteritis in the United States. In the United Kingdom, these viruses are more often reported in connection with endemic disease. Sporadic cases may go unrecognized as many affected individuals will not seek medical attention and those who do may not have a stool specimen examined; moreover, if they have, it may not be examined for Norwalk-like viruses. Analyses of electron microscopic reporting of gastrointestinal viruses in the United Kingdom between 1985 and 1987 showed that SRSVs are frequently found in adults as well as in children (19). Gray and co-workers (20) reported that out of 3250 serum specimens collected in England in 1991 and 1992, 73.3% were positive for Norwalk virus by testing with an indirect enzyme-linked immunosorbent assay (ELISA) for antibody to Norwalk virus using baculovirus-expressed capsid antigen. The prevalence of Norwalk virus antibody differed with age and region. The titers rise with age. Similar results were found in other countries (20). The less developed a country the earlier in age individuals acquire antibodies (see Chapter 3).

D. Hepatitis A

Hepatitis A infection occurs worldwide as sporadic cases as well as outbreaks and epidemics. In contrast to the situation in developed countries, adults in developing countries are often immune to the infection and epidemics are therefore uncommon. Where environmental sanitation is lacking or poor, infection is common and occurs at an early age. However, with improved sanitation in different developing countries young adults are left susceptible and therefore outbreaks are increasing.

In developed countries infection occurs in settings with children (e.g., day care center) and in traveler returning from areas where the disease is endemic. Epidemics often evolve slowly and may involve a wide area for many months. Epidemics caused by the food- or waterborne route may evolve more explosively. The disease is most common among school-aged boys and girls as well as young adults.

E. Rotaviruses

Symptomatic and asymptomatic rotavirus infections together with high prevalence of serum antibodies after 3 years of age are frequently described in all age groups and in different countries. This implies that sequential rotavirus infections occur throughout life (see also Chapter 4).

III. OUTBREAKS

The WHO defines a foodborne or waterborne outbreak as an incident in which two or more persons experience a similar illness after ingestion of the same food, or after ingestion of water from the same source, and where epidemiological evidence implicates the food or water as the source of the illness (6).

In viral foodborne disease outbreaks, where different transmission routes may be possible at the same time, the distinction between food- and waterborne transmission and any person-to-person spread is difficult.

Reporting of food- and waterborne outbreaks tends to be biased by different factors: Large outbreaks as well as multistate or international outbreaks are more likely to be reported. Despite international efforts there are different reporting and surveillance systems present. Even in the United States, systems are different in different states. Mild illnesses such as diarrhea or vomiting in families tend not to be reported, in contrast to outbreaks with severe or even life-threatening outcomes in residential settings. It is also more likely that outbreaks with commercial products are more likely to be reported than home-made food. In contrast to viral diseases, outbreaks with etiology known to the general public and general practitioners, and well-publicized transmission routes such as *Salmonella* and eggs, are more likely to be reported.

Further difficulties in reporting food- and waterborne disease outbreaks are the generally long incubation period (e.g., hepatitis A), the complexity of modern food production and distribution practices, and a lack of knowledge on the part of the average consumer. Viral diseases, in contrast to those by parasitic worms, are somewhat more readily acquired by the general public. People therefore do not tend to demand further investigations by their general practitioner. Therefore, a considerable number of cases might not be recognized because no samples are taken and investigated. These cases then tend to be reported as outbreaks of unknown etiology. Lack of specific treatment as well as limited resources in terms of cost does not motivate the general practitioner to take samples and have them investigated further. Nevertheless, a critical look at the available data can provide much useful information on the role of viruses as agents of foodborne disease.

A general overview in this regard is given by using data from the United States and Europe. Selected outbreaks will be used to highlight the major types of viruses and their role in causing foodborne disease.

A. Viral Foodborne Disease Outbreaks in the United States

In the United States, Bean and co-workers (21) reviewed data collected by the CDC through a collaborative surveillance program on the occurrence of foodborne disease outbreaks (FBDOs). In this report 2423 outbreaks were reported. They concluded that bacterial pathogens caused the largest percentage (79%) of these events as well as the highest percentage (90%) of cases. However, the responsible pathogen was not identified in more than half of the FBDOs reported. Table 2 is a summary of viral foodborne disease outbreaks (VFBDOs).

Bean et al. (21) concluded that there was significant underreporting of VFBDOs. There were no reports other than hepatitis A or Norwalk viruses. Of the total of 2423 outbreaks between 1988 and 1992, only 45 (1.9%) were listed as VFBDOs. Of the remaining 1422 FBDOs with unknown etiology, 35% were also thought to have been caused by viruses (22). Based on these data, the estimated number of VFBDOs is believed to be as high as 542 (22%) in the United States.

B. Viral Foodborne Disease Outbreaks in Europe

In Europe national reporting systems for foodborne diseases vary widely. Therefore, the data summarized in Table 3 do not allow a direct comparison of the situation in various

Table 2 Viral Foodborne Disease Outbreaks in the United States between 1988 and 1992[a]

Factor	1988	1989	1990	1991	1992
VFBDO					
HAV	12 (2.7%)	7 (1.4%)	9 (1.7%)	7 (1.3%)	8 (2%)
Norwalk	0	1 (0.2%)	0	0	1 (0.2%)
Unknown etiology	268 (59.4%)	284 (56.2%)	295 (55.5%)	314 (59.5%)	261 (64.1%)
Cases					
HAV	795 (5.1%)	329 (2.1%)	452 (2.3%)	114 (0.8%)	419 (3.8%)
Norwalk	0	42 (0.3%)	0	0	250 (2.3%)
Unknown etiology	7608 (48.4%)	8750 (55.1%)	9925 (49.9%)	8218 (55.2%)	5982 (54.3%)
Death					
HAV	3 (15.8%)	0	3 (20%)	0	0
Norwalk	0	0	0	0	0
Unknown etiology	0	0	3 (20%)	1 (10%)	0

[a] The given numbers represent the number of outbreaks, cases, and deaths and the percentage of these compared with the total reported.
Source: Modified after Ref. 21.

countries or permit the drawing of a common picture for Europe as a whole. However, looking at the given outbreaks in a critical manner, it allows one to estimate the reported causative agents and trends within a given country (6).

Hepatitis A virus was the main agent reported in terms of VFBDOs. Other etiological causes were reported in England and Wales and Germany only. It is not clear whether

Table 3 Viral Foodborne Disease Outbreaks in Europe between 1989 and 1992[a]

Country	Agent reported	VFBDO (%)			
		1989	1990	1991	1992
France	Hepatitis A		0	1 (0.2%)	0
	Unknown		85 (14.3%)	82 (12.6%)	138 (18.8%)
Germany	Hepatitis A	1 (2.8%)	4 (6.1%)	2 (2.2%)	2 (0.9%)
	Rotavirus	1 (2.8%)	0	0	0
	Norwalk	1 (2.8%)	0	0	0
	"Virus"	0	0	0	1 (0.3%)
	Unknown	5 (13.9%)	11 (16.7%)	22 (23.9%)	33 (14.5%)
Spain	Hepatitis A		0	2 (0.2%)	1 (0.1%)
	Unknown		351 (38.4%)	358 (39%)	386 (39.1%)
Sweden	Hepatitis A		1 (3.2%)	0	0
	Unknown		7 (22.6%)	11 (31.4%)	14 (53.8%)
England & Wales	Hepatitis A	1 (2.7%)	1 (1.8%)	1 (2.2%)	
	SRSV	6 (16.2%)	12 (22.2%)	11 (24.4%)	
	Unknown	23 (62.2%)	32 (59.3%)	24 (53.3%)	
U.K. Scotland	"Virus"		1 (0.6%)	2 (1.3%)	0
	Unknown		18 (10.3%)	7 (4.7%)	1 (0.5%)

[a] The given numbers represent the number of outbreaks and the percentage of these compared with the total reported.
Source: Modified after Ref. 6.

this is due to methodological difficulties or because these agents are not found in other European countries.

In England and Wales, 1280 general outbreaks of infectious intestinal disease were reported to the Communicable Disease Surveillance Centre (CDSC) in London between 1992 and 1994. A virus was confirmed as the cause for 389 (31%) of these, and 47 of those were attributed to foodborne transmission (12%). SRSVs were identified in 41 (87%), small round viruses (SRVs) in 3, astroviruses in 2, and rotavirus in 1 outbreak. However, SRVs are generally regarded as incidental findings in ill as well as healthy people and may be confused with SRSVs. The three outbreaks attributed to SRVs reported during this period conform to the pattern of illness described for SRSV infections and were included with the SRSVs (23).

SRSVs are known to cause epidemic gastroenteritis in institutions, accounting for 54% of outbreaks due to person-to-person spread and 6% of foodborne outbreaks in England and Wales (24).

C. Incriminated Foods and Settings of Viral Foodborne Disease Outbreaks

In the United States the settings of VFBDOs between 1988 and 1992 were reported as follows: 24 delicatessen, cafeterias, restaurants; 2 private homes; 4 schools; 1 church; 11 other; and 3 unknown settings.

The majority of the settings were outside private homes. Foods involved were seafood, fruits and vegetables, and desserts, as well as multiple and processed foods. Contributing factors of the 45 reported VFBDOs were poor personal hygiene (76%), food from unsafe source (9%), improper holding temperature (4%), and unknown or other, unspecified reasons (11%) (21).

In England and Wales, the reported outbreaks between 1989 and 1990 caused by hepatitis A virus were traced to locally produced bread, rolls, sandwiches, and filled rolls from one shop. The other two outbreaks were associated with consumption of molluscan shellfish or prawn vol-au-vents during a private party. Salads, buffet meals, or sandwiches, all requiring a high degree of handling during preparation, were implicated in 10 outbreaks caused by SRSVs. Seafood was the suspected vehicle of infection in four outbreaks. Eleven outbreaks were associated with the consumption of oysters, five were due to SRSVs; in the remaining six, no pathogen was identified. However, the pattern of illness as well as the incubation periods for the outbreaks with no identified pathogen suggests SRSVs as the likely cause. Other foods suspected in VFBDOs included chicken, burgers, barbecued meat, and pizza (25).

The infectious intestinal disease outbreaks reported to CDSC during the period 1992 to 1994 identified in 57% of the VFBDOs the settings in the commercial catering sector (canteens, restaurants, hotels, shops, and mobile retailers), 28% in institutional settings (army, navy camps, prisons, homes for the elderly, hospitals, schools), and 11% were associated with catering for large numbers from private homes. The remaining 4% were not specified.

Oysters were identified in 17% of the reported outbreaks caused by SRSVs. Other edibles included mixed foods (assorted Chinese food, fish dinner, raw mushroom, Stilton cheese, turkey, meat pies, pasta salad, prawn cocktail, sandwiches, vegetable soup, margarine), desserts (raspberry syllabub, custard slices, peach and raspberry pie), salads or vegetables and fruits (watercress, green salad, carrots, melon, and papaya cocktail).

Astrovirus outbreaks were associated with drinking water from an unsafe source. The only reported foodborne rotavirus outbreak was linked to the consumption of chicken marsala.

It should be noted here that of all these outbreaks only in one SRSV were detected in the food specimens (oysters) by reverse transcriptase polymerase chain reaction (RT-PCR). In all other outbreaks, food was implicated by statistical association with illness or on the basis of strong circumstantial evidence alone.

In Table 4 examples of various VFBDOs are given, as well as the number of cases involved and the suspected item(s) of food.

D. Difficulties

Laboratory confirmation of food in a particular outbreak caused by or thought to be caused by a virus is rare. There are no methods to test for viruses contained in foods on a routine basis (Chapter 6). The amount of virus contained in foods is, in contrast to stool samples, too low to be seen by electron microscopy directly. Therefore, molecular biological methods, such as RT-PCR, are necessary to identify these agents.

Outbreaks of SRSV caused by oysters show that RT-PCR has not always been successful in detecting the etiological agent. This may be because the virus detection limit of RT-PCR in oysters is probably much higher than the infectious dose of the virus for

Table 4   Examples of VFBDOs

Country	Year	Causative agent	No. of cases	Food involved	Agent detected in food/ water?	Ref.
China	1988	HAV	300,000	Raw clams	Yes	26
USA	1988	HAV	61	Raw oysters	Yes	27
USA	1988	Norwalk	1440	Celery/chicken salad	No	28
Japan	1988	Rotavirus group C	676,383	School lunch	No	29
USA	1990	Norwalk	217	Fresh-cut fruit	No	30
Japan	1991	Astrovirus	>4,700	School lunch	No	31
Brazil	1992	Rotavirus group A	132	Drinking water	No	32
Denmark	1992	''Virus''	1455	Drinking water	No	33
USA	1992	Norwalk	201	Ice	No	34
USA	1993	Norwalk	70	Raw oysters	No	35
New Zealand	1994	Norwalk	36	Oysters	No	36
Finland	1994	Norwalk	>1500	Drinking water	No	37
Germany	1994/95	HAV	49	Bakery products	No	38
USA/Canada	1995	Norwalk-like	108	Drinking water	Yes	39
USA (Florida)	1995	Norwalk	131	Oysters, raw and heat-treated	No	40
New Zealand	1995	HAV	36	Delicatessen	No	41
Italy	1996	HAV	5620	Raw seafood (e.g., oysters)	No	42

Source: Modified after Ref. 3.

humans. In addition, circulating strains can differ in different countries. In Malta, for example, an SRSV gastroenteritis outbreak in 1995, thought to be caused by direct person-to-person contact, was tested with a primer pair that detects most prevalent strains in the UK and gave only weak positive reactions. The genetic and antigenic divergence among different strains highlights the need for more broadly reactive diagnostic assays.

Occurrence of secondary cases caused by person-to-person transmission after an exposure to a particular food may mislead investigators. In outbreaks of SRSV secondary cases are frequent and are used in epidemiological investigations to characterize VFBDOs.

Cases of illness in persons who handle or serve food and ice may make it difficult to distinguish between the possible roles of contaminated raw food items, food handler contamination, or contaminated water sources as etiological factors. It is questionable as to whether this distinction is useful in controlling an outbreak. Regardless of whether food or water is identified as a primary source for that outbreak, ill food handlers can prolong the outbreak (43).

Another difficulty is that samples of suspected food item(s) are not always available. In outbreaks of SRSV the relatively short incubation period increases the chance of being able to collect such samples. On the other hand, the generally long (up to 50 days) incubation period for hepatitis A renders it virtually impossible to obtain samples of suspected foods and this makes it necessary to rely on epidemiological methods to investigate suspected foodborne outbreaks.

E. Role of Food Handlers

A significant number of cases of hepatitis A are food- or waterborne. A proportion of these cases are due to foods contaminated by food handlers, particularly when their work involves touching unwrapped foods that are to be consumed raw or without further cooking. It is the asymptomatic preicteric food handler, often with poor hand washing habits, who is usually the source of outbreaks involving food. A vaccine is available, and food handlers can be immunized. Operators of food and catering service establishments could thus consider investing in the vaccination of their employees as a complementary measure to personal hygiene and safe food handling practices (44).

Presymptomatic food handlers (43,45) as well as symptomatic food handlers (43) were described in various foodborne outbreaks as the most likely source of VFBDO.

F. Identification of Viral Foodborne Disease Outbreaks
Using Alternative Methods of Detection

In the absence of specific laboratory results, outbreaks of food- and waterborne gastroenteritis can be characterized by the use of epidemiological, clinical, and microbiological criteria. These criteria may help to identify an outbreak presumably caused by Norwalk or Norwalk-like viruses. Especially where sophisticated laboratory facilities are not easily available these criteria might help to prevent any spread of the agent by recognizing the viral origin of the outbreak.

Proposed criteria for considering an outbreak due to Norwalk-like virus are given in Table 5 and were described and modified by various authors (43,46,47).

Table 5 Norwalk-like Pattern of Illness in Outbreaks of Acute Gastroenteritis

Criteria	Demand
Stool samples	Negative for bacterial pathogens
Secondary cases	Occurrence
Percentage of patients vomiting	$\geq 50\%$
Ratio of patients vomiting to patients with fever	>1
Type of vomiting	Projectile, explosive
Median incubation period	24–48 h
Median duration of illness	12–60 h
Ratio of patients vomiting to patients with diarrhea in children and adolescents	>1
Ratio of patients vomiting to patients with diarrhea in adults	<1

Source: Data from Refs. 43, 46, and 47.

G. Outbreak Control

The first aspect of managing a foodborne or waterborne outbreak must be to prevent its further spread. In respect to the nature of the agents person-to-person transmission is the most likely ongoing transmission route.

Especially in settings with cohorts living close together (e.g., elderly homes, cruise ships, hotels, air craft carriers, etc.) the importance of personal hygiene must be stressed. The removal of any potentially contaminated food and cleaning of the contaminated environment (e.g., toilet facilities, kitchen environment) is absolutely necessary. As soon as possible the source of the outbreak must be traced, whether it is primary foodborne, secondary foodborne, or person-to-person spread. Interviewing the kitchen staff about their health status (and their relatives in close contact, such as children), analyzing the foods eaten and by whom it was prepared, and identifying possible risks (e.g., no heat process) should help to trace the source.

Primary contaminated food (e.g., oysters) must be traced back to their origin and steps to withdraw the same batch of products from the market considered. Preventive interventions to avoid future outbreaks must be introduced by hazard analysis critical control point analysis (HACCP).

Secondary contamination of food must be investigated by identifying any breach in hygienic practices. Meanwhile nonheated food must not be served unless it was produced outside the suspected area. Potentially contaminated food must be destroyed once the needed samples have been collected. The decontamination of kitchen surfaces and other suspected areas in the home must be performed using hot water and a general purpose yet safe disinfectant such as 500 ppm hypochlorite (48).

The most difficult part in managing a foodborne outbreak is to identify the infected food handler(s) followed by their exclusion from work until the risk of spreading the virus is minimized. It must not be forgotten that staff must be sent home and must not stay in areas (e.g., common rooms) where other staff may be infected consequently. This includes also managerial staff and not only staff working with food or patients directly. The guidance of the management of outbreaks of foodborne illness, published by the Department of Health in the United States recommends an exclusion of persons who pose a special risk until 48 hours after clinical recovery and 72 hours for children (49). Hedberg and Osterholm (43), however, recommended in outbreak settings in Minnesota exclusion of ill food handlers for 72 hours. When there is evidence of transmission among food handlers and transmission to patrons on multiple days, these authors recommend closing the restaurant for 72 hours to provide an opportunity for the virus to "burn itself out." However, even when it is highly desirable to exclude food handlers from work, the financial realities of the catering industry mean that this is unlikely to be achievable routinely (48). Individual solutions have to be found for each outbreak separately while balancing all relevant aspects.

To prevent a person-to-person spread, movement of patients and staff must be restricted. Whenever possible, the infected and therefore infectious people should be isolated and social events that lead to crowding avoided. All people should be encouraged to practice good personal hygiene (48).

Outbreaks of viral gastroenteritis that are waterborne demand investigations of the reasons of the contamination and disinfection of the drinking water to prevent further spread within the community. For individual households boiling of water for 10 min is sufficient to inactivate viral agents of gastroenteritis (18).

H. Conclusions

Unlike many food- and waterborne bacterial infections, viral infections may also spread from person to person directly. In outbreak situations it might therefore be difficult to trace a particular source. In general, once an epidemic has started, mixed transmission such as water- and foodborne as well as person-to-person spread are likely.

The identification of viruses directly from patients is possible but for some agents (e.g., SRSV) not available on a routine base. Isolation and identification of the etiological agent in food or water are not necessarily available in a routine lab. Alternatively, for viral gastroenteritis methods that use epidemiological, microbiological, and clinical criteria may be helpful to identify an agent and trace a possible source.

Oysters remain an important vehicle of VFBDO and continued vigilance is needed to ensure that oyster beds are not contaminated with human feces or inadequately treated sewage.

IV. TRANSMISSION

A. General

Most foodborne viruses infect their host per orally by ingestion and are subsequently shed. It is important to note that there is no replication of viruses in foods. However, viruses are remarkably stable and may survive for prolonged periods in the environment. Processes such as heating, boiling, and cooking inactivate viruses as they inactivate pathogenic bacteria and are therefore also critical control points for viruses. Depending on the origin and source of the viruses, two main transmission routes might be identified.

The only significant viral zoonosis transmissible via milk is a tickborne encephalitis found in certain parts of Europe (CEE). Infected animals shed the virus in the milk, which then may infect humans who consume it raw (50).

In general, transmission routes may be divided into the direct route, the indirect route, and other routes.

1. Direct Transmission

Foods that are contaminated at source with viruses and are eaten raw or after inadequate heat treatment serve as the main sources of foodborne outbreaks. Shellfish harvested from fecally contaminated seawaters and fruit and vegetables grown in and/or irrigated with fecally polluted waters are good example of such foods.

2. Indirect Transmission

Contamination of foods by an infected food handler during preparation or serving appears to be one of the most important sources for foodborne outbreaks. Characteristically, items of food that require intensive manual preparation (e.g., sandwiches, salads, fruit salads, etc.) and no further processing are most likely to be incriminated as the source of viral disease outbreaks (see Sec. III of this Chapter). Fecally contaminated water used to wash vegetables (52) and fruit and contaminated surfaces and utensils could also act as a source of viruses.

3. Other Transmissions

It is believed that human caliciviruses may spread also by the inhalation of aerosols from vomitus or feces (17). Such airborne particles could contaminate environmental surfaces as well and lead to virus spread by hand/mouth transmission (53). Experimental evidence for both of these points of view is still lacking.

    An overview of the most important foodborne viruses and their transmission routes is given in Table 6.

B. Secondary Spread

Food- and waterborne viruses can give rise to secondary cases more often than many such bacterial pathogens. Therefore, the epidemic potential of foodborne viral diseases remains higher than that of equivalent bacterial infections.

C. Conclusion

Transmission routes of food- and waterborne viruses vary with the type of virus. However, fecal–oral transmission, either direct or indirect, remains the main means of spread. Food

Table 6 Mode of Transmission of Various Viruses[a]

Mode of transmission	HastV	HuCV	HAdV	ROTAV	SRSV	HAV	HEV
Direct fecal–oral (person-to-person)	Yes	Yes	Yes	Yes	Yes	Yes	Yes
Water, food	Yes	Yes	?	Yes	Yes	Yes	Yes
Indirect, e.g., fomites	Yes	Yes	?	Yes	Yes	Yes	?
Airborne	?	Yes	?	?	Yes ?	?	?

[a] Transmission routes marked with an (?) indicate either not known or under discussion.
HastV, human astrovirus; HuCV, human calicivirus; HAdV, adenovirus, enteric adenoviruses; ROTAV, rotaviruses A, B, C; SRSV, small round-structured viruses, caliciviruses genotype I and II; HAV, hepatitis A; HEV, hepatitis E.

at risk are all items that are exposed to contamination by fecal matter or vomitus and that are consumed without any subsequent heat treatment.

V. RISK ASSESSMENT

A. Fecal Shedding

Norwalk virus infection of volunteers showed that the peak of viral shedding was between 25 and 72 h. Virus first appeared in stool at 15 h. Samples collected 7 days after infection were still positive for virus (54).

Volunteers who were repeatedly challenged with Norwalk virus showed that ill volunteers were significantly more likely than well volunteers to have virus antigen in their stool. After challenge, antigen shedding was detected on days 1–13. Ill volunteers shed the antigen longer than healthy individuals (55).

The pattern of shedding of the SRSV particles in the stools of patients who suffered from a food poisoning due to raw oysters was investigated by using electron microscopy. Haruki et al. (56) showed that fecal shedding occurred within 5 days of illness. During the course of illness, after 5 days the concentration of SRSV particles in feces decreased rapidly. Detection rate was 43% among those who were ill during the 5-day course. After this time, SRSV count generally was low and particle clusters were rarely found. The result indicated that the SRSV shedding reaches a peak during the duration of symptoms. Foodborne outbreaks have been described where asymptomatic food handlers with an ill child at home were responsible for preparing a salad (45), the suspected item of food.

In contrast to SRSV, infectious HAV is found in feces reaching peak levels the week or two before the onset of symptoms and diminishing rapidly after liver dysfunction or symptoms appear. However, when a food handler in a commercial setting is diagnosed with hepatitis A, the probability of such an individual giving rise to a recognizable outbreak in the consuming public is less than 1%. On the other hand, untrained, infected food handlers handling food for many other people have caused some significant outbreaks (57).

Astrovirus excretion usually continues for the duration of the diarrhea only, but chronic infections in immunocompromised individuals have been reported (58).

Rotaviruses are transmissible during the acute stage of the disease when large numbers of particles are shed in the feces. Virus shedding may last for 3–7 days, but longer periods of secretion are seen in immunocompromised patients.

In general, there are two main patterns of virus shedding. Some viruses (e.g., astroviruses, SRSVs, rotavirus) are shed only for a short period, mainly in parallel with clinical symptoms. Hepatitis A virus, on the other hand, is shed before onset of illness but most cases are probably noninfectious after the first week of jaundice. Prolonged excretion of HAV has been described only in infants born prematurely (59).

An understanding of these different patterns of viral shedding may be helpful in determining the period of exclusion of symptomatic food handlers and optimal time for collection of fecal samples for laboratory diagnosis.

B. Vomiting

Outbreaks of SRSV due to foods contaminated by vomiting food handlers have been recorded. It has been estimated that 20–30 million virus particles with an infectious dose of 10–100 particles are discharged in the vomitus of such individuals (60). Assuming a kitchen 10 m long, 10 m wide, and 2.0 m high, and assuming that 30 million virus particles are discharged homogeneously by an ill food handler, there will be 150,000 virus particles per cubic meter, or more than 1500 times the viral infectious dose!

The SRSVs are virulent and relatively resistant to environmental disinfection and decontamination. An outbreak was described by Patterson et al. (60) in which a kitchen assistant vomited into a kitchen sink after becoming ill. This sink was used after cleaning and disinfection to prepare a potato salad the next morning. It was subsequently identified as the source of infection in a cohort study of guests who became ill after consumption of the potato salad.

C. Viruses in the Environment

Viruses like SRSVs, rotaviruses, astroviruses, and HAV are remarkably stable in the environment and can survive for months (e.g., hepatitis A) in water; hence the sewage contamination of water supplies, and farms growing oysters and clams. Others, like HuCVs, can remain infectious for several years (17).

1. Surface Water

In Switzerland, for example, surface water that also infiltrates groundwater was tested by RT-PCR for enteroviruses, HAV, and SRSVs. Of 27 water samples tested 15% were positive for enteroviruses and 59% for SRSVs. Hepatitis A virus was not detected. The presence of *Escherichia coli* in the tested water samples did not correlate with the presence or absence of the investigated viruses (61).

2. Irrigation

For the irrigation of crops, river water as well as sewage water is used. It has been shown that viruses may persist for days on vegetation. However, the type of virus and type of crop are clearly important, but the dominant factors are temperature, sunlight, humidity, and radiation and desiccation. Exceptional survival times of up to 2 months are possible. Under normal circumstances almost complete elimination will occur in 5 days (62). When such virus-contaminated water is used on vegetables and fruits that are consumed raw, the result may be VFBDO (63). It has been clearly demonstrated that drip irrigation, particularly when combined with soil covered with plastic sheets, is a method of effluent application that minimizes the risk of crop contamination (62).

3. Drinking Water

Depending on the source of water (groundwater, well water) the risk of drinking water containing viruses varies with the exposure of the water to human sewage. Groundwater contamination of wells may occur after heavy rain fall, contamination by runoff of surface water, leakage of septic tanks, or cross-contamination with sewage water. Groundwater is usually of high microbiological quality; however, aquifers surrounded by permeable or fissured rock are likely to be more highly contaminated.

Rotaviruses may be found in different environmental areas and are known to be resistant. Levels of chlorine used in the terminal disinfection of sewage effluents or drinking water are not sufficient to inactivate these viruses. Some authors even suggest that public water supplies act as reservoirs for the viruses between seasonal epidemics (64,65). However, rotaviruses in drinking water are less resistant to inactivation with chlorine than

Norwalk viruses. Exposure for 30 min to 3.75 mg/L active chlorine is necessary to inactivate the agent (18).

An example of a multiple viral waterborne epidemic took place in a Finnish municipality in April 1994. Some 1500–3000 people (25–50% of the population) had symptomatic acute gastroenteritis. Laboratory findings identified adenovirus, a Norwalk-like agent, SRVs, and group A and C rotaviruses as causative agents. However, Norwalk virus was believed to be the cause of the infections. This epidemic was most probably associated with contaminated drinking water. The groundwater well, situated in the embankment of a river, was contaminated by polluted surface water by a nearby river during the spring flood. A backflow from the river to the well had occurred via a forgotten drainage pipe (66).

Water treatments at the point of use with pressed activated carbon block filter followed by ultraviolet (UV) treatment showed that remove and/or inactivation greater than 99.99% of hepatitis A, poliovirus type 1, and simian rotavirus might be achieved. These findings suggest that waterborne viruses might be removed from drinking water by UV light as well as by chlorination (67).

Drinking water may also lead to illness when contaminated water is used to prepare foods that will not be heat-treated further. An outbreak was described by Warner et al. in 1991 (52) whereby celery was exposed during processing to contaminated water and consequently used in a chicken salad. Of 3000 exposed cadets, approximately 1440 became ill.

4. Seawater

It is well recognized that consumption of raw or improperly cooked oysters harvested from fecally contaminated estuaries can give rise to foodborne outbreaks. One such outbreak of Norwalk virus is believed to have been caused by the contamination of the oyster beds from feces of an infected oyster harvester (68). It is important to note that fecal coliforms and vibrios used as indicators of fecal pollution of oyster-growing waters are inadequate to determine the virological quality of such waters (68).

5. Depuration of Oysters

Oysters may be able to rid themselves of 99% of enteric viruses in 25 days under laboratory conditions (68). Under commercial depuration conditions, rotavirus and HAV could be recovered from bivalves after 96 h of immersion in a continuous flow of ozonated marine water (69).

6. Heat Inactivation of Oysters

Heat can render many viruses noninfectious. There is little information on the ability of cooking oysters to inactivate Norwalk and related viruses. However, outbreaks have been reported whereby consumption of thoroughly cooked (40) or steamed oysters (70) still resulted in cases of infection.

Cooking experiments (69) with experimentally contaminated mussels showed that rotaviruses as well as HAV could still be recovered 5 min after the opening of the mussel valves.

Current RT-PCR techniques in oysters and hard-shelled clams seem to have a limit of detection of Norwalk virus particles when shellfish are seeded with a known virus in the laboratory. This lower limit of detection seems higher than the infectious dose of Norwalk viruses (68,71).

7. Fruit and Vegetables

On fruit, enteric viruses may be present as a result of fecal contamination either before or after harvest. More contamination of the crop may occur at subsequent stages of the processing and handling. Sources of contamination are infected workers handling crop or process steps that involve contaminated water. Viruses of most concern are HAV and SRSVs. The transmission of HAV by fruit juices was described for a variety of different fruits.

Viruses cannot grow on contaminated vegetables and fruits. However, they can survive there and can cause illness in humans upon the ingestion of such items. An increase in the consumption of fresh and lightly processed fruits and vegetables, combined with increased trade in produce from regions where standards of hygiene are poor, enhances the risk of foodborne outbreaks.

Hernandez et al. (72) assessed the presence of rotavirus and HAV in lettuce bought in farmers' markets in San Jose, Costa Rica. These authors were able to detect rotavirus as well as HAV during the period of high prevalence of diarrhea in pooled lettuce samples. In almost all cases fecal coliforms were detected as well.

8. Raw Milk

Raw milk has only occasionally been implicated in outbreaks of human illness with transmission of hepatitis A, poliomyelitis, and encephalitis only having been documented (50,73,74). Educating consumers to avoid bacterial infections by heating raw milk before consumption will result in inactivation of most if not all pathogenic viruses at the same time.

D. People at Risk

Gerba et al. (75) have identified four groups of individuals who are at greatest risk of serious illness and mortality from water- and foodborne enteric pathogens. These include the very young, the elderly, pregnant woman as well as immunocompromised individuals. In the elderly the greater risk from foodborne disease may be due to a combination of reduced gastric acidity, underlying medical conditions, and a weakening of the immune system (76). In view of the growing numbers of the elderly in developed countries, foodborne diseases will continue to be an important source of illness and death in this age group.

E. Conclusions

Application of the principles of HACCP for different types of foods (e.g., fruits, vegetables, seafood, etc.) will allow identifying possible risks caused by viruses to the consumer. In most cases, steps undertaken to inactivate foodborne bacteria may also inactivate viruses. However, much has to be done in educating food personnel to avoid any contamination of food by symptomatic food handlers. Only when food handlers as well as managers are aware of the epidemic potential of foodborne viruses, adequate precautions can be introduced to prevent foodborne illness especially among the most vulnerable.

ACKNOWLEDGMENTS

The author thanks Mrs. Sharon Hunter for the correction of this manuscript, as well as Paul Svoboda and Mike Hobbins for critical discussion.

ABBREVIATIONS

AIDS	acquired immunodeficiency syndrome
CDC	Centers for Disease Control and Prevention (Atlanta, GA)
CDSC	Communicable Disease Surveillance Centre (London, United Kingdom)
ELISA	enzyme-linked immunosorbent assay
EU	European Union
FAO	Food and Agriculture Organization
FBDO	foodborne disease outbreak
HACCP	hazard analysis critical control point
HastV	human astrovirus
HAV	hepatitis A virus
HIV	human immunodeficiency virus
HuCV	human calicivirus
RT-PCR	reverse transcriptase polymerase chain reaction
SRSV	small round structured viruses
SRV	small round viruses
VFBDO	viral foodborne disease outbreak
WHO	World Health Organization

REFERENCES

1. C Bern, RI Glass. Impact of diarrheal diseases worldwide. In: AZ Kapikian, ed. Viral Infections of the Gastrointestinal Tract. New York: Marcel Dekker, 1994, pp. 1–26.
2. J Zahorsky. Hyperemesis hiemis or the winter vomiting disease. Arch Pediatr 46:391–395, 1929.
3. AZ Kapikian, RG Wyatt, R Dolin, TS Thornhill, AR Kalica, RM Chanock. Visualization by immune electron microscopy of a 27-nm particle associated with acute infectious nonbacterial gastroenteritis. J Virol 10:1075–1081, 1972.
4. I Adler, R Zickl. Winter vomiting disease. J Infect Dis 119:668–673, 1969.
5. World Health Organization. WHO Surveillance programme for control of foodborne infections and intoxications in Europe. FAO/WHO Collaborating Centre for Research and Training in Food Hygiene and Zoonoses, First Report, Berlin: 1981.
6. World Health Organization. WHO Surveillance programme for control of foodborne infections and intoxications in Europe. Federal Institute for Health Protection of Consumers and Veterinary Medicine. FAO/WHO Collaborating Centre for Research and Training in Food Hygiene and Zoonoses, Sixth Report 1990–1992, K. Schmidt, ed. Berlin: 1995.
7. AMM Hoogenboom-Verdegaal, JC de Jong, M. During, R. Hoogenveen, JA Hoekstra. Community based study of the incidence of gastrointestinal diseases in the Netherlands. Epidemiol Infect 112:481–487, 1994.
8. P Sockett. Social and economic aspects of food-borne disease. Food Policy 18:110–119, 1993.
9. LA Jaykus. Epidemiology and detection as options for control of viral and parasitic foodborne disease. Emerg Infect Dis 3:529–539, 1997.
10. DE Archer, DA Kessler. Foodborne illness in the 1990s. JAMA 269:2737, 1993.
11. TM Lüthi. Food and waterborne viral gastroenteritis: a review of agents and their epidemiology. Mitt Gebiete Lebensm Hyg 88:119–150, 1997.
12. DO Cliver. New issues in food and environmental virology. J Food Protect 58 (Suppl): 59, 1995.
13. A Stolle, B Sperner. Viral infections transmitted by food of animal origin: the present situation in the European Union. Arch Virol 13 (Suppl):219–228, 1997.

14. RI Glass, J Noel, D Mitchell, JE Herrmann, NR Blacklow, LK Pickering, P Dennehy, G Ruiz Palacios, ML deGuerrero, SS Monroe. The changing epidemiology of astrovirus-associated gastroenteritis: a review. Arch Virol 12 (Suppl):287–300, 1996.

15. S Kriston, MM Willcocks, MJ Carter, WD Cubitt. Seroprevalence of astrovirus types 1 and 6 in London, determined using recombinant virus antigen. Epidemiol Infect 117:159–164, 1996.

16. GS Grohmann, RI Glass, HG Pereira. Enteric viruses and diarrhea in HIV-patients. N Engl J Med 329:14–20, 1993.

17. WD Cubitt. Caliciviruses. In: AZ Kapikian ed. Viral Infections of the Gastrointestinal Tract. New York: Marcel Dekker, 1994, pp 549–569.

18. Centers for Disease Control and Prevention (CDC). Viral agents of gastroenteritis: Public health importance and outbreak management. MMWR 39, RR-5:1–24, 1990.

19. SS Monroe, RI Glass, N Noah, TH Flewett, EO Caul, CI Ashton, A Curry, AM Field, R Madeley, PJ Pead. Electron microscopic reporting of gastrointestinal viruses in the United Kingdom, 1985–1987. J Med Virol 33:193–198, 1991.

20. JJ Gray, X Jiang, P Morgan-Carpner, U Desselberger, MK Estes. Prevalence of antibodies to Norwalk virus in England: detection by enzyme linked immunosorbent assay using baculovirus-expressed Norwalk virus capsid antigen. J Clin Microbiol 31:1022–1025, 1993.

21. NH Bean, JS Goulding, MT Daniels, FJ Angulo. Surveillance for foodborne diseases outbreaks, United States, 1988–1992. J Food Protect 60:1265–1286, 1997.

22. CG Helmick, PM Griffin, DG Addiss, RV Tauxe, DD Juranek. In: JE Everhardt, ed. Digestive diseases in the United States: epidemiology and impact. NIH Publication No. 94-1447. Washington DC: U.S. Department of Health and Human Services, Public Health Service, National Institutes of Health, National Institute of Diabetes and Digestive and Kidney Diseases, 1994, pp 85–123.

23. TM Lüthi, PG Wall, HS Evans, GK Adak, EO Caul. Outbreaks of foodborne viral gastroenteritis in England and Wales: 1992 to 1994. Commun Dis Rep CDR Rev 6:R131–R136, 1996.

24. T Dijuretic T, PG Wall, MJ Ryan, HS Evans, GK Adak, JM Cowden. General outbreaks of infectious intestinal disease in England and Wales: 1992 to 1994. Commun Dis Rep CDR Rev 6:R57–R63, 1996.

25. PN Sockett, JM Cowden, S LeBaigue, D Ross, GK Adak, H Evans. Foodborne disease surveillance in England and Wales: 1989–1991. Commun Dis Rep CDR Rev 3:R159–R173, 1993.

26. ZY Xu, ZH Li, JX Wang, ZP Xiao, DH Dong. Ecology and prevention of a shellfish associated hepatis A epidemic in Shanghai, China. Vaccine (Suppl 1):S67–S68, 1992.

27. JCA Desenclos. KC Klontz, MH Wilder, OV Nainan, HS Margolis, RA Gunn. A multisate outbreak of hepatitis A caused by the consumption of raw oysters. Am J Publ Health 81: 1268–1272, 1991.

28. RD Warner, RW Carr, FK McClesky, PC Johnson, LM Goldy-Elmer, VE Davison. A large nontypical outbreak of Norwalk virus: gastroenteritis associated with exposing celery to nonpotable water and with Citrobacter freundii. Arch Intern Med 151:2419–2424, 1991.

29. K Matsumoto, M Hatano, K Kobayashi, A Hasegawa, S Yamazaki, S Nakata, S Chiba, Y Kimura. An outbreak of gastroenteritis associated with acute rotaviral infection in schoolchildren. J Infect Dis 160:611–615, 1989.

30. BL Herwaldt, JF Lew, CL Moe, DC Lewis, CD Humphrey, SS Monroe, EW Pon, RI Glass. Characterization of a variant strain of Norwalk virus from a foodborne outbreak of gastroenteritis on a cruise ship in Hawaii. J Clin Microbiol 32:861–866, 1994.

31. I Oishi, K Yamazaki, T Kimoto, Y Minekawa, E Utagawa, S Yamazaki, S Inouye, GS Grohmann, SS Monroe, SE Stine, C Carcamo, T Ando, RI Glass. A large outbreak of acute gastroenteritis associated with astrovirus among students and teachers in Osaka, Japan. J Infect Dis 170:439–443, 1994.

32. MDST Timenetsky, V Gouvea, N Santos, ME Alge, JJ Kisiellius, RCC Carmona, H Tanaka, LTM Souza, DF Souza, M Ueda, CB Nascimento, AMG Dias, E Kano, SA Fernandes, AT

Tavecchio, CT Calzada, LK Nakahara, O Araujo, D Dimitrov, MTF Castro, FC Sanda, EA Noce, RF Boni, SN Neme, VA Domenegueti, IAZ Castanheiro, AB Rocha, M Raskin, MM Rocha, SR Baraldi, AY Tanaka, LD Ribeiro, MRN Esper, NR Reis, KMS Carraro. Outbreak of severe gastroenteritis in adults and children associated with type G3 rotavirus. J Diarrh Dis 14:71–74, 1996.

33. E Laursen, O Mygind, B Rasmussen, T Ronne. Gastroenteritis: a waterborne outbreak affecting 1600 people in a small Danish town. J Epidemiol Community Health 48:453–458, 1994.

34. AS Khan, CL Moe, RI Glass, SS Monroe, MK Estes, LA Chapman, X Jiang, C Humphrey, E Pon, JK Iskander, LB Schonberger. Norwalk virus-associated gastroenteritis traced to ice consumption aboard a cruise ship in Hawaii: comparison and application of molecular method based assays. J Clin Microbiol 32:318–322, 1994.

35. MA Kohn, TA Farley, T Ando, M Curtis, SA Wilson, Q Jin, SS Monroe, RC Baron, LM McFarland, RI Glass. An outbreak of Norwalk virus gastroenteritis associated with eating raw oysters: implications for maintaining safe oyster beds. JAMA 273:467–471, 1995.

36. Anonymous. Norwalk-like virus most likely cause of gastroenteritis associated with oysters. Newsletter. WHO Surveillance Programme for Control of Foodborne Infections and Intoxications in Europe 45:3, 1995.

37. M Kukkula, P Arstila, ML Klossner, L Maunula, CHV Bonsdorff, P Jaatinen. Waterborne outbreak of viral gastroenteritis. Scand J Infect Dis 29:415–418, 1997.

38. B Becker. Bakery products causing hepatitis A epidemic in Germany. Newsletter. WHO Surveillance Programme for Control of Foodborne Infections and Intoxications in Europe 49/50: 4, 1996.

39. M Beller, A Ellis, SH Lee, MA Drebot, SA Jenkerson, E Funk, MD Sobsey, OD Simmons III, SS Monroe, T Ando, J Noel, M Petric, JP Middaugh, JS Spika. Outbreak of viral gastroenteritis due to a contaminated well. International consequences. JAMA 278(7):563–568, 1997.

40. S McDonnell, KB Kirkland, WG Hlady, C Aristeguieta, RS Hopkins, SS Monroe, RI Glass. Failure of cooking to prevent shellfish-associated viral gastroenteritis. Arch Intern Med 157: 111–116, 1997.

41. J O'Hallahan. Outbreak of Foodborne hepatitis A associated with a Wellington delicatessen. Newsletter. WHO Surveillance Programme for Control of Foodborne Infections and Intoxications in Europe 49/50:5, 1996.

42. PL Lopalco, P Malfait, S Salmaso, C Germinario, M Quarto, S Barbuti, R Cipriani, A Mundo, G Pesole. A persisting outbreak of hepatitis A in Puglia, Italy, 1996: epidemiological follow up. Eurosurveillance 2:31–32, 1997.

43. CW Hedberg, MT Osterholm. Outbreaks of foodborne and waterborne viral gastroenteritis. Clin Microbiol Rev 6:199–210, 1993.

44. Anonymous. Prevention of foodborne hepatitis A considerations on the vaccination of food-handlers. Newsletter. WHO Surveillance Programme for Control of Foodborne Infections and Intoxications in Europe 41:5, 1994.

45. SV Lo, AM Connoly, SR Palmer, D Wright, PD Thomas, D Joynson. The role of presymptomatic food-handler in a common source outbreak of foodborne SRSV gastroenteritis in a group of hospitals. Epidemiol Infect 113:513–521, 1994.

46. JE Kaplan, R Feldman, DS Campell, C Lookabaugh, GW Gary. The frequency of a Norwalk-like pattern of illness in outbreaks of acute gastroenteritis. Am J Publ Health 72:1329–1332, 1982.

47. TM Lüthi, PG Wall, HS Evans, EO Caul. Applying epidemiological criteria to test for Norwalk-like pattern of illness in outbreaks of acute viral-gastroenteritis in the absence of virological results. Public Health Laboratory Service, 21st Annual Scientific Meeting, University of Warwick, 1996, p 117.

48. Viral Gastroenteritis Subcommittee. Outbreaks of gastroenteritis associated with SRSV's. PHLS Microbiol Dig 10:2–8, 1993.

49. Anonymous. Management of outbreaks of foodborne illness. Guidance produced by a Depart-

ment of Health working group. Department of Health. BAPS, Health Publication Centre, Heywood, UK, 1994, pp 95–96.

50. M Gresikova. Tickborne encephalitis. In: YH Hui, JR Gorham, KD Murell, DO Cliver, eds. Foodborne Disease Handbook, Vol. 2. Diseases Caused by Viruses, Parasites, and Fungi. New York: Marcel Dekker, 1994, pp 113–135.

51. Lee JV, Dawson SR, Ward S, Surman SB, Neal KR. Microbiological indicators of water quality as predictors of illness rates among users of the canoe slalom course at the National Water Sports Centre, Holme Pierrepont, Nottingham. PHLS 21st Annual Scientific Conference, University of Warwick, 1996, p 59a.

52. RD Warner, RW Carr, FK McClesky, PC Johnson, LM Goldy Elmer, VE Davison. A large nontypical outbreak of Norwalk-virus: gastroenteritis associated with exposing celery to non-potable water and with Citrobacter freundii. Arch Intern Med 151:2419–2424, 1991.

53. Caul EO. Small round structured viruses: airborne transmission and hospital control. Lancet 343:1240–1241, 1994.

54. DY Graham, X Jiang, T Tanaka, AR Opekun, HP Madore, ME Estes. Norwalk virus infection of volunteers: new insights based on improved assays. J Infect Dis 170:34–43, 1994.

55. PC Okhuysen, X Jiang, L Ye, PC Johnson, ME Estes. Viral shedding and faecal IgA response after Norwalk virus infection. J Infect Dis 171:566–569, 1995.

56. K Haruki, Y Seto, T Murakami, T Kimura. Pattern of shedding of small, round structured virus particles in stools of patients of outbreaks of food-poisoning from raw oysters. Microbiol Immunol 35:83–86, 1991.

57. DO Cliver. Epidemiology of foodborne viruses. In: YH Hui, JR Gorham, KD Murell, DO liver, eds. Foodborne Disease Handbook, Vol. 2. New York: Marcel Dekker, 1994, pp 159–175.

58. JB Kurtz. Astroviruses. In: AZ Kapikian, ed. Viral Infections of the Gastrointestinal Tract. New York: Marcel Dekker, 1994, pp 569–580.

59. AS Benenson. Control of Communicable Disease Manual. Washington, DC Amercan Public Health Association, 1995, pp 217–220.

60. W Patterson, P Haswell, PT Fryers, J Green. Outbreak of small round structured virus gastroenteritis arose after kitchen assistant vomited. Common Dis Rep 7:R101–R103, 1997.

61. M Gilgen, J Lüthy, D Häfliger, HP Bühler, U Müller, D Germann, P Hübner. Mikrobiologische Untersuchung von See- und Flussbädern des Kantons Bern auf ausgewählte enteropathogene Viren und Escherichia coli. Mitt Gebiete Lebensm Hyg 88:321–334, 1997.

62. RG Feachem, DJ Bradley, H Garelick, DD Mara. Sanitation and Disease. World Bank Studies in water supply and sanitation, Vol. 3. Chichester: John Wiley and Sons, 1983, pp 138–151.

63. BK Ward, LG Irving. Virus survival on vegetables spray-irrigated with wastewater. Water Res 21:57–63, 1987.

64. VC Rao, Metcalf TG, JL Melnick. Development of a method for concentrating of rotavirus and ist application to recovery of rotavirus from estuarian waters. Appl Environ Microbiol 52:484–488, 1986.

65. SA Ansari, VS Springthorpe, SA Sattar. Survival and vehicular spread of human rotaviruses: possible relation to seasonality of outbreaks. Rev Infect Dis 13:448–461, 1991.

66. M Kukkula, P Arstila, ML Klossner, L Maunula, CHV Bonsdorff, P Jaatinen. Waterborne outbreak of viral gastroenteritis. Scand J Infect Dis 29:415–418, 1997.

67. M Abbaszadegan, MN Hasan, CP Gerba, PF Roessler, ER Wilson, R Kuennen, E Van Dellen. The disinfection efficacy of a point-of-use water treatment system against bacterial, viral and protozoan waterborne pathogens. Water Res 31: 574–582, 1997.

68. MA Kohn, TA Farley, T Ando, M Curtis, S Wilson, Q Jin, SS Monroe, RC Baron, LM McFarland, RI Glass. An outbreak of Norwalk virus gastroenteritis associated with eating raw oysters: implications for maintaining safe oyster beds. JAMA 273:466–471, 1995.

69. FX Abad, RM Pinto-RM, R Gajardo, A Bosch. Viruses in mussels: public health implications and depuration. J Food Protect 60:677–681, 1997.

70. KB Kirkland, RA Meriwether, JK Leiss, WR MacKenzie. Steaming oysters does not prevent Norwalk-like gastroenteritis. Public Health Reports 111:527–530, 1996.
71. AB Dix, LA Jakus. Virion concentration method for the detection of human enteric viruses in extracts of hard shelled clams. J. Food Protect 61:458–465, 1998.
72. F Hernandez, R Monge, C Jimenez, L Taylor. Rotavirus and hepatitis A virus in market lettuce (Latuca sativa) in Costa Rica. Int J Food Microbiol 37:221–223, 1997.
73. FL Bryan. Epidemiology of milk-borne diseases. J Food Protect 46:637–649, 1983.
74. G Schagemann. Viruses In: The significance of pathogenic microorganisms in raw milk. Monograph of International Dairy Federation, Brussels, 1994.
75. CP Gerba, JB Rose, CN Haas. Sensitive populations: who is at the greatest risk? Int J Food Microbiol 30:113–123, 1996.
76. KC Klontz, WH Adler, M Potter. Age-dependent resistance factors in the pathogenesis of foodborne infectious disease. Aging Clin Exp Res 9:320–326, 1997.

9

Environmental Considerations in Preventing the Foodborne Spread of Hepatitis A

Syed A. Sattar
University of Ottawa, Ottawa, Ontario, Canada

Sabah Bidawid
Health Canada, Ottawa, Ontario, Canada

I. INTRODUCTION

Hepatitis A remains a major public health problem (1). It is the fifth most commonly reported infectious disease in the United States and ranks sixth among the top 10 causes of foodborne disease (2). It is an important cause of morbidity and results in significant economic losses in many parts of the world (3). Hepatitis A virus (HAV), the causative agent of hepatitis A, is transmitted mainly through the fecal–oral route (see Chapter 2). Whereas ingestion of HAV-contaminated water or food can result in the transmission of the disease (5), this chapter will not deal with the role of water in the transmission of hepatitis A.

In recent years there has been an increase in the number of cases of hepatitis A in Canada (6,7) and the United States (2,7,8), and more frequent reports of its foodborne spread; the outbreaks recorded in Colorado in 1993 (9) and in Michigan in 1997 (10) are cases in point. It should be noted here that at the present time hepatitis A is the only reportable foodborne viral disease in North America (5). It is also one of the more severe foodborne infections and as a result is on the list of severe hazards in Appendix V of the U.S. Food and Drug Administration (5). In view of this and the increasing threat from the disease, we propose a set of recommendations for reducing HAV transmission via foods instituting environmental control measures at various stages of the food production, processing, and marketing chain. Whereas the emphasis here is on hepatitis A, the measures listed should apply equally well in preventing and controlling the spread of foodborne infections in general.

II. BACKGROUND

A. The Virus and the Disease

Hepatitis A virus belongs to the genus *Hepatovirus* in the family Picornaviridae (11,12) and is shed in the feces of infected individuals. The virus is transmitted primarily through the ingestion of fecally contaminated material (13). Infections with HAV are endemic worldwide but cases occur much more frequently in regions with lower standards of hygiene. Outbreaks caused by it occur more frequently in settings such as hospitals (14,15), day care centers (16), schools (17,18), and in association with foods and food service establishments (19–22).

Hepatitis A has an incubation period of 15–50 days (average: 28 days). Virus excretion begins during the incubation period and continues for 7–10 days after the onset of clinically recognizable illness (11). After that time, excretion starts to decline and almost disappears with the appearance of jaundice. Nearly 90% of children under 5 years of age and 30–50% of adults infected with HAV remain asymptomatic (23,24). Those who become clinically ill may exhibit symptoms such as fever, chills, headache, fatigue, and malaise. These may be followed by anorexia, nausea, vomiting, diarrhea, right upper abdominal pain, passage of dark urine, and jaundice. The disease lasts for a few weeks, and complete recovery in adults usually takes a few months.

Infection with HAV confers lifelong immunity. Detection of IgG antibodies in the serum of an individual indicates a previous infection with HAV, whereas detection of IgM antibodies is indicative of a recent or current infection with the virus (25,26). Temporary protection for a few months is possible by passive immunization against HAV by intramuscular injection of pooled human immunoglobulin (Ig) either prior to exposure or within 2 weeks of exposure to the virus, but not later (27). Proper passive immunization protects 80–90% of immunized individuals against clinical disease but may not stop the fecal shedding of the virus. The World Health Organization (WHO) and the U.S. Centers for Disease Control and Prevention (CDC) recommend a dose of 0.02 mL/kg body weight to be administered intramuscularly within 2 weeks of suspected or anticipated exposure (3); such a dose may confer protection for 5–8 months.

Vaccines are now available for active prophylaxis against hepatitis A. Active immunization confers good (>90%) protection that may last for several years (28,29).

B. Transmission by Foods

Many types of food have been implicated in the transmission of HAV (5). These include shellfish, salads, sandwiches, vegetables, fruits, reconstituted frozen orange juice, ice cream, cheese, rice pudding, iced cake, custard, milk, bread, cookies, and other raw or undercooked foods. While the minimal infective dose for HAV in humans is not known, it is believed that as little as one infectious virus particle may be sufficient to infect susceptible hosts by the oral route. Contamination of foods with HAV may occur in several different ways: fruits and vegetables cultivated in and/or irrigated with fecally contaminated materials, shellfish grown in and harvested from fecally polluted waters (see Chapter 2), processing and preparation of food in fecally soiled equipment, and handling of ready-to-eat items of food by individuals with poor personal hygiene (30). Food establishments with poor sanitary conditions and inadequate waste disposal systems, along with unsatisfactory manufacturing practices, may also contribute to food contamination.

C. Environmental Factors

Hepatitis A virus is quite resistant to many environmental conditions, as well as physical and chemical agents (31–34). The virus survives for hours on human hands (35), for days on articles of common use and environmental surfaces (36), and for a few days to weeks in dried feces (37). Hands play an important role in the spread of HAV, and the transfer of infectious virus can occur readily on contact between animate and inanimate surfaces (35).

Hepatitis A virus can readily survive freezing temperatures, and persist in freshwater or salt water for up to 12 months (31). The virus may remain infectious for several days or longer in contaminated foods, and can be found in exceedingly high titers in shellfish, which are capable of concentrating the virus from contaminated water.

Heat is the most effective measure in the inactivation of HAV (38–41). Complete inactivation of HAV in the meat of shellfish can be achieved after heating (steaming) shellfish to an internal temperature of 85–90°C for 1.5 min (42). Lower temperatures have been reported to inactivate only a portion of the virus particles in suspension or feces. Recent studies with respect to heat inactivation of HAV in dairy products (skim and homogenized milk, and table cream) indicate that at 71°C a significant 4 $\log_{10}$ reduction in HAV titer was achieved in both skim and homogenized milk when exposed for 8.5 and 9.4 min, respectively. A 4 $\log_{10}$ reduction in HAV titer in table cream was attainable at 71°C after 13.5 min of exposure (43). These studies clearly indicate that milk HTST (high-temperature short-time) pasteurization is not sufficient to destroy HAV.

Only a few germicidal products are effective in inactivating HAV when used at appropriate concentrations (32,44,45). Ordinary soap and water primarily remove the virus from hands, where it was found that approximately 15–20% of the virus may still remain after casual washing (34). Ultraviolet irradiation of HAV-contaminated phosphate-buffered saline at 1.1-W output for 1 min could inactivate the virus (46,47). γ-Irradiation of sewage sludge only partially (<2.5 $\log_{10}$) inactivated HAV even at a dose of 1200 krad (45,47). A recent study on the effect of various doses of γ-irradiation on the inactivation of HAV indicated that at 10 kGy, only a 2 $\log_{10}$ and a 3 $\log_{10}$ reduction in virus titer in strawberries and lettuce, respectively, could be achieved (48).

III. SUMMARY OF FACTS TO BE CONSIDERED IN ENVIRONMENTAL CONTROL

The following well-established facts about hepatitis A and the virus that causes it have been used as the basis in the design of the recommendations:

1. Hepatitis A continues to take its toll on human health. In fact, recent years have seen an increase in the incidence of this disease in the United States, Canada, and elsewhere. The reasons for this increase are many but perhaps the important among them are (a) increased travel to and from areas where the disease is more common; (b) changes in our lifestyle including increases in high-risk sexual practices and the use of illicit drugs; (c) increasing importations of fresh fruits and vegetables from regions with poorer standards of hygiene and less stringent environmental controls; (d) ever larger numbers of children in day care where there is a higher risk of exposure to HAV with a correspondingly greater chance of passing the infection onto others in the family and the community.

2. Only one serotype of HAV is known and infection with the virus gives long-lasting immunity.

3. The average incubation period of hepatitis A is about 28 days and the disease itself may last for 2–6 weeks. Fewer than 1% of those who get the disease die from it. Therefore, the economic losses arise mainly from school and work days lost.

4. Persons with hepatitis A do not become chronic carriers of the virus. In view of this, individuals who have completely recovered from the disease pose no risk as food handlers.

5. The principal source of HAV is the feces of infected humans and virus spread occurs mainly through the ingestion of fecally contaminated material.

6. The virus can survive for several hours on human hands (35) and for several days on environmental surfaces indoors (36). Causal contact between contaminated and clean surfaces can readily lead to transfer of infectious virus (35). Therefore, handwashing alone without regular and proper cleaning and decontamination of surfaces and objects handled by many may not be sufficient in interrupting the spread of HAV.

7. Virus excretion in feces begins well before the appearance of clinical signs and symptoms of the disease and continues for several days thereafter. Many individuals, particularly children, may become infected with the virus without becoming sick. Such "silent" disseminators of the virus may pass the virus on to others around them. These factors make it virtually impossible to identify and isolate infected cases *before* they begin to excrete the virus; hence, the emphasis on good personal and environmental hygiene and the importance of washing hands properly and frequently.

8. Individuals who may contract the disease through the ingestion of HAV-con-

taminated food or water can give rise to ''secondary cases'' by passing the virus on to others around them.

9. Shellfish can concentrate human pathogens such as HAV from water when they filter-feed. Therefore, the levels of such pathogens in them may be several fold higher than those in the surrounding waters. HAV and other viruses of human origin cannot grow in shellfish but can survive there for prolonged periods (5). Depuration of shellfish cannot be relied on to rid shellfish of viral pathogens.

10. Hepatitis A virus can also remain infectious for days to weeks in dairy foods and on vegetables and fruits that are generally consumed raw (49). This highlights the need for avoiding food contamination in the first place. Cross-contamination of foods during any handling or preparation must be avoided for the same reason.

11. Hepatitis A virus is relatively resistant to inactivation by many common physical and chemical agents (33). This may explain its propensity for causing outbreaks through foods and other vehicles. It also reemphasizes the importance of avoiding food contamination to start with because reprocessing of foods with suspected HAV contamination is often not feasible without damaging the nutritive and aesthetic qualities of the suspect item.

12. Laboratory tests have shown many germicidal soaps to be more active against foodborne bacteria and less so against viruses such as HAV (34). Therefore, a given handwash agent per se should not be relied on to render hands safe. Instead, proper decontamination of hands must be regarded as the combined outcome of frequency of washing, application of the handwash agent with rubbing over both hands for at least 10 s, thorough rinsing in running water, and proper drying.

13. Hand rubs that contain >60% ethanol show good activity against HAV (34). However, such hand rubs may contain other germicidal chemicals and perfumes rendering them unsuitable for use by food handlers.

14. If and when a germicidal soap is to be used, it must be selected with care to ensure that it is nonirritating to the skin and safe for repeated use. Any product that is perceived as harmful by the staff will interfere with handwashing compliance.

15. Regular cleaning and proper decontamination of settings where food is processed and served are important adjuncts to reducing the risk of spread of HAV and other foodborne pathogens. In most situations, washing with soap and warm water may be sufficient. However, where chemical germicides are considered necessary, they must be selected with great care, ensuring that they are (a) demonstrated to be effective against common foodborne pathogens, (b) safe for humans, (c) environmentally benign, (d) fast acting, (e) compatible with the materials to be treated, and (f) easy to use.

16. All employees must be trained properly in the use of chemical germicides and all such products must be stored properly and applied in accordance with label directions.

17. Safe and effective vaccines are now available against hepatitis A. In the food industry the vaccination of food handlers is the best means of eliminating the risk of HAV spread through them. However, currently available vaccines are

expensive and require at least two injections over a period of several weeks. This makes them unsuitable for seasonal and short-term employees.

IV. RECOMMENDATIONS FOR ENVIRONMENTAL CONTROL

This guide is based on a consideration of data from our own research on HAV, an extensive review of the literature on foodborne outbreaks of the disease, and discussions with experts in food processing. The guide provides information and advice to (a) primary food production personnel, (b) retail and food service managers, (c) food handlers, including unpaid food handlers and servers, and (d) regulatory authorities.

A. Food Production, Transportation, and Storage

1. As far as possible, do not use untreated human feces or raw/inadequately treated sewage sludge to grow foods. This is particularly important for the cultivation of vegetables and fruits that are consumed raw or that do not require peeling. Note that cross-contamination of foods with HAV and other pathogens can readily occur during harvesting, storage, and transportation.
2. Do not grow shellfish in or harvest them from waters known to be contaminated with human fecal material or sewage. Remember that bacterial indicators of the sanitary quality of water are not always reflective of the presence or absence of infectious agents such as HAV. If such items of food are suspected to be contaminated, advise consumers to properly heat-treat them before consumption.
3. Do not use fecally polluted waters for the irrigation or washing of any foods that are consumed raw. Avoid the use of such waters also for the spraying of pesticides.
4. Protect the harvested items of food from contamination by human and animal waste during storage, any further handling, and transportation. This should include programs to limit access of vermin and insects to the produce.
5. Protect all food contact surfaces from fecal contamination and thoroughly clean containers and equipment used for food transportation and storage.
6. Adhere to ''Good Manufacturing Practices'' in the processing of foods. Further details in this regard have been published (50,51).
7. Educate all farm workers on the importance of good personal hygiene and means by which fecal contamination of foods can be avoided.

B. Washrooms

1. Provide employees at all stages of food industry with adequate, proper, and conveniently located facilities for handwashing. This is of utmost importance in preventing the spread of foodborne pathogens. Also ensure that such facilities are supplied at all times with sufficient quantities of clean water (preferably warm water), suitable handwash agents and items for drying washed hands (see below).

2. Installation of knee-, foot-, or wrist-operated faucets or electronically triggered ones is highly recommended. Such devices avoid the need to touch contaminated faucet handles with washed hands. When faucet handles and door knobs/handles must be touched with washed hands, train employees to do so with the paper or cloth towel used for drying the hands.

3. Wherever possible, the exit door to the washroom should be designed to open without the necessity of a door knob or handle being touched. This is essential to avoid the immediate recontamination of washed hands while exiting the washroom.

4. Institute and regularly monitor a program for the frequent and proper cleaning of washrooms and other handwashing facilities.

C. Handwashing and Hand Decontamination

1. Instruct all members of the staff on the importance of good personal hygiene and the need for proper and frequent handwashing to prevent the spread of foodborne infections.

2. Train all employees in the proper technique for washing hands. The recommended method for this purpose is as follows: (a) wet hands with water; (b) apply the handwash agent and rub it thoroughly for no less than 10 s on the surfaces of both hands; (c) rinse hands well in running tap water; and (d) dry them well.

3. In most situations, frequent and thorough washing of hands with an ordinary soap and water and their proper drying may be enough to render hands safe for handling foods.

D. Environmental Decontamination

1. Clean all food contact surfaces regularly and thoroughly. This is especially important in settings where foods to be consumed raw are handled and served.

2. If and when using chemical disinfectants, follow label directions for the product.

3. Rinse the product well to avoid contamination of foods.

E. Food Handlers

1. Prepare foods on properly cleaned and decontaminated surfaces; use clean utensils and avoid the use of the same utensils for preparing raw, partially cooked, and cooked foods; use the appropriate temperature and time combination for cooking foods; wash fruits and vegetables thoroughly before serving; clean and disinfect equipment, surfaces, and facilities in which food is handled to an adequate extent.

2. Consider wearing gloves to avoid fecal contamination of hands, particularly when handling foods served raw such as salads, certain types of shellfish, and fruits. Remember that fecally contaminated gloves can also spread pathogens through cross-contamination.

3. Food handlers with children in day care are at a higher risk of acquiring HAV because such children may contract the disease from others and the infection in young children is often subclinical. Therefore, maintain a close watch on

interfamilial spread and take appropriate preventive measures such as vaccination.

F. Retail and Food Service Managers

1. Provide adequate and conveniently located facilities for hand washing and drying in the food preparation area as required. Appropriate disinfectants and washing/cleaning products should be available, as well as an adequate supply of hot water.

2. Monitor staff for safe food handling techniques and encourage the practice of good personal hygiene. Also encourage employees to report illness, particularly diarrhea, without fear of loss of income or loss of job. Fear of loss of income while ill and away from work has led to a good deal of ''at-risk'' food handling over the years.

3. As a preventive measure, exclude any food handler(s) from work for 7–10 days after the appearance of jaundice. If a food handler is suspected of exposure to HAV, seriously consider excluding the individual from handling any foods until past the incubation period.

4. Place posters in various areas of the establishment directing and reminding employees of safe food handling techniques and the need for good personal hygiene, with particular emphasis on frequent and thorough washing and drying of hands. Make employees aware of their crucial role in preventing cases of HAV and other foodborne infections in customers, co-workers, and family contacts.

5. Give serious consideration to the fate of food items that may have been handled by an employee suspected of suffering from hepatitis A. Measures such as reprocessing or disposal of potentially contaminated foods depend on a number of considerations. These include the type of food in question, its quantity, stage of potential contamination, feasibility and success of treatment or reprocessing, and the safety of the food to consumers after the reprocessing treatment. HAV can retain its infectivity for several days in many types of foods, especially when they are refrigerated or frozen (49). The only general guide that can be given in such situations is to heat-treat the item, if and when possible, to inactivate the virus and make the food safe for human consumption. There may be situations where economic losses due to such contamination of food become unavoidable in order to protect the health of the consumer and safeguard the reputation of the company.

6. Educate and train food supervisors and managers to help implement the safety rules and to identify hazards and problems. Further details in this regard are available (50,51).

7. Implement the hazard analysis critical control point (HACCP) (52) system in the handling and processing of foods. The basic criteria for the implementation of HACCP include (a) identification of hazards; (b) determination of location or process that needs to be controlled, i.e., critical control point (CCP); (c) establishing critical limits for each CCP; (d) monitoring the efficacy of established control measures; (e) taking required corrective action to maintain efficacy of established control measures; and (f) maintaining adequate and relevant records of the operation.

8. Provide sufficient education and training to food handlers in the basic principles of food safety, including HACCP. Particular emphasis should be on time and temperature control; personal hygiene; sources of food pathogens; potential for cross-contamination of foods during transportation, handling, preparation, storage, serving, etc.; factors determining the survival and growth of pathogens in general in foods and in premises where items of food are processed and stored; and the need to report illness immediately (particularly those with gastrointestinal upset or jaundice). Teach food handlers proper methods of hand washing and drying, and of cleaning and decontaminating surfaces and equipment used in food preparation.

G. Regulatory Authorities

1. Offer specific educational/training programs to apprise regulatory agency personnel, including public health inspectors, of potential means of contamination of foods with HAV.
2. Develop and implement inspection programs incorporating the HACCP approach for food processing, food service, and retail food establishments.
3. Brief owners and managers of food producing, processing, and marketing establishments on their responsibilities with regard to food hygiene and the training and monitoring of their staff.
4. Develop and distribute as widely as possible pamphlets written in simple language about what employees should do to prevent the spread of hepatitis A and other diseases through foods.

V. CONCLUDING REMARKS

The experimental evidence generated in the past decade clearly shows the ability of HAV to survive well in the indoor environment, on human hands, as well as in or on many items of food (see Chapter 2). It has also been demonstrated that transfer of infectious virus can occur readily on casual contact between hands and environmental surfaces (35) and that HAV is relatively resistant to inactivation by chemical germicides and other chemical and physical agents (53). In view of all these considerations, it is considered ever more urgent to apprise all those connected with the food industry on the importance of environmental and other control measures based on the most recent scientific data available on HAV.

All aspects of the guide apply equally to other foodborne diseases except the recommendation with regard to vaccination against HAV. The advent of active immunization against hepatitis A should be regarded as an important advance in public health because its proper and sustained use could see the eventual eradication of this disease. The injectable nature of the available vaccines makes them somewhat expensive and difficult to administer. But the vaccination of targeted individuals, particularly those who handle food and work in day care and health care settings, should herald a noticeable reduction in the number of cases of hepatitis A in North America and other industrialized parts of the world.

No amount of published information on hepatitis A and its causative agent is likely to reduce the threat of spread of the disease through foods unless it reaches all those who work in the food industry in an understandable and meaningful format. This is particularly important for workers who actually handle foods at various stages of its production and sale on a regular basis. While it would not be feasible for a book chapter such as this to educate all those concerned, it is hoped that the information it contains will be helpful in the creation of instructional material to address the specific needs of the different components of the food industry.

ABBREVIATIONS

CDC   Centers for Disease Control and Prevention (Atlanta, GA)
HACCP   hazard analysis critical control point
HAV   hepatitis A virus
IgM   immunoglobulin G
WHO   World Health Organization

REFERENCES

1. RS Koff. Hepatitis A. Lancet 351:1643–1649, 1998.
2. Centers for Disease Control and Prevention. Ten leading nationally notifiable infectious diseases—United States, 1995. MMWR 45(41):883–884, 1996.
3. World Health Organization. Public health control of hepatitis A: memorandum from a WHO meeting. Bull WHO 73:15–20, 1995.
4. T Cromeans, M Favorov, O Nainan, H Margolis. Hepatitis A and E viruses. In: Foodborne Disease Handbook, 2nd ed. Marcel Dekker, New York, 2000, pp. 23–76.
5. DO Cliver. Virus transmission via food. World Health Sta Q 50:91–104, 1997.
6. Morbidity and Mortality Weekly Report (MMWR). Hepatitis A among homosexual men— United States, Canada, and Australia. 41(9):155–164, 1992.
7. Canada Communicable Disease Report. Vol. 24S7 (Oct., 1998). Laboratory evidence of human viral and selected non-viral infection in Canada—1989 to 1996.
8. Centers for Disease Control and Prevention. Summary of Notifiable Diseases, United States, 1996. MMWR 45:3, 1996.
9. Is food hepatitis on the rise? Food Protect Rep V 1(1), 1993.
10. FE Shaw. HAV-tainted frozen strawberries top national news: tale of the outbreak. Hepatitis Control Rep 1–12, 1997.
11. JL Melnick. History and epidemiology of hepatitis A virus. J Infect Dis 171 (Suppl 8), 1995.
12. SM Feinstone. Hepatitis A: epidemiology and prevention. Eur J Gastroenterol Hepatol 8:300– 305, 1996.
13. S Krugman, JP Giles, J Hammond. Infectious hepatitis: evidence for two distinctive clinical, epidemiological, and immunological types of infection. JAMA 200:365–373, 1967.
14. LS Rosenblum, ME Villarino, OV Nainan, AL Melish, SC Hadler, PP Pinsky, WR Jarvis, CE Ott, HS Margolis. Hepatitis A outbreak in a neonatal intensive care unit: risk factors for transmission and evidence for prolonged viral excretion among preterm infants. J Infect Dis 164:476–482, 1991.
15. JN Hanna, MR Loewenthal, P Negel, DJ Wenck. An outbreak of hepatitis A in an intensive care unit. Anaesth Intensive Care 24:440–444, 1996.
16. SC Hadler, L MFarland. Hepatitis in day care centres: epidemiology and prevention. Rev Infect Dis 8:548–557, 1986.
17. G Levinthal, M Ray. Hepatitis A: from epidemic jaundice to a vaccine-preventable disease. Gastroenterologist 4:107–117, 1996.

18. SM Lemon. Type A viral hepatitis: epidemiology, diagnosis, and prevention. Clin Chem 43: t-9, 1997.

19. A Mele, T Stroffolini, F Palumbo, G Gallo, P Ragni, E Balocchini, ME Tosti, R Corona, A Marzolini, A Moiraghi. Incidence of and risk factors for hepatitis A in Italy: public health indications from a 10-year surveillance. SEIEVA Collaborating. J Hepatol 26:743–747, 1997.

20. RG Pebody, T Leino, P Ruutu, L Kinnunen, I Davidkin, H Nohynek, P Leinikki. Foodborne outbreaks of hepatitis A in a low endemic country: an emerging problem? Epidemiol Infect 120:55–59, 1998.

21. CA Reed, J Hollingsworth, B Kaplan. Hepatitis A infrequent food link, but of concern. J Am Vet Med Assoc 210:1398, 1997.

22. AC Weltman, NM Bennett, DA Ackman, JH Misage, JJ Campana, LS Fine, AS Doniger, GJ Balzano, GS Birkhead. An outbreak of hepatitis A associated with a bakery, New York, 1994: the 1968 ''West Branch, Michigan'' outbreak repeated. Epidemiol Infect 117:333–341, 1996.

23. WM Lednar, SM Lemon, JW Kirkpatrick, RR Redfield, LM Fields, PW Kelly. Frequency of illness associated with epidemic hepatitis A virus infection in adults. Am J Epidemiol 122: 226–233, 1985.

24. SA Locarnini, AA Ferris, NI Lehmann, I Gust. The antibody response following hepatitis A infection. Intervirology 8:309–318, 1977.

25. HW Kao, M Aschavai, AG Redeker. The persistence of hepatitis A IgM antibody after acute clinical hepatitis A. Hepatology 4:933–936, 1984.

26. Centers for Disease Control and Prevention. Prevention of hepatitis A through active or passive immunization: recommendations of the Advisory Committee on Immunization Practices. MMWR 45 (RR-15):1–30, 1996.

27. A Werzberger, B Mensch, B Kuter, L Brown, J Lewis, R Sitrin, W Miller, D Shouval, B Wiens, G Calandra, J Ryan, P Provost, D Nalin. A controlled trial of formalin-inactivated hepatitis A vaccine in healthy children. N Engl J Med 327:453–457, 1992.

28. TF Bader. Hepatitis A vaccine. Am J Gastroenterol 91:217–222, 1996.

29. G Yao. Clinical spectrum and natural history of viral hepatitis A in the 1988 Shanghai epidemic. In: Viral Hepatitis and Liver Disease (FB Hollinger, SM Lemon, and HS Margolis, eds), Williams and Wilkins, Baltimore, 1991, p 76.

30. PW Lowry, R Levine, DF Stroup, RA Gunn, MH Wilder, C Konigsberg Jr. Hepatitis A outbreak on a floating restaurant in Florida. Am J Epidemiol 129:155–164, 1986.

31. MD Sobsey, PA Shields, FS Hauchman, AL Davis, VA Rullman, A Bosch. Survival and persistence of hepatitis A virus in environmental samples. In: Viral hepatitis and liver disease (AJ Zuckerman, ed). Alan R. Liss, New York, 1988, pp 121–124.

32. O Thraenhart. Measures for disinfection and control of viral hepatitis. In: Disinfection, sterilization, and preservation (SS Block, ed). Lea & Febiger, Philadelphia, 1991, pp 445–471.

33. JN Mbithi, VS Springthorpe, SA Sattar. Chemical disinfection of hepatitis A virus on environmental surfaces. Appl Environ Microbiol 56:3601–3604, 1990.

34. JN Mbithi, VS Springthorpe, SA Sattar. Comparative in vivo efficiencies of handwashing agents against hepatitis A virus (HM-175) and poliovirus type 1 (Sabin). Appl Environ Microbiol 59:3463–3469, 1993.

35. JN Mbithi, VS Springthorpe, JR Boulet, SA Sattar. Survival of hepatitis A virus on human hands and its transfer on contact with animate and inanimate surfaces. J Clin Microbiol 30: 757–763, 1992.

36. JN Mbithi, VS Springthorpe, SA Sattar. Effect of relative humidity and air temperature on survival of hepatitis A virus on environmental surfaces. Appl Environ Microbiol 57:1394–1399, 1991.

37. KA McCaustland, WW Bond, DW Bradley, JW Ebert, JE Maynard. Survival of hepatitis A in feces after drying and storage for one month. J Clin Microbiol 16:957–958, 1982.

38. JV Parry, PP Mortimer. The heat sensitivity of hepatitis A virus determined by a simple tissue culture method. J Med Virol 14:277–283, 1984.

39. B Fleming, A Billing, A Vallbracht, K Botzenhart. Inactivation of hepatitis A virus by heat and formaldehyde. Water Sci Technol 17:43–45, 1985.

40. S Krugman, JP Giles, J Hammond. Hepatitis virus: effect of heat on the infectivity and antigenicity of the MS-1 and MS-2 strains. J Infect Dis 122:432–436, 1970.

41. DA Peterson, LG Wolfe, EP Larkin, FW Deinhardt. Thermal treatment and infectivity of hepatitis A virus in human faeces. J Med Virol 2:201–206, 1978.

42. J Millard, H Appleton, JV Parry. Studies on heat inactivation of hepatitis A virus with special reference to shellfish. Part 1. Procedures for inactivation and recovery of virus from laboratory-maintained cockles. Part 2. Heat inactivation of hepatitis A virus in artificially contaminated cockles. Epidemiol Infect 98:397–414, 1987.

43. S Bidawid, JM Farber, SA Sattar. Heat inactivation of hepatitis A virus (HAV) in dairy foods. J Food Prot (in press).

44. PJ Provost, BS Wolanski, WJ Miller, OL Itensohn, WJ McAleer, MR Hilleman. Physical, chemical, and morphologic dimensions of human hepatitis A virus strain CR326. Proc Soc Biol Med 148:532–539, 1975.

45. R Scheid, F Deinhardt, G Abb J Frosner, R Zachoval, and G Siegl. Inactivation of hepatitis A and B viruses and risk of iatrogenic transmission. In: Viral hepatitis: 1981 International Symposium (W Szmuness, HJ Alter, JE Maynard, eds.). Franklin Inst. Press, Philadelphia, 1981, pp. 627–628.

46. AJ Zuckerman. Hepatitis A and non-A, non-B hepatitis, hepatitis C, hepatitis E. In: AJ Zukkerman, JE Banatvala, JR Pattison, eds. Principles and Practice of Clinical Virology, 2nd ed. John Wiley and Sons, London, 1990, pp 3–151.

47. G Siegl. The biochemistry of hepatitis A virus. In: Hepatitis A (RJ Gerety, ed). Academic Press, New York 1984, pp 9–32.

48. S Bidawid, JM Farber, SA Sattar. Gamma irradiation of hepatitis A virus in fruits and vegetables. Int J Food Microbiol (in press).

49. S Bidawid, JM Farber, SA Sattar. Survival of hepatitis A virus in food stored at modified atmosphere packaging environment. Abstract presented at the 66th Cojoint Meeting on Infectious Diseases, Newfoundland, 1998.

50. World Health Organization. Food and Agriculture Organization of the United Nations. Codex Alimentarius General Principles of Food Hygiene. ALINORM 95/13.

51. Code of Practice: General principles of food hygiene for use by the food industry in Canada, A Health and Welfare Canada Publication, 1983.

52. Sanitation Code for Canada's Foodservices Industry. Published by the Canadian Restaurant and Foodservices Association, 1992.

53. International Commission on Microbiological Specifications for Food (ICMSF) (B Simonsen, FL Bryan, JHB Christian, TA Roberts, JH Silliker, RB Tompkin). Prevention and control of foodborne salmonellosis through application of hazard analysis critical control point (HACCP). Int J Food Microbiol 4:227–247, 1987.

10

Taeniasis and Cysticercosis

Zbigniew S. Pawlowski
University of Medical Sciences, Poznan, Poland

K. D. Murrell
U.S. Department of Agriculture, Beltsville, Maryland

I. INTRODUCTION

Taeniasis is caused by the presence of adult cstodes or tapeworms in the intestines of humans and other mammals (primarily meat eaters). *Taenia saginata* (beef tapeworm) and *T. solium* (pork tapeworm) are the two classically known causes of taeniasis in humans. Cysticercosis is a tissue infection with a larval or metacestode stage of (cattle, humans, pigs). Larval stages of *T. saginata* occur only in beef and larval *T. solium* only in pigs and humans. Both *T. saginata* and *T. solium* are cosmopolitan tapeworm parasites with a high endemicity in Africa, Asia, Latin America, and some other countries, where humans and domestic animals live together under poor sanitary conditions and where raw/undercooked pork or beef is consumed (1). Because of the increase in tourism and immigration, taeniasis and cysticercosis have become widely disseminated worldwide. Estimated economic losses from medical care for neurocysticercosis, such as hospitalization, chemotherapy, neurosurgery, and radiological imaging, are great. It has been calculated that U.S. $14.5 million was spent in Mexico during 1980 to treat the 2700 new hospitalized patients (2). Because of the condemnation of infected carcasses at slaughter, more than U.S. $43 million was lost in 1980 in Mexico, which is equivalent to two-thirds of the total investment in pig production (2). In Africa, the loss to the cattle industry from *T. saginata* infection may be as high $1–2 billion per year.

Recently *T. saginata*-like tapeworms, obtained from Taiwan aborigines, were found to be transmitted from pigs rather than from cattle (3). After extensive morphological study, including molecular genetic analyses, this "*Taiwan Taenia*" was identified as a subspecies of *T. saginata* (3). *This Taiwan Taenia is also found in other Asian countries.*

II. DISTRIBUTION

Taenia saginata taeniasis/cysticercosis occurs in many cattle breeding regions and especially where beef is eaten raw or undercooked. Objective epidemiological data are not available from most of the countries as the collection of information is not standardized and originates from such various incomparable sources as helminthic surveys in various human populations, laboratory and hospital data from a few countries with obligatory notification of the cases, sporadic description of the human cases in countries where the infection is rare, and, finally, from meat inspection official reports, which may differ in quality. In the United States, the prevalence of bovine cysticercosis, as registered by Federal Meat Inspection, always remains low: 0.14% in 1912, 0.6% in 1942, and 0.04–0.08% in the 1960s. Bovine cysticercosis is concentrated in California and Texas, mainly due to

Table 1 *Taenia solium* and *T. saginata*: Incidence in Selected Countries

Country	Estimated % of population infected	1992 population (millions)	Estimated case (no.)[a]
Taenia solium			
USA		2553.6	122
Chile	0.3	13.6	40,800
Ecuador	0.9	10.0	90,000
Guatemala	1.1	9.7	106,700
Mexico	1.1	87.7	964,700
Taenia saginata			
USA	0.0	255.6	519
South America	0.3	300.0	1,000,000
Cuba	0.1	10.8	10,800
Guatemala	1.7	9.7	164,900
Chile	1.9	13.6	258,400
Asia	0.4	3207.0	15,000,000
Europe	2.1	511.0	11,000,000
Africa	2.7	654.0	18,000,000

[a] Except for the United States, these are rough estimates based on computations from incidence estimates, not reported cases.
Source: Data from Ref. 38.

cattle feedlot outbreak attributed to Mexican immigrant workers commonly infected with *T. saginata* (4). Large epizootic outbreaks of *T. saginata* cysticercosis in animal feed lots were described first in the United States but have been reported from other countries as well. Distribution of *T. solium* taeniasis/cysticercosis is usually confined to poor countries as it is mainly related to the low sanitation in pig husbandry. Human migration and increased consumption of pork increases the spread of taeniasis/cysticercosis from the endemic rural areas to urban areas. In the United States, between 1857 and 1954 only 42 cases of human cysticercosis could be traced, whereas in four hospitals in Los Angeles during 11 years (1973–1983), 497 patients were identified as having been treated for cysticercosis (5). Over 90% of these patients were Hispanic, the majority with a Mexican nationality. Cysticercosis was diagnosed in 12 U.S. citizens who had no history of travel to countries traditionally considered endemic for *T. solium* infections. Since 1981, approximately 80 cases of cysticercosis have been diagnosed in the United State annually; cysticercosis is not yet a notifiable disease in the United States.

The rate of infection of adult stage of both species of *Taenia* varies widely globally, but in some countries the number of cases may exceed 1 million (Table 1).

III. PUBLIC HEALTH IMPORTANCE

Information about the public health importance and economic costs of taeniasis/cysticercosis is lacking in hard data. Taeniasis itself causes little if any disability and morbidity. In the 1970s, 10% of French and 20% of Polish *T. saginata* carriers were hospitalized (6); now, with improved taeniacides, this proportion is lower. The diagnosis and treatment of taeniasis in outpatients is a low medical services cost. However, as humans are the only disseminators of infection to animals, human taeniasis plays an important role in sustaining the existence of the parasite; this also includes the risk of cysticercosis in the *T. solium* carrier, members of his or her family, or other people.

Bovine and pig cysticercosis is a veterinary problem that absorbs much of a meat inspector's time; causes financial losses to producers due to preventive handling and condemnation of infected carcasses (marking, transportation, refrigeration, boiling, burying, incineration, processing); and causes a loss of valuable animal protein. The total animal losses enter the range of millions of U.S. dollars per country with even a moderate incidence rate (7). The cost of a feed lot epidemic of bovine cysticercosis in the United States may exceed $500,000 (1). Bovine cysticercosis greatly reduces the chance of some African countries (Kenya, Botswana) to be leading beef meat producers and exporters (6). In 1980 in Mexico, the estimated economic losses due to pig cysticercosis were U.S. $43,310,524 (2).

Human cysticercosis has a definite public health importance, measurable in disability, morbidity, and mortality of infected people and a considerable burden to the health services. Disability may be expressed best by the fact that in endemic areas 30–60% of cases of epilepsy are caused by neurocysticercosis. Morbidity may be calculated roughly from the proportion of the cysticercosis patients among the patients of neurological wards in the areas where national services are available to the majority of the population; in endemic areas, 10–20% of such patients are treated because of cysticercosis. The total expenditures for medical care are enormous: computed tomography (CT) scanning alone to confirm brain cysticercosis cost over $100. Mortality due to neurocysticercosis is considerable and in endemic areas may range from 1% to 2% of all causes of death. There is an urgent need to evaluate the medical, social, and economic costs of taeniasis/cysticercosis and especially of *T. solium* cysticercosis in humans in order to better justify the need of having an effective national control program in endemic countries.

IV. LIFE CYCLES AND EPIDEMIOLOGY

These tapeworms live in the small instestine. They lack an intestinal tract, and nutrients are absorbed actively through their tegument. They have a simple nervous and excretory system extending along the strobila.

Most of the biological potential is focused on survival and reproduction. Not all of the factors are known that allow a tapeworm to survive in the intestine, a rather hostile environment that is permanently mobile, anaerobic, with periodic changes in pH and in digestive enzyme activities. The living tapeworms do resist digestion by the human small intestine enzymes but they may disintegrate quickly after exposure to an antithelminthic. The reproductive system develops in immature segments and fills in much of the mature and gravid proglottids. After a cross-fertilization, the embryos develop within the uterus, which is a closed system of tubules. The eggs can be liberated only when the proglottid is detached or broken. The gravid proglottid of *T. saginata* contains an average of 80,000 *Taenia* eggs, and that of *T. solium* an average of 60,000 eggs. Multiplying these numbers by an average number of proglottids produced daily (6–9 for *T. saginata* and 5 for *T. solium*) one can calculate the reproductive potential of *T. saginata* as on the order of 600,000 eggs daily and *T. solium* as 300,000 eggs daily. The almost unlimited access to food and energy by the tapework tegument, which participates actively in digestive/absorptive functions, facilitates such a high reproductive potential.

Usually one tapeworm parasitizes humans but multiple or mixed infections are not uncommon. The possibility of superinfection with *T. saginata* cysticerci has been proved in humans (8). In Poland, multiple infections occurred approximately in 1 of every 200 *T. saginata* carriers (9).

Large *T. saginata* tapeworms are usually located in the jejunum; they are bent a few times and have a scolex attached most frequently 40–50 cm below the duodenojejunal flexure. When a mature *T. saginata* tapeworm proglottid detatches, it is active. The detatched mobile gravid proglottids frequently leave the intestinal tract by crawling out of the anus at any time of the day or night independently of defecation. The proglottids of *T. solium* usually do detach in groups of several segments and leave the host during defecation.

The contamination of the environment by *Taenia* eggs in human feces or proglottids is restricted by sanitation. In some areas, free-roaming pigs feed on fecal matter deposited indiscriminately by people. This may protect the environment somewhat against contamination, but it leads to infected pork.

Field experiments have shown that *Taenia* eggs can be dispersed rapidly within 100 m around a site of contamination (10). The dispersion pattern suggests that water, wind, or ground slope may play a role in spreading of *Taenia* eggs. Birds may be involved in transmission of *Taenia* eggs from sewage works to pastures or open waters (11). Eggs may also be carried on shoes or a car's wheels and on agricultural machines, but this has not been carefully documented.

Cysticercosis, in contrast to taeniasis, is the systemic disease caused by the invasion of tissue by the infective larvae (cysticerci) of the tapeworms. In the case of *T. saginata* infection, humans are definitive hosts and cattle are intermediate hosts; therefore, *T. saginata* cysticercosis is only a disease of cattle (''measly beef'') and has veterinary importance in beef and dairy production (6).

In the case of *T. solium* infection, humans can be both the definitive and the intermediate hosts (pigs are the normal intermediate hosts) (Fig. 1).

When humans or pigs ingest *T. solium* eggs, the cysticerci localize in the muscles, the viscera, and often the brain. Thus, both *T. solium* taeniasis and cysticercosis are diseases of humans. The most frequently affected sites are the muscles and the skin. Patients are typically found with multiple subcutaneous nodular lesions without pain or inflammatory signs. Migration of the cysticerci into the central nervous system or into other strategic organs, such as the eyes or heart, however, often results in serious and life-threatening disease (6).

After the ingestion of contaminated human waste by the intermediate host (cattle, pigs), the hexacanth embryo is released from the eggs and penetrates the intestinal mucosa, and it then travels via the circulation to such tissues as the muscles, the viscera, and, in the case of *T. solium*, the brain. At these sites, the larvae develop into the cysticercus stage. When humans ingest undercooked beef or pork meat containing cysticerci, the larvae are freed from the tissue and attach to the intestinal mucosa and mature into adult tapeworms. Gravid proglottids begin to appear in the feces about 3 months after ingestion of the cysticerci.

Because humans are the only definitive hosts for *T. saginata* and *T. solium*, outdoor defecation, free access of animals to latrines, and use of sewage effluent for the irrigation of fruits and vegetables and pastures are the most important risk factors of taeniasis/cysticercosis in humans and in domesticated animals in developing countries (11). The number of eggs shed from an infected person per day may be very high, and massive

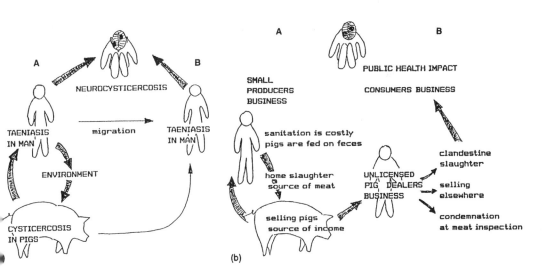

Figure 1   Cycles of *T. solium*: (a) biological (A = rural areas; B = urban areas); (b) economic.

outbreaks called ''cysticercosis storms'' have occurred in feed lot cattle in North America. In Danish herds, cysticercosis storms occur from the inappropriate handling of sewage sludge (12). *Taenia saginata* eggs can survive on soil for about 5–6 months and are regularly recovered from sewage sludge (12). Such risk factors implicate taeniasis/cysticercosis as also a waterborne parasitic zoonosis.

Since the 1980s, neurocysticercosis has been increasingly seen in the United States. Some of these cases were acquired locally by contact with a *T. solium* carrier. A survey for intestinal parasitic infections in North Carolina demonstrated that 3% of migrants from Central America are infected with *Taenia*, but not necessarily *T. solium*. Three cases of neurocysticercosis from the eastern United States recently were reported in the households where immigrants from Latin American endemic areas were hired to assist seasonally with work at home (13).

Taenia solium may spread epidemically, as was documented by the observation of an epidemic of neurocysticercosis in West Irian (Indonesia) (14). The first unusual sign noted was an epidemic of fits and burns reported in a few localities of the Wissel Lakes area. The primary source of the epidemic was *T. solium*-infected pigs, imported from Bali Island (Indonesia) where the prevalence of cysticercosis in humans and pigs is rather high. *T. solium* infections are spreading easily to other parts of West Irian (14). In Papua New Guinea, at the border with West Irian there are several camps for Irian Jaya refugees. The serological examination of 221 refugees revealed the first human case of highly probable *T. solium* in Papua New Guinea (15). The study concluded that West Irian refugees should be regarded as potential carries of *T. solium* and as such may introduce taeniasis/cysticercosis into Papua New Guinea. It also stresses the necessity of pay more attention to *Taenia* infections when transferring animals to ''virgin'' areas.

V. CLINICAL ASPECTS

A. Taeniasis

The presence of such a large parasite such as *Taenia* in the small intestine is likely to cause some pathomorphological and pathophysiological changes in the host. In 20 of 25 *T. saginata* carriers, a jejunal biopsy demonstrated deformation of villi, pronounced proliferation of enterocytes at the margins of the villi, crypt hyperplasia, and mononuclear cell infiltration of the lamina propria; in 5 patients, signs of fibrosis in the mucosal sublayer were observed (16). Participation of the lymphocytes and eosinophils in lamina propria infiltrations, moderate blood eosinophilia in onethird of patients, and elevated serum immunoglobulin E (IgE) (17) suggest an immunological nature of the clinical pathology in part of the cases.

In a study of 2200 human *T. saginata* taeniasis cases in Poland, the most common symptom noted by patients (98%) was an unpleasant sensation caused by active discharge of *Taenia* segments from the anus (6). Fewer than half of patients, especially long-term carriers, reported no other symptom; in the remaining patients, abdominal pain and nausea were the most common complaints.

Taenia solium taeniasis causes similar symptoms. In endemic regions it should be suspected in any individuals suffering from cysticercosis, as cysticercosis coexists with taeniasis in about 25% of the cases. For treatment of taeniasis, the drug of choice is praziquantel. Praziquantel is administered orally after a light meal in a single dose of 5 mg per kg of body weight. It may be unnecessary to use a higher dose of 10 mg/kg because even a low dose of $2^{1}/_{2}$ kg was found effective in *T. saginata* taeniasis (18). In a field study, doses between 3.4 and 7.4 mg/kg were found equally effective against *T. solium* taeniasis. It has been reported that a single dose of 20 mg/kg precipitated seizures and severe headache in an adult patient with asymptomatic cysticercosis (19). This is an additional argument for using the lowest effective dose of praziquantel (or, alternatively, niclosamid) in the regions endemic for *T. solium* cysticercosis.

B. Cysticercosis (*T. solium*)

Neurocysticercosis deserves separate discussion as it is the most pathogenic form of *T. solium* infection in humans. Pathogenesis of neurocysticercosis results from three major mechanisms: (a) mechanical compression and damage of tissues by cysticerci (e.g., in ventricular system) and /or host reaction to the cysticerci (e.g., granulomas, calcifications); (b) immunopathological effects such as cellular infiltration around cysticerci and endarteritis; and (c) late secondary sequelae of cysticerosis such as basal fibrosis, hydrocephalus, and brain atrophy. The clinical pathology processes end with fibrotic scars, especially in the meninges, replacing the destroyed cysticerci or, less frequently, in the form of calcifications of necrotic foci (20).

Epilepsy is the most common sign of neurocysticercosis. Typical is an onset of epileptic attacks in young adults (22). Depending on the epidemiological situation, cysticercosis is the cause of epilepsy in 0.0005–40% of cases (21). The epilepsy is either focal or generalized. The attacks may increase quickly, leading to status epilepticus or, what is more characteristic, may have long intervals between them or even disappear spontaneously. The focal neurological deficit signs refer, in order of their decreasing frequency, to pyramidal tract and brainstem dysfunction, cerebellar ataxia, involuntary movements, and sensory deficits (23).

Mental disorders occur in about a third of symptomatic cases and are common especially in diffuse neurocysticercosis or in the later stage of the disease. The disorders have a form of characteropathia, affective changes, confusion, hallucinations, delusions, apathy, depression, and dementia. Dementia is due to brain atrophy caused by vasculitis. Some minor psychiatric disorders may be the only symptoms of neurocysticercosis.

There are some indications that a subcutaneous location of *T. solium* cysticerci is more common in Southeast Asia than in Latin America. In an analysis of cysticerci in southern India, 18 of 38 patients examined had subcutaneous cysticercosis, especially in the neck and back regions. There was also a high incidence in individuals younger than 30 years. Ocular cysticercosis occurred frequently in the younger age group, possibly due to early detection, and only three cases of neurocysticercosis were detected in adults (24).

The antihelminthics praziquantel and albendazole are commonly used in human neurocysticercosis (18,19). The indications for a specific chemotherapy treatment are (a) cysticerci in the parenchyma, (b) multiple cysticerci in the brain or located at the optical chiasma or close to brain arteries. Cysticerci seen as calcified totally or partially or hypodense in CT scanning should not be treated by chemotherapy. Chances of a successful treatment are also small in cysticerci located in the ventricular system. Some sequelae of neurocysticercosis, such as calcification of the lesions, arachnoiditis, and intracranial hypertension, respond poorly to chemotherapy. Praziquantel usually is administered per os for 2 weeks at the daily dose of 50 mg per kg body weight. This may not be the optimal dose in all cases; some good results were achieved using higher doses for a shorter time with either praziquantel or albendazole.

Albendazole is administered per os in a daily dose of 10–15 mg per kg body weight for 8–15 days. Repetition of a treatment depends on the results of the initial chemotherapy (25).

Both drugs have similar cure rates of between 60% and 85%; however, there are ample individual variations in the drug's level in plasma and cerebrospinal fluid that may affect the efficacy of both drugs. The efficacy can be evaluated objectively only 6 months after treatment and depends on the status of the parasite before the treatment and the aggressiveness of the host response, which usually is lower in age-advanced patients. The positive effect of chemotherapy is evaluated by the disappearance of clinical symptoms and signs and by the results of a control CT scan.

VI. DIAGNOSIS

Detection of the adult worm (intestinal) infection is based primarily on identification of proglottids, eggs, or coproantigen in the faces (40). Serological diagnosis of *T. solium* cysticercus infection is becoming more reliable with the introduction of an enzyme-linked immunotransfer blot (EITB) using purified worm glycoprotein antigen, and a new enzyme-linked immunosorbent assay (ELISA) (26,27). The isolation of a 10- to 26-kDa genus-specific antigen has made serology testing for cysticercosis highly sensitive (28,29). These glycoproteins are highly promising for both diagnosis, seroepidemiology, and the monitoring of prognosis in the treatment of neurocysticercosis (30,41).

In general, clinical manifestations of cysticercosis have a weak correlation with immune responses. Close cooperation of the clinicians with the local serological laboratory usually helps to evaluate the diagnostic value of the tests objectively, whereas for any epidemiological surveys in communities, the antigens, reagents, and technologies have to be standardized.

VII. PREVENTION AND CONTROL

Current food safety control strategies for these parasitic zoonoses include detecting infections at slaughter in order to remove them from the food chain. Meat inspection procedures are not very efficient at present. But regardless of the success of efforts to improve the detection of infected animals at slaughter, effective control will always be best achieved by the minimization of risk factors. Governments must take steps to ensure that all beef and pork is marketed through channels that allow for meat inspection. On-farm management must ensure that animals are protected from ingesting feed or water contaminated with human feces. Whenever possible, farm workers should be examined for parasites and treated if warranted. The use of sewage sludge and effluents for agricultural purposes should receive more careful scrutiny. For example, cattle should be screened from direct contact with streams carrying untreated wastewater and effluent from plants (31). Legislation may also be needed to regulate the agricultural use of sewage sludge, especially on grazing and pasture land (32). Research is urgently needed on the effects of composting and alternative chemical treatment in the production of safe sewage sludges and effluents for agricultural use.

In countries where inspection for cysticercosis is not mandatory, it is generally recommended that consumers cook their meat to at least 60°C. The control of porcine cysticercosis in developing countries is especially urgent where the obstacles to change are poverty, tradition, and vested interests (33). The World Health Organization and the Pan American Health Organization have developed two alternative strategies for the control of human *T. solium* infections (34): (a) comprehensive long-term intervention and (b) short-term intervention based on mass treatment of adult worm (intestinal) infections (42). The comprehensive long-term program includes appropriate legislation, modernization of swine production systems, improvement of meat inspection efficiency and coverage, provision of adequate sanitary facilities, and adoption of measures to identify and treat human tapeworm carriers. As pointed out by Schantz et al. (33), several of these features will be difficult to achieve because of social, political, and economic realities in endemic areas. Therefore, the short-term program of identifying foci and treating all diagnosed or suspected human cases was developed (34). The prospects for eradication of *T. solium* are considered good because of the vulnerability of the life cycle to improved sanitation, the introduction of confined (intensive) swine housing, and rigorous meat inspection (33); therefore, the International Task Force for Disease Eradication has accepted taeniasis/cysticercosis as potentially eradicable.

Other strategies for control of porcine cysticercosis include vaccines for pigs. Homologous antigens from *T. solium* cysticerci are able to induce a high level of protective immunity (71%) against a challenge infection (35). Current efforts are aimed at identifying the protective antigens; whether a practicable vaccine can be produced will depend on a complex set of issues (efficacy, cost effectiveness, compatibility with other control efforts, etc.). Recently, a promising protovaccine for bovine cysticercosis was announced (36).

In prevention of neurocysticercosis, the major role is played by early detection and treatment of *T. solium* taeniasis and strict personal hygiene. The easy spread of *T. saginata* eggs was demonstrated by the results of an examination of 10 *T. saginata* carriers in Germany; as many as 68 embryophores were found in fingernail dirt, up to 700 *Taenia* eggs in water used for handwashing, and over 50,000 eggs per liter of water used to soak underwear of infected persons (37).

The individual prophylaxis of taeniasis is accomplished by avoiding consumption of raw pork or beef or semiraw meat products. Freezing of infected pork or beef can kill the cysticerci when the temperature inside a piece of meat or a carcass is below $-5°C$ (23°F) for at least 4 days or around $-20°C$ for at least 12 h. Cysticerci are killed best by cooking the meat until it loses its pink color.

REFERENCES

1. Dewhirst, L. W. (1975). Parasitologic and economic aspects of cysticercosis in the Americas. PAHO Scientific Publication 295, Pan American Health Oganization, Washington, D.C., p. 133.
2. Acevedo-Hernandez, A. (1982). Economic impact of porcine cysticercosis. In: Cysticercosis: Present State of Knowledge and Perspectives, A. Flisser, K, Willms, J. P. Lacletter, C. Larralde, C. Ridaura, and F. Beltran, eds., Academic Press, New York, p. 63.
3. Zarlenga, D. S., McManus, D. P., Fan, P. C., and Cross, J. H. (1991). Characterization and detection of a newly described Asian taeniid using cloned ribosomal DNA fragments and sequence amplification by the polymerase chain reaction. Exp. Parasitol. 72:174.
4. Schultz, M. G., Hermos, J. A., and Steele, J. H. (1970). Epidemiology of beef tapeworm infection in the United States. Publ. Health Rep. 85:169.

5. Richards, F. O., Jr., Schantz, P. M., Ruiz-Tiben, E., and Sorvillo, F. J. (1985). Cysticercosis in Los Angeles County. JAMA 254:3444.
6. Pawlowski, Z. S., and Schultz, M. G. (1972). Taeniasis and cysticercosis (Taenia saginata). Adv. Parasitol. 10:269.
7. Loaharanu, P., and Murrell, K. D. (1994). A role for eradication in the control of foodborne parasites. Trends Food Sci. Technol. 5:190–195.
8. Sterba, J., and Dykova, I. (1979). Symptomatology of taeniasis caused by Taenia saginata. Folia Parasitol. (Praha) 26:281.
9. Pawlowski, Z. S., and Rydzewski, R. (1958). Observations on taeniarhynchosis in the Poznan Province. Wiad. Parazytol. 4:509.
10. Gemmell, M. A., and Johnstone, P. D. (1976). Factors regulating tapeworm populations. Dispersion of eggs of Taenia hydatigena on pasture. Ann. Trop. Med. Parasitol. 70:431.
11. Murrell, K. D., and Nawa, Y. (1998). Animal waste: risk of zoonotic parasite transmission. Rev. Environ. Health 13:169–178.
12. Nansen, P., and Henriksen, Sv. Aa. (1986). The epidemiology of bovine cysticercosis (C. bovis) in relation to sewage and sludge application on farmland. In Epidemiological Studies of Risks Associated with the Agricultural Use of Sewage Sludge: Knowledge and Needs, J. C. Block, A. H. Havelaar, and P. L'Hermite, eds. Elsevier, London, p. 76.
13. Moore, A. C. (and 18 authors). (1995). Seroprevalance of cysticercosis in an Orthodox Jewish community. Am. J. Trop. Med. Hyg. 53:439–442.
14. Muller, R., Lillywhite, J., Bending, J. J., and Catford, J. C. (1987). Human cysticercosis and intestinal parasitism amongst the Ekari people of Irian Jaya. J. Trop. Med. Hyg. 90:291.
15. Fritzsche, M., Gottstein, B., Wigglesworth, M. C., and Eckert, J. (1990). Serological survey of human cysticercosis in Irianese refugee camps in Papua New Guinea. Act. Trop. 47:69.
16. Kociecka, W., Gustowska, L., and Blotna-Filipiak, M. (1985). Jejunal mucosa biopsy in patients with T. saginata taeniasis. Wiad. Parazytol 31:263.
17. Nepote, K. H., Pawlowski, Z. S., and Soulsby, E. J. L. (1974). Immunoglobulin levels in patients infected with Taenia saginata. In: Parasitic Zoonoses: Clinical and Epidemiological Studies, E. J. L. Soulsby, ed., Academic Press, New York, p. 241.
18. Pawlowski, Z. S. (1991). Efficacy of low doses of praziquantel in taeniasis. Acta. Trop. 48: 83.
19. Torres, J. R., Noya, O., Noya, B. A., and Mondolfi, A. G. (1988). Seizures and praziquantel. A case report. Rev. Inst. Med. Trop. Sao Paulo 30:433.
20. DeVilliers, J. C. (1983). Cysticercosis of the nervous system. South Afr. Med. J. 63:769.
21. Chopra, J. S., Kaur, U., and Mahajan, R. C. (1981). Cysticerciasis and epilepsy: a clinical and serological study. Trans. R. Soc. Trop. Med. Hyg. 75:518.
22. Medina, M. T., Rosas, E., Rubi-Donnadieu, F., and Sotelo, J. (1990). Neurocysticercosis as the main cause of late-onset epilepsy in Mexico. Arch. Intern. Med. 150:325.
23. Del Brutto, O. H., and Sotelo, J. (1988). Neurocysticercosis: an update. Rev. Infect. Dis. 10: 1075.
24. Veliath, A. J., Ratnakar, C., and Thakur, L. C. (1985). Cysticercosis in South India. J. Trop. Med. Hyg. 88:25.
25. Sotelo, J., and Jung, H. (1998). Pharmacokinetic optimisation of the treatment of neurocysticercosis. Clin. Pharmacokinet. 34:503–515.
26. Tsang, V. C. E., Pilcher, J. A., Zhou, W., Boyer, A. E., Kamango-Sollo, E. I. P., Rhoads, M. L., Murrell, K. D., Schantz, P. M., and Gilman, R. H. (1991). Efficacy of the immunoblot assay for cysticercosis in pigs and modulated expression of distinct IgM/IgG activities in Taenia solium antigens in experimental infections. Vet. Immunol. Immunopathol. 29:69.
27. Hayunga, E. G., Sumner, M. P., Rhoads, M. L., Murrell, K. D., and Isenstein, R. S. (1991). Development of a serologic assay for cysticercosis using an antigen isolated from Taenia. Am. J. Vet. Res. 52:462.
28. Yang, H. J., Chung, J. Y., Yun, D. H., Kong, Y., Ito, A. Ma, L., Liu, Y. H., Lec, S. C., Kang,

S. Y., and Cho, S. Y. (1988). Immunoblot analysis of a 10kda antigen in cyst fluid of Taenia solium matacestodes. Parasite Immunol. 20:483–488.

29. Tsang, V. C., Brand, J. A., and Boyer, A. E. (1989). An enzyme-linked immunotransfer blot assay and glycoprotein antigens for diagnosing cysticercosis. J. Clin. Microbiol. 33:3124–3128.

30. Ito, A., Plancarte, A., Ma, L., Kong, Y., Flisher, A., Cho, S. Y., Liu, Y. H., Kamhawi, S., Lightowlers, W., and Schantz, P. M. (1998). Novel antigens for neurocysticercosis diagnosis: simple method for preparation and evaluation for serodiagnosis. Am. J. Trop. Med. Hyg. 59: 291–294.

31. Ilsoe, B., Kyusgaard, N. C., Nansen, P., and Henriksen, S. A. (1990). Bovine cysticercosis in Denmark. Acta Vet. Scand. 31:159–168.

32. Barbier, D., Perrine, D., Duhamel, C., Doublet, R., and Georges, P. (1990). Parasitic hazards with sewage sludge applied to land. Appl. Environ. Microbiol. 56:1420–1422.

33. Schantz, P. M., Cruz, M., Sarti, E., and Pawlowski, Z. (1993). Potential eradecability of taeniasis and cysticercosis. Bull. PAHO 397–403.

34. Pawlowski, Z. S. (1990). Prospectives on the control of Taenia solium. Parasitol. Today 6: 371.

35. Nascimento, E., Costa, J. O., Guimares, M. P., and Tarares, C. A. P. (1994). Effective immune protection of pigs against cysticercosis. Vet. Immunol. Immunopathol. 45:127–137.

36. Lightowlers, M. W., Rolfe, R. and Gavci, C. G. (1996). Taenia saginata: vaccination against cysticercosis in cattle with recombinant oncosphere antigens. Exp. Parasitol. 84:330–338.

37. Ockert, G., and Obst, J. (1973). Dissemination of encapsulated oncospheres by tapeworm carriers. Monatshefte fur Vet. 28:97.

38. Roberts, T., Murrell, K. D., and Marks, S. (1994). Economic losses caused by foodborne parasitic diseases. Parasitology Today 10:419–423.

39. Escobedo, F., Penagos, P., Rodriguez, J. and Sotelo, J. (1980). Albendazole therapy for neurocysticercosis. Arch. Intern. Med. 147: 738.

40. Allan, J. C., Velasquez, T. M., Torres-Alvarez, R., Yurrita, P., Garcia-Noval, J. (1996). Field trail of the coproantigen-based diagnosis of Taenia solium taeniasis by enzyme-linked immunosorbent assay. Am. J. Trop. Med. Hyg. 54:352–356.

41. Garcia, H. H., Parkhouse, R. M., Montenegro, T., Martines, S. M., Tsang, V. C., and Gilman, R. H. (1998). Specific antigen detection ELISA for the diagnosis of human neurocysticercosis. Trans. R. Soc. Trop. Med. Hyg. 92:411–414.

42. Allan, J. C., Velasquez, T. M., Fletes, C., Torres-Alvarez, R., Lopez-Virula, G., Yurrita, P., Soto-de-Alfaro, H., Rivera, A., Garcia-Noval, J. (1997). Mass chemotherapy for intestinal Taenia solium infection: effect on prevalence in humans and pigs. Trans. R. Soc. Trop. Med. Hyg. 91:595–598.

11

Meatborne Helminth Infections: Trichinellosis

William C. Campbell
Drew University, Madison, New Jersey

I. INTRODUCTION

A. Definition and Distribution

Trichinellosis is a foodborne disease caused by infection with parasitic worms of the genus *Trichinella*. Infection results from ingestion of meat harboring the infective larvae of the parasite. The disease has long been known as trichinosis because the parasite was formerly called *Trichina*. The term *trichinellosis* is preferred, and has gained wide acceptance internationally. Many books and comprehensive reviews on trichinellosis have been published, including several in recent decades (1–9).

The disease may be severe and even fatal. It is cosmopolitan in distribution and occurs in frigid, temperate, and torrid zones (see Sec. IV).

B. History

Trichinella has enjoyed a long history, first as a zoological curiosity and later as a pathogen commanding a great deal of attention in the fields of public health and biomedical research (10). The parasite was first discovered during a routine autopsy in England in 1835 and was originally known as *Trichina spiralis*. In 1895, the generic name *Trichinella* was proposed and was adopted slowly by the medical community.

In 1860, *T. spiralis* was found in the tissues of a patient who had died of an extremely painful muscle disease, and this important case revealed both the pathogenicity of the organism and the link between the disease and the ingestion of undercooked pork. The muscle-dwelling larva was thus one of the first microorganisms to be recognized as a systemic pathogen of humans. During the latter part of the nineteenth century, the essentials of the life cycle were worked out, and the pathogenicity of the parasite was shown dramatically by a series of severe outbreaks in Germany.

The painfulness of the disease and its high mortality rate called for preventive measures. From the beginning, it was considered more feasible to seek out and discard the infected swine carcasses than to change traditional methods of preparing pork and sausage dishes. While consumers everywhere, and especially in the United States, were urged to cook their porcine meats thoroughly, the thrust of control measures in Europe was the microscopic inspection of small pieces of muscle from swine carcasses (trichinoscopy). Restriction on the importation of pork also was employed, the most celebrated example being the "pork war" between Germany, which espoused trichinoscopy, and the United States, which did not. The control of trichinellosis gradually came to rest on multiple measures (see Sec. VIII) and on an increasing public awareness of the desirability of cooking or freezing meats derived from swine or wild animals.

II. PUBLIC HEALTH IMPORTANCE

A. Prevalence and Significance of Clinical Disease

In the years following World War II, some 300–500 clinical cases per year were reported in the United States (trichinellosis became reportable in 1947). Since then, the incidence has declined markedly (11) and in recent years deaths have rarely been attributed to the disease. In 1955 only 28 cases were reported, and in 1997 only 9 (12,13). The decline in clinical infections undoubtedly is attributable to the same factors as were responsible for the decline in subclinical infections (see Sec. II.B).

Similarly, there has been a striking decline in clinical trichinellosis in the European Union. Outbreaks such as those occurring in Germany in the late nineteenth century no longer occur and, indeed, trichinellosis is recorded very rarely except when caused by consumption of nonporcine flesh. Usually, the culprit has been the flesh of game animals (14). Surprisingly, horse meat also has been implicated (in outbreaks in France and Italy). Given the opportunity, horses will eat dead mice with relish, and natural infections in horses have been reported (15). Horses probably become infected by intentionally or inadvertently eating fodder contaminated with the bodies of dead rats or mice. Trichinellosis outbreaks of varying severity continue to occur in eastern Europe, the former USSR, and Asia. Recent reports from China suggest that trichinellosis may be more common there than had been supposed. In Southeast Asia, the distribution of trichinellosis is spotty, and uncooked (sometimes fermented) pork sausages are eaten with relative impunity except among certain groups, such as the hill tribes of northern Thailand (see Sec. VII.C). In southern Thailand *T. pseudospiralis* has been identified as the cause of a severe clinical outbreak, with one fatality (16).

B. Subclinical Infection

Because the larval parasites become calcified in human flesh, microscopic infection of cadaver tissue provides good evidence of the number of people who have had trichinellosis sometime during their life. Surveys in the United States in the 1930s indicated an astonishingly high prevalence, with rates of 16% for the population at large, and local rates as high as 26%. By 1970, the national rate had fallen to 4%, and now is believed to be less than 1%.

The validity of the prevalence rates naturally is related to the magnitude of the surveys. The earlier surveys involved large numbers of cadavers (more than 5000 in the 1930s) and the decline in prevalence undoubtedly is real. There is little motivation for new surveys that would provide a sound current figure. The decline in the United States is attributable to (a) an increased awareness of the danger of eating uncooked pork, (b) an incidental increase in the use of home freezers, with resultant destruction of larvae that may be present in meat stored in such appliances, (c) the imposition of regulations for the processing of meat products that are likely to be eaten without further cooking, and (d) a reduction in the infection rate in swine resulting from the prohibition of the feeding of garbage to swine and the widespread practice of raising swine in confinement housing.

In other countries, especially in western Europe, the dramatic decline in clinical trichinellosis (Sec. II.A) presumably has been accompanied by a corresponding decline in subclinical human infection, and certainly by the virtual elimination of the infection from domestic swine. Control of the infection in swine in such countries has been achieved primarily by massive and well-supervised trichinelloscopy of swine carcasses at slaughter.

C. Significance for Industry

For the livestock industry, trichinellosis has been of concern ever since the disease was first linked to the eating of pork. Outbreaks of the disease, and the attendant publicity, discourage the consumption of pork. Media reports on the disease, even in the absence of a current outbreak, result in a transient depression in the demand for pork and therefore have an adverse economic impact on the industry. Measures to control the disease through inspection of carcasses, processing of meat products, or other means impose a financial burden that must be reckoned in the cost of meat production—regardless of how that cost is distributed among the taxpayer, the producer, the merchant, and the consumer.

D. Significance for Consumer

For the consumer trichinellosis is a hazard of uncertain dimension. In most parts of the world the probability of contracting the disease is low, but its painfulness and potential lethality have engendered an understandable, if often exaggerated, fear of the disease. Pork, in consequence, is often overcooked. To the extent that the retail price of the pork may reflect the cost of disease prevention, there is, as stated, a direct financial effect on the consumer.

III. THE PARASITE

A. Morphology

1. General

Trichinella exhibits the basic nematode characteristics of elongated, colorless cylindrical, fluid-filled (pseudocelomate) body, with sexual dimorphism. It is one of the smallest parasitic nematodes known and has a smooth annulate cuticle. The esophagus is surrounded for much of its length by highly distinctive cells, the stichocytes, which form the stichosome and are of immunological significance.

2. Male and Female Adult

The male *T. spiralis* measures 1.0–1.5 mm long and approximately 0.03 mm wide. It has an elongate tubular testis that widens into a vas deferens and further into a seminal vesicle before opening into the terminal cloaca. The posterior tip of the worm bears small sensory papillae and a pair of large lateral extensions, the copulatory appendages. A delicate bell-shaped structure, the copulatory bell, occasionally may be seen at the cloacal opening and is thought to be an extension of the cuticular lining of the rectum.

The female *T. spiralis* measures approximately 3 mm long and 0.05 mm wide. It has a single hologonic ovary in the posterior portion of the worm, and from this a short oviduct leads to a wide uterus that occupies much of the body until it opens at a vulva situated near the distal end of the stichosome. The portion of the uterus adjacent to the oviduct constitutes a seminal receptacle. The surface of the female worm does not have specialized appendages.

3. First-Stage Larva

The first-stage larva of *T. spiralis*, shed from the adult female, measures almost 0.1 mm long and, apart from possession of a penetrating stylet, essentially is undifferentiated. This form, known as a newborn larva (NBL), undergoes irregular growth and differentiation in a host muscle cell, reaching a length of 1.0 mm. It has a well-developed stichosome and genital primordium. There are no sexual appendages, but certain fine details of anatomy make it possible to differentiate male and female larvae by microscopic examination.

B. Life Cycle

The life cycle of *Trichinella* is shown diagrammatically in Fig. 1. In mice, the host for which the life cycle is best known, the worms grow to adulthood in the epithelial lining

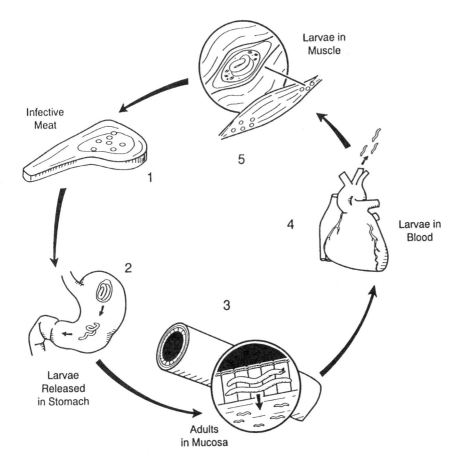

Figure 1 Life cycle of *Trichinella* (see text for more detail). (1) Meat harboring encysted larvae is ingested. (2) Larvae are released by digestion of cyst walls in gastric fluid. (3) Larvae pass to small intestine, invade epithelial cells of mucosa, grow to maturity, and mate. Female worm sheds newborn larvae, which enter vascular system. (4) Larvae are carried to right side of heart, then to the lungs and left side of heart, and are disseminated via the aorta. (5) Larvae reaching skeletal muscle invade individual muscle cells and induce transformation into nurse cells, which gradually become thick-walled infective cysts. In humans, transmission from one individual host to another does not occur ordinarily.

of the host's intestinal mucosa and mate when about 36 h old (i.e., 36 h after entering the host as larvae). Starting at about day 5 of infection, and ending about day 10, the female worm gives birth to numerous (several hundred to a few thousand) first-stage larvae (NBLs). These larvae probably are deposited within the series of epithelial cells in which the parent resides, but almost nothing is known of this process or the process by which the larvae reach the vascular system of the host. It is known that they can migrate through solid tissues, but most of them are transported quickly to the right side of the heart, either via the lymphatics that drain into the vena cava or, as now seems more likely, via the veins (17).

The period of larviposition ends rather abruptly because the worms are expelled abruptly by the host's immune response (leaving a small residuum of worms, the majority of them male). In the absence of such response, the worms may remain alive and fecund for long periods. It therefore must be emphasized that this account of the life cycle applies to primary infections in the typical laboratory mouse and to infections of sufficient magnitude to induce a typical immune response. In other hosts, the timing of events may vary because of differences in host susceptibility, immune responsiveness, parasite load, parasite species or strain, and so on. In humans, gravid female worms have been found a couple of months after exposure to infection (so treatment may be indicated even late in the course of infection).

The NBLs, having been swept to the heart, reach the arterial blood flow and are disseminated throughout the body. Those reaching striated muscle penetrate individual muscle cells and induce those cells to transform into nurse cells. Dissemination of larvae from gut mucosa to musculature is essentially instantaneous, but approximately 2 weeks are required for the larvae to grow to the infective stage (still first-stage larvae, but now 10 times longer and almost 300 times greater in volume) and for the surrounding nurse cell to mature into a thick-walled living "capsule." Thus, in a typical murine infection, the musculature harbors infective encapsulated larvae beginning about 3 weeks (and as early as 17 days) after the host has become infected. The larvae remain infective for months, even years. In some hosts, especially humans, they eventually die and become calcified, and their remains serve indefinitely as evidence of former infection.

When the muscle-dwelling larvae are ingested by another host, the capsule wall is digested in the new host's stomach and the larvae are liberated. Within minutes after reaching the duodenum, the larvae enter the columnar epithelium of the mucosa. Threading themselves through a series of adjacent epithelial cells, they molt four times and become young adults in little more than 24 h.

C. Special Biological Features

Adult *T. spiralis* have little storage of carbohydrate, but the first-stage larvae accumulate considerable quantities of glycogen during their development in the host musculature. There is evidence that larval worms, while developing and molting in the host intestine, switch from anaerobic to aerobic metabolism. Little is known about the nutrition of the species, but adult worms presumably derive nutrients from the cytoplasm of the surrounding epithelial syncytium. The first-stage larva presumably depends on nutrients provided by the surrounding nurse cell, at least during its phase of growth and differentiation.

One of the most striking features of *T. spiralis* is its ability to induce in the host musculature a series of changes that render it suitable as a milieu for the larval parasite (18,19). The most important of these are (a) loss of normal muscle form and function,

(b) enlargement of nuclei, (c) modification of the glycocalyx to form a thick collagen wall, and (d) induction of local angiogenesis resulting in a network of blood vessels around the infected muscle cell.

The vascular network (rete), presumably serving to bring nutrients and dispose of waste products, is made up of vessels that are not typical capillaries. Formation of the network appears to involve two concurrent processes (20). Larval penetration of the host muscle cell results in mitochondrial damage that induces localized hypoxia. This event (occurring about a week after infection) activates the synthesis of vascular endothelial growth factor (VEGF). The parasite thus exploits the host genome to produce a peptide (per VEGF mRNA) that causes nearby endothelial cells to divide, migrate, and form a network of blood vessels around the developing nurse cell. Concurrently, the parasite induces the nurse cell to synthesize glycogen, which provides a supporting framework for the network. In mice the network begins to form at about day 12, and the whole process of nurse cell formation is completed by day 20. The parasite probably continues indefinitely to secrete the protein molecules necessary to maintain the nurse cell in the hypoxic state needed by the essentially anaerobic larva.

Early reports of structural, biochemical, and physiological alterations in parasitized muscle tissue have been reviewed (21) and later reports have been summarized (22).

The row of large periesophageal cells constituting the stichosome is an exceptional morphological feature and has remarkable immunological significance, yet its primary function is unknown. Ducts lead from the stichocytes to the esophageal lumen, but the composition of the secretion has not been determined. In the stichocytes themselves, two kinds of granules may be seen. These have been designated *alpha* and *beta*, and each is the source of a specific antigen. These are by no means the only antigens isolated from *Trichinella*. The parasite was one of the first helminths to be shown to induce protective immunity, and the ease with which it can be passaged in mice has made it a popular subject for immunological studies. The immunology of trichinellosis is reviewed in Chapter 16.

Trichinella is remarkable among parasitic nematodes in having neither a free-living stage between individual hosts nor an intermediate host. The dissemination of worm progeny throughout the host musculature permits passage to another individual host, but only after the death of the first host.

D. Infraspecific Variation

Until recently, *T. spiralis* was accepted as the sole representative of the genus *Trichinella*. It now has become clear that not all populations of the parasite are the same. The differences between them are in some cases small, and cross-breeding experiments to establish reproductive isolation are difficult and seldom undertaken. The criteria used in characterizing a population include:

Geographic distribution
Fertility of cross-bred organisms
Morphology of cyst
Morphology of adult or larva
Appearance on scanning electron microscopy
Signs and symptoms in human infections
Induction of capsule in host
Infectivity and pathogenicity in mice, rats, pigs

Distribution and duration in intestine
Ratio of larval burden to inoculum size
Larval output in vitro
Chemical attraction in vitro
Antibody characteristics
Antigen characteristics
Protein composition
Behavior in response to cold, heat
Survival after freezing
Sensitivity to antihelminthics
Enzyme and isoenzyme characteristics
Chromosome form and number
DNA profile

Molecular techniques have now been used to clarify the situation, and the following five species are currently recognized (14).

1. *Trichinella spiralis* (Owen, 1835) Railliet, 1895; with cosmopolitan distribution in domestic swine; pathogenic in humans.
2. *Trichinella nativa* Britov and Boev, 1972: found in wildlife in northern parts of the Old World and New World (Holarctic distribution); has remarkable resistance to freezing; pathogenic in humans.
3. *Trichinella britovi* Pozio et al., 1992; with Palearctic (Old World) distribution in wildlife; little pathogenicity in humans.
4. *Trichinella pseudospiralis* Garkavi, 1972; characterized by absence of capsule formation around the muscle-dwelling larvae; cosmopolitan distribution in birds and animals; rare but pathogenic in humans.
5. *Trichinella nelsoni* Britov and Boev, 1972; common in wildlife in equatorial Africa; infective for humans but pathogenic only in very heavy infections.

The biological properties of these species have been described in considerable detail. Three other populations that are distinguishable genetically have been designated T5, T6, and T8 pending resolution of their taxonomic status.

For expert characterization of newly isolated *Trichinella*, the larval forms may be sent to a reference center, designated as such by the International Commission on Trichinellosis. For optimal characterization, larvae should be preserved in the prescribed manner (24).

IV. EPIDEMIOLOGY

A. Geographic Distribution and Prevalence

The distribution of *Trichinella* infection (like that of the clinical disease; see Sec. II.A) is cosmopolitan. The parasite appears to have a very limited distribution in Australia and New Zealand, but is widely distributed throughout the rest of the world. It is found in the Arctic, in temperate lands, and in the tropics, though some species predominate in particular geographic regions (Sec. III.D).

In some regions, prevalence in humans and swine has been brought to an extremely low level, and this is especially true where swine raising is an industry rather than a rural farm operation (Secs. II.A and B). The parasite generally persists at higher levels in wildlife. Prevalence rates tend to be higher in large carnivorous and scavenging species than in rodents, herbivores, or other types. Bears, for example, are a significant source of human infection, with prevalence being particularly high (up to 50%) in polar bears. In the United States, prevalence rates of 5% and 11% have been recorded for mink and coyotes, respectively. In Europe, the red fox (*Vulpes vulpes*) is the prime wildlife host, though many other species may also be infected. Prevalence in wild boars is low, except in areas where the domestic and sylvatic cycles overlap; see Sec. IV.B (14,25). In Thailand, wild pig meat was responsible for the only known outbreak of disease attributed to *T. pseudospiralis*, while an anomalous single case in New Zealand may have been acquired from marsupial meat (16,26).

In China, *T. spiralis* isolated from dogs differed in isoenzyme and DNA characteristics from *T. spiralis* isolated (in a different part of the country) from pigs (27). In Japan, *T. britovi* is found in wildlife (28).

B. Transmission Patterns

It has long been recognized that trichinellosis is transmitted in two epidemiological cycles, the domestic and the sylvatic, which are quite distinctive but not absolutely distinct from each other. The *domestic cycle* consists of the cyclic transmission of *Trichinella* from pig to pig, or pig to rat to pig, with the potential for noncyclic transmission of the parasite from pig to humans. Although this is the classical source of human trichinellosis, humans must be regarded as an offshoot of the domestic cycle, not an integral part of it and certainly not the central focus of it. The cycle is perhaps best thought of as operating between live pigs and dead pigs (Fig. 2a). There are at least three ways in which the parasite is conveyed: (a) by the feeding of uncooked pork scraps (e.g., in garbage) to swine being raised in commercial piggeries; (b) by cannibalism among pigs (i.e., scavenging the carcasses of pigs that have died in the piggeries, and possibly the biting of tails of live pigs; (c) by the disposal of uncooked pork (e.g., in garbage) in dumps to which rats have access, and the subsequent ingestion of *Trichinella*-infected rat flesh by swine in rat-infected piggeries. Transmission to humans, the very important offshoot of this cycle, occurs when the flesh of slaughtered swine is eaten by humans without cooking it enough to kill the parasite.

The *sylvatic* cycle (Figs. 2b–d) consists of transmission of *Trichinella* among individuals and among species of wildlife. Again, human beings are an offshoot of the cycle, becoming infected by consumption of game meat, but not passing the infection on to other animals.

The lack of absolute separation between the two transmission cycles derives from the fact that swine may eat the flesh of wild animals, and wild animals may eat the flesh of domestic swine. Rats are probably of special significance in this regard because of their scavenging habits. If they feed at town dumps and acquire *Trichinella* from discarded scraps of pig flesh, they may in turn be eaten by wildlife predators or die and be eaten by wild carrion feeders. If, on the other hand, rats acquire the infection by scavenging wildlife carcasses, they subsequently may enter commercial piggeries and be eaten by swine.

The crossover of *Trichinella* from one transmission cycle to another probably is infrequent, but is of significance in relation to the distribution of particular strains or types of the parasite. Wildlife strains may have low infectivity for rats or swine. Crossover in the other direction (domestic to sylvatic) may be more likely, and there is considerable evidence that this in fact has occurred (29).

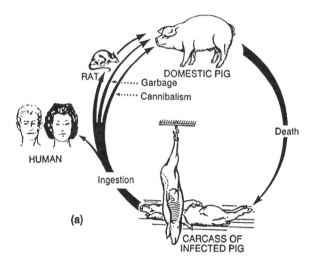

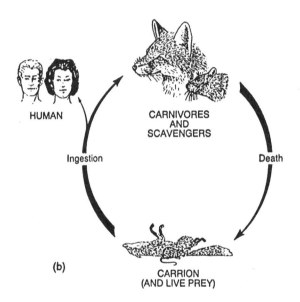

Figure 2 The transmission of trichinellosis: (a) the domestic cycle; (b) the sylvatic cycle in temperate zones; (c) the sylvatic cycle in torrid zones; (d) the sylvatic cycle in frigid zones. [Modified from Ref. 4 and Campbell, W. C. (1988). Trichinosis revisited—another look at modes of transmission, Parasitol. Today 4:83–86.]

V. PATHOGENESIS AND CLINICAL FEATURES

A. Intestinal Phase

Parasitism of the intestinal mucosa by adult *Trichinella* results in a variety of structural and functional disturbances readily observed in experimental animal infections. These include inflammation, flattening of the villus profile, thickening of the smooth muscle layer of the intestinal wall, accumulation of fluid in the gut lumen, alteration of intestinal

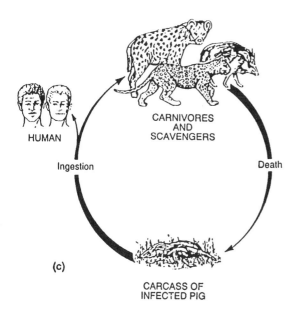

(c)

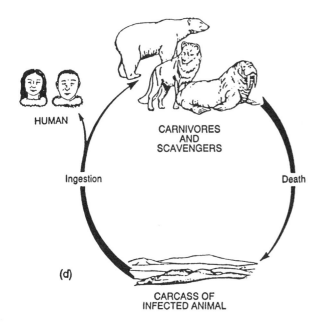

(d)

Figure 2 Continued

motility, and endocrine disturbances. It would be expected that similar effects occur in humans and that the severity of the damage is roughly proportional to the number of worms.

Numerous signs and symptoms have in fact been attributed to intestinal trichinellosis, but the connection is very doubtful except in the cases of diarrhea and abdominal discomfort (perhaps accompanied by nausea and vomiting). The symptoms are likely to be mild and therefore not diagnosed (or misdiagnosed). Indeed, it is believed that intestinal phase symptoms are not apparent in at least 85% of the cases that subsequently are diagnosed as trichinellosis on the basis of muscle phase symptoms. However, trichinellosis among Inuit people of the Canadian Arctic region was characterized by prolonged diarrhea and only mild muscle pain (30). It is not known whether the intestinal phase of the parasitism was prolonged correspondingly, or whether a conventional intestinal phase induced a clinical effect of unusual duration. (The infection was attributed to walrus meat.)

B. Muscle Phase

The muscle phase of clinical trichinellosis corresponds to the invasion of muscle tissue by *Trichinella* larvae shed by intestinal female worms. In humans, the illness typically begins 1–4 weeks after the patient has ingested infective meat and is characterized by myalgia, fever, edema, and eosinophilia. The incubation period and severity depend on several factors.

The severity of helminth diseases typically depends on the number of worms acquired because, unlike bacteria or protozoa, the worms do not multiply within the host body. Trichinellosis is unusual in that, while there is no buildup in number of adult worms, their progeny remain within the body. Thus, the reproductive capacity of the worms as well as their initial number affects the severity of the disease. The number of larvae produced is affected by the fecundity of worms and the duration of their sojourn in the gut, and largely is determined by the immune status of the host. The severity of the muscle phase also may depend on strain differences in parasite (or host). Patients in Kenya, for example, infected with sylvatic *T. nelsoni* have been known to tolerate intensities of parasitism that would ordinarily be fatal.

Because the larvae invade many organs of the host body, they elicit signs and symptoms of great variety (31). Invasion of striated muscle tissue is a key element of the life cycle and, not surprisingly, muscle pain is an outstanding feature of the muscle phase of trichinellosis. The pain may be severe and incapacitating. The jaw muscles are one of the predilection sites for larval invasion, and pain in this region may be intense. In *T. pseudospiralis* infections in humans, in which larvae do not become encapsulated, myositis may persist for several months (16). The thoracic diaphragm is another predilection site; and animal studies suggest that the function of diaphragmatic muscle is impaired during the early phase of infection. The impairment may be the result of inflammation induced by metabolic products released from nonencapsulated larvae (22). Although the larvae do not survive in heart muscle, they may induce severe, even fatal, myocarditis. Death from cardiac failure, when it occurs, usually occurs late in the illness when most of the acute symptoms have subsided. Death also may result from encephalitis or pneumonitis. Eosinophilia, which is of diagnostic value (Sec. VI), appears at about the same time as invasion of the musculature, but there is evidence from infections in mice that it is in fact triggered by the intestinal phase.

Persistent fever, resembling typhoid fever, also is characteristic. Strangely, this has not been demonstrable in mice or swine even in heavy infections. Another outstanding clinical feature is facial edema, especially in the periorbital region. It is believed that the leakage of fluid from blood vessels into surrounding tissues reflects an allergic vasculitis that develops early in the illness. The edema, like the fever, usually subsides within a few days or a few weeks. Extravasation of whole blood also may occur, and hemorrhage of the ocular conjunctiva or the capillary beds of the finger nails, while not common, are very characteristic of trichinellosis.

In addition to the neovascularization around the muscle-dwelling larva and the other changes associated with nurse cell induction (Sec. III.B), some degree of inflammation also may develop in the affected muscle. The degree of inflammation varies enormously. In rats and mice, common laboratory strains of *T. spiralis* often incite heavy infiltration of inflammatory cells in the connective tissue of the surrounding muscle. In pigs, the same strains may elicit little or no inflammation of muscle. In humans, heavy infections generally cause severe myositis, with intense infiltration of connective tissue and loss of striation in muscle fibers. In humans and other animals, the infiltration of inflammatory cells typically is concentrated around the poles of the encapsulated larva but is diffuse in severe cases. The variable nature of the inflammation probably is related to immunological factors as well as to strain and species.

VI. DIAGNOSIS

A. Clinical

1. History

A clinical history that reveals the prior consumption of raw or undercooked meat is of evident importance in the diagnosis of trichinellosis. This information often will not be proffered until the examining physician has some reason to suspect trichinellosis and to inquire about food habits. In diagnosing trichinellosis during the intestinal phase, a history of raw meat consumption is likely to be the only basis for considering such a diagnosis, but this generally becomes feasible only when a particular case is part of an outbreak in which other patients, with more advanced disease, already have been diagnosed. In diagnosing the disease during the muscle phase, a history of raw meat consumption will be strongly supportive of a diagnosis based on signs and symptoms.

2. Symptomatology

Because of the potential involvement of many organs (Sec. V) and consequent wide variety of signs and symptoms, trichinellosis may simulate many other diseases. Nevertheless, there is a cluster of signs and symptoms that, when taken together, are virtually pathognomic. These are the clinical features of fever, myalgia, and periorbital edema, accompanied by the laboratory finding of eosinophilia. Less common, but nonetheless indicative, are splinter hemorrhages under the fingernails.

B. Parasitological

The definitive diagnosis of trichinellosis is made by finding the larvae of the parasite in biopsy specimens from skeletal muscle. (The adult worms are not seen ordinarily in the course of antemortem or postmortem procedures.) When clinical and serological findings point clearly to trichinellosis, there is little justification for biopsy procedures. If parasitological confirmation is needed, the biopsy may be taken from one of the large muscles (e.g., the biceps or gastrocnemius). The intercostal muscles are probably a predilection site, although not the primary ones (Sec. V.B). The tissue specimen may be fixed and sectioned for examination, but it often is more desirable to examine it directly by squash procedures or to digest the tissue (1% pepsin, 1% HCl) and search microscopically for liberated larvae.

Despite the definitive nature of a positive parasitological finding, the value of biopsy is limited. In examining a large skeletal muscle, larvae are unlikely to be found in a minute fragment unless the infection is heavy, in which case other diagnostic methods probably would suffice. In some cases, larvae may not be seen in histological sections, but pieces of cyst material or associated inflammatory infiltrate may enable a diagnosis of trichinellosis to be made. It usually would be pointless to do a biopsy during the first 2–3 weeks of a suspected infection because the larvae would not yet have reached the characteristic muscle-dwelling encysted stage.

C. Serological

Immunodiagnostic methods for trichinellosis have been available for many years and gradually have become more and more reliable. Testing services are provided by governmental agencies in many countries. For information on this subject, see Chapter 15.

VII. TREATMENT

A. Principles

Trichinellosis is not often diagnosed in the intestinal phase (Sec. V.A) and so treatment rarely is considered at that time. When under exceptional circumstances the disease is diagnosed or suspected during the intestinal phase, it is important to initiate treatment so that the production of larvae, and invasion of muscle, can be prevented or minimized.

The muscle phase of trichinellosis in humans is usually a self-limiting disease. The larval population reaches a plateau and the parasites eventually die in situ. The question of treatment then becomes one of acceptable degrees of disease severity. Treatment may be indicated to alleviate pain and discomfort, and this must be decided on a case-by-case basis. When the disease is life threatening, treatment is imperative but it is not always clear as to whether this should include not only symptomatic treatment but also specific antihelminthic medication. Published case reports suggest that specific therapy may have been crucial to the successful outcome of some severe cases. Withholding specific therapy in such cases would be justified only if there were equally weighty concerns about drug toxicity or hypersensitivity reactions to antigens released by dead parasites.

B. Symptomatic Treatment

Symptomatic treatment is centered on the use of analgesic and antipyretic medications and bed rest. Other supportive measures are used as clinical circumstances dictate. To alleviate inflammation and counter shock-like symptoms in severe trichinellosis, corticosteroids are employed (e.g., prednisolone at 50 mg/day). Such therapy may be needed to avert a fatal outcome but in itself is complicated by the danger of exacerbating the disease. This is because steroidal suppression of immunity may prolong the intestinal phase of the infection and greatly increase the number of larvae invading the host tissues. Adjunctive antihelminthic therapy therefore should be used together with the steroid, especially early in the infection when substantial numbers of adult worms are likely to be in the intestine. (The dynamics of the intestinal phase and significance of immune responses are not well known in human trichinellosis, but there is evidence that adult parasites may persist in small numbers for at least a couple of months.)

C. Specific Therapy

1. Intestinal Stages

Although trichinellosis is rarely diagnosed in the intestinal phase, treatment directed against the intestinal stages of *Trichinella* (preadults and adults) may be indicated under two circumstances: (a) in prepatent infections, i.e., when a person is known to have eaten infective meat but symptoms associated with larval dissemination have not yet appeared; and (b) in patent infections, when larval dissemination has occurred but larviposition by adult worms may be continuing.

a. Prepatent Infection. Recognition of prepatent infection is obviously extremely rare. The purpose of treatment in such cases is to prevent overt disease by killing preadult and adult worms in the gut or by suppressing the reproductive function of the adults. If infective meat has been ingested within the previous 24 h, it is likely (chiefly on the basis of animal studies) that the preadult worms could be destroyed by treatment with pyrantel, oxantel, thiabendazole, mebendazole, or albendazole. If the interval between ingestion of meat and treatment is more than 24 h, pyrantel or levamisole could, at least in theory, be used to destroy the adult worms. For pyrantel, a dosage of 10 mg/kg/day for 5 days has been suggested on the basis of clinical experience (32). Alternatively, a benzimidazole (see Sec. VII.C.2) could be used to sterilize the worms and so prevent larval invasion of the musculature.

b. Patent Infection. In patent infections, the situation is similar to prepatent infections except that larval invasion of the musculature already has occurred. Treatment options are the same, theoretically, as for prepatent infections of more than 24 h duration, the intent now being to prevent further muscle damage. In practice, however, a benzimidazole would be the drug of choice because (depending on the regimen used) it may have some effect on the muscle-dwelling larvae (see Sec. VII.C.2).

It has been argued that all patent infections should be treated with antihelminthics because it is not currently possible to determine whether adult worms are surviving and shedding larvae at any given time during patency, and the assumption therefore should be made that the disease may get worse if left untreated. On the other hand, it has been contended that light infections are better left untreated because the disease is by definition mild (there being no other way to tell that the infection is light), will be self-limiting, and may confer a measure of immunity. Clinical monitoring of the illness should preclude any confusion between mild symptoms as indicators of low worm numbers and as forerunners of severe disease. The putative acquisition of immunity naturally would be of value primarily under such conditions of frequent exposure as those encountered by polar Eskimos in Greenland (33).

2. Muscle Stages

When trichinellosis is deemed sufficiently severe to warrant specific medication, a benzimidazole should be used. In rodents, and probably in people, these drugs are especially efficacious against the young, muscle-dwelling larvae (34). Beginning about 3 weeks after ingestion of infective larvae, the benzimidazoles become progressively less effective as encapsulation takes place and as the capsule wall becomes progressively thicker. Thiabendazole was the first of the series but has been superseded, first by mebendazole and then by albendazole. The new drugs are more potent and better tolerated than thiabendazole, which no longer is recommended for this use.

It must be realized that trichinellosis does not occur in circumstances that allow large-scale controlled clinical trials of candidate drugs. Dosage recommendations emerge from isolated cases or small outbreaks. They are therefore approximate guidelines and the regimen should be individualized for each patient. Nevertheless, clinical experience suggests that the following regimen can be expected to be satisfactory: mebendazole 200–300 mg (2.9–4.3 mg/kg) given t.i.d. (three times daily) for 5–10 days. Albendazole has shown promise at 400 mg (5.7 mg/kg) daily for 3 days, followed by 800 mg for 15 days (35). Albendazole, at 800 mg/day in four divided doses for 4 weeks, appeared to be effective in severe and prolonged clinical cases of *T. pseudospiralis* infection, with adjunctive corticosteroid treatment to relieve respiratory difficulty and increased myalgia in the immediate posttreatment period (16). Both animal studies and clinical experience suggest that albendazole is the current drug of choice.

The cited dosages represent much more intensive therapy than the regimens used for conventional antihelminthic treatment with these drugs. They should therefore be used prudently and modified as needed. Studies in laboratory animals strongly suggest that it is more important to maintain plasma levels of drug over a prolonged period than to achieve a high peak plasma concentration. Thus, in adjusting the regimen for an individual patient, maintenance of plasma drug concentration should be an objective.

If attempts are made to monitor the success of treatment by means of biopsy, it should be borne in mind that neither the absence of live larvae in a particular specimen nor the presence of an occasional dead larva can be assumed to be drug effects. Furthermore, the finding of live larvae only shows that treatment did not kill all of them and does not show that treatment was a failure. The best measure of treatment success is clinical improvement in the patient's condition, but unfortunately it usually is not possible to make a scientifically rigorous assessment of the role played by chemotherapy in that improvement.

VIII. PREVENTION AND CONTROL

A. Swine Production

To prevent human trichinellosis, swine herds should be kept free of *Trichinella* infection. Because pork scraps may be present in domestic or commercial garbage (and slaughterhouse offal), steps should be taken to ensure that raw garbage is not fed to pigs. Unless extremely heavy, *Trichinella* infection does not cause significant ill health or unthriftiness in pigs. Producers therefore have little incentive to guard against the infection, and legislation may be necessary. Laws requiring only that garbage be cooked before being fed to swine are difficult to enforce, so it is prudent to prohibit garbage feeding in general. In the United States, garbage feeding in fact is prohibited in every state.

Another factor in controlling *Trichinella* in swine is the ratproofing of piggeries. It is now well documented that rats may play a role in *Trichinella* transmission (16). Pigs may eat rats that harbor sylvatic species or strains of the parasite but obviously are more likely to consume rats that have acquired infection in town dumps or in piggeries, and therefore are carrying domestic or "synanthropic" types.

The importance of good husbandry recently has become apparent because of new evidence of "cannibalism" among pigs kept under poor conditions. Crowded pigs may be capable of transmitting the infection through tail biting, but pigs are much more likely to become infected as a result of eating the flesh of other pigs that have died but have not been removed from the piggery.

Vaccines, made from secretory/excretory antigens, are protective against *Trichinella* under experimental conditions (see Chapter 16 for a review of immunology). Such vaccines do not protect absolutely and, in any case, the need for such measures under conditions of routine swine production is questionable. Similarly, chemical compounds are known that could be used for chemoprophylaxis of trichinellosis, but there is no justification for their use in swine production.

The industrialization of pig production, with its reliance on confinement housing, has largely obviated the need for control of *Trichinella* transmission. Under rural farm conditions, control measures are more needed but less attainable.

B. Meat-Packing Level

1. Trichinelloscopy

Ever since the first outbreaks of clinical trichinellosis were recognized in the late nineteenth century, attempts have been made to prevent human infection by ensuring that the flesh of infected animals is excluded from the food supply or is rendered harmless before being eaten. The earliest and most widely used method has been to search microscopically for the larval parasites. This inspection, trichinelloscopy, has been carried out primarily by direct microscopic examination of squashed fragments of muscle tissue taken from pigs at slaughter.

To achieve more reliable information, larger pieces of muscle may be digested for subsequent microscopic examination of the residue for larvae. To conduct surveillance on a large scale, samples from several animals may be pooled and digested together, and inspection of individual carcasses then needs to be done only if a batch is found positive. These methods are described in Chapter 16. The pooled digestion method is officially approved in the United States for the certification of pork. There is no requirement that pork destined for sale within the United States be inspected for *Trichinella* larvae. Pork destined for export may be subject to inspection, depending on the requirements of the importing country. The United States, however, carries out an inspection of all horse meat that it exports. [This is required by importing countries because of the rare reports implicating horse meat in outbreaks of human trichinellosis (see Sec. II.A).]

In countries or regions where *Trichinella* has been shown to be essentially absent from commercial swine, it may be difficult to justify the cost of inspecting all slaughtered pigs. In the European Union, hundreds of millions of dollars could be saved annually if the examination of slaughtered swine were to be based on epidemiological data, with corresponding targeting of areas of known risk (14).

2. Immunodiagnosis

Immunological methods for the detection of *Trichinella* in swine have been developed to a high degree of sophistication. They have been applied both to epidemiological survey work and to the surveillance of swine at slaughter. These methods are described in Chapter 16.

3. Processing of Meats and Meat Products

Some meats and meat products are likely to be eaten without consumer cooking. This applies particularly to sausage products designed to be eaten cold (e.g., summer sausage). Such food items can be made safe, with respect to trichinellosis, by freezing, freeze-drying, cooking, irradiating, and by certain methods of curing (36). In the United States, specific methods of freezing, cooking, and curing are approved for use in the commercial production of these foods. Compliance with the relevant federal law is monitored by the Food Safety and Inspection Service of the U.S. Department of Agriculture.

Trichinella larvae in meat can be killed by irradiation of the meat at approximately 5 kilograys (kGy) and can be rendered noninfective by dosages as low as 0.1 kGy. In the United States, the irradiation of pork at 0.3–10.0 kGy has been approved since 1985 but is not in common use.

C. Consumer Level

1. Education

Awareness of the cause and mode of transmission of trichinellosis gives the consumer the ability to avoid infection by the simple expedient of eschewing meat that has not been frozen or cooked thoroughly. (Sometimes it is not simple for the consumer to tell the status of a particular dish that is served.) In the United States, education in this matter has been prolonged, pervasive, and apparently persuasive. Popular knowledge of the disease is probably a major factor, although by no means the only factor, in the current low prevalence of clinical trichinellosis in this country. In such countries as Germany, where trichinelloscopy has been applied intensively for over a century, the role played by consumer education is not self-evident and indeed is the subject of some disagreement among observers.

Ethnic or cultural practices that entail special risk may call for special educational measures. Some European dishes, especially traditional types of sausage that receive little cooking or that are sampled for flavor during preparation, have long been associated with trichinellosis. Immigrants from Southeast Asia to the United States have experienced several outbreaks of trichinellosis as a result of eating traditional uncooked sausages made of American pork. Education could be used to call attention to the need to cook or freeze such foodstuffs. Freezing may be more acceptable because meat is less altered by freezing than by cooking.

If educational campaigns are carried out in a way that engenders an exaggerated fear of infection, the result may be an excessive cooking of pork dishes. Overdone warnings lead to overdone pork.

2. Cooking

The thermal death point of *T. spiralis* is approximately 57°C (135°F) and it is unlikely that there are appreciable differences among species or strains. Cooking pork so that it reaches a temperature of 77°C (170°F) gives a margin of safety without destroying the gastronomic appeal of the meat. Slightly lower temperatures may be adequate provided the temperature is achieved uniformly throughout the piece of meat being cooked (22). In a controlled study, frying in deep fat at 162°C for 6 min raised the temperature at the center of pork chops to 71.8°C, yet the meat remained infective, suggesting that other parts of the chops had not reached such a temperature. Cooking chops in a conventional oven at 162°C for 43 min raised the temperature of the center to a similar temperature, 71.7°C, but the meat was not infective. With both methods of cooking, prolonging the cooking to achieve a temperature of 77°C (170°F) at the center was sufficient to make the chops noninfective (37). As a practical matter, it can be assumed that pork will be noninfective if it has been cooked until it is gray throughout, with no pink or red coloration. Thorough cooking similarly is advisable in the preparation of game meat (especially bear or walrus).

In the case of microwave cooking, special care must be taken to ensure that the meat is heated evenly. There is evidence that, in some ovens, larvae may survive in the interstices between intersecting microwaves in which the meat does not reach the prescribed temperature. In one study, a microwave oven operating at 619 W was used to raise the temperature at the center of pork chops to 81.2°C (178.2°F) in a period of 3 min, yet even at this temperature the meat was infective (22). In using microwave ovens, therefore, it may be advisable to divide the cooking period into two or more time periods and to turn (reposition) the meat in the oven between heating periods.

3. Freezing

It generally is not convenient, at the consumer level, to use freezing for the express purpose of killing *Trichinella* larvae in pork. Nevertheless, the widespread use of home freezers provides incidental protection against trichinellosis. To be effective, the meat must be of such a size, and must be stored in the freezer for such a time, as to ensure that it is frozen throughout. Under ordinary domestic conditions, meats usually are stored for prolonged periods and are likely to be frozen thoroughly.

Some Arctic strains of *Trichinella* have a considerable degree of resistance to freezing (see Sec. III.D). This means that frozen wildlife carcasses in the Arctic may be infective for humans, and outbreaks of human trichinellosis in Alaska have in fact been attributed to the eating of frozen bear meat.

REFERENCES

1.   Bessanov, A. S. (1972). Epizootiology and Control of Trichinellosis. Mintis, Vilnius, U.S.S.R. (now Lithuania).
2.   Bessanov, A. S. (1975). Diagnostics of Trichinellosis. Mintis, Vilnius, U.S.S.R. (now Lithuania).
3.   Britov, V. A. (1982). Wozbuditjeli Trichinelljeza. Nauka Press, Moscow.
4.   Campbell, W. C. (1983). Trichinella and Trichinosis. Plenum Press, New York.
5.   Campbell, W. C., Griffiths, R. B., Mantovani A., and Pawlowski, Z. S. (1988). Guidelines on Surveillance, Prevention and Control of Trichinellosis. Instituto Superiore di Sanita, Rome.
6.   Gould, S. E. (1970). Trichinosis in Man and Animals. Charles C Thomas, Springfield, IL.
7.   Soulé, C. and Dupouy-Camet, J. (1991). La trichinellose, une zoonose enévolution. CNEVA-OIE, Paris.
8.   Lupascu, G. (1970). Trichineloza. Editura Academeci Republicii Socialiste Romania, Bucharest, Romania.
9.   Dupouy-Camet, J., Ancelle, T., Fourestié, V., Boireau, P., and Soulé, C. (1998). Trichinelloses. Encyclopédie Médico-Chirurgicale. 8-517-A-10. Elsevier, Paris.
10. Campbell, W. C. (1979). History of trichinosis: Paget, Owen and the discovery of Trichinella spiralis. Bull Hist Med 53:520–553.
11. Bailey, T. M., and Schantz, P. M. (1990). Trends in the incidence and transmission patterns of trichinosis in humans in the United States, Rev. Infect Dis. 12:5–11.
12. Anon. (1996). MMWR 45(1), Jan. 12.
13. Anon. (1997). MMWR 46(52), Dec. 20.
14. Pozio, E. (1998). Trichinellosis in the European Union: epidemiology, ecology and economic impact. Parasitol. Today 14:35–38.
15. Arriaga, C., Yepez-Mulia, L., Viveros, N., Arturo, L., Zarlenga, D. S., Lichtenfels, J. R., Benitez, E., and Ortega-Pierres, M. G. (1995). Detection of *Trichinella spiralis* in naturally infected horses. J Parasitol. 81:781–783.

16. Jingwutiwes, S., Chantachum, N., Kraivichian, P., Siriyastien, P., Putaporntip, C., Tamburrini, A., La Rosa, G., Sreesunpasirikul, C., Yingyourd, P., and Pozio, E. (1998). First outbreak of human trichinellosis caused by *Trichinella pseudospiralis*. Clin. Infect. Dis. 26:111–115.

17. Wang, C. H., and Bell, R. G. (1986). *Trichinella spiralis*: newborn larval migration route in rats reexamined, Exp. Parasitol. 61:76–85.

18. Despommier, D., Symmans, W. F., and Dell, R. (1991). Changes in nurse cell nuclei during synchronous infection with *Trichinella spiralis*. J. Parasitol. 77:290–295.

19. Despommier, D. D. (1976). Musculature. In: Ecological Aspects of Parasitology. C. R. Kennedy, ed., North-Holland, Amsterdam, pp. 269–285.

20. Capo, V. A., Despommier, D. D., and Polvere, R. I. (1998). *Trichinella spiralis*: vascular endothelial growth factor is up-regulated within the nurse cell during the early phase of its formation. J. Parasitol. 84:209–214.

21. Stewart, G. L. (1983). Pathophysiology of the muscle phase. In: Trichinella and Trichinosis. W. C. Campbell, ed. Plenum Press, New York. pp. 241–264.

22. Harwood, C. L., Young, L. S., Lee, D. L. and Altringham, J. D. (1996). The effect of *Trichinella spiralis* infection on the mechanical properties of the mammalian diaphragm. Parasitology 113:535–543.

23. Pozio, E., LaRosa, G., Murrell, K. D., and Lichtenfels, J. R. (1992). Taxonomic revision of the genus Trichinella. J. Parasitol. 78:654–659.

24. Pozio, E. (1990). Trichinosis. Parasitol. Today 5:169–170.

25. Pozio, E., La Rosa, G., Serrano, F. J., Barrat, J., and Rossi, L. (1996). Environmental and human influence on the ecology of *Trichinella spiralis* and *Trichinella britovi* in Western Europe. Parasitology 113:527–533.

26. Andrews, J. R. H., Bandi, C., Pozio, E., Gomez Morales, M. A., Ainsworth, R., and Abernethy, D. (1995). Identification of *Trichinella pseudospiralis* from a human case using random amplified polymorphic DNA. Am. J. Trop. Med. Hyg. 53:185–188.

27. Zhang, Y., Wang, H., Weide, L., Zhou, M., Wu, Z., Liu, H., and Shi, H. (1996). Analysis of Chinese isolates of *Trichinella spiralis* by molecular biotechnology. Chinese Med. J. 109: 665–669.

28. Pozio, E., La Rosa, G., Yamaguchi, T., and Saito, S. (1996). *Trichinella britovi* from Japan. J. Parasitol. 82:847–849.

29. Murrell, K. D., Stringfellow, F., Dame, J. B., Leiby, D. A., Duffy, C., and Schad, G. A. (1987). *Trichinella spiralis* in an agricultural ecosystem. J. Parasitol. 73:103–109.

30. Viallet, J., MacLean, J. D., Goresky, C. A., Standt, M., Routhier, G., and Law, C. (1986). Arctic trichinosis presenting as prolonged diarrhea. Gastroenterology 91:938–946.

31. Pawlowski, Z. S. (1983). Clinical aspects in man. In: Trichinella and Trichinosis W. C. Campbell, ed. Plenum Press, New York, pp. 367–401.

32. Pawlowski, Z. S. (1985). Trichinosis. In: Current Diagnosis, 7th ed. R. B. Conn, ed. W. B. Saunders, Philadelphia, pp. 297–300.

33. Bohm, J. (1985). Epidemiology of trichinellosis in Greenland. In: Trichinellosis. C. W. Kim, ed. State University of New York Press, Albany, pp. 268–273.

34. Campbell, W. C., and Cuckler, A. C. (1964). Effect of thiabendazole upon the enteral and parenteral phases of trichinosis in mice. J. Parasitol. 50:481–488.

35. Fourestie, V., Bougnoux, M. E., Ancelle, T., Liance, M., Roudot-Thoraval, F., Naga, H., Pairon-Pennachioni, M., Rauss, A., and Lejone, J. L. (1988). Randomized trial of albendazole versus tiabendazole plus flubendazole during an outbreak of human trichinellosis. In: Trichinellosis. (C. E. Tanner, A. R. Martinez-Fernandez, and F. vanKnapen, eds. CSIC Press, Madrid, pp. 366–369.

36. Murrell, K. D. (1985). Strategies for the control of human trichinosis transmitted by pork. Food Technol. 39:65–68, 110–111.

37. Kotula, A. W., Murrell, K. D., Acosta-Stein, L., and Douglass, L. (1983). Destruction of *Trichinella spiralis* during cooking. J. Food Sci. 48:765–768.

12

Fish- and Invertebrate-Borne Helminths

John H. Cross

Uniformed Services University of the Health Sciences, Bethesda, Maryland*

I. INTRODUCTION

Many parasites are acquired by humans from the ingestion of fish, mollusks, and arthropods; and the parasites' endemnicity commonly is associated with cultural and eating habits that have been in practice in populations for generations. In most cases, infections are acquired by eating intermediate hosts that are raw or incompletely cooked, partially pickled or smoked, or poorly preserved. The infections are preventable if the food is prepared sufficiently to destroy the infective stages of the parasite carried by the intermedi-

---

*The opinions or assertions contained herein are the private ones of the author and are not to be construed as official or reflecting the views of the U.S. Department of Defense or the Uniformed Services University of the Health Sciences.

ate or paratenic hosts. It is extremely difficult, however, to change cultural and eating habits; therefore, these parasitoses will continue to prevail. There are a large number of foodborne parasitoses, but this chapter deals with a few selected nematodiases, cestodiases, and trematodiases.

II. FISHBORNE HELMINTHIASES

People's desire to eat fish uncooked has perpetuated myriad helminthic infections. While most of these parasitoses are reported in Asians, other population groups worldwide acquire infections when they eat fish raw.

A. Intestinal Capillariasis

1. Introduction

Intestinal capillariasis is caused by *Capillaria philippinensis*, a tiny nematode that was unknown until it was recovered from a Filipino who died in a Manila hospital in 1963 after suffering from diarrhea, cachexia, emaciation, and recurrent ascites (1). An epidemic of the disease was recognized in northern Luzon in the Philippines in 1965–1966; by 1967 over 1000 cases were documented and 77 had people died. The disease later was reported from many of the coastal provinces of northern Luzon and later in the central islands of the Philippines. It was recognized in 1973 in Thailand (2) and eventually in Japan, Iran, Egypt, Taiwan (3), and Korea (4). Single cases have also been reported from India, as well as from Italy and Spain. The latter two may have been acquired in Indonesia and Colombia, respectively (5).

Infection with *C. philippinensis* causes serious illness and usually leads to death if not treated in time. Deaths have been reported from the Philippines and Thailand, while in other countries the disease was identified before the severe manifestations of the infection developed.

2. Etiology

Capillaria philippinensis is a trichuriid nematode with characteristics of the group, including a narrow anterior end containing the esophagus, stichocytes, and stichosome, and a wider posterior end with the intestinal tract and reproductive organs. Males (Fig. 1) are 1.5–3.9 mm in length, and have widths of 3–5 μm at the head, 23–28 μm at the stichosome, and 18 μm at the cloaca. The male spicule is 230–330 μm in length; the spicular sheath is without spines and extended measures 440 μm. Females (Fig. 2) measure 2.3–5.3 mm in length, with widths at the head of 5–8 μm, 25 μm at the stichosome, 28–36 μm at the vulva, and 29–47 μm at the postvulva.

The worms are found in the small intestines, where the females produce thick-shelled eggs, thin-shelled eggs, and larvae. The eggs, when found in the feces, are peanut-shaped with a striated shell and flattened plugs at each end. The eggs measure 36–45 μm by 20 μm (Fig. 3). All stages of the parasite (eggs, larvae, and adults) are found in the feces.

The life cycle of the parasite has been determined experimentally in monkeys and Mongolian gerbils (2). Eggs pass in the feces and embryonate in the soil or water in 5–10 days. Embryonated eggs are swallowed by fish, hatch in the intestines of the fish, and the released larvae double in size and become infectious in 3 weeks. Larvae fed to gerbils develop into adults in 10–11 days; at 14 days, the female worms produce larvae. These

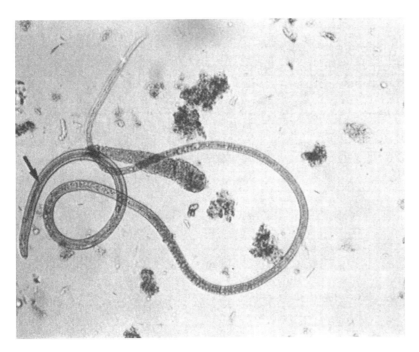

Figure 1 Male *Capillaria philippinensis* measures 1.5–3.9 mm in length; note spicule (arrow) (×160).

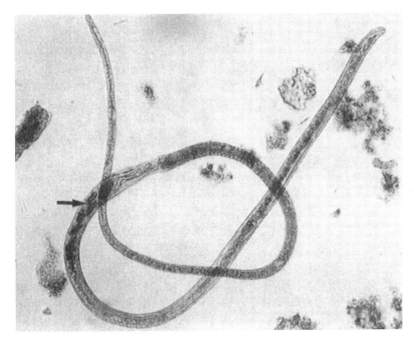

Figure 2 Female *Capillaria philippinensis* measures 2.3–5.3 mm in length. This is an old female with only a few eggs in the uterus (arrow) (×160).

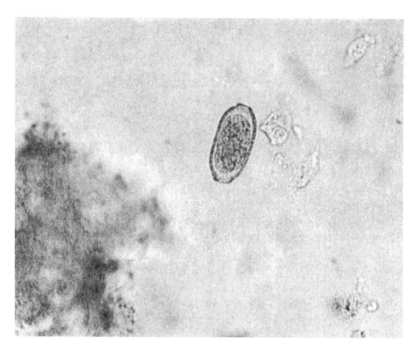

Figure 3 Eggs of *Capillaria philippinensis* in feces. Eggs have flattened bipolar plugs, a striated shell, and measure 36–45 μm by 20 μm (×400).

larvae mature into adults in another 12–14 days and begin producing thick-shelled eggs that pass in the feces.

While most female worms produce eggs, there are always a few that continue to produce larvae. These larvae replace adult worms that die sometime after reproducing; however, the number of worms continues to increase until the gerbil dies as a result of the infection. Internal autoinfection also occurs in monkeys (2), but these animals do not succumb to the infection. Humans and gerbils, on the other hand, die if not treated and at autopsy massive numbers of worms have been recovered.

Nearly 2000 cases of intestinal capillariasis have been recorded in the Philippines, while in Thailand there have been only a few hundred cases documented. Over 100 people have died as a result of the infections in the Philippines. Deaths also have occurred in Thailand, but the numbers are not known entirely.

It has been shown experimentally that freshwater fish can transmit the parasite in the Philippines and Thailand. In the Philippines, gerbils became infected when fed fish (*Hypseleotris bipartita*) taken from the lagoons in northern Luzon. A number of species of freshwater fish have been infected experimentally and it is suspected that many may be able to serve as intermediate hosts for the parasite. The fish are usually small and are eaten whole by both Filipinos and Thais.

The fish are eaten whole by fish-eating birds and several species of these birds also have been infected experimentally with the parasite (6,7). On one occasion, a yellow bittern (*Ixobrychus*) was found infected naturally with a *C. philippinensis* male. Birds now are believed to be a natural host for the parasite and are able to transmit the infection along their natural flyways.

Sanitation in rural areas of Southeast Asia is considered poor. Toilets may be present but in many households are not used because of improper maintenance. Defecation in the fields is a common practice; during the monsoon rains, the feces are pulverized, washed away, and carried to freshwater streams and ponds. Drinking water usually is from natural sources and is consumed without boiling or chemical treatment. In addition to fish, many other foods are eaten uncooked, especially in the endemic areas of the Philippines and in Thailand.

3. Pathogenesis and Pathology

The parasite is found predominantly in the jejunum of the small intestines. At autopsy, the villi are flattened and denuded. The mucosal glands are dilated and there is cellular infiltration into the lamina propria (Fig. 4). Worms in all stages are found in the intestinal lumen along with cellular debris (8) (Fig. 5). Electron microscopy studies of gerbil intestines revealed microulcers in the epithelium, compressive degeneration of the cells, mechanical compression, and a homogeneous material was seen at the anterior end of larval parasites (9).

In humans, biopsied jejunal tissue showed a loss of adhesion specialization and separation of epithelial cells. These changes in the intestinal tissue are thought to be the cause of clinical manifestations of fluid, protein, and electrolyte loss. At one human autopsy, over 200,000 worms were recovered in 1 L of bowel fluid. In other organs, such pathological changes as fatty metamorphosis of the liver, vacuolization of the renal proximal convoluted tubular cells, vacuolization of myocardial cells, and lipochrome pigment concentrations in the myocardium were also seen (8).

4. Clinical Manifestations

Clinical symptoms first experienced by infected persons are abdominal pain, diarrhea, and a gurgling stomach. As the weeks pass, diarrhea increases, with 8–10 voluminous stools

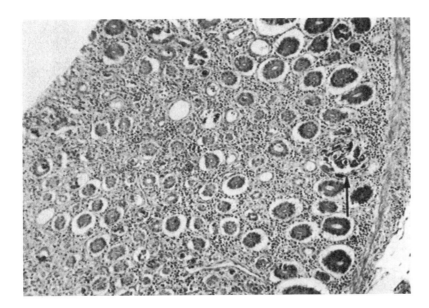

Figure 4 Intestinal mucosa taken at autopsy from a patient who died with intestinal capillariasis. Note cellular infiltration, dilatation of mucosal glands, and sections of worm (arrow) (×160).

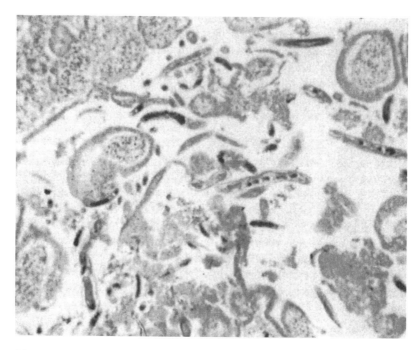

Figure 5 Intestinal tissue from patient who died with intestinal capillariasis. Note sections of worms and cellular debris in lumen of the intestine (×100).

passed each day. The patients lose a great deal of body weight and suffer from malaise, anorexia, and vomiting. Studies by Whalen and associates reported borborygmi, muscle wasting and weakness, distant heart sounds, hypotension, edema, gallop rhythm, pulsus alterans, abdominal distention and tenderness, and hyporeflexia (10).

Laboratory findings were a protein-losing enteropathy, malabsorption of fats and sugars, decreased excretion of xylose, and low serum levels of potassium, sodium, calcium, carotene, and total protein. Low levels of immunoglobulin G (IgG), immunoglobulin M (IgM), and immunoglobulin A (IgA) with elevated levels of immunoglobulin E (IgE) were seen at the time of illness, but all reverted to normal on follow-up several months later (11).

If treatment is not initiated soon enough, patients die because of the irreversible effects of the electrolyte loss, heart failure, or from septicemia.

5. Diagnosis

In endemic areas such as the Philippines, a clinical diagnosis can be made in patients presenting with abdominal pain, diarrhea, and gurgling stomach. In chronic infections, the patients will have experienced weight loss, wasting, and an intractable diarrhea.

The parasitological diagnosis is made by detecting *C. philippinensis* eggs, larvae, and adults microscopically in the feces (Fig. 3). Direct and concentration methods of examining the stools usually will reveal the infection. Multiple stool examinations may be necessary, or small intestinal biopsy or intubation also will demonstrate the parasite or eggs. On a few occasions in the Philippines, unconfirmed cases were treated with an antihelminthic and the parasite was found in the feces after medication.

6. Treatment

The recommended treatment is the oral administration of mebendazole, 400 mg/day in two divided doses for 20 days, or albendazole, 400 mg in two divided doses for 10 days. Relapses are known to occur and the treatment should be repeated with mebendazole for 30 days or albendazole for 20 days. Relapses result from incomplete elimination of larval parasites that grow into adults and reestablish infections. Relapses were common when thiabendazole was used (25 mg/kg body weight per day for 30 days) and with lower doses of mebendazole or albendazole. Patients with chronic infections should be given electrolyte replacement, an antidiarrheal, and a high-protein diet.

7. Prevention and Control

Intestinal capillariasis can be prevented by educating indigenous populations on the hazards of eating small freshwater fish uncooked. It is difficult to change eating habits that have been traditional for generations, but it is the only way to control this disease. Human-to-human transmission through fish most likely occurred during the epidemics in Thailand and the Philippines since indiscriminate disposal of feces is common in these countries. In such cases, improvement of sanitary conditions would have been beneficial. The zoonotic aspects of the parasitoses impede realistic control measures.

Severe illness and death can be prevented by early recognition, diagnosis, and treatment. Untreated infections invariably lead to death.

B. Gnathostomiasis

1. Introduction

Gnathostomiasis is a disease caused by infection with the larval or immature adult stages of nematodes of the genus *Gnathostoma*. Adult parasites are reported in canines, felines, and other carnivorous animals worldwide, but only two, *G. spinigerum* and *G. hispidum*, are reported from humans. In recent years, however, *G. doloresi* (12) and *G. nipponicum* (13) have been found in Japanese individuals. Human gnathostomiasis has been reported sporadically in Southeast Asia, China, and Mexico, but is seen most often in Thailand and Japan. *Gnathostoma* species have been reported in animals in the Middle East, Africa, and the former Soviet Union.

Gnathostoma spinigerum was described first in gastric tumors found in a tiger that died at the London Zoological Garden. Miyazaki reviewed the genus and reported 19 species (14), but Daengsvang considers only 12 to be valid species (15). Most of the scientific information relating to these parasites is credited to the work of Miyazaki and Daengsvang.

Larval and immature *G. spinigerum* adults cause migratory, often transitory, subcutaneous swellings in humans. The worms occasionally enter the internal organs and central nervous system (CNS), but they never reach sexual maturity in humans.

2. Etiology

Gnathostoma spinigerum is a short, stout nematode with a subglobose head armed with 7–9 transverse rows of hooklets (Fig. 6). Spines also extend halfway down the body. Males measure 11–25 mm and females 25–54 mm in length. Adult worms live in tumors in the stomach wall of fish-eating mammals. Females produce brownish ovoid eggs with a mucoid plug at one end; they measure 56–79 μm by 35–43 μm. The eggs pass in the animal feces, reach water, and embryonate in 7–10 days.

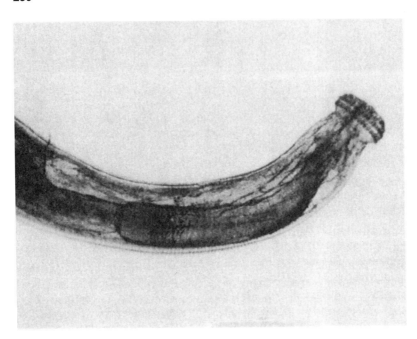

Figure 6   Larva of *Gnathstoma* species recovered from a fish in Thailand; note spiny head (×100).

The first-stage larva hatches from the egg and is eaten by a freshwater copepod of the genus *Cyclops*, in which it develops into a second-stage larva. When the copepod is eaten by a second intermediate host, including fish, birds, amphibians, reptiles, and mammals, the parasite enters the tissue and becomes a third-stage larva. Paratenic, or transport, hosts may eat a second intermediate host and the infective stage larva becomes encapsulated in the tissue and does not develop further. When a second intermediate host, or paratenic host, is eaten by a definitive host, the parasite is digested from the tissue, penetrates the stomach wall, and migrates to the liver and then to other organs, eventually returning to the peritoneal cavity and penetrating the stomach to form a tumor-like mass. The worms reach maturity and produce eggs in approximately 6 months.

Felines and canines are definitive hosts for *G. spinigerum*, and domestic and wild pigs are natural hosts for *G. hispidium*. Infections are acquired by eating one of the many intermediate hosts. In Thailand, Daengsvang listed 44 species of vertebrates naturally infected with *G. spinigerum* third-stage larvae, including freshwater fish, frogs, snakes, chickens, ducks and other birds, rats, mongooses, and tree shrews (14). The main source of human infection in Thailand is the snake-headed fish *Ophicephalus*, which is permitted to ferment for a few days and is eaten raw (*som fak*). Human infections also are reported from eating raw or poorly cooked catfish, eels, frogs, chickens, ducks, and snakes. In Japan, freshwater *Ophicephalus* species are eaten raw as *sashimi*. The ingestion of raw or inadequately cooked fish is the major source of infection in other areas reporting the disease. Infections in Mexico are attributed to eating raw cycloid fish as ceviche (16).

Infections occur in all age groups and both sexes. A 7-day-old Thai baby was found infected and prenatal transmission was suspected (17). The parasitosis is seen with regularity throughout Thailand, especially in medical centers in Bangkok and Chiang Mai; most are subcutaneous infections.

3. Pathogenesis and Pathology

Humans are abnormal hosts for gnathostomiasis and the parasite once acquired does not reach sexual maturity. After ingestion of the second intermediate host, the larva migrates through the tissues and into any organ. It eventually enters subcutaneous tissue and becomes a larval migrans. Lesions develop along the worm's migratory track, causing inflammation, necrosis, and hemorrhage. Toxic products, immunological responses, and mechanical injury cause inflammatory reactions, eosinophilia, swelling, pain, and edema. A reaction is transient and disappears in a few days only to reappear sometime later at a new location. The parasite may enter an eye, causing subconjunctival edema, hemorrhage, and retinal damage. Invasion into the CNS leads to encephalitis, myelitis, radiculitis, and subarachnoid hemorrhage. Intracerebral hematoma and transitory obstructive hydrocephalus also have been described in Thailand (18).

4. Clinical Manifestations

In the natural host, *G. spinigerum* causes tissue damage as it migrates. Adult worms cause cavities to form in the stomach wall that become filled with serosanguinous fluid and eventually cause nodular formulation. In humans, the parasite rarely becomes sexually mature and the larva migrates to various sites. Usually only a single larva is involved in human infections.

The first clinical symptoms occur shortly after the ingestion of infected food; the parasite enters the intestinal tissue, causing epigastric pain, nausea, vomiting, pruritus, urticaria, and low-grade fever. There is a leukocytosis with a marked eosinophilia. Acute pain is experienced as the larva migrates through abdominal and thoracic organs. The worm eventually makes its way to the subcutaneous tissue in about a month. The larva moves through the tissues, causing edematous swellings up to 15 cm, usually on the chest, abdomen, or extremities. The lesion is red and warm with pruritus and lasts about a week. It may recur weeks or months later. At times, the larva may cause serpiginous tracks that are similar to but bigger than those caused by animal hookworm larvae.

The gnathostome larvae in the eye cause exophthalmus, vitreous, corneal, or lenticular damage, impaired vision, and blindness. CNS invasion causes headache, nuchal rigidity, drowsiness, coma, and symptoms of a cerebral vascular accident. In the spinal cord, the worm may migrate along the nerve trunk, causing root pain and paralysis.

5. Diagnosis

The definitive diagnosis is made by the recovery and identification of larvae from surgical specimens, urine, sputum, or vaginal discharge. A presumptive diagnosis in endemic areas is based on history, symptoms, or serology. Specific antigens are available for immunodiagnosis (19).

In the differential diagnosis, such other parasitic infections as hookworm, cutaneous larva migrans, myiasis, sparganosis, cutaneous paragonimiasis, loiasis (calabar swelling), visceral larval migrans (toxocariasis), and meningitis caused by *Angiostrongylus cantonensis* should be considered.

6. Treatment

There is presently no specific chemotherapeutic agent for the treatment of gnathostomiasis. Trials with mebendazole have not been successful but long-term treatment with albendazole has shown promise (19). Surgical removal is indicated when the parasite is accessible in the skin, eye, or abdominal mass. Supportive, symptomatic, and antiinflammatory treatment is recommended.

The prognosis is good in cutaneous gnathostomiasis, although the infection may persist intermittently for months or years. CNS involvement may resolve spontaneously or may lead to death. The fatality rate for cerebral gnathostomiasis in Thailand is estimated to be 40% (19).

7. Prevention and Control

Health education programs would be beneficial for control in endemic areas of Asia. Infections could be prevented if people would eat only well-cooked intermediate or paratenic hosts (e.g., fish, eels, snakes, frogs, and poultry), and drink only boiled or treated waters that are potentially copepod-infested.

C. Anisakiasis

1. Introduction

Anisakiasis is a gastrointestinal parasitosis caused by the larval stages of an anisakid nematode. The adult worms are common parasites of marine mammals and the larval stages are found in marine fish and squid and possibly scallops and shrimp. Humans acquire infections by eating these marine animals raw or improperly cooked or preserved. The disease was recognized first in Holland, where it was associated with pickled herring, and later in Japan, with infections acquired from eating raw fish prepared as sashimi. The disease also is reported from the Americas, the Pacific islands, and other areas of northern Europe.

There are many species of anisakine nematodes, but those most often associated with human illness are *Anisakis simplex* and *Pseudoterranova decipiens*. There are two types of disease: invasive and noninvasive. *P. decipiens* often is noninvasive. Geographic distribution is associated with eating habits.

2. Etiology

Adult anisakids are located in the stomach of marine cetaceans and pinnipeds but do not develop in humans. It is only the third larval stage of the parasite that is harmful to humans. These are elongated and tapered at both ends. *A. simplex* larvae are 20–35 mm in length and 0.3–0.06 mm in width, and are white or milky in color. *P. decipiens* larvae measure 25–50 mm in length and 0.3–1.22 mm in width; they are yellowish brown.

The mouth of anisakids has three lips and a boring tooth. The digestive tract consists of a postesophageal expansion called a ventriculus (Fig. 7) and in other species there may be an intestinal cecum and/or an intestinal appendage. The position of the excretory pore, usually anterior, is helpful in identification. A mucron is located at the posterior end (Fig. 8). Cross-sections of worms in tissue reveal polymyrian and cyclomyarian muscles, thick cuticle, thick-walled intestine, and Y-shaped lateral cords (Fig. 9).

Female worms deposit eggs that pass in the feces of the marine mammals onto the ocean floor. First-stage larvae then second-stage larvae develop and eventually hatch from the egg and are eaten by small marine crustaceans (*Euphausia* and *Thysanoessa*) and develop into third-stage larvae. These larvae are infective to squid and marine fish. The larvae may pass from squid to fish and from smaller fish to larger fish along the predatory chain. The larvae penetrate the gut to the peritoneal cavity and musculature, then transfer to the definitive host when the intermediate or paratenic host is eaten. Humans become infected when the intermediate host is eaten uncooked.

Whales, dolphins, and porpoises serve as definitive hosts for *A. simplex*, while seals,

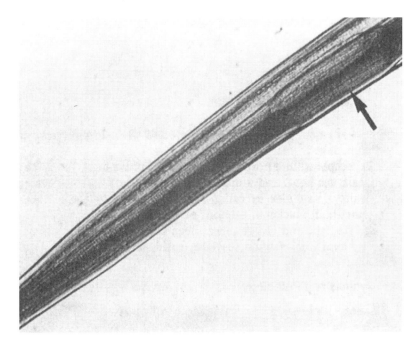

Figure 7 Anterior end of *Anisakis simplex* larva recovered from a marine fish from Taiwan; arrow indicates ventriculus (×100).

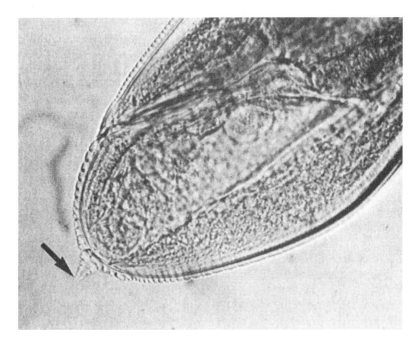

Figure 8 *Anisakis simplex* larva recovered from a marine fish from Taiwan; note mucron (arrow) (×63).

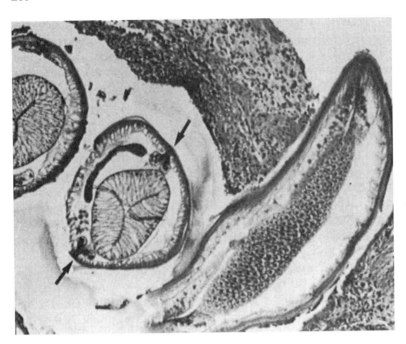

Figure 9 Sections of *Anasakis simplex* larva in stomach of a Mongolian gerbil; note Y-shaped lateral cord (arrow).

sea lions, and walruses are natural hosts for *P. decipiens*. Humans acquire *A. simplex* by eating raw or poorly salted, pickled, or smoked herring, cod, mackerel, salmon, or squid, and *P. decipiens* from cod, halibut, flatfish, and red snapper. Such traditional preparations as green herring, lomi lomi salmon, ceviche, sushi, and sashimi are major sources of infections.

3. Pathogenesis and Pathology

Anisakid larvae are released in the stomach upon digestion of the intermediate host tissue and penetrate the gastric or intestinal mucosa. Worms may penetrate the gut wall and enter the peritoneal cavity. The parasite elicits a foreign body reaction with cellular infiltration with neutrophils, a few eosinophils, and giant cells. In the gastrointestinal tract, an abscess develops with massive infiltration of eosinophils, lymphocytes, monocytes, neutrophils, histiocytes, and plasma cells, and eventually necrosis and hemorrhage. Worms are found in early lesions but in time are destroyed by the inflammatory reaction. Tunnels and burrows in the stomach mucosa also have been reported and are thought to be caused by protease secreted from the worms (22). Larvae entering the peritoneal cavity may migrate into lymph nodes, the mesentery or omentum, the pancreas, and other organs. Some ingested larvae do not penetrate the gut mucosa and are vomited, migrate up the esophagus, or pass in the feces.

4. Clinical Manifestations

Clinical symptoms may appear within a few hours or a few days after ingestion of infected intermediate hosts. In stomach anisakiasis, there is epigastric pain, nausea, and vomiting. The acute symptoms subside in a few days, but a vague abdominal pain with intermittent nausea and vomiting persists for weeks. Occult blood may be present in gastric aspirates.

Anasakis simplex, more often than *P. decipiens*, invades the intestinal mucosa, usually the distal ileum, and symptoms appear within a week of an infective meal. There is abdominal pain, nausea, vomiting, diarrhea, and possibly fever. Symptoms may resemble peptic ulcer and occult blood may be found in the feces. Parasites in the peritoneal or pleural cavities may cause peritonitis or pleurisy, respectively.

5. Diagnosis

Anisakiasis is misdiagnosed easily because of its similarity to such other gastrointestinal conditions such as acute abdomen, gastric ulcer or neoplasm, appendicitis, acute regional ileitis, diverticulitis, Crohn's disease, tuberculous enteritis, cholelithiasis, or intestinal obstruction. A parasitological diagnosis is made only when the worm is recovered or sections seen histologically after surgical removal of tissue indicates the presence of the parasite. The finding of Y-shaped lateral cords of the worm is helpful. Double-contrast X-ray shadowing or ultrasonic echo analysis of the stomach is also of value. Radiology also may be useful when it shows luminal narrowing of the small intestines. Such serological tests as immunofluorescence and radioallergosorbent tests may provide a presumptive diagnosis. A microenzyme-linked immunosorbent assay using monoclonal antibodies was recently described (23).

6. Treatment

Antihelminthics have not been reported effective in the treatment of anisakiasis. Fiberoptic endoscopy and removal of the larvae with forceps is the easiest treatment for stomach anisakiasis. Surgical removal of the parasite may be necessary in some cases of gastric and intestinal disease, but in some chronic as well as acute cases only symptomatic treatment is recommended. The prognosis is good once the parasite is removed or passes from the body. In chronic cases, the parasite eventually will be destroyed by the host reaction.

7. Prevention and Control

The best prevention and control of anisakiasis is eating only well-cooked marine fish and squid. Larvae are killed by cooking at 60°C or above. Fish frozen at −20°C for 24 h will kill the larvae. Salt curing, marinating, microwaving, and smoking temperatures are insufficient to kill the parasite. Fish infections are highest in areas in which there are high marine mammal populations, such as coastal Japan and Alaska. In recent years, fish infections have increased with the rise of the seal population along the California coast.

D. Diphyllobothriasis

1. Introduction

There are a number of tapeworms that humans acquire through their eating practices, but only a few are acquired by eating fish. Although there are various species of fish tapeworm reported in humans, the most important is *Diphyllobothrium latum*. Two less important species are *D. pacificum*, reported from Peru, Chile, and Japan, and *D. nikonkaiense* from Japan. Some closely related diphyllobothriid species cause sparganosis in humans, but these usually are not fish-borne.

The distribution of the fish tapeworm or broad tapeworm *D. latum* is widespread in the temperate and sub-Arctic regions of the Northern Hemisphere where freshwater fish are eaten. The Baltic region in northern Europe, especially among the Finnish populations, once was highly endemic, but the prevalence has decreased in recent years. This also has occurred in northern Italy and Switzerland; however, in the Danube delta in Romania and in the Volga basin of Russia, the prevalence rate remains high. Freshwater lake regions in Africa, Asia, and South America report the parasitosis. In North America, diphyllobothriasis has been reported in the Eskimos of Alaska and Canada and, in the Great Lakes region, in people of Scandinavian origin. Russians and Scandinavian immigrants to North America introduced the disease into these areas.

The disease usually is asymptomatic but may cause anemia and/or low serum vitamin B_{12} levels. Although fish-eating animals may be infected, humans are considered the major source of infection because of poor sanitary practices.

2. Etiology

The fish tapeworms are among the largest worms to infect humans. They are located in the small intestine and range from 2 to 15 m in length with a maximum width of 20 mm. The scolex or head is 2 mm in length and 1 mm in width and has a dorsal-ventral sucking groove or bothrium that serves as a holdfast organ. The neck is thin and unsegmented, and is followed by segmented immature, mature, and gravid proglottids. There may be as many as 3000 proglottids. A mature proglottid is broader than it is long and has a rosette-shaped, egg-filled uterus; at the posterior end, it is square. The parenchyma is filled with male and female genital organs and vitelline follicles. Eggs are shed from the uterine pore and, after all eggs are expelled, the proglottids disintegrate.

Eggs of *D. latum* passed in the feces must reach freshwater; in 10–14 days, a ciliated hexacanth embryo called a *coracidium* develops. This emerges from the egg through an opened operculum and swims freely in the water until ingested by a copepod. The coracidium develops into an elongated procercoid larva in the hemocoel of the copepod. Further development occurs when the first intermediate host is eaten by a second intermediate host, a fish. The procercoid passes from the fish's intestine to the musculature and develops into the pleurocercoid larva or sparganum. When an infected fish is eaten uncooked, the pleurocercoid emerges from the fish tissue and develops into an adult tapeworm in 3–5 weeks. The worms are known to live for as long as 10 years.

Although a variety of fish-eating mammals can serve as a definitive host for *D. latum*, humans are considered the major reservoir host and source of infection. Dissemination of the parasite is thought to have been through immigration of infected persons from endemic areas into lake regions in which susceptible copepods and fish were present. Russians may have introduced the parasite into Alaska, and Scandinavian immigrants are blamed for the introduction of the parasite into the Great Lakes region of North America. Wild animals are able to maintain the infection but they are poor hosts compared to humans. A high prevalence of infection in freshwater lakes is attributed to the practice of resort hotels along lake shores of pumping raw sewage into the lakes.

Important animal hosts are the dog, fox, mink, cat, pig, bear, seal, and sea lion. Many species of fish serve as second intermediate hosts (e.g., pike, perch, turbot, salmon, and trout). The preparation of fish varies among different ethnic populations. Some prefer smoked or pickled fish while others, like the Japanese, eat raw fish prepared as sashimi or sushi. Jewish housewives often become infected by tasting fish being prepared for gefilte fish.

3. Pathogenesis and Pathology

Although *D. latum* is a large worm that occupies the small intestine, its presence produces little pathology. Usually, only one worm is involved, but with more than one worm there may be intestinal obstruction. When the worm is located in the jejunum, it may compete with the host for vitamin B_{12} and, after 2–3 years of infection, the host may develop a megaloblastic anemia. This is found primarily in Scandinavian countries, especially Finland. Hematological changes associated with the anemia include increases in mean corpuscular volume and hemoglobin content, and nucleated erythrocytes with megaloblasts and normalblasts occurring in the peripheral blood. The number of platelets are reduced, the coagulation time is prolonged, and a mild leukopenia may occur. A mild hemolysis occurs in association with ineffective erythropoiesis. Toxic substances excreted by the worm may affect the CNS, causing peripheral and spinal nerve degeneration.

4. Clinical Manifestations

Diphyllobothriasis usually is asymptomatic and patients are unaware of the infection until the passage of spent proglottids in the stool or vomit. Clinical symptoms reported in Finns are fatigue, weakness, the desire for salt, diarrhea, epigastric pain, and fever. Anemia develops in very few people, but many may experience pallor, glossitis, dyspnea and tachycardia, and low serum vitamin B_{12} levels. Bloating, sore gums, and a sore tongue also are reported. When the CNS is affected, there are symptoms of weakness, numbness of extremities, dizziness, paraesthesia, and disturbances in mobility and coordination. There may be ocular atrophy causing central scotoma.

5. Diagnosis

The parasitological diagnosis of diphyllobothriasis is made by detecting characteristic *D. latum* eggs in the feces. The eggs are a yellowish color, thin-shelled, operculated, and approximately 60 × 40 μm in size (Fig. 10). Stool concentration methods may have to be used in early infections, but in later infections the eggs are abundant. Proglottids found

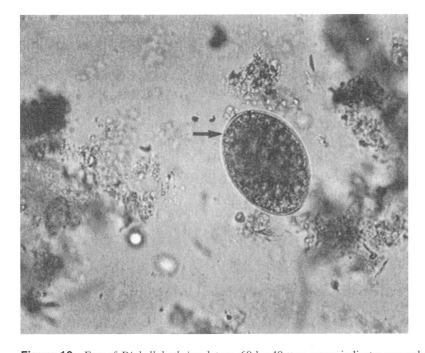

Figure 10 Egg of *Diphyllobothrium* latum, 60 by 40 μm; arrow indicates operculum (×400).

in feces or vomitus can be identified as a diphyllobothriid by the rosette-shaped uterus. Tapeworm anemia can be differentiated from true pernicious anemia by the presence of free hydrochloric acid in gastric juice, which is associated with achlorhydria.

6. Treatment

Two drugs are effective in the treatment of fish tapeworms: niclosamide and praziquantel. Niclosamide is given in a single oral dose of 4 tablets (2-g chewable tablets) on an empty stomach. The dosage is reduced to 3 tablets for children who weigh more than 34 kg and 2 tablets for those who weigh less than 34 kg. Praziquantel is given in a single oral dose of 10–20 mg/kg body weight for adults and children. Side effects are minimal. Supplements of vitamin B_{12} and folic acid are recommended following treatment in cases of anemia. If the scolex is not removed, the worm may regenerate and follow-up examinations should be made after 6 weeks to determine the presence of eggs in the feces. The prognosis is good, with rapid clinical improvement following treatment.

7. Prevention and Control

Infection with *D. latum* is prevented easily by the eating of only well-cooked freshwater fish. Cooks should avoid tasting fish while it is being cooked. The changing of dietary habits through education would reduce infection. If fish is to be eaten raw, smoked, or pickled, freezing at −10°C for 1–3 days should kill pleurocercoid larvae in the fish tissue. Sewage from lakeside hotels and from ships sailing in the lakes should be treated before release into the lakes. It would be difficult to control contamination of water by infected wildlife populations, but pets can be dewormed periodically.

E. Clonorchiasis

1. Introduction

Clonorchis sinensis, the Chinese liver fluke, was found first in a Chinese person who died in Calcutta, India, in 1874. It also is called *Opisthorchis sinensis* as it is similar to other liver flukes of the genus *Opisthorchis*, but most parasitologists prefer to retain the original name.

   The parasite is endemic in China, Korea, Japan, Taiwan, and Vietnam. Close to 4 million people are infected in China, nearly 1 million in Korea, and 1 million in Vietnam (24). Reports of infection in persons outside of Asia are considered cases imported from Asia. Infection rates have decreased in Japan in recent years but the parasitosis remains widespread in China, Taiwan, and Korea. The infections are acquired by eating freshwater fish raw or poorly cooked or preserved. Cyprinoid species of fish are the major source of infection.

2. Etiology

Clonorchis sinensis is a flat, slender trematode, attenuated at the anterior end and with a rounded posterior (Fig. 11). It measures 10–14 mm in length and 3–5 mm at the widest area of the posterior end. The hermaphroditic adults inhabit the smaller biliary passages of humans and other fish-eating mammals. The oral sucker opens at the anterior end and immediately behind is a divided intestinal cecum. The branching testes are in the posterior third of the body and the ovary is rounded and anterior to the testes. The egg-filled uterus is in the middle of the anterior part of the body. The vitellaria are lateral to the uterus and intestinal ceca and extend from the ovary to the acetabulum. Eggs are small, ovoid,

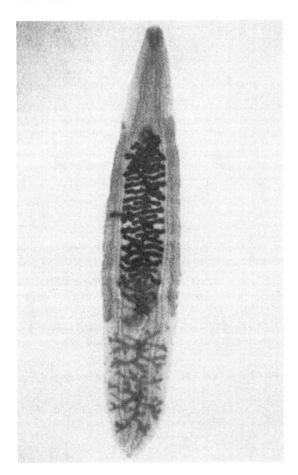

Figure 11 *Clonorchis sinensis* adult (×10).

and yellowish browns with an operculum on one end and a small knob at the aoperculated end. The yellow-brown eggs measure 26–30 μm by 15–17 μm and are passed by the worm down the bile duct to the intestine and out with the feces.

The eggs are embryonated when passed and must reach freshwater containing hydrobiid snails, the first intermediate host. The snail ingests the egg and the miracidium hatches and migrates to hemolymph spaces and develops into a sporocyst. Redia develop in the sporocysts, escape, and migrate to the snail digestive gland, in which cercariae are formed. The cercariae leave the snail and swim in the water in search of specific fish that are second intermediate hosts. The cercariae penetrate beneath the fish's scales and encyst in the muscle and subcutaneous tissue as metacercariae, becoming infective in 3–4 weeks.

When the fish is eaten raw, cooked inadequately or pickled poorly, metacercariae are digested from the fish tissue, excyst, migrate to and enter the ampulla of Vater, and pass through the common bile duct to the intrahepatic bile ducts. The young flukes reach the bile ducts in 1 or 2 days and mature into adults in 30 days. Some worms are known to live for as long as 30 years.

Important snail hosts are *Parafossarulus manchouricus*, *Bithynia fuchsiana*, and *Alocimna longicornis*, snails commonly found in fish culturing ponds. Important fish second intermediate hosts are members of the family Cyprinidae. Carp such as *Ctenopharyn-godon idellus* in China and *Pseudorasbora parva* in Japan and Korea are important hosts. Over 100 species of fish are reported hosts for the parasite in Asia.

The prevalence and distribution of clonorchiasis is associated with the presence of susceptible snail and fish intermediate hosts and habits of the indigenous populations. It has been estimated that close to 7 million people are infected in Asia. Southern Chinese are infected more often, but foci of infections are known in Beijing and other areas of northern China. In many parts of China, including Taiwan, fish often are raised in ponds fertilized with human feces. Susceptible snails also are present in the ponds and the life cycle is perpetuated. The people eat the fish raw in thin slices over a watery rice called congee. Many people in Hong Kong are infected by eating this preparation, but neither the fish nor snails are present in Hong Kong. Infected fish are imported from the Chinese mainland. In Korea, infections are seasonal since snails hibernate in winter and the larval stages often are damaged due to the cold weather. The snails become infected in the spring and shed cercariae. Korean men are infected more commonly than women since the men eat raw fish while drinking wine. Children in Korea are less likely to be infected, but in parts of China clonorchiasis occurs in children as well as in adults. It is interesting that in Vietnam infections are much more common in people in the North than in the South.

In most endemic areas, cats and dogs usually are infected in high numbers. At one time in Taiwan, clonorchiasis was endemic in only two areas, one an aborigine village in a central lake region and the other in a Hakka (originally from South China) village in southern Taiwan with many fish ponds. Dogs and cats around the island generally are infected. The eating of raw freshwater fish has become popular in Taiwan and consequently the prevalence of human infection is increasing. Carnivorous animals, including pigs, rats, mink, and other wild animal life, are reservoir hosts in endemic areas.

3. Pathogenesis and Pathology

Persons with clonorchiasis may be infected for many years essentially without disease. The pathogenesis, however, is associated with worm loads. Infection with a few thousand worms is not unusual and a record 20,000 flukes were recovered from a person at autopsy. There is an indication that mechanical irritation caused by the worm suckers and toxic substances produced by the parasite may contribute to the pathology; this becomes more significant with increased numbers of worms.

Worm presence causes proliferation of bile duct epithelium and an increase in goblet cells and mucus secretion. The ductal wall becomes thickened as a result of cellular proliferation and leukocyte infiltration. The lumen becomes dilated and tortuous, and there may be ductal and penductal fibrosis. There are no changes in the liver parenchyma. Cholangitis and cholecystitis may occur as a result of secondary bacterial infection during obstruction of the bile ducts. Calculi also may form around dead worms; gallbladder and bile duct stones were found in 70% of clonorchiasis patients autopsied in Hong Kong. The parasite also may enter the pancreatic ducts, causing dilatation and fibrosis and pancreaticoliths (25). There is strong evidence of an association between *C. sinensis* infection and cholangiocarcinoma. Throughout endemic areas, the cancer often is discovered in patients with the parasitosis. The tumors usually originate at sites at which the worms are located. The exact role of the parasite is unclear, but probably is associated with length and severity of infection. Host immune response and such variables as ingestion of dietary carcinogens may be factors related to the carcinogenic event (26).

4. Clinical Manifestations

Clinical symptoms of acute clonorchiasis can occur as early as 10–26 days after eating fish infected with large numbers of metacercariae. Patients experience fever, epigastric pain, anorexia, diarrhea, hepatosplenomegaly, leukocytosis, and eosinophilia. Most persons in endemic areas are subjected to continuous exposure to infection, and with progressive infection patients complain of irregular appetite, indigestion, epigastric discomfort, fullness of the abdomen, diarrhea, and fever.

Recurrent pyogenic cholangitis is a frequent complication, accompanied by fever, chills, abdominal pain, jaundice, and hepatomegaly. *Escherichia coli* commonly is responsible for secondary bacterial infections. Adult worms may invade the pancreatic duct and cause symptoms of acute pancreatitis, often 1–3 h after a meal that stimulates an excessive flow of pancreatic juices. Biliary and pancreatic stones may cause obstructions; the most serious consequence of chronic disease is the possibility of cholangiocarcinoma.

5. Diagnosis

Eggs of *C. sinensis* (Fig. 12) may be present in stool specimens as early as 1 month after an infected meal. Eggs also may be recovered from the bile duct drainage. The eggs are characteristic, but at times may be difficult to differentiate from eggs produced by such other trematodes as *Heterophyes heterophyes, Metagonimus yokogawai, Haplorchis taichui*, and *Opisthorchis* species. A definitive diagnosis is made after examination of adult worms passed in the feces following treatment. Several serologic tests have been developed, but most are nonspecific. An enzyme-linked immunosorbent assay (ELISA) is reported to be of value however (27). Computed tomography may be helpful in diagnosis in endemic areas (28) and M-mode sonograms may detect fluke movements.

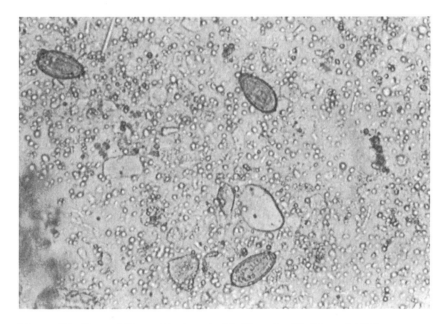

Figure 12   Eggs of *Clonorchis sinensis*, 26–30 μm by 15–17 μm (×400).

In the differential diagnosis, consideration must be given to other parasitoses, peptic ulcers, gallstones, and other causes of carcinoma, cholangitis, and cholecystitis.

6. Treatment

Praziquantel is the drug of choice in treating clonorchiasis, given at 75 mg/kg body weight daily for 2 days in three divided doses. Side effects of headache, dizziness, and nausea develop in some patients, but are mild and transient. In cases of biliary obstruction, surgical drainage may be required; in patients with cholangitis, antibiotics should be given. The prognosis is good for light infections of short duration, especially after praziquantel therapy. Chronic, long-term, untreated infections may lead to severe hepatobiliary disease, liver dysfunction, and carcinoma of the bile duct. Relapsing cholangitis of a long duration can lead to death.

7. Prevention and Control

Clonorchiasis can be prevented if populations in endemic areas would avoid eating raw, undercooked, or improperly pickled, salted, dried, or smoked fish. The eating of raw freshwater fish has been a practice among Asian groups for centuries and they would be unlikely to change their dietary habits. Some education projects have been successful for control, but sustainability is difficult. Freezing fish for a week at $-10°C$ would be beneficial, but even frozen fish has been incriminated with *Clonorchis* outbreaks in nonendemic areas.

In areas in which fish are raised in ponds, human and animal feces should be composted or sterilized before their application as fertilizer in the ponds. Snail control measures are unrealistic since most molluscicides are harmful to fish. Treatment of all infected persons and improved sanitation would help control infection, but in some treatment programs people often become reinfected. Thorough cooking of all freshwater fish is the most effective means of control.

F. Opisthorchiasis

1. Introduction

Two species of *Opisthorchis* infect humans and cause a disease similar to clonorchiasis. *O. viverrini* is found in humans and animals in Southeast Asia, especially northern Thailand and Laos, and *O. felineus* (the cat liver fluke) is found in similar hosts in eastern Europe, Poland, Germany, European Russia, and Siberia. There are also scattered reports from several other countries. Nearly all new arrivals in Siberia become infected during the first year of residence. It is estimated that over 8 million people are infected with *O. viverrini* and 1.6 million with *O. felineus* (24). Prevalence rates in highly endemic areas are over 90% for *O. viverrini* and 85% for *O. felineus*. In Thailand, Chinese populations rarely are infected, with most infections occurring in Laotians and Thais of Laotian ancestry. Opisthorchiasis has significant public health importance since it often causes serious disease and is associated more with cholangiocarcinoma than clonorchiasis.

2. Etiology

Opisthorchis species are similar morphologically to *C. sinensis* except that the testes of the opisthorchids are lobate. *O. viverrini* differs from *O. felineus* in that there are more notches in the testes, a greater proximity of ovary to testes, and the vitellaria are aggregated into clusters. Although there are variations in size between the opisthorchids and *C. sinensis*, it is very difficult to make specific identification from eggs.

The life cycle of the parasite is similar to that of *C. sinensis*. Adult opisthorchids reside in the distal and at times the proximal biliary ducts and gallbladder. Eggs pass to the intestines and out with the feces. The eggs are embryonated and, upon reaching freshwater, are ingested by snails, *Bithynia goniomplalos, B. funiculata*, and *B. siamensis* for *O. viverrini* and *B. leachi* for *O. felineus*.

Cyprinoid fish or carp serve as second intermediate hosts and are the main source of human and animal infections. Fish are eaten raw by Thais and Laotians, especially in the rainy season when the fish are most abundant. *Cyclocheilihthys siaja, Hampala despaor*, and *Puntius phoides* are fish commonly infected, with infection rates as high as 95%. In Europe and Siberia, metacercariae are common in the freshwater fish *Leuciscus rutilus, Blicca bjorkna, Tinca tinca*, and *Barbus barbus*.

In areas endemic for *O. felineus*, the freshwater fish is eaten dried, salted, or raw. In one European outbreak, salted fish was eaten the first day of salting. In Thailand, fish are raised in ponds and the ponds are fertilized with human and animal feces. Infants become infected when mothers feed them raw or partly cooked fish. Only a few worms are found in infants but over the years with constant exposure the worm load increases. The intensity of infection indicated by egg counts increases with age but becomes stable in groups over 20 years of age. Worm burden also increases, peaking between the ages of 31 and 40 (30). The parasites are known to live for 20 years.

Animal reservoirs include most fish-eating mammals in endemic areas. In some parts of Europe, 85% of the cat population is infected. Cats and dogs in Southeast Asia are important reservoirs for *O. viverrini*. Other opisthorchids are occasionally from humans; *O. guayaquilensis, Amphimerus pseudofelineus*, and *Metorchis conjunctus*. The latter was incriminated in an epidemic among Koreans in Canada who ate white suckerfish raw (31).

3. Pathogenesis and Pathology

The pathologic effects of *Opisthorchis* infections are not significantly different from those caused by *C. sinensis*. Worms usually mature in the distal part of the biliary tract. There is hyperplasia of the biliary ducts and thickening of the walls. Large numbers of flukes produce more changes. Biliary tract obstruction is not uncommon with large worm loads and marked dilatation of the bile ducts occurs at the distal ends. Worms usually are present in the gallbladder, common bile duct, and pancreatic duct in heavy infections.

Suppurative cholangitis occurs in some patients and the infection may extend to the liver parenchyma and cause hepatitis. The incidence of cholangiocarcinoma is high in cases of opisthorchiasis and hepatocellular carcinoma occurs more often with opisthorchid infections than in patients with clonorchiasis. The causes of carcinoma in opisthorchid infections are not known, but recent studies speculate that low levels of nitrosamines in foods preserved by bacterial fermentation may be contributory. In addition, it has been hypothesized that N-nitro compound produced endogenously by activated macrophages in the chronically inflamed biliary tract may serve as the carcinogen for the fluke-associated carcinogenesis (32).

4. Clinical Manifestations

Opisthorchid infection causes the same clinical symptoms as *C. sinensis*. Most infections are asymptomatic, depending on the degree of parasitoses and duration of infection. Acute disease with fever, lymphadenopathy, myalgia, rash, and eosinophilia is reported shortly after exposure.

Right upper quadrant pain that radiates to the epigastrium appears in some patients, usually in late afternoon especially in chronic infections involving hundreds or thousands of worms. There may be diarrhea and flatulent dyspepsia after meals; the liver may be enlarged and painful. There may be jaundice and, in some individuals, fever. Micro- and macroabscesses are reported along with catarrhal inflammation of the pancreatic ducts. Papuloerythematous lesions are reported in some *O. felineus* infections.

5. Diagnosis

A parasitologic diagnosis is made by finding eggs in the feces. The eggs cannot be differentiated from those of *C. sinensis* or eggs from small intestinal trematodes. ELISA and DNA probes to detect antibodies and antigens have been developed in Thailand that could aid in diagnosis (33,34). Ultrasonography and liver scans show lesions compatible to opisthorchiasis and peritoneoscopic findings have been shown to be helpful in the diagnosis of infection and frequently encountered cholangiocarcinoma (35).

6. Treatment

Praziquantel is the drug of choice for treatment, in a dose of 25 mg/kg body weight, which is given three times in 1 day after meals for adults and children. In mass treatment campaigns for a village population, the drug may be given as a single dose of 50 mg/kg body weight. Mebendazole, 30 mg/kg/day for 3–4 weeks, and albendazole, 400 mg twice daily for 3–7 days, may be used if praziquantel is not available. Antibiotics are given in cases of cholangitis and obstructive jaundice, and surgery may be required in some cases of obstructive jaundice. The prognosis is good following treatment, but reinfection is common. Cholangiocarcinoma may be fatal.

7. Prevention and Control

Education of the population to avoid eating raw or poorly preserved fish could prevent opisthorchiasis. Habits are difficult to break; therefore, to control infection, mass treatment with praziquantel must be used. Reinfection after treatment is common in endemic areas; in a study in Thailand, 87.7% of the treated population was reinfected after 1 year (36). Sanitary practices should be changed and fish ponds should not be fertilized with night soil. Snail control and eradication of reservoir hosts is not feasible. It has been suggested that community-based control programs should be directed toward treatment of heavily infected individuals (37).

G. Heterophyiasis

1. Introduction

There are a large number of tiny flukes of the family Heterophyidae that are parasites of humans and fish-eating mammals. More than 10 species have been reported in humans, but the most important are *Heterophyes heterophyes* and *Metagonimus yokogawai*. These are intestinal parasites acquired by eating freshwater fish raw or poorly cooked or preserved. *H. heterophyes* is found in the Middle East, Turkey, the Balkans, Spain, and parts of Asia. It is highly prevalent in the Nile delta of Egypt. *H. heterophyes* or a closely related species is endemic to areas of the Philippines, Indonesia, Thailand, China, Japan, and Korea. *M. yokogawai* is found primarily in Asia, particularly in China, Japan, and Korea. There are sporadic reports in Siberia, Spain, and the Balkans. They cause little intestinal disease, but ectopic eggs may cause serious disease in vital organs (29).

2. Etiology

The integument of heterophyid trematodes characteristically is covered with scalelike spines, which are more pronounced at the anterior end. There are anterior and ventral suckers, with the anterior sucker one-third the size of the ventral, and the ventral either median or submedian. *H. heterophyes* is 1–1.8 mm by 0.3–0.7 mm in size and the anterior is rounded broadly (Fig. 13). *M. yokogawai* measures 1–2.5 mm by 0.4–0.8 mm (Fig. 14). The ventral sucker is deflected to the right posterior toward the genital atrium.

Eggs are small, 28–30 μm by 15–17 μm, yellow to brown in color, and similar to those of *C. sinensis* and the opisthorchids. The eggs are embryonated when passed and, reaching water, hatch after being ingested by a snail intermediate host. In Egypt, *H. heterophyes* uses the snail *Perinella conica* as a host, while in Asia *Cerithidae cingulata* is the important snail host. The first intermediate hosts for *M. yokogawai* are *Semisulcospira libertina* and *Thiara granifera*. Second intermediate hosts for *H. heterophyes* include the mullet species *Mugil cephalus, M. capito, Talapia nilotica,* and *Aphanius fasciatus* in Egypt, and *Acanthogobius* in Japan. Fish hosts for *M. yokogawai* include the trout *Plectoglossus altivelis* and *Odontobulis obscurus* and the salmon *Salmo perryi* and *Tribolodon*

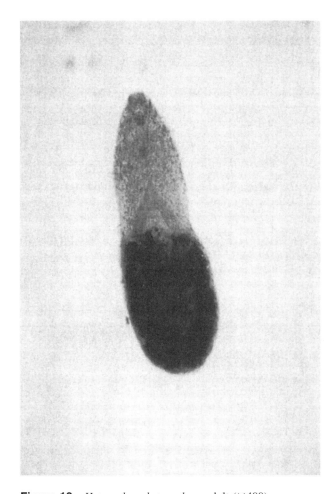

Figure 13 *Heterophyes heterophyes* adult (×400).

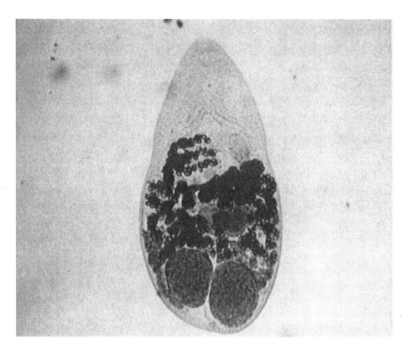

Figure 14 *Metagonimus yokogawi* adult (×400).

hakonensis. The metacercariae encyst under the scales in subcutaneous tissue and on the gills.

Infection is acquired by eating the fish uncooked. The metacercariae excyst in the small intestine and develop into adults in 7–10 days. The parasites move into the crypts and are associated closely with the mucosa (Fig. 15). Fish-eating birds and mammals serve as reservoir hosts and infections rarely persist for more than a year. Marine fish and shrimp also may be sources of infection.

3. Pathogenesis and Pathology

When a large number of *H. heterophyes* are present in the small intestine, there may be ulceration, inflammation, or superficial necrosis. Some worms may penetrate the mucosa and produce a granulomatous response that encapsulates the parasite. Worms deep in the crypts release eggs that may enter the lymphatics and circulation and be carried to such ectopic locations as the spleen, lungs, liver, spinal cord, brain, and heart. Eggs are known to cause fibrosis in the myocardium, and fibrosis and calcification in heart valves. Adult worms also may enter ectopic locations. Similar pathology may be caused by *M. yokogawai*, but less often. Ulceration, inflammation, and granulomatous infiltration into the duodenum and jejunum are associated with *M. yokogawai*.

4. Clinical Manifestations

Chronic, intermittent, often bloody diarrhea is associated with heterophyiasis, along with dyspepsia, upper abdominal pain, anorexia, nausea, vomiting, and weight loss. Light infections are asymptomatic, however. Eggs or worms in the CNS may cause clinical symptoms similar to those for cerebral hemorrhage or manifestations of Jacksonian epilepsy. In the

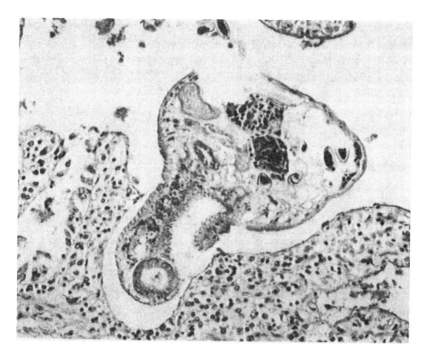

Figure 15 Heterophyid trematode in the intestinal mucosa of a Filipino who died with intestinal capillariasis (×400).

heart, there may be indications of myocarditis or chronic congestive heart failure. Death can occur in cases of ectopic heterophyiasis.

5. Diagnosis

The confirmed diagnosis is made by finding and identifying characteristic eggs in the feces. The eggs of these trematodes are alike and specific identification is difficult. The eggs resemble those of *C. sinensis*, except for the opercular shoulder. Heterophyid eggs in feces of Asians living outside of Asia probably would be from *C. sinensis* or *O. viverrini* if they were out of the endemic areas for more than 2 years; the heterophyid infections rarely persist for more than 2 years. A confirmed diagnosis is made by recovery of adult flukes at autopsy or from the feces following treatment. Recovery of these parasites is very difficult and requires tedious microscopic searching through the fecal material.

Symptoms associated with ectopic infections may be diagnostic if there is evidence of intestinal heterophyiasis. A confirmed diagnosis, however, is made at autopsy. Serological tests are not available.

6. Treatment

Treatment consists of praziquantel, either as a single dose of 20–25 mg/kg of body weight or 75 mg in three doses in one day. Niclosamide in a single 2-g dose also has been used but is not as satisfactory as praziquantel. It is not known how effective these drugs are on adult ectopic heterophyiasis.

7. Prevention and Control

Heterophyiasis could be prevented if populations in endemic areas would avoid eating infected brackish water or freshwater fish raw, improperly cooked, salted, or pickled. Education in regard to the hazards of eating raw fish and the practice of good sanitation would be helpful for control. Raw fish also should not be fed to dogs or cats.

H. Nanophyetiasis

Nanophyetes salmincola (*Troglotrema salmincola*) is a well-recognized intestinal trematode of dogs, foxes, raccoons, mink, and other fish-eating mammals of eastern Siberia and North America. It causes "salmon poisoning" in these animals. The parasite is not responsible for producing salmon poisoning; the etiological agent is a rickettsia, *Neorickettsia helmintheca*. *N. salmincola* was not considered an important parasite of humans until a number of human infections were reported from a clinic in Oregon (38).

The parasite is small, approximately 1 by 0.4 mm, and lives in the intestinal wall of the host. Eggs pass in the feces. The snail host in the United States is *Oxytrema silicula* and in Siberia *Semisulcospera laevigata*. The cercariae from snails encyst in the kidneys and other organs of salmon and trout. Other species of fish can serve as second intermediate hosts. Animals and humans acquire the infection by eating the fish uncooked, although there is one report of the infection being acquired by handling salmon (39).

Humans infected with the fluke do not develop salmon poisoning; this only occurs in dogs and other animals. The parasite causes a gastroenteritis in humans, resulting in frequent bowel movements or diarrhea, abdominal discomfort, nausea, vomiting, weight loss, fatigue, and peripheral blood eosinophilia.

The diagnosis is made from eggs recovered from feces or adult worms following treatment. Niclosamide and bithionol have been used to treat some patients, but the drug of choice is praziquantel, 60 mg/kg body weight in 1 day in three divided doses (40). Praziquantel also has been used in animals but had no effect on metacercariae when used in salmon.

III. MOLLUSCAN-BORNE HELMINTHIASES

Snails, clams, oysters, and a variety of other mollusks are part of the diets of many people worldwide. The eating of oysters on the half shell is a culinary delight popular among Americans. Other cultural groups also enjoy fresh raw oysters. Helminthic parasites, however, are not acquired naturally from oysters. Clams, on the other hand, are known sources for echinostome infections, and terrestrial and freshwater snails and other gastropods are vectors of metastrongylid nematodes, especially *Angiostrongylus* species. In all cases of molluscan-borne helminthiases, the animals are eaten raw or poorly cooked, pickled, or smoked.

A. Angiostrongyliasis

1. Introduction

Angiostrongylus cantonensis and *A. costaricensis* are two snail-borne nematodes that can cause disease in humans. Two other species, *A. malaysiensis* and *A. mackerrasse*, are potentially pathogenic, but as yet no human infections have been documented. *A. costaricensis* is indigenous to the Americas and causes abdominal angiostrongyliasis or granulomatous eosinophilic lesions in the intestines. Infections are not attributed to specific eating habits but through accidental ingestion of infected slug intermediate hosts. Since it is not a specific foodborne disease, it is not considered further here.

The snail host for *A. cantonensis*, on the other hand, is part of the diet of cultural groups in Asia and the Pacific Islands. Infective third-stage larvae are in the snail meat; when eaten uncooked, the larvae migrate to the CNS and cause eosinophilic meningitis or eosinophilic meningoencephalitis.

The parasite is widespread in Asia and the Pacific Islands, Australia, India, Sri Lanka, Africa, the Caribbean, and in Louisiana in the United States. Infections in humans have been reported nearly everywhere the parasite occurs. Most infections are acquired by eating snails, but accidental infections occur through tiny snail or slug contamination of raw vegetables.

Most human infections and deaths have been reported from Taiwan, Thailand, and some Pacific islands. It is a public health problem in those areas in which snails are an important part of the diet.

This is a relatively new human parasitosis, with the first case reported from Taiwan during World War II. It gained international recognition, however, when the parasite was recovered at autopsy in a man who died in Hawaii (41). Since then, the parasite and disease have been found internationally and appear to be spreading (42).

2. Etiology

Adult male and female *A. cantonensis* live in the pulmonary arteries of rats (*Rattus* species) and bandicoots (*Bandicota* species). Female worms lay eggs that lodge in small lung capillaries and develop first-stage larvae. The larvae enter the alveolar spaces, migrate up the trachea, epiglottis, and esophagus to the alimentary tract, and pass in the rodent feces. The larvae enter the first intermediate host by either direct penetration or ingestion; a variety of gastropods such as terrestrial snails and slugs or aquatic snails serve as hosts. The larvae enter the molluscan muscular tissue, molt into the second and third stages, and are in the infective stage after 12–16 days.

When the intermediate host is eaten by rodent hosts, the larvae are digested out of the tissue, enter the intestinal tissue, and are carried by the circulatory system to the brain; this can occur in 1–2 days. They migrate to the olfactory lobes and cerebral hemispheres and molt to the fourth stage in 4–6 days. They finally move to the subarachnoid spaces and molt to the fifth, young adult stage in 7–9 days postinfection. After 10 days, these young worms migrate to the pulmonary arteries and become sexually mature and reproduce. The prepatent period is 42–45 days (43).

The adult worms are filariform and taper at both ends. The cephalic tip is simple and the cuticle smooth with transverse striae. Females measure 25–30 mm in length and 0.3–0.5 mm in diameter. The milky white uterine tubules spiral around the blood-filled intestine, giving a ''barber pole'' appearance. Males measure approximately 18 mm in length and possess a well-developed caudal bursa and spicule (1–1.5 μm). The arrangement of the bursal rays and spicule size are important in species identification (43).

Humans become infected with *A. cantonensis* by intentionally or accidentally eating infected snails or slugs. In most endemic areas of Asia, snails are eaten with regularity. The Thais are fond of aquatic *Pila* species of snails, while on Taiwan infections are attributed to the ingestion of the giant African land snail, *Achatina fulica*. Infections in the Taiwanese are more severe than infections in Thais. It has been suggested that *A. fulica* is a better intermediate host than *Pila* species, with the presence of more third-stage larvae in *A. fulica* than in *Pila* species. More infections occur in children on Taiwan, while in Thailand more adults are infected. Males more than females are infected in Thailand, but no sex differences are found on Taiwan.

Snails and slugs are used by some Asians for medicinal purposes and such paratenic hosts as toads and frogs, used for medicinal purposes, have also caused infection in Okinawans. Larvae released from dead snails in water may cause infection when the untreated water is ingested. In some areas of the Pacific, shrimp or sauces prepared from shrimp juices have been considered a source of infection, as has the eating of fresh vegetation containing infected planaria or tiny slugs. Land crabs serve as paratenic hosts and, when eaten by Pacific Islanders, infections occur.

The parasitosis continues to be reported from new areas of the world. The spread has been attributed to the dissemination of the snail, but it is more realistic to consider stowaway rats on ships as the disseminators. Terrestrial and aquatic snails are universally susceptible to infection, and when rats arrive into a new area the local mollusks can be readily infected.

3. Pathogenesis and Pathology

The pathogenesis of angiostrongyliasis follows the migration of the infective stage larvae; however, most of the evidence has been with experimental animals (44). The larvae invade the gastric mucosa, pass to the blood vessels, and are carried to the liver. This may cause gastroenteritis and hepatomegaly. The larvae leave the liver and are carried to the right heart, then to the lungs, from which they reenter the blood through the walls of the pulmonary veins. In rats, the larvae are carried to the CNS or to skeletal muscles; they destroy the muscle fiber, then find a small nerve and travel to the CNS. These observations in animals explain myalgia, pain, and paresthesia in humans. In the CNS, the larvae invade neural tissue; some die and become surrounded by a granulomatous reaction. Others survive, emerge onto the surface, and become preadults.

In humans, the worms are present on the brain surface, surrounded by an inflammatory reaction (Fig. 16), and others cause burrowing tracts in the brain. The inflammation and tracts are responsible for encephalitis; in the spinal cord, the worms may cause myelitic symptoms. Cranial nerves and spinal roots are affected by the inflammatory reaction from the infiltration of lymphocytes, plasma cells, and eosinophils. This may lead to cranial nerve palsies and optic neuritis. Worms may migrate into the eye chamber from the brain through spaces between the optic nerve and its sheath. The worms usually die in the CNS of humans, but adult worms have been recovered from the lungs of three humans at autopsy (44–46).

At autopsy in humans, the brain is found congested and there is inflammation of the leptomeninges. The inflammatory reaction consists of leukocytes, including eosinophils and plasma cells. Isolated giant cells may be present throughout the subarachnoid spaces. Dead worms and fragments of worms may be found in the brain and spinal cord surface. Tracts may contain debris, glitter cells, and Charcot-Leyden crystals.

4. Clinical Manifestations

The incubation period varies from 1 to 36 days, but averages about 2 weeks. A few hours after eating infected snails, there may be vomiting, abdominal pain, and diarrhea, probably in response to larvae penetrating the intestinal mucosa. Urticarial skin rashes also have been observed in Thailand after the eating of snails. Such flu-like symptoms as mild fever, malaise, cough, sneezing, and rhinorrhea are seen early. In most patients, clinical symptoms are abrupt, with the development of headache, nausea, vomiting, fever, and stiff

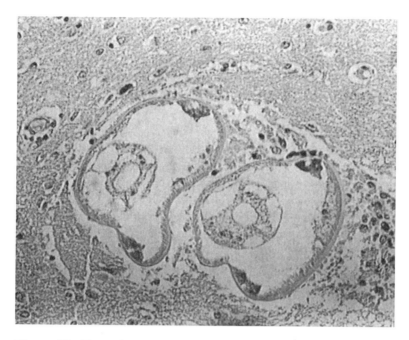

Figure 16 Human brain with sections of *Angiostrongylus cantonensis* in the leptomeninges (×400).

neck. Paresthesia of the skin on the trunk and extremities is common and may persist for weeks. Cranial nerve palsies also have been observed. Coma has been reported in more patients in Taiwan than in Thailand and probably is associated with large numbers of parasites. Moving worms have been observed in the eye chamber, but in such cases meningitis symptoms are not seen. Eosinophilic pleocytosis is common, with worms being recovered in spinal fluid. This also is found most often in Taiwanese cases. Death is not common and occurs only with heavy infections.

5. Diagnosis

In endemic areas, the diagnosis is made by the symptoms, eosinophilic pleocytosis, and a history of eating or exposure to infected snails or paratenic hosts. A parasitological diagnosis is made by finding worms in the cerebrospinal fluid or eye chamber. Most confirmed cases have been from Taiwan.

Serological tests have been improving. An ELISA has some value (47,48), but more recently the tests have been improved with the use of purified antigens (49) and monoclonal antibodies (50). Magnetic resonance imaging (51) and computed tomographic myelography (52) have aided in the diagnosis of the parasitoses.

6. Treatment

There is presently no specific therapy for angiostrongyliasis. Antihelminthics have been shown to be effective in early infections in animals but not in humans. Mebendazole and albendazole have been shown to be effective in children in Taiwan (53). Many physicians do not recommend specific treatment since mass destruction of the parasites may cause more serious disease. Treatment remains supportive. Lumbar puncture provides some relief. Corticosteroids and analgesics may alleviate the headache and radicular symptoms. Surgical removal of the worm is necessary in cases of ocular angiostrongyliasis.

The disease is self-limiting in most cases. Symptoms may continue for several weeks but generally resolve spontaneously. There are few deaths reported; most have been in Taiwan and the patients died in a coma after a few weeks.

7. Prevention and Control

Angiostrongyliasis can be prevented by educating populations in endemic areas to avoid eating uncooked mollusks, particularly land snails. Education of children not to play with snails is important as hands may become contaminated. Freezing will kill larvae if the snails are frozen at −15°C for 12–24 h. Paratenic hosts (shrimp, prawn, crabs) should be cooked before eating and vegetables washed before eating raw to help control the disease.

B. Echinostomiasis

1. Introduction

Trematodes of the family Echinostomatidae are intestinal parasites of birds and mammals; approximately 15 species have been reported from humans. Most infections occur in Asians and involve such species as *Echinostoma ilocanum, E. lindoensis, E. malayanum, E. hortense, E. cinetorchis, E. revolutum, E. recurvatum,* and *Hypoderaeum conoidium.* Infections are widespread throughout the Far East, with the highest prevalences occurring in the Philippines, Thailand, Indonesia, Taiwan, and Korea (54). Although many people in Asia are infected with echinostomes, very few have serious illness. An outbreak of gastroenteritis caused by echinostome infection recently was reported in a group of Americans returning from a tour in Africa (55).

One of the earliest reports of echinostomiasis in humans was from the Philippines. Philip Garrison found eggs in the feces of prisoners at Bilibid Prison in Manila in April 1907. Five prisoners from northern Luzon were found to be passing eggs and after treatment with male fern, 21 worms were recovered from the feces (56).

2. Etiology

The Echinostomatidae are small flukes 2–6 mm in length, 1–1.5 mm in width, and with both ends attenuated (Fig. 17). The anterior oral sucker is surrounded by spines and smaller spines are present in the anterior tegument. There is a large ventral sucker with the genital pore anterior to it. Two round testes are located centrally, with the ovary anterior to the testes. The uterus is between the testes and ventral sucker. The number and arrangement of the circumoral spines are considered of taxonomic importance.

The parasites are located in the small intestines and produce large (86–117 μm by 53–82 μm), thin-shelled, operculated eggs that pass with the feces. The eggs reach water in which they embryonate in 14 days, hatch, and release a ciliated miracidium. The larval trematodes then find the first intermediate snail host (*Pila, Viveparum, Lymnaea* species). After a cycle of several weeks in the snail, cercariae are released that enter a second intermediate host (the same snail, other species of mollusks, fish, or tadpoles). Humans and other animals acquire infections after eating the second intermediate host raw or partially cooked. The metacercariae are released after digestion of the infected tissue and the worms mature into adults in 2 weeks.

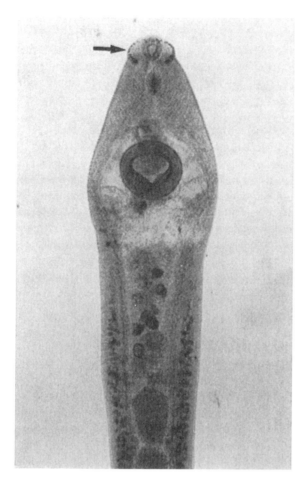

Figure 17 *Echinostoma ilocanum* anterior end showing circumoral spines around the anterior sucker (arrow) (×160).

In the Philippines, infections of *E. ilocanum* and *E. malayanum* are acquired, respectively, by eating *Pila luzonica* and *Lymnaea cumingiana*. In Indonesia, *E. lindoensis* infections were attributed to eating freshwater clams of the *Corbicula* species. The infection was once highly endemic in the Lindu Lake Valley of Sulawesi but it nearly vanished because of the introduction of fish (*Talapia mossambica*). The fish and the clams competed for zooplankton for food. The fish also fed on the clam larval stages (54).

In Thailand, *H. conoideum* may infect as many as 50% of the population. Raw snails, such as *Pila* species and *Indoplanorbis exustus*, frogs, and tadpoles are eaten raw or undercooked. In Korea, echinostome infection is obtained by eating raw or insufficiently cooked fish.

3. Pathogenesis and Pathology

There is little pathology associated with echinostomid infections unless a large number of worms are involved. Infections are usually short lasting. Inflammatory lesions with shallow ulcers in the mucosa may develop at the site of the fluke attachment. There may be cellular infiltration and necrosis. Endoscopic studies in Japan demonstrated multiple intestinal erosions in association with infections (57).

4. Clinical Manifestations

Heavy infections may cause clinical symptoms of abdominal pain, flatulence, and diarrhea. Some patients also experience eosinophilia, abdominal pain, soft stools, anorexia, and edema.

5. Diagnosis

For diagnosis, eggs present in the feces must be differentiated from those produced by *Fasciolopsis buski* or *Fasciola hepatica*. Echinostome eggs are ellipsoidal, thin-shelled, operculated, unembryonated, and yellow to yellowish brown. The eggs of different species vary in size, with *E. ilocanum* 83–116 μm by 53–82 μm (Fig. 18), *E. lindoensis* 97–107 μm by 65–73 μm, and *H. conoideum* 137 by 75 μm. Species identification can be made only after the recovery of adult worms following antihelminthic treatment.

6. Treatment

Praziquantel, 40 mg/kg body weight in a single dose, is the drug of choice. Albendazole, 400 mg/day for 3 days, is an alternate drug. Mebendazole, used in the treatment of *Capillaria philippinensis*, was found to eliminate *E. ilocanium* in patients with concomitant infections. The fluke eggs were not seen after 4 days of treatment.

7. Prevention and Control

Echinostomid infections are preventable if such second intermediate hosts as snails, clams, tadpoles, frogs, and fish are well cooked before eating. Education programs informing indigenous populations of the hazards of eating raw animal life should be implemented. Since there is little illness associated with echinostomiasis, it is unlikely that health authorities will implement a program to control this parasitosis.

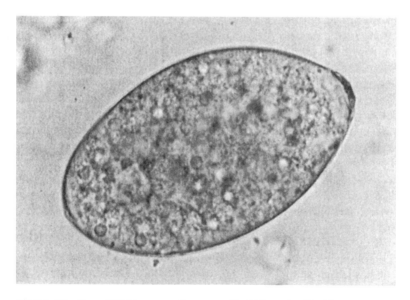

Figure 18 Egg of *Echinostoma ilocanum*, 83–116 μm by 53–82 μm (×400).

IV. ARTHROPOD-BORNE HELMINTHIASIS

A. Paragonimiasis

1. Introduction

There are over 40 species of *Paragonimus* described from around the world, but only 8 are considered important. *P. westermani* is widespread in the Far East and is responsible for most cases of pulmonary paragonimiasis. *P. heterotremus* is reported from China and Southeast Asia, *P. skrjabini* from China, *P. miyazakii* from Japan, *P. uterobilateralis* and *P. africanus* from Africa, and *P. mexicanus* from Latin America. *P. kellicotti* is a North American species reported from animals and only twice in humans.

The first report of *Paragonimus* was in a Brazilian otter in 1850, and the second finding was by Cobbold in 1859 in an Indian mongoose. In 1877, a Bengal tiger died in the Amsterdam zoo; *Paragonimus* was found in the lung and eventually named after the zoo director, G. F. Westerman. In 1879, the first human infection was from Taiwan; the patient was a Portuguese sailor who had been treated previously by Patrick Manson in Amoy, China. After learning of these findings, Manson concluded that the hemoptysis seen in his Chinese patients was due to this fluke. Workers in Japan found the worm at autopsy and Braun named it *Paragonimus westermani* (58).

Paragonimiasis has a sporadic distribution with foci in China, Korea, Japan, Taiwan, the Philippines, Thailand, Vietnam, Laos, Cambodia, Nepal, Burma, and scattered areas of Africa and Latin America. Over 20 million infections are estimated worldwide (24).

2. Etiology

The adult *P. westermani* usually is located in pairs in granulomatous capsules in the lungs. It is a plump, oval, reddish brown trematode that is often shaped like a coffee bean. It is 7–16 mm in length, 4–8 mm in width, and 5 mm thick. The tegument is spiny and the suckers are of equal size. The esophagus divides into two ceca that extend to the posterior end of the body. The testes are lobed irregularly and situated side by side on the posterior half of the body. The lobated ovary is located on one side, anterior to the testes and opposite the egg-filled uterus. The vitellaria are lateral along the entire length of the body (Fig. 19).

The thick-shelled operculated eggs pass from the worms, through the cyst, into the bronchi, where they are coughed up in the sputum or are swallowed and pass in the feces. The eggs are unembryonated and must reach water, where they become embryonated in 3 weeks. The miracidium hatches from the egg and swims in the water to find specific snails of the families Hydrobiidae, Thiaridae, and Pleurocercidae. Important snails in Southeast Asia are *Semisulcospira libertine*, *Thiara* species, and *Botria* species.

The miracidium enters the snail tissue, goes through polyembryony, and produces stubby-tailed cercariae. The cercariae are poor swimmers and must crawl to a suitable crustacean second intermediate host in which they form into metacercariae in the gills, muscles, legs, and viscera. Freshwater crabs of the genera *Erlocheir*, *Potamon*, and *Sundathelphusa* and crayfish of the genus *Cambaroides* are important second intermediate hosts. In Africa and Latin America, other species of snails and crustaceans serve as first and second intermediate hosts for *Paragonimus* species endemic in these areas.

Humans and other animals acquire lung fluke infections by ingesting infected crustaceans raw or poorly cooked, pickled, or salted. In some areas of China, crabs are soaked in rice wine for a period of time before eating; these are called *drunken crabs*. In other parts of China, people eat crayfish curd, raw crab sauce, or jam. The Thais eat raw freshwa-

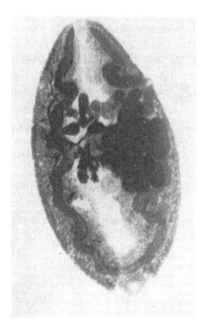

Figure 19 *Paragonimus westermani* adult (×10).

ter shrimp salad and crab sauce. Koreans eat uncooked crabs immersed in soy sauce. Crabs are eaten roasted or raw in the Philippines, and crab juice is used in preparing some meat dishes. Some Asians use crabs and crayfish for medicinal purposes; the juices are used to treat measles, diarrhea, and urticaria. Contaminated utensils are also a source of infection and, since metacercaria can live for weeks outside the animals, water also may be a source of infection.

Infections also are made possible by the eating of animal meat containing immature flukes. In Japan, infections followed the ingestion of uncooked slices of wild boar meat (59). In times of famine, especially in Africa, people will eat raw crab and crayfish and acquire paragonimiasis. Also in Africa, women believe that raw crab will increase fertility.

After the infected second intermediate hosts are ingested, the metacercariae are released from the animal tissue and excyst in the duodenum. The tiny worms penetrate the intestinal tissue, enter the body cavity, and either enter the abdominal wall or remain in the body cavity. After several days, the young worms return to the body cavity, penetrate the diaphragm, enter the pulmonary cavity and then the lungs. The worms become sexually mature and produce eggs in about 10 weeks. Some infections are reported to persist for 10 or even 20 years.

Infections generally are more common in children; children in Liberia become infected by chewing on raw crab legs (60).

3. Pathogenesis and Pathology

The pathology associated with paragonimiasis depends on the number of worms involved with the infection and the location of the parasite in the tissue.

Young worms passing through the intestinal and abdominal wall, peritoneal cavity, and diaphragm cause little damage. The passage is assisted by lytic substances that digest the tissue. The worms settle in the lung parenchyma close to the large bronchioles or

bronchi beneath the pleural surface. An inflammatory reaction with an infiltration of eosinophils and neutrophils develops around the worm. A fibrous cyst develops containing one
or usually two worms, cellular debris, a brownish pus, Charcot-Leyden crystals, and eggs.
When the cyst ruptures, the necrotic material is released into the surrounding lung tissue.
Eggs in the tissue may provoke pseudotubercles with eosinophils, lymphocytes, and eventually giant cells and fibroblasts. Cysts usually develop in the right lung, with no more
than 20 cysts in both lungs. Cyst contents may rupture and discharge into the bronchus,
giving rise to a blood-tinged sputum containing eggs. Surrounding lung tissue may become
atelectatic and walls of old cysts become fibrosclerotic and calcified.

Erratic paragonimiasis is not uncommon in endemic areas. Cerebral or CNS involvement occurs frequently with either immature or mature worms. In CNS involvement, the
parasites probably migrate from the body cavity directly to the brain. When mature worms
are present, it is believed that they first develop into adults in the lung and then migrate
to the CNS. In the brain, the parasite lies in a cavity surrounded by a membrane produced
by toxic substances. A cyst eventually forms that becomes necrotic and filled with eggs,
cellular debris, and Charcot-Leyden crystals. Egg granulomata may develop and finally
calcification may occur.

Wandering worms, usually immature forms, may encyst in any organ or tissue of
the body and provoke an inflammatory reaction. *Paragonimus* species also may enter
subcutaneous tissue and form cysts, eosinophilic infiltrations, and abscesses. The parasites
often are responsible for cutaneous larval migrans and swellings.

4. Clinical Manifestations

Early clinical symptoms are associated with the migration of the young worms in the
digestive tract and abdominal cavity. Patients may experience diarrhea and abdominal
pain. When the parasites migrate to the lung, there may be allergic manifestations, urticaria, and eosinophilia, as well as fever, chills, and chest pains. Once the worms are
established, there may be general malaise, cough, dyspnea, and night sweats.

In chronic infections, pleurisy with hemoptysis, pleural effusion, persistent rales,
clubbed fingers, and pneumothorax may develop. Paroxysmal coughing, pronounced in
the morning, produces a gelatinous, tenacious, rusty brown, blood-streaked sputum. Ex-
pectoration produces 25–50 mL/day. Radiographic examination may demonstrate the appearance of nodules or ring shadows similar to lesions caused by tuberculosis. Most cases
involving a few worms generally are asymptomatic.

Extrapulmonary paragonimiasis usually involves the brain and spinal cord. Cerebral
symptoms begin about 10 months after the appearance of pulmonary symptoms. The
symptoms are similar to those caused by brain tumor, abscess, or Jacksonian epilepsy.
Headache, fever, vomiting, seizures, and visual disturbances may be seen. With spinal
involvement, there could be paraplegia, monoplegia, limb weakness, and sensory deficits.
Cerebral paragonimiasis has been documented in Korea, Japan, and the Philippines, and
Paragonimus species other than *P. westermani* may be involved.

Abscesses with worms and eggs generally are associated with paragonimiasis in the
extrapulmonary locations. The parasites have been found in the abdominal cavity and
wall, liver, spleen, lymph nodes, omentum, pericardium, myocardium, male and female
genital organs, bone marrow, urinary bladder, and skin. These infections may be symp-
tomless or cause a great deal of tenderness and pain. *P. skrjabini* in Chinese patients has
been associated with liver and skin infections and *P. mexicanus* and *P. miyasakii* with
cutaneous lesions.

5. Diagnosis

In patients with tenacious bloody sputum, tuberculosis must be considered. Microscopic diagnosis of sputum and feces usually will reveal the presence of typical *Paragonimus* eggs. The eggs are thick-shelled with a flattened operculum, a pronounced operculum rim, and a thickening at the aoperculated end. The eggs are immature when passed, are yellow to golden brown, and measure 80–120 μm by 50–60 μm (Fig. 20). Worms may be found at autopsy or, on rare occasion, in sputum. Several sputum and stool examinations may be required and sodium hydroxide solution may be needed to dissolve tenacious sputum. Species identification cannot be made by the eggs since all species are similar in appearance.

Serological testing is of value. The intradermal test developed in Japan using a veronal-buffered saline (VBS) extract of *P. westermani* will indicate past or present infections. It has been used extensively in surveys and when positive is followed up by a complement fixation test or ELISA (61,62). Immunoblot also has been found to be sensitive and specific (63). Immunodiagnostic tests have particular value in the presumptive diagnosis of extrapulmonary paragonimiasis.

Chest X-rays will show abnormalities, but a specific diagnosis is difficult. However, ring shadows with a crescent-shaped opacity along one side of the border is considered pathognomonic. Cerebral angiography, computed tomographic scanning, and possibly magnetic resonance imaging may be of value for cerebral paragonimiasis.

6. Treatment

Praziquantel, 25 mg/kg body weight three times a day for 3 days, is the drug of choice for treating paragonimiasis. Eggs cease to appear in the sputum in a few weeks and chest

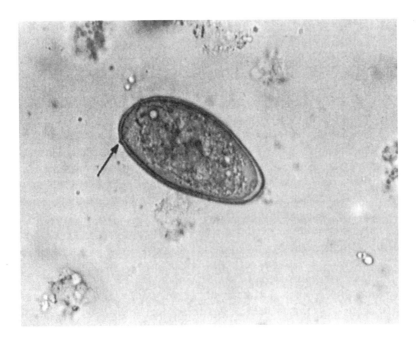

Figure 20   Egg of *Paragonimus westermani*, 80–120 μm by 50–60 μm; note operculum (arrow) (×400).

X-rays clear in several months. In cases of cerebral paragonimiasis, corticosteroids also should be given since dead parasites may provoke a local reaction. Such transient side effects as headache, dizziness, urticaria, fever, and gastrointestinal disturbances may be experienced with the use of praziquantel. The dosage for children is the same as that for adults.

Niclofolan has been used successfully in Korea in doses of 25 mg/kg body weight for 3 days. All patients should be reexamined after 6 months and treatment repeated if infection remains. Surgery may be required for extrapulmonary paragonimiasis, especially cerebral and cutaneous disease.

The prognosis in cases of pulmonary paragonimiasis is good. In light infections, the disease is self-limiting and spontaneously cured after 5–10 years. Reinfection, however, can occur in endemic areas. The prognosis is not good in heavy infections and repeated treatment may be necessary. Ectopic paragonimiasis, especially foci in the CNS, pericardium, or visceral organs, may lead to death.

7. Prevention and Control

Paragonimiasis can be prevented by changing people's eating habits and practices. The cooking of crustaceans at 55°C for 5 min will kill metacercariae. Education and improved sanitation would be beneficial, as would the avoidance of using crustaceans for medicinal purposes. In Taiwan, school children were taught the hazards of eating raw crabs and infection disappeared in many localities. Patients with bloody sputum who do not respond to antituberculosis treatment should have the sputum examined for *Paragonimus* eggs. Mass chemotherapy campaigns would benefit highly endemic areas. Control by attempts to eliminate snails and crustaceans in nature, however, would not be realistic.

REFERENCES

1. Chitwood, M. B., Velasquez, D., and Salazar, N. G. (1968). Capillaria philippinensis sp. n. (Nematode: Trichinellida) from intestine of man in the Philippines, J. Parasitol., 54:368–371.
2. Cross, J. H., and Bhaibulaya, M. (1983). Intestinal capillariasis in the Philippines and Thailand. In: Human Ecology and Infectious Diseases (N. Croll and J. Cross, eds.), Academic Press, New York, pp. 103–136.
3. Cross, J. H. (1992). Intestinal capillariasis, Clin. Microbiol. Rev., 5:120–129.
4. Lee, S. H., Hong, S. T., Chai, J. Y., Kim, W. H., Kim, Y. T., Song, I. S., Kim, S. W., Choi, B. I., and Cross, J. H. (1993). A case of intestinal capillariasis in Korea, Am. J. Trop. Med. Hyg., 48:542–546.
5. Cross, J. H. (1998). Capillariosis. In: Zoonosis: Biology, Clinical Practice, and Public Health (S. R. Palmer, Lord Soulsby, and D. I. H. Simpson, eds.), Oxford University Press, New York, pp. 759–772.
6. Bhaibulaya, M., and Indra-Ngarm, S. (1979). Amaurornis phoenicurus and Ardeola bacchus as experimental definitive hosts for Capillaria philippinensis in Thailand, Int. J. Parasitol., 9: 321–322.
7. Cross, J. H., and Basaca-Sevilla, V. (1983). Experimental transmission of Capillaria philippinensis to birds, Trans. R. Soc. Trop. Med. Hyg., 77:511–514.
8. Fresh, J. W., Cross, J. H., Reyes, V., Whalen, G. E., Uylangco, C. V., and Dizon, J. J. (1972). Necropsy findings in intestinal capillariasis, Am. J. Trop. Med. Hyg., 21:169–173.
9. Sun, S. C., Cross, J. H., Berg, H. S., Kau, S. L., Singson, C., Banzon, T., and Watten, R. H. (1974). Ultrastructural studies of intestinal capillariasis Capillaria philippinensis in human and gerbil hosts, Southeast Asian J. Trop. Med. Public Health, 5:524–533.

10. Whalen, G. E., Rosenberg, E. B., Strickland, C. T., Gutman, R. A., Cross, J. H., Watten, R. H., Uylangco, C., and Dizon, J. J. (1969). Intestinal capillariasis—a new disease in man, Lancet, 1:13–16.

11. Rosenberg, E. B., Whalen, G. E., Bennich, H., and Johansson, S. G. O. (1970). Increased circulating IgE in a new parasitic disease—human intestinal capillariasis, N. Engl. J. Med., 83:1148–1149.

12. Nawa, Y., Imai, J.-I., Ogata, K., and Otsuka, K. (1989). The first record of a confirmed human case of Gnathostoma doloresi infection, J. Parasitol., 75:166–169.

13. Taniguchi, Y., Hashimoto, K., Ichikawa, S., Shimizu, M., Ando, K., and Kotani, Y. (1991). Human gnathostomiasis, J. Cutan Pathol., 18:112–115.

14. Miyazaki, I. (1960). On the genus Gnathostoma and human gnathostomiasis with special reference to Japan, Exp. Parasitol., 9:338–370.

15. Daengsvang, S. (1980). A Monograph on the Genus Gnathostoma and Gnathostomiasis in Thailand. SEAMIC (Southeast Asian Medical Information Center), Tokyo.

16. Martinez-Cruz, J. M., Bravo-Zamudio, R., Aranda-Patraca, A., and Martinez-Maranon, R. (1989). La gnatostomiasis en Mexico [Gnathostomiasis in Mexico], Salud Publica Mex., 31: 541–549.

17. Radomyos, P., and Daengsvang, S. (1987). A brief report on Gnathostoma spinigerum specimens obtained from human cases, Southeast Asian J. Trop. Med. Publ. Health, 18:215–217.

18. Schmutzhard, E., Boongird, P., and Vejjajiva, A. (1988). Eosinophilic meningitis and radiculomyelitis in Thailand, caused by CNS invasion of Gnathostoma spinigerum and Angiostrongylus cantonensis, J. Neurol. Neurosurg. Psychiatry, 51:80–87.

19. Nopparatana, C., Setasuban, P., Chaicumpa, W., and Tapchaisri, P. (1991). Purification of Gnathostoma spinigerum specific antigen and immunodiagnosis of human gnathostomiasis, Int. J. Parasitol., 21:677–687.

20. Kravichian, P. Kulkumthorn, M., Yingygourd, P., Akarabovcorn, P., and Paireepai, C. (1992), Albendazole for the treatment of human gnathostomiasis. Trans. R. Soc. Trop. Med. Hyg. *6: 418–421.

21. Thanogsak, B. (1991). Gnathostomiasis. In: Hunter's Tropical Medicine, 7th ed. (G. T. Strickland, ed.), W. B. Saunders, Philadelphia, pp. 764–767.

22. Sakanari, J. A., and McKerrow, J. H. (1989). Anisakiasis, Clin. Microbiol. Rev., 2:278–284.

23. Yagihashi, A., Sato, N., Fakahashi, S., Ishikura, H., and Kikuchi, K. (1990). A serodiagnostic assay by microenzyme-linked immunosorbent assay for human anisakiasis using a monoclonal antibody specific for Anisakis larvae antigen, J. Infect. Dis., 161:995–998.

24. WHO (1995) Control of foodborne trematode infections. WHO Tech. Rep. Ser. 849:1–57.

25. Ona, F. V., and Dytoc, J. N. (1991). Clonorchis-associated cholangiocarcinoma: a report of two cases with unusual manifestations, Gastroenterology, 101:831–839.

26. Schwartz, D. A. (1980). Review: Helminths in the induction of cancer: Opisthorchis viverrini, Clonorchis sinensis and cholangiocarcinoma, Trop. Geogr. Med., 32:95–100.

27. Chen, C. Y., Hsieh, W. C., Shih, H. H., and Chen, S. N. (1987). Evaluation of enzyme-linked immunosorbent assay for immunodiagnosis of clonorchiasis, Chung Hua Min Kuo Wei Sheng Wu Chi Mien I Hsueh Tsa Chih (Taiwan), 20:241–246.

28. Choi, B. I., Kim, H. J., Han, M. C., Do, Y. S., Han, M. H., and Lee, S. H. (1989). CT findings of clonorchiasis, Am. J. Roentgenol., 152:281–284.

29. Lloyd, S., and Soulsby, J. L. (1998). Other trematode infections. In: Zoonoses: Biology, Clinical Practice, and Public Health. (Lloyd, S. R., Lord Soulsby, Simpson, D. I. H., eds. Oxford University Press, New York, pp. 731–746.

30. Sithithaworn, P., Tesana, S., Pipitgool, V., Kaewkes, S., Thaiklar, K., Pairojkul, C., Sripa, B., Paupairoj, A., Sanpitak, P., and Aranyanat, C. (1991). Quantitative post-mortem study of Opisthorchis viverrini in man in northeast Thailand, Trans. R. Soc. Trop. Med. Hyg., 85:765–768.

31. MacLean, J. D., Arthur, J. R., Ward, B. J., Gyorkos, T. W., Curtis, M. A., and Kokosin E.

(1996). Common-source outbreak of acute infection due to the North American liver fluke Metorchis conjunctus. Lancet 347:154–158.

32. Haswell-Elkins, M. R., Sithithaworn, P., and Elkins, D. (1992). Opisthorchis viverrini and cholangiocarcinoma in Northeast Thailand, Parasitol. Today, 8:86–89.

33. Elkins, D. B., Sithithaworn, P., Haswell-Elkins, M., Kaewkes, S., Awacharagan, P., and Wongratanacheewin, S. (1991). Opisthorchis viverrini: relationships between egg counts, worms recovered and antibody levels within an endemic community in northeast Thailand, Parasitology, 102:283–288.

34. Sirisinha, S., Chawengkirttikul, R., Sermswan, R., Amornpant, S., Mongkolsuk, S., and Panyim, S. (1991). Detection of Opisthorchis viverrini by monoclonal antibody-based ELISA and DNA hybridization, Am. J. Trop. Med. Hyg., 44:140–145.

35. Hitanant, S., Trong, D. T., Damrongsak, C., Chinapak, O., Boonyapisit, S., Plengvanit, U., and Viranuvatti, V. (1987). Peritoneoscopic findings in 203 patients with Opisthorchis viverrini infection, Gastrointest. Endosc., 33:18–20.

36. Upatham, E. S., Viyanant, V., Brockelman, W. Y., Kurathong, S., Lee, P., and Kraengraeng, R. (1988). Rate of re-infection by Opisthorchis viverrini in an endemic northeast Thai community after chemotherapy, Int. J. Parasitol., 18:643–649.

37. Haswell-Elkins, M. R., Elkins, D. B., Sithithaworn, P., Treesarawat, P., and Kaewkes, S. (1991). Distribution patterns of Opisthorchis viverrini within a human community, Parasitology, 103:97–101.

38. Eastburn, R. L., Fritsche, T. P., and Terhune, C. A., Jr. (1987). Human intestinal infection with Nanophyetus salmincola from salmonid fishes, Am. J. Trop. Med. Hyg., 36:586–591.

39. Harrell, L. W., and Deardorff, T. L. (1990). Human nanophyetiasis: Transmission by handling naturally infected coho salmon (Oncorhynchus kisutch), J. Infect. Dis., 161:146–148.

40. Fritsche, T. R., Eastburn, R. L., Wiggins, L. H., and Terhune, C. A., Jr. (1989). Praziquantel for treatment of human Nanophyetus salmincola (Troglotrema salmincola) infection, J. Infect. Dis., 160:896–899.

41. Rosen, L., Chappell, R., Laqueur, G. L., Wallace, G. D., and Weinstein, P. P. (1962). Eosinophilic meningitis caused by a metastrongylid lungworm of rats, JAMA, 179:620–624.

42. Cross, J. H. (1987). Public health importance of Angiostrongylus cantonensis and its relatives, Parasitol. Today, 3:367–369.

43. Bhaibulaya, M. (1979). Morphology and taxonomy of major Angiostrongylus species of eastern Asia and Australia. In: Studies in Angiostrongyliasis in Eastern Asia and Australia (J. H. Cross, ed.), U.S. Naval Medical Research Unit #2, Taipei, Taiwan, NAMRU-2-SP#44, pp. 4–13.

44. Hung, T.-P., and Chen, E.-R. (1988). Angiostrongyliasis (Angiostrongylus cantonensis). In: Handbook of Clinical Neurology, Vol. 8(52) (Microbial Disease) (A. A. Harris, ed.), Elsevier, Amsterdam, pp. 545–562.

45. Yii, C. Y., Chen, C. Y., Fresh, J. W., Chen, T., and Cross, J. H. (1968). Human angiostrongyliasis involving the lungs, Chinese J. Microbiol., 1:148–150.

46. Sonakul, D. (1978). Pathological findings in four cases of human angiostrongyliasis, Southeast Asian J. Trop. Med. Publ. Health, 9:220–227.

47. Cross, J. H., and Chi, J. H. C. (1982). ELISA for the detection of Angiostrongylus cantonensis antibodies in patients with eosinophilic meningitis, Southeast Asian J. Trop. Med. Publ. Health, 13:73.

48. Jaroonvesama, N., Charoenlarp, K., Buranasin, P., Zaraspe, G. G., and Cross, J. H. (1985). ELISA testing in cases of clinical angiostrongyliasis in Thailand, Southeast Asian J. Trop, Med. Publ. Health, 16:110–112.

49. Yen, C.-M., and Chen, E.-R. (1991). Detection of antibodies to Angiostrongylus cantonensis in serum and cerebrospinal fluid of patients with eosinophilic meningitis, Int. J. Parasitol., 21:17–21.

50. Shih, H. H., and Chen, S. N. (1991). Immunodiagnosis of angiostrongyliasis with monoclonal

antibodies recognizing a circulating antigen of mol. wt. 91,000 from Angiostrongylus cantonensis, Int. J. Parasitol., 21:171–177.

51. Hsu, W.-Y., Chen, J.-Y., Chien, C.-T., Chi, C.-S., and Han, W.-T. (1990). Eosinophilic meningitis caused by Angiostrongylus cantonensis, *Pediatr. Infect. Dis. J.*, 9:443–445.

52. Wood, G., Delamont, S., Whitby, M., and Boyle, R. (1991). Spinal sensory radiculopathy due to Angiostrongylus cantonensis infection, *Postgrad. Med. J.*, 67:70–72.

53. Hwang, K. P. and Chen, E. R. (1991) Clinical studies on angiostrongyliasis cantonensis among children in Taiwan. In: Proceedings of the 33rd Seminar on Emerging Problems in Food-borne Parasitic Zoonoses, November 1990, Chiang-Mai, Thailand, pp. 194–199.

54. Carney, W. P., Sudomo, M., and Purnomo. (1980). Echinostomiasis: a disease that disappeared, Trop. Geogr. Med., 32:106–111.

55. Poland, G. A., Navin, T. R., and Sarosi, G. A. (1985). Outbreak of parasitic gastroenteritis among travelers returning from Africa, Arch. Intern. Med., 145:2220–2221.

56. Garrison, P. E. (1909). A new intestinal trematode of man (Fascioletta ilocana gen. nov., sp. nov.), Phil. J. Sci., 3:385–393.

57. Harada, H., Fukumoto, K., and Yamaguti, K. (1983). Echinstoma hortense by endoscopical examination, Gastroenterol. Endosc., 25:622–627.

58. Cross, J. H. (1982). Liver and lung fluke infections. In Clinical Medicine (J. A. Spitter, Jr., ed.), Harper and Row, Philadelphia, pp. 1–13.

59. Miyasaki, I., and Hirose, H. (1976). Immature lung fluke first found in the muscle of a wild boar in Japan, J. Parasitol., 62:835–837.

60. Sachs, R., and Cumberlidge, N. (1990). Distribution of metacercariae in freshwater crabs in relation to Paragonimus infection of children in Liberia, West Africa, Ann. Trop. Med. Parasitol., 84:277–280.

61. Indrawati, I., Chaicumpa, W., Setasuban, P., and Ruangkunaporn, Y. (1991). Studies on immunodiagnosis of human paragonimiasis and specific antigen of Paragonimus heterotremus, Int. J. Parasitol., 21:395–401.

62. Ikeda, T., Oikawa, Y., and Nishiyama, T. (1996). Enzyme-linked immunosorbent assay using cystine-proteinase antigen for immunodiagnosis of human paragonimiasis. Am. J. Trop. Med. Hyg., 55:435–437.

63. Slemenda, S. B., Maddison, S. E., Jong, E. C., and Moore, D. D. (1988). Diagnosis of paragonimiasis by immunoblot, Am. J. Trop. Med. Hyg., 39:469–471.

13

Waterborne and Foodborne Protozoa

Ronald Fayer

U.S. Department of Agriculture, Beltsville, Maryland

I. INTRODUCTION

Protozoa are single-celled eukaryotic microorganisms that do not form a natural group but are placed together essentially for convenience. They are ubiquitous in natural waters and moist soil, and are found as symbionts, endocommensals, or parasites of most animals and many plants. They are separated into major taxa based primarily on microscopic features such as flagella, cilia, and pseudopods, and then further defined by ultrastructural morphology, uniqueness of life cycle, host specificity, biochemistry, and molecular genetics.

Of nearly 40 species of parasitic protozoa infectious for humans, this chapter describes only six genera. These include the ameba *Entamoeba histolytica*, the ciliate *Balantidium coli*, the flagellate *Giardia duodenalis* (synonymous with *G. lamblia, G. intestinalis*), and five apicomplexans, *Cyclospora cayetanensis, Cryptosporidium parvum, Sarcocystis suihominis, Sarcocystis bovihominis*, and *Toxoplasma gondii*.

All of the protozoan parasites discussed in this chapter are at some stage excreted in the feces of a host. Routes of transmission through the environment to other susceptible hosts can be extremely complex. The protozoa can enter water in a variety of ways and can then be transmitted directly by drinking contaminated water or indirectly by contamination of food, utensils, food preparation surfaces, hands of food handlers, or other surfaces exposed to contaminated water. The protozoa can also be be transmitted directly by poor hygiene or by mechanical vectors. A study in the city of Hyderabad, India illustrates some of the complexity (1). Of 232 samples of stored water collected from social welfare hostels, restaurants, households, railway stations, and bus depots, as well as hand washings from street food vendors and food handlers, and vegetable washings, 61 samples contained cysts of *Entamoeba histolytica, Balantidium coli*, and *Giardia*.

II. PROBLEMS WITH DETERMINING PREVALENCE OF HUMAN INFECTION WITH PROTOZOAN PARASITES

The problem of determining prevalence exists for all protozoa discussed in this chapter. Reports of foodborne or waterborne disease due to protozoa are relatively rare. Most data are derived from individual cases or outbreaks reported in scientific or medical journals. Occasionally, surveys of specific population groups are reported. Even then the data are often incomplete or confusing and therefore of limited use. Of all the Apicomplexa infecting humans (*Cryptosporidium, Isospora, Sarcocystis*, and *Toxoplasma*), none are included as reportable diseases in the National Notifiable Diseases Surveillance System maintained by the U.S. Centers for Disease Control (CDC) (2). Giardiasis is not a reportable disease. After giardiasis became a notifiable disease in Wisconsin the annual number of cases increased by more than 20-fold from 2.2 cases per 100,000 population to 49.1 cases per 100,000 in 1988 (3). Amebiasis is a reportable disease, but it is not clear if *E. histolytica* is the only intended agent or if all intestinal amebae or *Acanthamoeba* and *Naegleria* that cause meningoencephalitis also are included under this heading.

CDC Surveillance Summaries (1990) contain two major topics: "Waterborne Disease Outbreaks, 1986–1988" (4) and "Foodborne Disease Outbreaks, 5-Year Summary, 1983–1987" (5). In Ref. 4, only 9 outbreaks involving 1169 cases were reported for *Giardia* and only 1 outbreak involving 13,000 cases was reported for *Cryptosporidium*. The authors state that data in this report should be interpreted with caution because the number of waterborne outbreaks reported to the CDC and the U.S. Environmental Protection Agency (EPA) represent only a fraction of the total number that occur and therefore should not be used to draw firm conclusions about the true incidence of waterborne disease outbreaks or the relative incidence of waterborne diseases of various etiologies.

Similarly, only 1 outbreak of *Giardia* involving 13 cases was reported in 1985 and only 2 outbreaks involving 28 cases were reported for foodborne *Giardia* in 1986, whereas no other protozoa were reported to be involved in foodborne infections (5). The authors of this report also state that the number of outbreaks of foodborne disease reported by this surveillance system clearly represents only a small fraction of the actual outbreaks and that less commonly observed pathogens and those associated with milder illness are even more underrepresented.

Distortions in reporting are compounded by two other factors (6): (a) many physicians who encounter reportable infectious diseases do not know the requirements or methods for reporting, and (b) other physicians assume that clinical laboratories and not the physicians are responsible for reporting. Epidemics such as acquired immunodeficiency syndrome (AIDS) that increase the workload of health care departments have resulted in a decrease in the reporting of infectious disease outbreaks (5). Further distortion in prevalence data results from the difficulty of diagnosing many protozoan diseases. Direct detection of organisms in food or water is often difficult or impossible because, unlike bacterial, fungal, or viral culture methods that permit the expansion of an initially small number of organisms, there are no equivalent culture methods for any of the Apicomplexa (7).

Problems associated with determining the prevalence of human infections with protozoa apply to virtually all organisms discussed in this chapter but especially to the Apicomplexa.

III. AMEBIASIS

There is current acceptance of two distinct species that may not have been clearly differentiated until recently because they are morphologically identical but biologically distinct. The distinction between the two species is based on immunological, biochemical, and molecular data needed to identify the species. *Entamoeba dispar* is a nonpathogen, and *Entamoeba histolytica* is a pathogen of concern to public health.

The primary route of transmission of *E. histolytica* is from human to human. It was estimated to infect half a billion people annually with disease in 10–15% of those infected resulting in 50,000–100,000 deaths worldwide (8). In 1990 it was the most common pathogenic intestinal parasite in Mexico, varying in prevalence from place to place but being higher in areas of poor sanitation, in households and institutions, and among male homosexuals (9). A pattern was observed with a high incidence of infection in the spring (9). Closed communities where a common source or close person-to-person contact is most likely implicated are exemplified by outbreaks at a rehabilitation center in Saudi Arabia (10) and among schoolchildren in Ethiopia (11).

Entamoeba histolytica lives primarily in the intestinal lumen as a trophozoite, a motile feeding ameboid stage that resembles a macrophage. Trophozoites are spherical or oval with a large single nucleus that contains a central karyosome and minute, irregularly shaped masses of peripheral chromatin. Cytoplasm generally contains vacuoles and particulate material and sometimes extends as irregular finger-like projections (pseudopods) from the margin (12). Trophozoites can transform to an environmentally resistant cyst stage when they pass from the body in the feces. Mature cysts often have four, and rarely eight, nuclei, but uni- and binucleate cysts are also observed. Each has a delicate nuclear membrane with condensed chromatin on the inner surface and a tiny karyosome in the center. The cytoplasm frequently contains highly refractive, cigar-shaped structures (12).

Transmission from human to human by the fecal–oral route usually is associated with poor hygiene or poor water quality. Cysts can be introduced into the mouth by soiled hands of food handlers, family members, hospital personnel, and other close personal contacts; by food contamination via flies; and by water contaminated with sewage.

Despite documentation of naturally infected nonhuman primates and experimental infection of such other animals as cats and dogs with *E. histolytica*, human infection from any animal source has not been documented. However, knowledge that *E. histolytica* can infect animal hosts serves as warning that under the right conditions animals might be a source of *E. histolytica* infecting humans.

Infections can range from asymptomatic to mild bowel discomfort to diarrhea or dysentery with and without blood and mucus to tissue invasion with ulcerative lesions in such extraintestinal sites as the liver and the skin, or even the brain, and can last from a few days to several months.

Presumptive diagnosis of *E. histolytica* can be made by identification of trophozoites or cysts in stool specimens or trophozoites in tissues. Periodic acid Schiff (PAS) staining helps to locate the trophozoites by staining the cytoplasm intensely, but hematoxylin-eosin or iron-hematoxylin stains are necessary for identifying nuclear features. Because *E. histolytica* and the nonpathogenic *E. dispar* are morphologically indistinguishable, there is presently a need for simple and reliable tests to differentiate these species in clinical samples, especially for use in impoverished countries where invasive amebiasis is a health problem. Differences in surface antigens, isoenzyme patterns, and gene sequence differences are helpful in differentiating these species in sophisticated laboratories. Treatment varies with the extent of infection. Metronidazole is often the drug of choice for acute colitis and abscesses. Emetine hydrochloride plus chloraquine phosphate also is effective (12).

Prevention is by elimination of fecal contamination of food and water. Good personal hygiene (especially thorough washing of hands by food handlers), good sanitation, and high-quality water treatment are essential for preventing transmission. Knowledge that the incidence is higher in certain seasons provides warning that special care may be needed in the handling and processing of water and food.

IV. BALANTIDIASIS

Rarely reported and not usually considered a waterborne or foodborne zoonotic, *Balantidium coli* is the largest protozoan and the only ciliate parasite of humans (8); it is the only member of its genus of medical importance (13).

A. Prevalence and Transmission

Found worldwide in humans, the prevalence of *B. coli* is based primarily on case reports and appears low. Only 722 cases were reported by 1960 and few have been reported since then. The low prevalence may be in part a combination of lack of recognition of the organism and inappropriate diagnostic procedures. Trophozoites are found in the cecal and sigmoidorectal portion of the bowel. They are oval with a slightly pointed anterior end, have a thin membrane covered with cilia, foamy cytoplasm with a large macronucleus and a tiny micronucleus, and measure about 40–80 by 25–45 μm (14). They multiply in the tissues by binary fission and under certain conditions transform into cysts that pass from the body in the feces. Cysts are round to oval, surrounded by a thick wall, contain a macro- and micronucleus, and measure about 40–65 μm.

Transmission via the fecal–oral route is from human to human. Organisms indistinguishable from *B. coli* have been found in the pig, peccary, chimpanzee, orangutan, macaque, dog, rat, guinea pig, and buffalo, although it is not known if all of these organisms are the same species (13). Transmission is facilitated in confined environments in which group hygiene may be poor and in some communities in which public hygiene is poor. Although people in tropical areas are those most commonly affected, infections have been reported in villagers in semiarid regions of rural peninsular India (15) and in Indian children in the Bolivian Altiplano at 11,000 feet above sea level (16). Flies are possible mechanical vectors. *Balantidium coli* is a widespread asymptomatic commensal in the intestine of pigs and pigs have been implicated as a major source of human infection, but their role remains controversial. Based on a review of the literature (13), over 50% of persons with balantidiasis reported contact with pigs. In an unusual case, a family claimed that infection followed ingestion of raw pork sausage. Surveys throughout New Guinea revealed prevalence rates of balantidiasis as high as 29% in some areas in which people lived in intimate contact with pigs. In Truk, in the Caroline Islands, where people live in close association with a variety of domestic animals, especially pigs, an outbreak of balantidiasis followed a typhoon. Because the outbreak began and ended spontaneously, a common source was indicated. Pig feces frequently were found near affected families, suggesting the outbreak resulted from widespread contamination of drinking water by pig feces. In a study designed to determine if *B. coli* was transmissible between humans and animals, both piglets and monkeys became infected and developed diarrhea after experimental infection with cysts from human feces (17).

B. Clinical Signs

Most *B. coli* infections in humans are either asymptomatic or characterized by intermittent diarrhea and constipation (14). Severe clinical disease also has been reported for nonhuman primates, especially the great apes. Clinical infections in humans resemble amoebiasis with pain, nausea, vomiting, and tenesmus. Diarrhea occasionally contains mucus, blood, or pus. Chronic infections are characterized by alternate episodes of diarrhea and constipation with cramps, abdominal pain, and tenesmus. Some cases last for years. *Balantidium coli* can invade the wall of the bowel and cause ulcers resembling amebic ulcers (14). Infection with *B. coli* has presented as acute appendicitis (18). Rare complications include perforation of the bowel wall and extraintestinal infection of the liver, vagina, ureter, and bladder (14). A South African patient infected with *B. coli* had a perforated terminal ileum that led to fecal peritonitis and death from septicemia (19).

C. Diagnosis, Treatment, and Prevention

Diagnosis is by identification of the parasite in stool specimens or biopsies taken by sigmoidoscope. Trophozoites are 50–200 by 40–70 μm, ciliated, with two nuclei (one large and one small), and a characteristic mouth opening called a cytostome (8). Cysts are usually spherical, thick-walled, measure 50–70 μm, and have a visible large nucleus (8). Patients have been treated for balantidiasis with paromomycin, carbarsone, diiodohydroxyquin, tetracycline (14), and metronidazole (20). Prevention is the same as for amebiasis. Under severe environmental conditions, water thought to be contaminated should be boiled before use.

V. GIARDIASIS

A. Taxonomy

This protozoan, described first by Antony van Leeuwenhoek (1632–1723) when examining his own stool, remains a problem with respect to the validity of a species name. Names used for this protozoan described from humans include *Lamblia intestinalis*, *Giardia lamblia*, *Giardia intestinalis*, *Giardia enterica*, and *Giardia duodenalis*. *Giardia duodenalis* was accepted by Meyer (21) and will be used in this chapter to refer to the human parasite.

B. Life Cycle

Giardia duodenalis is a flagellate of unique appearance that has a simple, direct life cycle (Fig. 1) similar to the intestinal amoebae. After ingestion and passage to the duodenum, the organism leaves the cyst (excysts) and undergoes binary fission to form two trophozoites that continue to proliferate and establish infection in the new host (22). The binucleate, motile feeding stage, the trophozoite, attaches to the epithelium in the upper part of the small intestine by a ventral disk-like organelle. Trophozoites swept into the lumen undergo mitotic division or transform into the cyst stage that passes from the host in the feces.

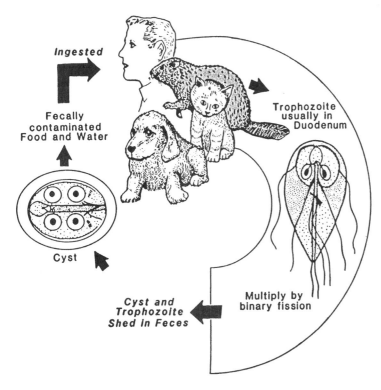

Figure 1. Transmission and life cycle of *Giardia*. *Giardia* is transmitted primarily by ingestion of water contaminated with *Giardia* cysts in feces from humans and possible other animals. Upon ingestion, encysted parasites transform into trophozoites that inhabit the mucosal surface of the duodenum and jejunum. They multiply by binary fission and encyst when swept into the lumen and down the gastrointestinal tract.

The morphology of both the trophozoite and cyst stage are reviewed in detail at both the light and electron microscope levels (22). When observed with the light microscope, iron-hematoxylin-stained trophozoites have a face-like appearance. They are pear-shaped and, when viewed with the narrow end down, the paired nuclei look like eyes, the anterior flagella that loop above them appear like eyebrows, and the median body resembles a downturned mouth (refer to Fig. 1). Trophozoites measure 9–21 μm long by 5–15 μm wide (23). Similarly prepared cysts have a visible but not prominent wall, are ellipsoid, contain two or four nuclei and paired axonemes, and measure about 8–12 μm long by 7–10 μm wide (23).

C. Prevalence

Giardia is found in humans and animals worldwide. In a 1976 survey of 55 state and territorial laboratories in the United States, *Giardia* was found in 3.8% of nearly 415,000 stools; in 1980, it was found to infect 2–20% of various populations in the United States; 58,504 cases were reported from 1976 to 1980 (24). Giardia is the most frequently reported intestinal parasite in Canada with more than 2000 cases annually in British Columbia alone (25). In the United States rates of giardiasis were highest in children less than 5 years of age and women of childbearing age (26). (See Sec. D.2, ''Waterborne Outbreaks'').

D. Transmission and Reservoir Hosts

Transmission of the dormant, environmentally resistant cyst stage is via the fecal–oral route through contaminated water, contaminated food, or person-to-person contact. Cysts can remain infectious for several months under moist and cool conditions, but lose infectivity under dry and hot conditions.

Because *Giardia* is so widespread among domestic and wild-animal hosts, its zoonotic potential has aroused interest and concern. Human-to-human transmission has been well documented from outbreaks in nurseries, schools, and day care centers extending to family members and also from person-to-person contact between male homosexuals.

Animal-to-human transmission of giardiasis remains controversial. The close association between humans and companion animals provides opportunity for transmission but unequivocal evidence of transmission is lacking. From six studies conducted in the United States, up to 1982 12% of 3523 dogs were found infected with *Giardia* (27). More recently, from six studies conducted in the United States and Canada, 1–68% of 6815 dogs were found infected (28). Although the information from these studies and anecdotal information from veterinarians indicate that dogs, especially puppies, could serve as sources of human infection, there has not been a single documented case of dog-to-human transmission (28). Cats also might be a source of human infection, but reported cases of feline giardiasis are rare and sometimes circumstantial (28). *Giardia* from pigs, cattle, sheep, and goats was infectious for dogs and cats under experimental conditions. No human infection related to these animals has been reported.

Based on reports of giardiasis in hikers and campers who obtained drinking water from untreated surface waters in areas in which few humans are present, wild animals have been suspect reservoir hosts. A statewide telephone survey in Colorado reported on 256 cases and matched controls (29). A higher proportion of cases than controls visited Colorado mountains (69% vs. 47%), camped out overnight (38% vs. 18%), and drank untreated mountain water (50% vs. 17%). Such wild animals as beavers, raccoons, bighorn, mouflon, and pronghorn have been infected with *Giardia* from humans (30). Wild animals associated with aquatic environments such as beavers and muskrats as well as such aquatic birds as egrets and green herons have been found infected with *Giardia* (28). However, until a marker such as a morphological feature, unique biochemical, antigen, or gene can be used to identify and distinguish *Giardia* from wild animals, domesticated animals, and humans, our ability to attribute outbreaks to specific sources remains circumstantial.

1. Foodborne Transmission

Evidence has been reviewed for foodborne transmission of *Giardia* due to fecal contamination (31). One especially well-documented outbreak of giardiasis in a farming community in Minnesota resulted after a person changed the diaper of a child with giardiasis and then prepared a salad (32). An outbreak of giardiasis involving 27 persons occurred after a meeting at a restaurant where circumstantial evidence suggested that ice contaminated by a food handler may have been the vehicle (33). Twenty-six persons acquired giardiasis after eating raw vegetables in an employee cafeteria served by an infected food handler (34). A fruit salad preparer who had diapered a child and had a pet rabbit at home, both positive for *Giardia*, was thought to be responsible for an outbreak of giardiasis following a family party attended by 25 persons (35).

2. Waterborne Transmission

In the United States, *Giardia* has been identified more often than any other pathogen in waterborne outbreaks resulting in illness. From 1971 to 1985, of 502 reported outbreaks, 251 were associated with specific etiological agents; *Giardia* was identified in 92 of these outbreaks and resulted in over 24,000 cases of illness (24).

A review of CDC reports from 1986 to 1988 indicated nine outbreaks resulting in 1169 cases due to waterborne giardiasis (4). These cases resulted from consumption of water intended for drinking or for domestic purposes as well as from consumption from nonpotable sources by swimmers, divers, hikers, and campers. Outbreaks were reported in 22 states, most (90%) in the mountains of 17 states in the Northeast, West, and Northwest. Water in these areas is primarily from surface water in streams or reservoirs uncontaminated by wastewater and in areas that are populated sparsely or restricted in large part from human activities. Treatment therefore has been minimal, consisting primarily of disinfection but not filtration. For 1989–1990, 16 states reported 26 outbreaks involving 4288 persons associated with the drinking of unfiltered surface water or surface-influenced groundwater (36). Similarly, community outbreaks have been reported in Canada from chlorinated, unfiltered surface water supplies (37,38).

In the United States in 1990, surface water was used by over 155 million people in 6000 community water systems; of these, 23% provided unfiltered water to 21 million people (24). Outbreaks have occurred mostly in small community water systems; such noncommunity systems as those supplying camps, parks, resorts, and institutions, or individual water systems including nonpotable sources; and water ingested while swimming or diving. Outbreaks have affected visitors as well as permanent residents. A seasonal trend, especially during the summer, implies either increased contamination of noncommunity water supplies or use by greater numbers of susceptible persons. Most cases were attributed to contaminated surface water but a few represent groundwater found to be contaminated by human waste or downstream from a beaver dam. In outbreaks involving unfiltered water, distinfection was inadequate for a variety of reasons, but primarily because of insufficient chlorine levels and too little contact time.

E. Clinical Signs

Infection with *Giardia* can range from asymptomatic to fatal. Following a communitywide outbreak in New Hampshire in which 213 symptomatic cases were diagnosed at a local hospital laboratory, a survey of city residents revealed that 76% of *Giardia* infections during the epidemic were asymptomatic and resolved without treatment (39). In contrast, although giardiasis is not generally thought to cause severe illness, a review of hospital data from the United States and the state of Michigan found that from 1979 to 1988, 4600 persons were hospitalized annually for giardiasis (26). Illness associated with giardiasis has been reviewed in detail (40,41). The most prominent initial signs of giardiasis are nausea, inappetence, discomfort in the upper intestine, and fatigue, followed by bursts of foul-smelling watery diarrhea, flatulence, and abdominal distention, usually lasting only a few days. Occasionally, this acute stage can last for months, causing malabsorption, weight loss, and debilitation. Chronic infection is marked by recurrent brief episodes or, to a lesser extent, persistent episodes of foul-smelling loose stools, flatulence, and abdominal distention. Lactose intolerance, common during active infection, may persist for a period thereafter. Mortality has been reported only twice.

F. Diagnosis

Acute diarrhea caused by giardiasis must be differentiated from other agents. Giardiasis should be suspected when upper abdominal cramps, distention, gas, and foul-smelling feces are present but blood and mucus are absent from the feces.

Chronic diarrhea caused by giardiasis must be differentiated from amebiasis, cryptosporidiosis, coccidiosis, strongyloidosis, inflammatory bowel disease, and irritable colon (41). When symptoms include upper intestinal discomfort, heartburn, belching, and intestinal rumbling, they must be differentiated from ulcers, hiatal hernia, gallbladder disease, and pancreatic disease.

Traditionally, diagnosis is based on identification of cysts or trophozoites in feces, or trophozoites in duodenal aspirates or biopsies. Most diagnoses can be made by stool examination. In formed stools found after the acute diarrheal period, cysts usually are present. In diarrheic stools during the acute period, trophozoites predominate.

Fresh stools must be examined immediately by wet smear or preserved in formalin, merthiolate-iodine-formalin (MIF), or polyvinyl alcohol (PVA) for later examination. At least three stools collected on alternate days should be examined before other diagnostic procedures are tried. When numerous, *Giardia* can be seen in direct fecal smears mixed with saline or Lugol's iodine or fixed in Schaudinn's fluid and stained with iron-hematoxylin (23). When stages in fecal smears are absent or scarce, zinc sulfate solution should be used for flotation and concentration; sugar and other salts should not be used because they distort the parasite and make it unrecognizable (23). When still no *Giardia* can be found in stools from suspected cases, duodenal or jejunal fluid should be obtained and examined by duodenal tube, endoscope, or Enterotest (Hedeco, Mountain View, CA). With the latter, a rubber-lined, weighted capsule attached to a nylon string is swallowed by the patient, left in place for several hours, and retracted. Then the lower portion of string is used to smear microscope slides to be examined for *Giardia* (41).

Because these traditional diagnostic methods lack sensitivity, are difficult to perform, or involve invasive procedures, immunological methods have been developed to aid diagnosis. Serodiagnosis and detection of anti-*Giardia* antibodies and *Giardia* antigen in clinical specimens has been reviewed (42). The serological assay with the greatest clinical utility detects immunoglobulin M (IgM) but its disadvantages include the variability of patient immune response, prolonged elevation of antibody, and cross-reactions. Antigen detection methods include immunodiffusion, immunofluorescence, counterimmunoelectrophoresis, and enzyme-linked immunosorbent assay (ELISA). A large commercial market for rapid and reliable immunodiagnostics now is driving the development of new and modified procedures. Molecular techniques are now being utilized for diagnosis and epidemiological investigations (43).

G. Treatment

Treatment for giardiasis is somewhat controversial in the United States. This subject has been reviewed recently by Davidson (44), who provides detailed description of the major drugs, risks, and benefits, and a critical review of efficacy trials. The treatment of choice recommended by the CDC is quinacrine (Atabrine, Sanofi Winthrop Pharmaceuticals, New

York, NY). Although metronidazole (Flagyl, G. D. Searle and Co., Chicago, IL) is used frequently, it is not approved by the Food and Drug Administration (FDA). Furazolidone (Furoxone, Roberts Pharmaceutical Corp., Eatontown, NJ) is recommended for children. Paromomycin (Humatin, ParkeDavis, Morris Plains, NJ) has been proposed for treatment of pregnant women.

H. Prevention

The most important steps to prevention of giardiasis are good personal hygiene (thorough washing of hands after changing diapers or having a bowel movement), proper sanitation of toilet seats, and treatment of drinking water. Because of the potential for virtually all surface waters to be exposed to *Giardia* contamination from human or animal sources, caution dictates that all cysts of *Giardia* be regarded as potentially infective for humans.

Filtration with diatomaceous earth, slow sand, and coagulation-filtration commonly is used by the water supply industry and can be designed to remove 99% or more of the *Giardia* cysts (45). Such chemical disinfectants as chlorine, chlorine dioxide, chloramine, and ozone can further reduce the number of infectious cysts in drinking water (45). These disinfectants can be reduced in effectiveness by changes in pH, water temperature, turbidity, concentration of disinfectant, and short contact time. A combination of filtration and disinfection appears best. Specific guidelines, recommendations, and regulations for controlling waterborne *Giardia* have been developed or proposed by various national and international organizations through 1998.

VI. CYCLOSPORIASIS

At this time cyclosporiasis is considered a newly emerging disease. More correctly, cyclosporiasis should be considered a newly recognized disease that is being detected with greater frequency in persons traveling from industrialized countries to countries where sanitation, treatment of drinking water, and sewage disposal are poor or nonexistent. It is also being detected in industrialized countries in association with the importation of fresh produce.

A. Taxonomy

Originally described as a cyanobacterium-like organism (CLB), *Cyclospora cayetanesis*, named and described by Ortega et al. (46,47), is the only member of the genus *Cyclospora* reported to infect humans. Other species have been described from snakes, moles, and rodents (47).

B. Life Cycle

The life cycle of *C. cayetanensis* has not been thoroughly studied but appears similar to that of other coccidian parasites. In the fecal stage, the oocyst is 8–10 μm in diameter and is excreted unsporulated; and it is not infectious until it sporulates. After 1 to 2 weeks in an aqueous medium at 20 to 30 °C oocysts sporulate, contain two sporocysts each with two sporozoites, and are infectious. After ingestion of fecal contaminated food or water oocysts release sporozoites, which penetrate duodenal or jejunal epithelial cells and develop asexually into schizonts that contain 4–6 or 10–16 merozoites approximately 1–2 μm wide by 5–6 μm in length (48,49). The number of asexual generations is not known and sexual stages have not been described.

C. Prevalence, Transmission, and Reservoir Hosts

Cyclospora cayetanensis oocysts or CLB have been found in human feces in Peru, Guatemala, Mexico, Haiti, the United States, Italy, France, the Netherlands, Republic of South Africa, Nepal, Bangladesh, Thailand, Malaya, and New Guinea. They have also been found in feces of travelers returning to western Europe, North America, or Australia from Turkey, Morocco, Gabon, Guatemala, Mexico, Haiti, Dominican Republic, Puerto Rico, Pakistan, India, Sri Lanka, Cambodia, Thailand, China, Indonesia, and the Solomon Islands. Infections have been reported in children, immunocompetent adults, and in persons with HIV infection.

In Nepal where the organism has been known for several years 350 symptomatic persons were diagnosed with cyclosporiasis between 1988 and 1993 (50). In Nepal and elsewhere, few cases have been diagnosed between November and April.

Contaminated drinking water is considered the major source of infection worldwide. In the United States an outbreak occurred in a hospital dormitory among 21 staff and employees who drank water within 24 h after the dormitory water pump failed and dispersed sludge from the bottom of a rooftop holding tank (51).

Foodborne cyclosporiasis has also been reported. Vegetables were collected in markets in an endemic area of Peru (see list in cryptosporidiosis section) and examined for *Cyclospora* and *Cryptosporidium* oocysts (52). Yerba buena and huacatay were found to be contaminated with *C. cayetanensis* oocysts.

In 1996 an outbreak of cyclosporiasis in adults in North America was reported in 20 states, the District of Columbia, and 2 Canadian provinces affecting 1465 persons associated with the consumption of Guatemalan raspberries (53). In the spring of 1997, 510 cases were reported in adults in 8 states and 2 provinces with raspberries from Guatemala as the probable vehicle of transmission in most cases and mesclun lettuce (a mixture of various baby leaves of lettuce) as the other (54). In the summer of 1997 at least 20 cluster cases of cyclosporiasis involving approximately 185 adults were associated with consumption of foods containing fresh basil prepared in a production kitchen for a chain of 8 retail food stores in the northern Virginia, Washington, D.C., and Maryland metropolitan area (55). One case of cyclosporiasis was reported in a woman experiencing loose stools beginning 48 h after eating fried clams (56), although the association was not proven.

Cyclospora oocysts resembling *C. cayetanensis* have been reported in the feces of a duck (57), chickens (57), dogs (58), and nonhuman primates (59). However, there is no evidence of transmission of *Cyclospora* from any of these hosts to humans.

D. Clinical Signs

Symptoms include diarrhea, nausea, vomiting, and abdominal cramps beginning 1–14 days after exposure (median incubation period: 8 days). In one outbreak in which 45 persons had diarrhea the median number of stools per day was 7 (range: 3–35) and the median duration of diarrheal illness was 5 days (range: 1–10) (54). Weight loss has been reported in immunocompetent as well as immunocompromised patients.

E. Diagnosis

Although invasive biopsies of duodenum and jejunum have revealed developmental stages in epithelial cells, the finding of oocysts in feces is the diagnostic method of choice. Oocysts are spheroidal 8.6 ± 0.6 μm (7.7–9.9 μm) in diameter with a bilayered wall (47). When viewed by bright field or interference contrast microscopy unsporulated oocysts contain a granular cytoplasm; sporulated oocysts contain 2 sporocysts each with 2 sporozoites. Oocysts autofluoresce when viewed with fluorescence microscopy using a filter for the wavelength of 340–380 nm (60). Oocysts usually appear red when stained by acid-fast staining methods (Kinyoun and Ziehl-Neelson staining procedures) but are variably acid-fast and sometimes refractory to staining. The use of fecal smears heated in a microwave oven followed by safranin staining was found superior to acid-fast staining methods (61).

F. Treatment

Trimethoprim-sulfamethoxazole (cotrimoxazole) has been effective for treatment of children, as well as immunocompetent and immunocompromised adults (62,63). This medication has also been used for prophylaxis.

G. Prevention

Avoid drinking water and eating raw fruit or vegetables in endemic areas. Even washing of contaminated fruits and vegetables does not completely remove oocysts (52). Drink only boiled or bottled beverages and eat only thoroughly cooked vegetables.

VII. CRYPTOSPORIDIOSIS

A. Taxonomy

Although over 20 species of *Cryptosporidium* have been named from fish, reptiles, birds, and mammals, not all are valid and only one species, *Cryptosporidium parvum*, infects humans (64). *Cryptosporidium parvum* also infects many other mammals, including companion, domesticated, and wild animals. Based on molecular data there appears to be two genotypes of *C. parvum* that are infectious for humans; genotype 1, which is transmissible from humans to humans, and genotype 2, which is zoonotic (65).

B. Life Cycle

The life cycle of *C. parvum* resembles that of such other coccidian parasites as *Eimeria*, *Isospora*, and *Cyclospora* in which all endogenous stages develop in a single host (Fig. 2) (64,66).

Within that host, infection begins when the oocyst stage is ingested and the four sporozoites within the oocyst are released. The sporozoites enter cells and all subsequent stages develop intracellularly at the surface of intestinal epithelial cells among the microvilli, primarily in the lower small intestine. In immunocompromised persons, parasites also have been found in the colon, rectum, gallbladder, pancreatic and bile ducts, and lungs.

Each stage is surrounded by the host cell membrane but remains separated from the

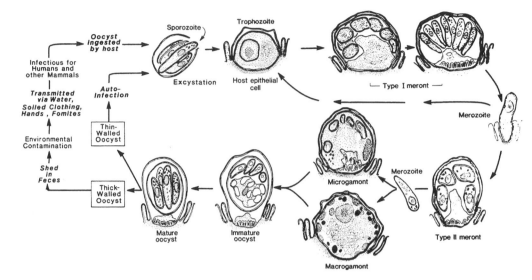

Figure 2. Transmission and life cycle of *Cryptosporidium parvum. Cryptosporidium* is transmitted by the oocyst stage in the feces contaminating food and water or by close personal contact with infected persons or animals or with soiled fomites. Ingestion of the oocyst followed by release of the sporozoite stage initiates a series of developmental sages in the microvilli of the epithelial cells lining the lower small intestine. In immunosuppressed hosts, the infection may not be confined to this location. The life cycle is completed with formation of the oocyst stage that passes in the feces.

host cell cytoplasm, apparently deriving essential nutrients for growth and development through a unique placenta-like organelle (see Fig. 2). Sporozoites lose their spindle shape and anterior organelles and become rounded or oval trophozoites that undergo nuclear division to form multinucleate schizonts. Each nucleus becomes incorporated within a merozoite, a stage that closely resembles the sporozoite. When fully mature, merozoites leave the schizont and enter other intestinal epithelial cells in which they initiate a second generation of schizont formation (schizogony).

Merozoites from the second generation leave the schizont and parasitize additional intestinal cells in which they develop into yet another schizont generation or form sexual stages—macrogamonts (females) or microgametes (males). Fertilization of macrogamonts is followed by their development into oocysts. Undifferentiated cytoplasm and nucleoplasm in the oocyst undergoes sporulation, the process in which four sporozoites are formed. The oocyst then passes in the feces from the body into the environment. The morphology of all of these stages has been reviewed in detail at both the light and electron microscope levels (64,66). The oocyst stage of *C. parvum* is notable for its small size, 4.5–6.0 μm in diameter.

C. Prevalence

Human infection with *C. parvum* was brought to international attention primarily as an opportunistic pathogen associated with AIDS in the early 1980s. Since its discovery in humans, surveys have reported human infections in 93 countries, on 6 continents, in industrialized and nonindustrialized countries, and in urban as well as rural locations. Based on selected populations of persons with diarrhea or other gastrointestinal disorders as well as stool specimens submitted for diagnosis, the following prevalence ranges were reported in the late 1980s: North America, 0.6–4.3%; Europe, 1–2%; Central and South America, Africa, Asia, and Australia, from 3% or 4% to 10% or 20% (67).

D. Transmission and Reservoir Hosts

Transmission from host to host is via the oocyst stage by the fecal–oral route through contaminated water, contaminated food, animal-to-person contact, or person-to-person contact. Oocysts of *C. parvum*, unlike many other coccidia, are sporulated when excreted and therefore are thought to be infectious immediately. They survive in wet or moist conditions outside the body for weeks or months. In the laboratory, they survive longer when stored at refrigeration temperature than at room temperature, suggesting a similar pattern in nature. Oocysts are rendered noninfectious within minutes at −20°C or colder (68), survive as long as 2 weeks at −10°C and 12 weeks at −5°C, 24 weeks or longer at 0–20°C, 16 weeks at 25 and 30°C, and 2 weeks at 35°C (69). At higher temperatures they survive only minutes and at 73.7°C they are killed in 5 s (70).

Over 130 species of mammals have been found naturally infected with *C. parvum* or species indistinguishable from it. These include wild mice, house mice, and rats (71); domestic cats, dogs, ferrets, raccoons, rabbits, and monkeys (72); pigs (73); cattle, sheep, goats, farmed red deer, wild ruminants including fallow deer, roe deer, sika deer, mule deer, Eld's deer, axis deer, and barasingha deer, water buffalo, Persian gazelles, blackbuck, sable antelope, scimitar horned oryx, fringe-eared oryx, addaxes, impalas, springbok, nilgai, gazelles, eland, and mouflon (74); and many others (75). Of these animals, cattle have been implicated as, but not proven to be, one of the most common sources of human infection (76).

1. Foodborne Transmission

Cryptosporidium parvum has been transmitted via apple cider, chicken salad, kefir, and milk. Oocysts were detected in mussels from the coast of Ireland (77) and in oysters from six Chesapeake Bay sites in Maryland (78). A selected sample of oocysts recovered from oysters was found to be infectious by testing in mice but there has been no report of human infection resulting from the eating raw shellfish (78). Oocysts have also been detected on fresh vegetables collected in markets in an endemic region of Peru (52). These included cabbage, celery, cilantro, green onions, ground green chili, leeks, lettuce, parsley, yerba buena, and basil.

2. Waterborne Transmission

References 79–81 are excellent reviews on the subject of waterborne infection. Oocysts have been found in recreational and drinking water as well as water from reservoirs, lakes, ponds, streams, rivers, ground water, and in marine waters. In the water surveys, the species of *Cryptosporidium* has not been identified often but other reports have documented human cryptosporidiosis resulting from ingestion of water from most of the sources listed above.

Despite the failure to clearly identify species or to determine if oocysts were infectious, as well as problems in recovering oocysts from environmental samples, surveys have yielded significant information. Oocysts have been found in small numbers in all water sources but are more prevalent in surface waters than in groundwaters. Accidental pollution and heavy rainfall, especially following the application of manure to agricultural land, appears to rapidly increase background levels. Most oocysts in these waters are thought to come from animal feces transported by postrainfall runoff and by products from sewage treatment that pass through treated effluent directly into water courses. Water samples collected from four locations on two rivers in Washington State over a 4-month period from pristine mountain locations to downstream agricultural sites showed that oocyst concentrations consistently were above the minimal detection level and were found continuously, not intermittently (82).

Outbreaks of cryptosporidiosis associated with contaminated drinking or recreational water have been reported from the United States, Canada, the United Kingdom, and Australia. In an outbreak in Georgia in 1987 an estimated 13,000 people became ill after consuming water from a filtered, chlorinated public water supply that met state and federal standards (4). The outbreak in Milwaukee, Wisconsin, affected an estimated 403,000 persons (83). Fifty-four deaths occurred during the 2-year postoutbreak period compared with 4 deaths in the 2 years before (84). AIDS was the underlying cause of death for 85% of the postoutbreak cryptosporidiosis cases (84).

E. Clinical Signs

Clinical signs of cryptosporidiosis from those most to least often observed include diarrhea, abdominal pain, nausea, vomiting, and fever (67). Nonspecific symptoms include loss of appetite, malaise, muscle pain, weakness, and headache. The onset of cryptosporidiosis often begins with an influenza-like illness. The major symptom, diarrhea, is more profuse in immunocompromised persons than in immunocompetent persons. In such severely immunocompromised persons as those with AIDS, infections are of long duration, often followed by death. When an immunosuppressant can be removed, the patient usually recovers. Other persons at risk include those at either end of the age spectrum and those with nutritional deficiencies.

The infective dose can be small, as few as 30 and an ID_{50} of 132, and the time from exposure until clinical illness can range from 5 to 21 days (67,85).

F. Diagnosis

In addition to typical symptoms of cryptosporidiosis, a good medical history can provide clues helpful for diagnosis. Cluster cases in families, association with day care centers, and personal contact with infected persons suggest person-to-person transmission. Exposure to young pets or farm animals suggests zoonotic transmission. Use of immunosuppressant medications or persistent diarrhea of long duration suggests impaired immunity.

A variety of techniques for processing stool specimens to obtain and identify oocysts have been reviewed (85). The simplest is the direct smearing of feces on a glass microscope slide, followed by drying and staining. Several stains have been used to differentiate oocysts from surrounding debris, including acid-fast stains, Giemsa, Gram stain, methenamine silver, PAS, and others. Of these, acid-fast stains are fast, relatively easy to use, and inexpensive. Most oocysts stain red, differentiating them from green-stained yeasts of the same size (4.5–5 μm diameter) and shape (round). Fluorescent stains alone or combined with specific anti–*C. parvum* antibody also have been used successfully and are available commercially in kits.

When oocysts are not found readily in smears, stool concentration techniques can

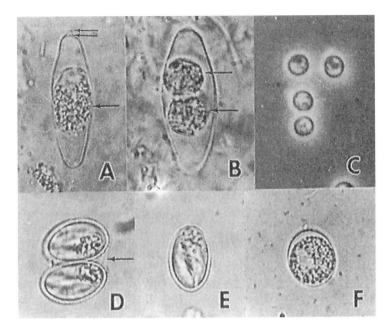

Figure 3. Comparative size and shape of coccidian oocysts; all are shown at the same magnification, illustrating the large size range: (A) unsporulated oocyst of *Isospora belli* in human feces (18 by 25 µm) showing micropyle cap on oocyst wall (double arrows) and unsporulated cytoplasm (arrow) in aqueous solution observed with bright field microscopy; (B) sporulated oocyst of *I. belli* containing two sporocysts (arrow) (11 by 28 µm); (C) oocysts of *Cryptosporidium parvum* (4–5 µm in diameter) in sugar flotation observed with phase contrast microscopy; (D) oocyst of *Sarcocystis* (13 by 16 µm) showing thin oocyst wall (arrow) surrounding two sporocysts, each of which contains four elongate sporozoites and a granular residual mass; (E) single sporocyst of *Sarcocystis* (8 by 13 µm); (F) unsporulated oocyst of *Toxoplasma gondii* (10 by 11 µm) from cat feces.

be used to decrease debris while increasing the relative number of oocysts (86). These include flotation of aqueous fecal slurries in saturated sucrose solution, zinc sulfate, or sodium chloride, or sedimentation in formalin ethyl acetate. Phase contrast microscopy greatly enhances visibility of oocysts in wet mounts from flotations compared with the more often used bright field microscopy. When viewed with phase contrast, oocysts in sucrose flotation appear distinct and luminescent against a darkened background (Fig. 3).

Diagnosis by examination of tissue biopsy specimens for stages in the epithelium is less desirable. It was used historically before oocyst identification was known, but it is an invasive process and the actual area that can be observed is quite restricted.

Use of molecular techniques, especially the polymerase chain reaction (PCR), for detection of oocysts in tap water and environmental samples has advanced rapidly and many of those techniques have been reviewed (87).

G. Treatment

Despite anecdotal success reported for a few drugs and reports of efficacy of polyether antibiotics in rodent models (88,89), of over 100 therapeutic and preventive modalities tested against cryptosporidiosis none has proven clearly effective in large-scale human studies. For a detailed review and listing of compounds, see Ref. 90.

H. Prevention

To prevent or control cryptosporidiosis, oocysts must be eliminated from the environment or rendered nonviable, or susceptible persons must avoid contact with such known sources of oocysts as infected animals and persons and contaminated environments.

Cold and hot temperature extremes kill oocysts (see Sec. D), so that boiling water and cooking vegetables in endemic areas or during outbreaks is helpful. However, temperature extremes cannot be applied practically in the environment or even in most large buildings. Few chemical disinfectants have been found effective within a relatively short period of exposure. Of those tested, the most effective are low molecular weight gases such as ammonia, ethylene oxide, methyl bromide, and formaldehyde (91). Because of their toxicity, general application of these chemicals is impractical.

Personal as well as environmental sanitation can reduce the number of oocysts. Thorough washing of hands and the removal of soil from fresh vegetables can reduce the risk of hand-to-mouth or foodborne ingestion of oocysts.

A variety of sequences are used for the treatment of water to render it safe for drinking; however, unless prepared specially, no water is totally clear of particles under 10 μm and the addition of chlorine as a final disinfectant for bacteria has no effect on oocysts. An exhaustive review of water purification methods for removal of oocysts is given by Ives (92). At the present time, slow sand filtration combined with chemical flocculation is the best method in widespread use in the water treatment process.

VIII. SARCOCYSTOSIS

A. Taxonomy

Parasites in the genus *Sarcocystis* derive their name from the intramuscular (*sarco* = "muscle") cyst stage of the life cycle. For over 100 years after the discovery of these parasites, this was the only stage known. Of *Sarcocystis* species infectious for humans, cysts of *S. hominis* are microscopic and those of *S. suihominis* are macroscopic in muscles of the respective intermediate hosts, cattle and swine.

B. Life Cycle

Cysts have been found in virtually all skeletal muscles of the body of infected hosts including the tongue, esophagus, and diaphragm, as well as in cardiac muscle and to a lesser extent in neural tissue including the brain. Within the intramuscular cysts are large numbers of spindle-shaped bradyzoites that appear dormant until the cyst is ingested (Fig. 4). Following digestion, bradyzoites liberated from the cyst become active and enter cells in the intestinal lamina propria in which they develop into the sexual stages of microgametes or macrogamonts. The sperm-like microgametes fertilize the egg-like macrogamonts, which then develop into the zygote and then the oocyst stage. Oocysts sporulate in the intestine, forming two sporocysts that contain four sporozoites each. Often, the thin cyst wall ruptures during passage in the feces, with individual sporocysts excreted by the infected person. These sporocysts are infectious only for the same species of intermediate

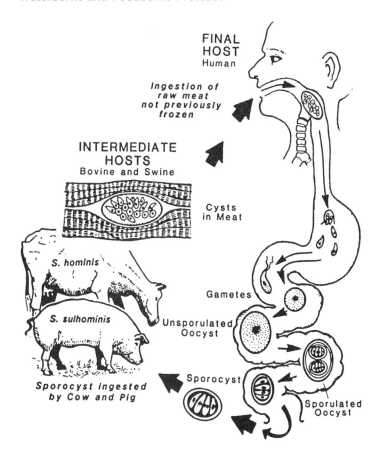

Figure 4. Transmission and life cycle of *Sarcocystis*. *Sarcocystis* is transmitted to humans by ingestion of raw or undercooked meat that contains mature cysts of *S. hominis* or *S. suihominis*. Bradyzoites liberated from the cysts enter cells lining the intestine, transform into sexual stages that undergo fertilization, and develop into oocysts. These sporulate in the gut and pass as oocysts or sporocysts in the feces. Human feces contaminating the environment in which livestock are raised are the only source of infection. Ingestion of oocysts/sporocysts leads to development of cysts in muscles and neural tissue.

host that served as the original source of infection. Thus, *S. hominis* sporocysts excreted approximately a week after eating raw or undercooked beef is infectious only for cattle and *S. suihominis* from pork is infectious only for pigs.

After ingestion of sporocysts, sporozoites liberated in the gut find their way to arterioles throughout the body from which they enter cells and undergo asexual multiplication, producing merozoites. These in turn leave the host cell and travel through the bloodstream to the capillaries, in which they enter new host cells and again undergo asexual multiplication to produce a second generation of merozoites. When these become mature and leave the host cell, they enter skeletal and cardiac muscle cells from which they give rise to the cyst stage.

The morphology of all stages has been reviewed in detail at both the light and electron microscope levels (93). Sporocysts in human feces are diagnostic. They are sporulated when excreted, contain four sporozoites and a granular residual body, and measure approximately 10 by 15 μm. They cannot be distinguished from sporocysts excreted by other definitive hosts but are programmed genetically to complete the life cycle only in specific hosts.

Cysts in the intermediate hosts have unique morphological features. The sarcocyst wall of *S. hominis* is up to 6 μm thick and appears radially striated from numerous villar protrusions that are up to 7 μm long and 0.7 μm wide; bradyzoites are 7–9 μm long and arranged in packets. The cyst of *S. suihominis* is up to 15 μm long and the wall is 4–9 μm thick with some villar protrusions up to 13 μm long; bradyzoites are 15 μm long.

C. Prevalence

Based on examination of tissues from abattoirs, most cattle in the United States and worldwide are infected with *Sarcocystis*, with *S. cruzi* (infectious from cattle to canines) being the most prevalent and easiest to identify histologically. Because *S. hominis* (infectious from cattle to humans) and *S. hirsuta* (infectious from cattle to felines) are difficult to distinguish except by electron microscopy, some prevalence data may be erroneous. However, there appears to be little or no *S. hominis* in the United States, whereas up to 63% of cattle in Germany have been found infected. The overall prevalence of *Sarcocystis* in pigs is low, 3–36% worldwide. *Sarcocystis suihominis* was found more prevalent in Germany than Austria, but information is not available from other countries.

Based on surveys conducted worldwide, more intestinal sarcocystosis in humans was reported in Europe than any other continent (93). The greatest prevalence was found in 10.4% of fecal specimens from children in Poland and 7.3% of samples from Germany.

D. Transmission and Reservoir Hosts

Humans acquire gastrointestinal sarcocystosis only by ingesting raw or undercooked meat from cattle or pigs harboring mature *S. hominus* or *S. suihominis* cysts. Other species of *Sarcocystis* such as *S. cruzi* in cattle muscle are not infectious for humans.

Based on a report from Thailand indicating intestinal lesions in persons having eaten undercooked meat from *Bos indicus* cattle (94), there may be other unnamed species of *Sarcocystis* infectious for humans. Other meat animals that harbor *Sarcocystis* with unknown life cycles include camels, llamas, water buffalo, yaks, and species of pigs other than the domesticated *Sus scrofa*.

Only about four dozen humans have been reported with *Sarcocystis* causing muscular infections. In such cases, humans harbor the cyst stage and therefore serve as the intermediate host. Based on all other *Sarcocystis* life cycles, infected human tissues must be eaten by a carnivore to complete the life cycle. Because there is no known predatory or scavenging cycle in nature in which human tissues are eaten regularly by carnivores, humans most likely become infected accidentally by ingestion of food or water contaminated with feces from a carnivore that participates in a primate–carnivore cycle involving an unknown species of *Sarcocystis*. Most human cases involving muscular infection have been in persons living or traveling to tropic areas.

E. Clinical Signs

The ingestion of raw beef containing *S. hominis* has been documented in studies involving human volunteers who became infected and excreted oocysts in their feces (95,96). Only one of these became ill. Signs that appeared 3–6 h after eating the beef included nausea, stomach ache, and diarrhea. In contrast, volunteers who ate raw pork infected with *S. suihominis* and became infected and excreted oocysts had much more dramatic signs that appeared 6–48 h after eating the pork (96,97). These clinical signs included bloat, nausea, loss of appetite, stomach ache, vomiting, diarrhea, difficulty breathing, and rapid pulse. Volunteers who ate well-cooked meat from the same pigs had no clinical signs (97).

F. Diagnosis

Presumptive diagnosis of human intestinal sarcocystosis is based on clinical signs and a medical history indicating ingestion of raw or undercooked meat. Definitive diagnosis requires identification of sporocysts in the feces. Sporocysts of *S. hominis* are excreted 14–18 days after the ingestion of beef. Sporocysts average 14.7 by 9.3 μm. Sporocysts of *S. suihominis* are excreted 11–13 days after ingestion of pork and average 13.5 by 10.5 μm. Sporocysts are seen readily by bright field microscopy of sucrose-fecal flotations by focusing the microscope at the uppermost optical plane of the wet mount just beneath the coverslip (see Fig. 3). Because of the overlap in size and shape, the two species are not discernable from one another when both are present at the same time.

Determination of *Sarcocystis* in meat can be made by direct observation of macroscopic cysts or microscopic examination of histological sections. Meat also can be ground and artificially digested with a mixture of pepsin and hydrochloric acid at 37°C for 1 h, centrifuged, and the pellet examined microscopically for the presence of bradyzoites.

G. Treatment

There is no known treatment for intestinal sarcocystosis. It is self-limiting and of short duration.

H. Prevention

To prevent human intestinal sarcocystosis, meat should be cooked thoroughly or frozen (temperature and time unknown) to kill bradyzoites in cysts. Chemoprophylaxis with the anticoccidial drugs amprolium and salinomycin has been effective in preventing sarcocystosis in experimentally infected cattle and sheep (98,99).

To interrupt the life cycle and prevent infection of livestock, carcasses, meat scraps, offal, or other, raw or undercooked tissues should not be available to domestic or wild carnivores in areas where livestock are raised.

IX. TOXOPLASMOSIS

A. Taxonomy

Toxoplasma gondii is the only member of its genus. The genera *Hammondia*, *Neospora*, and *Besnoitia* are closely related as determined by molecular studies but differ biologically.

B. Life Cycle

Toxoplasma gondii is a protozoan parasite of domestic and wild cats that potentially is capable of infecting all vertebrates (100). Cats can become infected from three sources: (a) by ingesting tissues from animals harboring cysts of *Toxoplasma* that contain the slowly developing organisms called cystozoites or bradyzoites, (b) by ingesting tissues from acutely infected animals with rapidly developing tachyzoites, or (c) by ingesting oocysts excreted in the feces of other cats.

The following sequence of development was determined by studies in which cats were infected with bradyzoites (Fig. 5). After ingestion of cysts, bradyzoites are liberated and enter the epithelium of the small and large intestines in which they develop through some or all of five asexual stages (types A–E). These stages differ slightly from one another in size, number of organisms, location in the intestine, and method of multiplication. Development of sexual stages (gametocytes) follows, mostly in the ileum. After fertilization, zygotes develop into oocysts that enter the lumen of the intestine and are excreted unsporulated in the feces.

The prepatent period, from ingestion of cysts until oocysts are excreted, varies from 3 to 10 days; the patent period usually lasts 7–21 days. Tens of millions of oocysts can be excreted by a single cat in a single day. Oocysts are not immediately infectious; they require exposure to air for about 48–72 h. The sporulated oocyst contains two sporocysts (cysts within the oocyst), each of which contains four sporozoites, the spindle-shaped

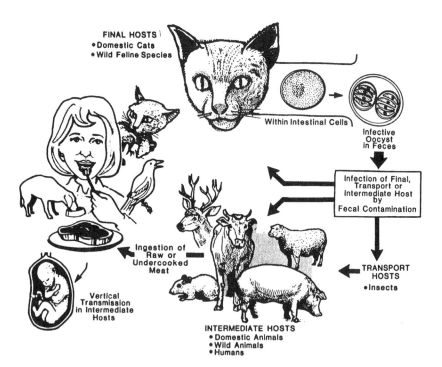

Figure 5. Transmission and life cycle of *Toxoplasma*. Oocysts in cat feces spread throughout the environment are infectious for all vertebrates including humans and other cats. Humans, cats, and other carnivorous animals also can become infected by ingested raw or undercooked meat or other animal tissues containing the cyst stage.

organism infective for cats and virtually all vertebrates, including humans. The number of times a cat might excrete oocysts is not known. Cats can repeat oocyst excretion after reinfection, although the number excreted is far less than during the primary infection. Older cats usually do not excrete oocysts unless severely ill, especially with enteritis from some other cause.

When either oocysts or tissue cysts are ingested, sporozoites or bradyzoites enter the tissues of a nonfeline host and transform into the tachyzoite stage. Tachyzoites undergo repeated, rapid multiplication by a process called *endodyogeny* (two progeny form within the parent tachyzoite) until, influenced by the host's immune system, the multiplication rate slows and the tachyzoites transform into bradyzoites. These become surrounded by a cyst wall and remain dormant but infectious in the tissue for long periods of time, sometimes for the life of the host. Bradyzoites transform back into tachyzoites when the cysts are ingested by another host or when the immune system of the current host is suppressed by malnutrition, disease, or medications.

C. Prevalence

Toxoplasma is among the most widespread and prevalent protozoan parasites on earth. Surveys in the 1970s indicated infections in an estimated half billion people (101). The range of prevalence reported in numerous surveys among and within countries and locales is great. In the United States, of the adult subpopulations examined for serum antibody to *Toxoplasma*, 0–39% were seropositive, indicating prior exposure or current infection (100). In other countries, the prevalence was much higher. In France, for example, 42–84% of the subpopulations examined were seropositive (100).

Summarizing 18 serological surveys in the United States (100), or 4871 cats examined, 25% were seropositive for *Toxoplasma* antibody. The prevalence of active excretion of oocysts by cats at any given time is less than 2%. This figure can lead to underestimation of the potential of cats to spread infection. The large number of oocysts excreted, the ability of oocysts to remain viable in the environment for a long time, and the intimate proximity of catas to humans and to human activities all contribute to the spread of toxoplasmosis.

Based on 38–50 reports, including serological testing as well as isolation of organisms from tissues, on the worldwide prevalence of *Toxoplasma* in meat animals, the following data were calculated (101). Of 16,293 cattle, 9654 sheep, and 17,499 pigs examined, the average rates of infection were 25%, 31%, and 29%, respectively. A study and review of *Toxoplasma* antibody in meat animals in the Netherlands showed a very high percentage of seropositive sheep (89–92%) and swine (50–86%) that declined significantly after farming practices changed to provide better hygiene (102). A study nearly 10 years later by the same group showed a very low seropositivity in continually housed pigs (1.8%) and calves (1.2%) whereas sows and dairy cows had seropositivity of 30.9% and 27.9%, respectively, demonstrating environmental infection pressure (103). A serological survey of 12,298 pigs in the United States indicated a similar percentage of seropositive swine, 24% of feeder pigs and 42% of breeder pigs (104). These data might lead one to believe that the percentage of infected beef, lamb, and pork is the same as the serological findings, but this has not been demonstrated. Factors responsible for these differences include differences in the infectivity and adaptability of strains of *Toxoplasma* that initiated infection, kinetics of the host's serological response to infection, and genetic or other unknown factors imparting resistance to persistent infection in the host. For example, despite seropositivity of 25% in cattle, *Toxoplasma* has rarely been isolated from tissues of naturally infected bovines and some of these isolations, upon reevaluation, probably were not *Toxoplasma* (100).

D. Transmission and Reservoir Hosts

Toxoplasma can be transmitted via several routes. Although a major source of infection is thought to result from contamination of the environment with oocysts excreted in cat feces, the extent of human infection resulting from this route is not known. Feces are deposited in gardens, fields, lawns, pastures, playgrounds, barns, stables, feed troughs, and almost any other location available to cats. The oocysts measure only 10–12 µm in diameter and are dispersed easily by wind, water, earthworms, slugs, insects, birds, the feet of humans and animals, tools, machinery, and assorted fomites. They can remain infectious for a year or more if protected from temperature extremes, intense sunlight, and drying. Virtually all warm-blooded animals can acquire infection from oocysts and relatively few oocysts are required to initiate infection. Thirteen of 14 pigs fed only one oocyst developed infections confirmed by bioassay in mice (105). Field studies on 47 swine farms in Illinois during 1992 and 1993 confirmed the possibility of transmission of *T. gondii* to swine via consumption of oocysts in feed, and soil (106). Of 188 cats examined on these farms, 67% had antibody to *T. gondii*. A study of over 11,000 human sera specimens from 10 Canadian cities collected over 13 years revealed a cycle of *Toxoplasma* prevalence based on environmental factors (107). The finding of fewer seropositive persons in the fall after a dry summer was interpreted as an indication that oocysts had not survived without moisture, which led to the hypothesis that the higher number of seropositive persons found at other times must be due to oocyst infection.

Oocyst infection also seemed to be the only explanation for the relatively high prevalence of *Toxoplasma* antibody in vegetarians in both India (108) and the United States (109). Oocyst infection was found to explain the high prevalence of *Toxoplasma* antibody in Costa Ricans whose homes were constructed with slotted ceilings, a frequently found building technique (110). In Costa Rica, cats are numerous, have easy access, and defecate in the ceilings, and fecal debris and oocysts drop through the slots, contaminating the living quarters below. Agricultural workers in California who worked in and handled soil were found at greater risk of acquiring *Toxoplasma* than nonagricultural or cattle feedlot employees (111).

Like vegetarians and workers close to the soil, herbivores must acquire *Toxoplasma* infection from oocysts. The prevalence of *Toxoplasma* antibody in herbivores such as cattle and sheep is a good indicator of the distribution of cat feces in the environment.

1. Foodborne Transmission

Transmission of *Toxoplasma* by ingestion of tissue cysts in raw or undercooked meat from a variety of livestock and game animals has been documented as another major source of human infection (100). At a university snack bar in New York City, 5 students ate undercooked hamburgers containing meat of undetermined origin and developed acute toxoplasmosis 8–13 days later (112). At a university in Brazil, 110 persons ate undercooked meat of undetermined origin and likewise developed toxoplasmosis (113).

Several outbreaks, however, have been associated with specific cultural preferences for undercooked meat rather than mishandling or accidental undercooking. In Russia, significantly more *Toxoplasma*-seropositive persons were identified among women who tasted raw minced meat during cooking than among those who did not taste raw meat (114). In Hawaii, a higher prevalence of infection was found among Hawaiians and Filipinos who ate raw meat (including pork) than among Caucasians and Japanese who ate thoroughly cooked meat (115). In France, 204 children became infected while they were hospitalized and fed meals of very rare or raw meat (116). In the United States, 19 people became infected from eating the Syrian recipe for kibee nayee containing raw beef (117). A similar dish, steak tartare, was thought to be responsible for the greater number of infected affluent whites than poor blacks in New York City (118). In Korea, 2 outbreaks were associated with eating raw spleen and liver from a wild pig and with eating raw liver from a domestic pig (119). In Brazil, 17 persons developed acute toxoplasmosis after eating raw mutton at a party (120). Although no reports have associated acquired toxoplasmosis from eating poultry, experimentally infected chickens and quail were shown to harbor infectious cysts in their muscle (121,122).

Less clear is the risk of general association with meat. Abattoir workers in Egypt and Ghana were not found to be at higher risk of acquiring *Toxoplasma* than the general population, whereas abattoir workers, butchers, and meat inspectors in Japan and Brazil were found to be at higher risk (101). One route of transmission to such workers is illustrated by the unusual case of a woman who died of toxoplasmosis after being stabbed with a butcher's knife (123).

In the United States, beef has not been found to be infected naturally with *Toxoplasma*, very little unfrozen lamb or horsemeat is eaten, and poultry is well cooked before being eaten (100). Of all the meat animals, a relatively high percentage of swine have *Toxoplasma* antibody (see Sec. IX.C, above) and the organisms have been isolated from commercial cuts of pork including arm picnic, Boston butt, ham, tenderloin, and spareribs from experimentally infected pigs (124). Therefore, infected pork could be a significant potential source of infection.

2. Waterborne Transmission

Waterborne transmission of *Toxoplasma* oocysts has been documented only twice but, based on prevalence of *Toxoplasma* antibody in herbivores and the opportunity for runoff water to contaminate drinking water supplies, it is likely that this route of infection is not as rare as the literature suggests. Shortly after U.S. military troops returned from training in Panama, toxoplasmosis was diagnosed (125). Only affected personnel obtained drinking water from a common source, a pool in the jungle. Water collected at that site weeks later and brought back to the United States for examination contained a variety of helminth egges, coccidian oocysts, and oocysts morphologically similar to *Toxoplasma*. One hundred people aged 6–83 years acquired acute toxoplasmosis in the city of Victoria and adjacent areas of Vancouver Island in Canada (126). The weight of evidence suggested that water from one of the local reservoirs was the source of the outbreak.

E. Clinical Signs

Clinical signs of toxoplasmosis can be highly variable because *Toxoplasma* can infect virtually every organ in the body. However, most cases probably are asymptomatic or resemble influenza. Acute infections have resulted in dermatomyositis, encephalitis, hepatitis, lymphadenitis, myocarditis, placentitis, pneumonitis, retinochoroiditis, skeletal myositis, tenosynovitis, tonsillitis, vasculitis, anemia, and fever (101).

The interval from exposure to clinical illness is quite variable. Generally, clinical signs are manifested from a week to a month after exposure. Duration, severity, and outcome generally vary according to the immunological state of the host. In cancer and AIDS patients, infections of long duration followed by death are more common than in immunocompetent patients.

A risk group of special concern is pregnant women. In Europe, congenital toxoplasmosis has been found in 10–60% of newborns and is probably the most serious form of *Toxoplasma* infection. The fetus is thought to become infected via the placenta as a consequence of parasitemia in the mother. Although repeated congenital infection is rare, it has been hypothesized that if cysts form in the uterine wall they could burst during a subsequent pregnancy, causing infection in the fetus. Infection acquired during pregnancy is transmitted to the fetus about 40% of the time; of those infected, 40% have clinical illness; of those with clinical illness, 40% of the illnesses are severe or lethal (101). The earlier in pregnancy infection is acquired, usually the greater the extent of lesions in the fetus.

In the United States, over 3000 babies are born with toxoplasmosis each year; of these, 5–15% die, 8–10% have brain and ocular lesions, 10–13% have visual damage, and the 58–72% who are normal at birth later develop active infections (100). In 1992 the annual cost of neonatal toxoplasmosis in the United States was estimated at $5256 million for income loss, special education, and medical costs (127). Assuming that pork causes 50% of U.S. cases, the foodborne cost was estimated at $2628 million (127).

F. Diagnosis

Diagnosis is based on medical history, tissue sections, or tissue smears that contain *Toxoplasma*, paired serological tests with rising antibody titers, and isolation of organisms from tissues or body fluids. Important aspects of the medical history would include ingestion of undercooked meat of any kind, ingestion of unpasteurized milk of any kind, ingestion of uncooked eggs, contact with cats, known chronic infection, blood transfusion, organ transplant, immunosuppressive medication, AIDS, cancer, and maternal infection during pregnancy.

Serological tests include agglutination tests, the Sabin-Feldman dye test, indirect fluorescent antibody, indirect hemagglutination, complement fixation, and ELISA. Agglutination tests and ELISA are available commercially and have been modified to detect M and G immunoglobulins. Serological tests can indicate that an animal has been exposed to *Toxoplasma*. To determine if an animal or meat actually is infected, serology is unreliable. Furthermore, because cysts can be few in number, they might not be detected easily by histological examination. Therefore, to determine infectivity, organisms must be isolated from specific tissues. The best method of isolating organisms is by digesting host tissue with pepsin-HCl and then inoculating the suspect debris intraperitoneally into mice known to be free of *Toxoplasma* antibody. To confirm transmission of organisms from suspect tissues, tachyzoites can be found in mouse peritoneal exudate 4–6 days after inoculation, antibody to *Toxoplasma* can be found in the mice 2–3 weeks after inoculation, and cysts will be present in the brain a month or more after inoculation.

G. Treatment

Chemotherapeutic treatment for toxoplasmosis in humans has been reviewed by Dubey and Beattie (100). Briefly, pyrimethamine with sulfonamides acts synergistically against the rapidly dividing tachyzoite stage and is usually the treatment of choice in the United States. Spiramycin and clindamycin, macrolide antibiotics, also have been found to be effective treatments. All of these compounds have serious side effects and must be used cautiously.

H. Prevention and Control

Any measures that reduce or eliminate contamination of the environment with oocysts from cat feces will help prevent toxoplasmosis. Cats should be prevented from hunting birds and rodents. They should be fed thoroughly cooked food, dry food, or canned food. Litter pans should be used in the home and emptied daily. Children's sandboxes should be covered when not in use. Gardeners handling soil should wear gloves and wash their hands thoroughly before eating or handling food. Cats on farms should be prevented from entering food storage facilities and animal quarters. Barns, stalls, feed bunkers, waterers, and other areas should be kept free of cat feces.

All meat should be cooked thoroughly before eating. Extensive studies have been conducted that demonstrate thermal death curves for cooking meat and indicate the necessary combination of time vs. temperature for rendering *Toxoplasma* noninfectious; 61°C or higher for 3.6 min or longer is one of many combinations (128). It is important to know the actual temperature at the core of the meat, not just the surface temperature. Microwave ovens should not be used for initial cooking because of the uneven distribution of heat resulting in hot and cold spots. Freezing also kills some strains of *Toxoplasma* but is less dependable.

Irradiation of meat at 50 kilorads or greater renders *Toxoplasma* noninfectious (129). However, irradiation is not used presently for treatment of meat sold at retail stores in the United States.

Finally, water from lakes, ponds, streams, or other untreated sources should be boiled before drinking or being used to wash food.

REFERENCES

1. Jonnalagadda, P. R., and R. V. Bhat, Parasitic contamination of stored water used for drinking/cooking in Hyderabad, *SE Asian J. Trop. Med. Publ. Health*, 26(4): 789–794 (1995).
2. Summary of notifiable diseases, United States, 1991, *Morbidity Mortality Weekly Report*, 40(53): 1–63 (1992).
3. Addiss, D. G., J. P. Davis, J. M. Roberts, and E. E. Mast, Epidemiology of giardiasis in Wisconsin: increasing incidence of reported cases and unexplained seasonal trends, *Am. J. Trop. Med. Hyg.*, 47(1): 13–19 (1992).
4. Levine, W. C., and G. F. Craun, Waterborne disease outbreaks, 1986–1988, *Morbidity Mortality Weekly Report*, 39(SS-1): I–13 (1990).
5. Bean, N. H., P. M. Griffin, J. S. Goulding, and C. B. Ivey, Foodborne disease outbreaks, 5-year summary, 1983–1987, *Morbidity Mortality Weekly Report*, 39(SS-1): 15–57 (1990).
6. Chorba, T. L., R. L. Berkelman, S. K. Safford, N. P. Gibbs, and H. F. Hull, Mandatory reporting of infectious diseases by clinicians, *Morbidity Mortality Weekly Report*, 39(RR-9): 1–17 (1990).
7. Casemore, D. P., Foodborne protozoal infection, *Lancet*, 336:1427–1432 (1990).

8. Orihel, T. C., and L. R. Ash, Parasites in Human Tissues, American Society of Clinical Pathologists, Chicago, 1995, p. 4.

9. Arroyave, R. J., D. E. Ayala, and R. C. Hermida, Differences in circannual characteristics of the incidences of amebiasis and giardiasis, *Prog. Clin. Biol. Res.*, 341B:717–727 (1990).

10. Omar, M. S., M. E. al-Awad, and A. A. al-Madani, Giardiasis and amoebiasis infections in three Saudi closed communities, *J. Trop. Med. Hyg.*, 94(1): 57–60 (1991).

11. Erko, B., H. Birrie, and S. Tedla, Amoebiasis in Ethiopia, *Trop. Geog. Med.*, 47(1): 30–32 (1995).

12. Connor, D. H., R. C. Neafie, and W. M. Meyers, Amebiasis. In: *Pathology of Tropical and Extraordinary Diseases, Vol. 1* (C. H. Binford and D. H. Connor, eds.), Armed Forces Institute of Pathology, Washington, DC, 1976, pp. 308–316.

13. Walzer, P. D., and G. R. Healy, Balantidiasis. In *CRC Handbook Series in Zoonoses, Section C, Parasitic Zoonoses*, Vol. 1 (L. Jacobs and P. Arambulo, eds.), CRC Press, Boca Raton, FL, 1982, pp. 15–24.

14. Neafie, R. C., Balantidiasis. In: *Pathology of Tropical and Extraordinary Diseases, Vol. 1* (C. H. Binford and D. H. Connor, eds.), Armed Forces Institute of Pathology, Washington, DC, 1976, pp. 325–327.

15. Bidinger, P. D., D. W. Crompton, and S. Arnold, Aspects of intestinal parasitism in villagers from rural peninsular India, *Parasitology*, 83(Pt2):373–380 (1981).

16. Basset, D., H. Gaumerais, and A. Basset-Pougnet, Intestinal parasitoses in children of an Indian community of Bolivian altiplano, *Bull. Soc. Pathol. Exot.* Filiales, 79(2): 237–246 (1986).

17. Yang, Y., L. Zeng, M. Li, and J. Zhou, Diarrhoea in piglets and monkeys experimentally infected with *Balantidium coli* isolated from human faeces, *J. Trop. Med. Hyg.*, 98(1): 69–72 (1995).

18. Dodd, L. G., *Balantidium coli* infestation as a cause of acute appendicitis, *J. Infect. Dis.*, 163(6): 1392 (1991).

19. Currie, A. R., Human balantidiasis. A case report, *South Afr. J. Surg.*, 28(1): 23–25 (1990).

20. Garcia-Laverde, A., and L. de Bonilla, Clinical trials with metronidazole in human balantidiasis, *Am. J. Trop. Med. Hyg.*, 24(5): 781–783 (1975).

21. Meyer, E. A. Introduction. In: *Giardiasis* (E. A. Meyer, ed.), Elsevier, New York, 1990, pp. 1–10.

22. Feely, D. E., D. V. Holbertson, and S. L. Erlandsen, *The biology of* Giardia. In: *Giardiasis* (E. A. Meyer, ed.), Elsevier, New York, 1990, pp. 11–50.

23. Levine, N. D., *Protozoan Parasites of Domestic Animals* and *of Man*, Burgess, Minneapolis, MN, 1961, pp. 119–120.

24. Craun, G. F., Waterborne giardiasis. In: *Giardiasis* (E. A. Meyer, ed.), Elsevier, New York, 1990, pp. 267–293.

25. Isaac-Renton, J. L., and J. J. Philion, Factors associated with acquiring giardiasis in British Columbia residents, *Can. J. Publ. Health*, 83(2): 155–158 (1992).

26. Lengerich, E. J., D. G. Addiss, and D. D. Juranek, Severe giardiasis in the United States, *Clin. Infect. Dis.*, 18(5): 760–763 (1994).

27. Kirkpatrick, C. E., and J. P. Farrell, Giardiasis, *Comp. Contin. Ed. Pract. Vet.*, 4:367–377 (1982).

28. Healy, G. R., Giardiasis in perspective: The evidence of animals as a source of human *Giardia* infections. In: *Giardiasis*, Vol. 3 (E. A. Meyer, ed.), Elsevier, New York, 1990, pp. 305–313.

29. Wright, R. A., H. C. Spencer, R. E. Brodsky, and T. M. Vernon, Giardiasis in Colorado: an epidemiologic study, *Am. J. Epidemiol.*, 105(4): 330–336 (1977).

30. Islam, A., Giardiasis in developing countries. In: *Giardiasis*, Vol. 3 (E. A. Meyer, ed.), Elsevier, New York, 1990, pp. 235–266.

31. Barnard, R. J., and G. J. Jackson, *Giardia lambia*, the transfer of human infections by food.

In: Giardia *and Giardiasis: Biology, Pathogenesis and Epidemiology* (S. L. Erlandsen and E. A. Meyer, eds.), Plenum Press, New York, 1984, pp. 365–378.

32. Osterholm, M. T., J. C. Forgang, T. L. Ristinen, A. G. Dean, J. W. Washburn, J. R. Godes, R. A. Rude, and J. C. McCullough, An outbreak of foodborne giardiasis, *N. Engl. J. Med.*, 304:24–28 (1981).

33. Quick, R., K. Paugh, D. Addiss, J. Kobayashi, and R. Baron, Restaurant-associated outbreak of giardiasis, *J. Infect. Dis.*, 166(3): 673–676 (1992).

34. Mintz, E. D., M. Hudson-Wragg, P. Mshar, M. L. Carter, and J. L. Hadler, Foodborne giardiasis in a corporate office setting, *J. Infect. Dis.* 167(1): 250–253 (1993).

35. Porter, J. D., C. Gaffney, D. Heymann, and W. Parkin, Food-borne outbreak of *Giardia lamblia, Am. J. Publ. Health* 80(10): 1259–1260 (1990).

36. Herwaldt, B. L., G. F. Craun, S. L. Stokes, and D. D. Juranek, Waterborne-disease outbreaks, 1989–1990, *MMWR CDC Surveill. Summ.*, 40(3): 1–21 (1991).

37. Moorehead, W. P., R. Guasparini, C. A. Donovan, R. G. Mathias, R. Cottle, and G. Baytalan, Giardiasis outbreak from a chlorinated community water supply, *Can. J. Publ. Health* 81(5): 358–362 (1990).

38. Isaac-Renton, J. L., L. F. Lewis, C. S. Ong, and M. F. Nulsen, A second outbreak of waterborne giardiasis in Canada and serological investigation of patients, *Trans. R. Soc. Trop. Med. Hyg.*, 88(4): 395–399 (1994).

39. Lopez, C. E., A. C. Dykes, D. D. Juranek, S. P. Sinclair, J. M. Conn, R. W. Christie, E. C. Lippy, M. G. Schultz, and M. H. Mires, Waterborne giardiasis: a communitywide outbreak of disease and a high rate of asymptomatic infection, *Am. J. Epidemiol.*, 112(4): 495–507 (1980).

40. Meyer, E. A., and E. L. Jarroll, Giardiasis. In: *CRC Handbook Series in Zoonoses, Section C.- Parasitic Zoonoses*, Vol. 1 (L. Jacobs and P. Arambulo, eds.), CRC Press, Boca Raton, FL, 1982, pp. 25–40.

41. Wolfe, M. S., Clinical symptoms and diagnosis by traditional methods. In: *Giardiasis* (E. A. Meyer, ed.), Elsevier, New York, 1990, pp. 175–185.

42. Engelkirk, P. G., and L. K. Pickering, Detection of *Giardia* by immunologic methods. In: *Giardiasis* (E. A. Meyer, ed.), Elsevier, New York, 1990, pp. 187–198.

43. Baruch, A. C., J. Isaac-Renton, and R. D. Adam, The molecular epidemiology of *Giardia lamblia*: a sequence based approach, *J. Infect. Dis.*, 174(1): 233–236 (1996).

44. Davidson, R. A., Treatment of giardiasis: the North American perspective. In: *Giardiasis* (E. A. Meyer, ed.), Elsevier, New York, 1990, pp. 325–334.

45. Jakubowski, W., The control of *Giardia* in water supplies. In: *Giardiasis* (E. A. Meyer, ed.), Elsevier, New York, 1990, pp. 335–353.

46. Ortega, Y. R., C. R. Sterling, R. H. Gilman, V. A. Cama, and F. Diaz, *Cyclospora* species—a new protozoan pathogen of human, *N. Engl. J. Med.*, 328:1308–1312 (1993).

47. Ortega, Y. R., R. H. Gilman, and C. R. Sterling, A new parasite (Apicomplexa: Eimeriidae) from humans, *J. Parasitol.*, 80(4): 625–629 (1994).

48. Nhieu, J. T., F. Nin, J. Fleury-Feith, M. T. Chaumette, A. Schaeffer, and S. Bretagne, Identification of intracellular stages of *Cyclospora* species by light microscopy of thick sections using hematoxylin, *Human Pathol.*, 27(10): 1107–1109 (1996).

49. Sun, T., C. F. Ilardi, D. Asnis, A. R. Bresciani, S. Goldberg, B. Roberts, and S. Teichberg, Light and electron microscopic identification of *Cyclospora* species in the small intestine; evidence of the presence of asexual life cycle in human host, *Am. J. Clin. Pathol.*, 105(2): 216–220 (1996).

50. Hoge, C. W., D. R. Shlim, and P. Escheverria, Cyanobacterium-like *Cyclospora* species, *N. Engl. J. Med.*, 329(20): 1505–1506 (1993).

51. Huang, P., J. T. Weber, D. M. Sosin, P. M. Griffin, E. G. Long, J. J. Murphy, F. Kocka, C. Peters, and C. Kallick, The first reported outbreak of diarrheal illness associated with *Cyclospora* in the United States, *Ann. Intern. Med.*, 123(6): 409–414 (1995).

52. Ortega, Y. R., C. R. Roxas, R. H. Gilman, N. J. Miller, L. Cabrera, C. Taquiri, and C. R. Sterling, Isolation of *Cryptosporidium parvum* and *Cyclospora cayetanensis* from vegetables collected in markets in an endemic region in Peru, *Am. J. Trop. Med. Hyg.*, 57(6): 683–686 (1997).

53. Herwaldt, B. L., and M. L. Ackers, An outbreak in 1996 of cyclosporiasis associated with imported raspberries, *N. Engl. J. Med.*, 336:1548–1556 (1997).

54. CDC, Update: Outbreaks of cyclosporiasis-United States and Canada, 1997, *Morbidity Mortality Weekly Report*, 46(23): 521–523 (1997).

55. CDC, Outbreak of cyclosporiasis-Northern Virginia-Washington, D.C.-Baltimore, Maryland, metropolitan area, 1997, *Morbidity Mortality Weekly Report*, 46(30): 689–691 (1997).

56. Ooi, W. W., S. K. Zimmerman, and C. A. Needham, *Cyclospora* species as a gastrointestinal pathogen in immunocompetent hosts, *J. Clin. Microbiol.*, 33(5): 1267–1269 (1995).

57. Garcia-Lopez, H. L., L. E. Rodriguez-Tovar, and C. E. Medina-De la Garza, Identification of *Cyclospora* in poultry, *Emerg. Infect. Dis.*, 2(4): 356–357 (1996).

58. Yai, L. E., A. R. Bauab, M. P. Hirschfeld, M. L. deOliveira, and J. T. Damaceno, The first two cases of *Cyclospora* in dogs, Sao Paulo, Brazil, *Rev. Inst. Med. Trop. Sao Paulo*, 39(3): 177–179 (1997).

59. Smith, H. V., C. A. Paton, R. W. Girdwood, and M. M. Mtambo, *Cyclospora* in non-human primates in Gombe, Tanzania, *Vet. Rec.*, 138(21): 528 (1996).

60. Dytrych, J. K., and R. P. Cooke, Autofluorescence of *Cyclospora, Br. J. Biomed. Sci.*, 52(1): 76 (1995).

61. Visvesvara, G. S., H. Moura, E. Kovacs-Nace, S. Wallace, and M. L. Eberhard, Uniform staining of *Cyclospora* oocysts in fecal smears by a modified safranin technique with microwave heating. *J. Clin. Microbiol.*, 35(3): 730–733 (1997).

62. Hoge, C. W., D. R. Shlim, M. Ghimire, J. G. Rabold, P. Pandey, A. Welch, R. Rajah, P. Gaudio, and P. Escheverria, Placebo-controlled trial of co-tri moxazole for cyclospora infections among travellers and foreign residents in Nepal, *Lancet*, 345:691–693 (1995).

63. Madico, G., R. H. Gilman, E. Miranda, L. Cabrera, and C. R. Sterling, Treatment of Cyclospora infections with co-trimoxazole, *Lancet*, 342:122–123 (1993).

64. Fayer, R., C. A. Speer, and J. P. Dubey, General biology of *Cryptosporidium. In: Cryptosporidiosis of Man and Animals* (J. P. Dubey, C. A. Speer, and R. Fayer, eds.), CRC Press, Boca Raton, FL, 1990, pp. 1–29.

65. Peng, M. M., L. Xiao, A. R. Freeman, M. J. Arrowood, A. A. Escalante, A. C. Weltman, C. S. L. Ong, W. R. MacKenzie, A. A. Lal, and C. B. Beard, Genetic polymorphism among Cryptosporidium parvum isolates: evidence of two distinct human transmission cycles, *Emerg. Infect. Dis.*, 3(4): 567–573 (1997).

66. Fayer, R., C. A. Speer, and J. P. Dubey, The general biology of *Cryptosporidium. In: Cryptosporidium and Cryptosporidiosis* (R. Fayer, ed.), CRC Press, Boca Raton, FL, 1997, pp. 1–42.

67. Ungar, B. L. P., Cryptosporidiosis in humans *(Homo sapiens). In: Cryptosoporidiosis of Man and Animals* (J. P. Dubey, C. A. Speer, and R. Fayer, eds.), CRC Press, Boca Raton, FL, 1990, pp. 59–82.

68. Fayer, R., T. Nerad, W. Rall, D. S. Lindsay, and B. L. Blagburn, Studies on cryopreservation of *Cryptosporidium parvum, J. Parasitol.*, 77:357–361 (1991).

69. Fayer, R., J. M. Trout, and M. C. Jenkins, Infectivity of *Cryptosporidium parvum* oocysts stored in water at environmental temperatures, *J. Parasitol.*, 84:1165–1169 (1998).

70. Harp, J. A., R. Fayer, B. A. Pesch, and G. J. Jackson, Effect of pasteurization in infectivity of *Cryptosporidium parvum* oocysts in water and milk, *Appl. Environ. Microbiol.*, 62:2866–2868 (1996).

71. Perryman, L. E., Cryptosporidiosis in rodents. In: *Cryptosporidiosis of Man and Animals* (J. P. Dubey, C. A. Speer, and R. Fayer, eds.), CRC Press, Boca Raton, FL, 1990, pp. 125–132.

72. Riggs, M. W., Cryptosporidiosis in cats, dogs, ferrets, raccoons, opossum, rabbits and non-human primates. In: *Cryptosporidiosis of Man and Animals* (J. P. Dubey, C. A. Speer, and R. Fayer, eds.), CRC Press, Boca Raton, FL, 1990, pp. 113–124.

73. Kim, C. W., Cryptosporidiosis in pigs and horses. In: *Cryptosporidiosis of Man and Animals* (J. P. Dubey, C. A. Speer, and R. Fayer, eds.), CRC Press, Boca Raton, FL, pp. 105–112.

74. Angus, K. W., Cryptosporidiosis in ruminants. In: *Cryptosporidiosis of Man and Animals* (J. P. Dubey, C. A. Speer, and R. Fayer, eds.), CRC Press, Boca Raton, FL, 1990, pp. 83–104.

75. O'Donoghue, P. Cryptosporidium and cryptosporidiosis of man and animals, *Int. J. Parasitol.*, 25(2): 139–195 (1995).

76. Shield, J., J. H. Baumer, J. A. Dawson, and P. J. Wilkinson, Cryptosporidiosis: an educational experience, *J. Infect.*, 21:297–301 (1990).

77. Chalmers, R. M., A. P. Sturdee, P. Mellors, V. Nicholson, F. Lawlor, F. Kenny, and P. Timpson, *Cryptosporidium parvum* in environmental samples in the Sligo area, Republic of Ireland: a preliminary report, *Lett. Appl. Microbiol.*, 25:380–384 (1997).

78. Fayer, R., T. K. Graczyk, E. J. Lewis, J. M. Trout, and C. A. Farley, Survival of infectious *Cryptosporidium parvum* oocysts in seawater and eastern oysters (*Crassostrea virginica*) in the Chesapeake Bay, *Appl. Environ. Microbiol.*, 64(3): 1070–1074 (1998).

79. Sterling, C. R., Waterborne cryptosporidiosis. In: *Cryptosporidiosis of Man and Animals* (J. P. Dubey, C. A. Speer, and R. Fayer, eds.), CRC Press, Boca Raton, FL, 1990, pp. 51–58.

80. *Cryptosporidium* in water supplies. *Report of the Group of Experts, Her* Majesty's Stationery Office, London, 1990, p. 230.

81. Rose, J. B., J. T. Lisle, and M. LeChevallier, Waterborne cryptosporidiosis: incidence, outbreaks, and treatment strategies. In: *Cryptosporidium and Cryptosporidiosis* (R. Fayer, ed.), CRC Press, Boca Raton, FL, 1997, pp. 93–110.

82. Hansen, J. S., and J. E. Ongerth, Effects of time and watershed characteristics on the concentration of *Cryptosporidium* oocysts in river water, *Appl. Environ. Microbiol.,* 57:2790–2795 (1991).

83. MacKenzie, W. R., W. L. Schell, KJ. A. Blair, D. G. Addiss, D. E. Peterson, N. J. Hoxie, J. J. Kazmierczak, and J. P. Davis, Massive outbreak of waterborne cryptosporidium infection in Milwaukee, Wisconsin: recurrence of illness and risk of secondary transmission, *Clin. Infect. Dis.*, 21(1): 57–62 (1994).

84. Hoxie, N. J., J. P. Davis, J. M. Vergeron, R. D. Nashold, and K. A. Blair, Cryptosporidiosis associated mortality following a massive waterborne outbreak in Milwaukee, Wisconsin, *Am. J. Publ. Health*, 87(12): 2032–2035 (1997).

85. Dupont, H., C. Chappell, C. Sterling, P. Okhuysen, J. Rose, and W. Jakubowski, The infectivity of *Cryptosporidium parvum* in healthy volunteers, *N. Engl. J. Med.*, 332:855–859 (1995).

86. Current, W. L., Techniques and laboratory maintenance of *Cryptosporidium*. In: *Cryptosporidiosis of Man and Animals* (J. P. Dubey, C. A. Speer, and R. Fayer, eds.), CRC Press, Boca Raton, FL, 1990, pp. 31–50.

87. Morgan, U. M., and R. C. A. Thompson, PCR detection of *Cryptoporidium*: the way forward? *Parasitol. Today*, 14(6): 241–245 (1998).

88. Blagburn, B. L., C. A. Sundermann, D. S. Lindsay, J. E. Hall, and R. R. Tidwell, Inhibition of *Cryptosporidium parvum* in neonatal Hsd:(ICR)BR Swiss mice by polyether ionophores and aromatic amides, *Antimicrob. Ag. Chemother.*, 35:1520–1523 (1991).

89. Rehg, J. E., Activity of azithromycin against cryptosporidiosis in immunosuppressed rats, *J. Infect. Dis.*, 143:1293–1296 (1991).

90. Blagburn, B. L. and R. Soave, Prophylaxis and chemotherapy: human and animal. In: *Cryptosporidium and Cryptosporidiosis* (R. Fayer, ed.), CRC Press, Boca Raton, FL, 1997, pp. 111–128.

91. Fayer, R., T. K. Graczyk, M. R. Cranfield, and J. M. Trout, Gaseous disinfection of *Cryptosporidium parvum* oocysts, *Appl. Environ. Microbiol.*, 62(10): 3908–3909 (1996).

92. Ives, K., *Cryptosporidium and water supplies: Treatment processes and oocyst removal. In: Cryptosporidium in Water Supplies. Report of the Group of Experts*. Her Majesty's Stationery Office, London, 1990, pp. 154–184.

93. Dubey, J. P., C. A. Speer, and R. Fayer, *Sarcocystosis of Animals and Man*, CRC Press, Boca Raton, FL, 1989, pp. 215.

94. Bunyaratvej, S., P. Bunyawongwiroj, and P. Nitiyanant, Human intestinal sarcosporidiosis: report of six cases, *Am. J. Trop. Med. Hyg.*, 31:36–41 (1982).

95. Aryettey, M. E., and G. Piekarski, Serologische *Sarcocystis-studien* an Menschen und Ratten, *Z. Parasitenk.*, 50:109–124 (1976).

96. Rommel, M., and A. O. Heydorn, Bietrage zum Lebenszyklus der Sarkosporidien. III. *Isospora hominis* (Railliet und Lucet, 1891) Wenyon, 1923, eine Dayerform der Sarkosporidien des Rindes und des Schweins, *Berl. Munch. Tierarztl. Wochen.*, 85:143–145 (1972).

97. Heydorn, A. O., Sarkosporidieninfiziertes Fleisch als mogliche Krankheitsurache fur den Menschen, *Arch. Lebensmittelhygiene*, 28:27–31 (1977).

98. Fayer, R., and A. J. Johnson, Effect of amprolium on acute sarcocystosis in experimentally infected calves, *J. Parasitol.*, 61:932–936 (1975).

99. Leek, R. G., and R. Fayer, Experimental *Sarcocystis ovicanis* infection in lambs: Salinomycin chemoprophylaxis and protective immunity, *J. Parasitol.*, SS9:271–276 (1983).

100. Dubey, J. P., and C. P. Beattie, *Toxoplasmosis of Animals and Man, CRC* Press, Boca Raton, FL, 1988, p. 220.

101. Fayer, R., Toxoplasmosis and public health implications, *Can. Vet. J.*, 22:344–352 (1981).

102. van Knapen, F., J. H. Franchimont, and G. van der Lust, Prevalence of antibodies to *Toxoplasma* in farm animals in the Netherlands and its implication for meat inspection, *Vet. Q.*, 4:101–105 (1982).

103. van Knapen, F., A. F. Kremers, J. H. Franchimont, and U. Narucka, Prevalence of antibodies to *Toxoplasma gondii* in cattle and swine in the Netherlands: towards an integrated control of livestock production, *Vet. Q.*, 17(3): 87–91 (1995).

104. Dubey, J. P., J. C. Leighty, V. C. Beal, W. R. Anderson, C. D. Andrews, and Ph. Thulliez, National seroprevalence of *Toxoplasma gondii* in swine, *J. Parasitol.*, 77:517–521 (1991).

105. Dubey, J. P., J. K. Lunney, S. K. Shen, O. K. Kwok, D. A. Ashford, and P. Thulliez, Infectivity of low numbers of *Toxoplasma gondii* oocysts to pigs, *J. Parasitol.* 82(3): 438–443 (1996).

106. Dubey, J. P., R. M. Weigel, A. M. Siegel, P. Thulliez, U. D. Kitron, M. A. Mitchell, A. Mannelli, N. E. Mateus-Pinilla, S. K. Shen, O. K. Kwok, Sources and reservoirs of *Toxoplasma gondii* infection on 47 swine farms in Illinois, *J. Parasitol.*, 81(5): 723–729 (1995).

107. Tizard, I. R., N. A. Fish, and J. P. Quinn, Some observations on the epidemiology of toxoplasmosis in Canada, *J. Hyg. Camb.*, 77:11–21 (1976).

108. Rawal, B. D., Toxoplasmosis. A dye-test survey of sera from vegetarians and meat eaters in Bombay, *Trans. R. Soc. Trop. Med. Hyg.*, 53:61–63 (1959).

109. Jacobs, L., J. S. Remington, and M. L. Melton, A survey of meat samples from swine, cattle, and sheep for the presence of encysted *Toxoplasma, J. Parasitol.*, 46:23–28 (1960).

110. Ruiz, A., and J. K. Frenkel, Isolation of *Toxoplasma* from cat feces deposited in false attics of homes in Costa Rica, *J. Parasitol.*, 63:931–932 (1977).

111. French, J.-G., H. B. Messenger, and J. McCarthy, A study of *Toxoplasma gondii* infection in farm and non-farm groups in the same geographic location, *Am. J. Epidemiol.*, 21:185–191 (1970).

112. Kean, B. H., A. C. Kimball, and W. N. Christenson, An epidemic of acute toxoplasmosis, *JAMA* 208:1002–1004 (1969).

113. Magaldi, C., H. Elkis, D. Pattoli, and A. L. Coscina, Epidemic of toxoplasmosis at a univer-

sity in Sao-Jose-dos-Campos, S. P. Brazil. I. Clinical and serologic data, *Rev. Latinoam. Microbiol. Parasitol.*, 11:5–13 (1969).

114. Melnik, M. N., N. A. Iova, L. N. Pukhteeva, and I. D. Netrebko, Raw meat as a factor in the transmission of toxoplasmosis, *Problemy Parazit. Materialy VIII Nauch. Konfer. Parazit.*, 22–24 (1975).

115. Wallace, G. D., The prevalence of toxoplasmosis on a Pacific Island and the influence of ethnic group, *Am. J. Trop. Med. Hyg.*, 25:48–53 (1976).

116. Desmonts, G., J. Couvreur, F. Alision, J. Baudelot, J. Gerbeaux, and M. Lelong, Etude epidemiologue sur la toxoplasmose: de l'influence de la cuisson des viandes de boucherie sur la frequence de l'infection humaine, *Rev. Fr. Etud., Clin. Biol.*, 10:952–958 (1965).

117. CDC, Toxoplasmosis Pennsylvania, *Morbidity Mortality Weekly Report*, 24:285–286 (1975).

118. Jones, S. R., Toxoplasmosis, a review, *J. Am. Vet. Med. Assoc.*, 163:1038–1042 (1973).

119. Choi, W. Y., H. W. Nam, N. H. Kwak, W. Huh, Y. R. Kim, M. W. Kang, S. Y. Cho, and J. P. Dubey, Foodborne outbreaks of human toxoplasmosis, *J. Infect. Dis.*, 175(5):1280–1282 (1997).

120. Bonametti, A. M., J. do N. Passos, E. M. daSilva, and A. L. Bortoliero, Outbreak of acute toxoplasmosis transmitted thru the ingestion of ovine raw meat, *Rev. Soc. Bras. Med. Trop.*, 30(1):21–25 (1996).

121. Dubey, J. P., M. D. Ruff, M. E. Camargo, S. K. Shen, G. L. Wilkins, O. C. Kwok, and P. Thulliez, Serologic and parasitologic responses of domestic chickens after oral inoculation with *Toxoplasma gondii* oocysts, *Am. J. Vet. Res.*, 54(10): 1668–1672 (1993).

122. Dubey, J. P., M. D. Ruff, O. C. Kwok, S. K. Shen, G. C. Wilkins, and P. Thulliez, Experimental toxoplasmosis in bobwhite quail (*Colinus virginianus*), *J. Parasitol.* 79(6): 935–939 (1993).

123. Ferre, F., J. C. Desport, P. Peze, P. L. Ferrat, H. Bertrand, H. Grandchamp, and P. Weinbeck, Toxoplasmose apres plaie par arme blanche, *Presse Med.*, 15:929–931 (1986).

124. Dubey, J. P., K. D. Murrell, R. Fayer, and G. A. Schad, Distribution of *Toxoplasma gondii* tissue cysts in commercial cuts of pork, *J. Am. Vet. Med. Assoc.*, 188(9): 1035–1037 (1986).

125. Benenson, M. W., E. T. Takafuji, S. M. Lemon, R. L. Greenup, and A> J. Sulzer, Oocyst-transmitted toxoplasmosis associated with ingestion of contaminated water. *N. Engl. J. Med.*, 307(11): 666–669 (1982).

126. Bowie, W. R., A. S. King, D. H. Werker, J. L. Isaac-Renton, A. Bell, S. B. Eng, and S. A. Marion, Outbreak of toxoplasmosis associated with municipal drinking water. The BC Toxoplasma Investigation Team, *Lancet*, 350(9072): 173–177 (1997).

127. Roberts, T., and K. D. Murrell, Economic losses caused by foodborne parasitic diseases, Int. Symp. Cost–Benefit Aspects of Food Irradiation Processing, Aix-en-PRovence, France, March 1–5, 1993, IAEA/WHO/FAO.

128. Dubey, J. P., A. W. Kotula, A. Sharar, C. D. Andrews, and D. S. Lindsay, Effect of high temperature on infectivity of *Toxoplasma gondii* tissue cysts in pork, *J. Parasitol.*, 76:201–204 (1990).

129. Dubey, J. P., R. J. Brake, K. D. Murrell, and R. Fayer, Effect of irradiation on viability of *Toxoplasma gondii* cysts in tissues of mice and pigs, *Am. J. Vet. Res.*, 47:518–522 (1986).

14

Medical Management

Paul Prociv
The University of Queensland, Brisbane, Queensland, Australia

I. INTRODUCTION

Apart from those infections transmitted by arthropod vectors, virtually all human parasitoses may be acquired orally. Even organisms that normally penetrate skin, such as schistosome cercariae and hookworm larvae, can establish infection after being ingested in drinking water or on contaminated food, although for them this portal hardly constitutes a significant route of transmission. Other fecally dispersed parasites, including the cysts of *Giardia* and *Entamoeba*, oocysts of *Cryptosporidium*, *Cyclospora* and other less common coccidia, and the eggs of geohelminths, may be transported mechanically on food or in drinking water, but they do not occur within the substance of unprocessed food.

In this chapter, only parasites that develop in food as part of their normal life cycle (i.e., those with infective stages that develop in or on the tissues of edible intermediate hosts) are considered. Examples of foodborne transmission, such as the ingestion of hypobiotic *Ancylostoma* larvae in meat, that constitute only a speculative or remote possibility are not discussed. Also excluded is *Dracunculus medinensis*, the Guinea worm, which is acquired by the accidental ingestion of infected cyclopoid copepods in drinking water.

The unequivocally foodborne parasites can be defined as those that utilize food animals (including insects, which are eaten either intentionally or accidentally) as intermediate or paratenic hosts. In addition, there are a few trematode species that utilize edible plants, even though their cercariae do not penetrate aquatic vegetation but simply encyst on its surface.

This means that all human foodborne parasitic infections are zoonoses in the wider sense that they have significant reservoirs of infection in other animal species. In most cases, humans function as paratenic, intermediate, or inadequate final hosts. As a result, the infections usually do not become patent and so cannot be diagnosed by the specific microscopic detection of parasitic stages in feces, urine, or blood. The converse of this is seen with spurious parasitism, in which individuals pass helminth eggs or protozoan cysts in their feces after having ingested animal tissues harboring the parasites; in such cases, it often can be very difficult to determine whether the infection is real or spurious (1). Diagnosis then depends on identifying the organisms in tissue samples or, less definitively, detecting reactive antibodies in the serum and other body fluids (and perhaps, eventually, parasite genetic material in these specimens).

Immunodiagnosis is covered in detail in Chapters 14–16 of this volume. For many of these infections, specific and sensitive serology either is not widely available or has not been tested adequately in appropriate epidemiological settings. The specificity of very few serological assays used in the diagnosis of helminthic infections has been adequately assessed by testing against a comprehensive range of related and unrelated parasitic and other infections.

Associated difficulties, common to all serology, include antibodies persisting after the infective organisms have perished or have been eradicated, and the incidental presence of antibodies to such ubiquitous parasites as *Toxoplasma*, which might not have been responsible for the presenting clinical problem in a particular case. Sporadic positive serological findings in the absence of supportive history and clinical manifestations often are meaningless and clinically unhelpful (they also confuse the results of seroepidemiological surveys). These uncertainties are compounded further by the possible existence of other latent human infections that at present cannot be diagnosed readily by other means. For these reasons, positive serology alone may not be sufficient evidence of an active, ongoing infection and rarely justifies therapeutic intervention.

Recent efforts to circumvent the drawbacks of serology by using monoclonal antibody and DNA probes have shown increasing promise in detecting traces of specific parasite material in blood, feces, sputum, urine, cerebrospinal fluid (CSF), and tissue samples from patients with a variety of infections. These analyses are technically sophisticated and not yet widely available; most are still in the early stages of experimental development and clinical evaluation. The expense of the reagents, and the necessary hardware, effectively will limit their availability to laboratories in developed countries and the range of infections sought to those that are more commonly encountered.

In the clinical setting, diagnosis requires the initial recognition of a suggestive presentation, which is reinforced by a history of exposure to the parasite and then established by specific, often specialized, laboratory investigations. Success rests entirely on the degree of suspicion among attending physicians, which depends on their awareness of the possibilities.

While a confident diagnosis is an essential prerequisite to specific therapy and important for patient reassurance, it does not guarantee that treatment will lead to the effective termination of infection and subsidence of disease. For example, available treatment for some helminthiases may be useless or, worse, frankly dangerous. Parasites living within tissues may elicit minimal local reactions, perhaps as a result of mechanisms that avoid or suppress immune and inflammatory responses; when the invaders perish, severe inflammation could ensue that precipitates clinical disease. Paradoxically, the continued survival of such parasites may be in the patient's interests; it might be wiser to adopt a policy of vigilant inactivity, leaving the disease to run its natural course, in the hope of an eventual satisfactory, spontaneous resolution.

Manifestations of some infections can be particularly disturbing to the patient. The physician must provide reassurance by explaining clearly what is happening and what is to be expected from treatment. It often can be just as important to explain how recurrence of infection should be avoided.

In this chapter, the clinical approach to specific parasitic infections is considered in groups according to their normal modes of acquisition. Some organisms can utilize a variety of routes; for instance, sparganosis may result from ingesting plerocercoids in undercooked meat, poultry, or fish, from drinking water that harbors infected copepods, or from topically applying raw animal tissues. These conditions are discussed under the presumed most frequent source of infection. As medical management includes first establishing a diagnosis, salient points in the history and clinical presentation are outlined that may help in this regard. Prevention is considered only as it applies to the individual; controlling infections is a public health challenge beyond the immediate clinical sphere and is covered in detail in other chapters of this volume.

II. MEATBORNE INFECTIONS

A. Toxoplasmosis

1. History of Exposure

Infection with the ubiquitous protozoan *Toxoplasma gondii* may be acquired by diverse routes, including ingesting oocysts with soil-contaminated food or water, eating tissue cysts in raw or undercooked meat, parenteral inoculation through blood transfusions and implanted organs, and transplacental infection. Once infected, most people continue to harbor latent tissue cysts and remain seropositive for toxoplasmosis for life. In societies in which raw meat is eaten traditionally, seroconversion occurs earlier in life and most adults are seropositive (2). Therefore, the precise origin of infection for any particular patient rarely is identifiable, and attempts to pinpoint this source usually are speculative and unhelpful clinically, often generating unjustified feelings of guilt and regret.

2. Clinical Manifestations

Acute infection is not diagnosed often in otherwise healthy individuals. When symptoms are recognized, they are nonspecific, resembling either an acute, viral, upper respiratory tract infection with lymphadenopathy, or the early stages of infectious mononucleosis (3). Very rarely, pneumonia, myocarditis, or encephalitis may develop during acute infection in apparently immunocompetent individuals. Chronic infections are asymptomatic but account for most cases of positive serology because a large proportion of the adult population in most parts of the world is infected.

Recrudescent or newly acquired acute infection in an immunosuppressed host is potentially disastrous. Illness may progress rapidly or evolve slowly, but eventually manifests as a combination of pulmonary, cardiac, and neurological abnormalities.

In congenitally infected neonates, the severity and extent of disease bear no relation to maternal symptoms; acute infections acquired during pregnancy usually are subclinical and detected only by maternal seroconversion. The most common manifestation of congenital toxoplasmosis is retinopathy, which usually is bilateral, may be pathognomonic, and presents as visual defects at any time in the first 20 years of life. The wide range of associated early complications includes spontaneous miscarriage, stillbirth, hydrocephalus, microcephaly, mental retardation, and epilepsy, but none of these is specific to toxoplasmosis.

3. Laboratory Diagnosis

The diagnosis of toxoplasmosis in the vast majority of cases rests entirely on serology (see Chapter 16, this volume), so it is imperative to distinguish chronic infection, which is asymptomatic and untreatable, from acute and congenital infections, which may be associated with significant disease. In acute or recrudescent immunosuppressed cases (i.e., those associated with parasitemia), the diagnosis is confirmed by finding tachyzoites in blood buffy coat smears, or in biopsy specimens of lymph nodes and other tissues. Should histology be negative, infection may be confirmed by inoculating blood or biopsy material into laboratory mice, which then are monitored for seroconversion or killed to identify the tachyzoites in peritoneal effusions. Cell cultures may give a more rapid result but are less sensitive than mouse inoculation (3). This is important particularly in acquired immunodeficiency syndrome (AIDS) patients with encephalitis, in whom serology may be unreliable and tachyzoites very difficult to identify in brain biopsy material.

In immunocompetent individuals, high levels of specific immunoglobulin M (IgM) and rapidly changing titers of immunoglobulin G (IgG) are consistent with recent infection. In chronic infections, anti-*Toxoplasma* IgG levels are raised but do not change rapidly with time, and specific IgM rarely is detectable. This means that two different types of tests must be done, and usually repeated after an interval of 2–3 weeks, to detect changing antibody titers plus specific IgG and IgM responses. Clinicians must be familiar with their local laboratory's serological repertoire and cutoff levels, as these vary widely among diagnostic centers. Detecting specific *Toxoplasma* antigens or DNA in clinical specimens may facilitate diagnosis, but reliable tests are not yet widely available.

Congenital infection can be detected only by finding organisms or specific antibody responses in the neonate. Not all pregnant women who seroconvert (i.e., develop acute infection) transmit the organism to their fetuses, and not all infected fetuses manifest clinical signs of disease.

In ocular toxoplasmosis, establishing the diagnosis in the absence of typical retinal lesions is fraught with difficulty (3).

4. Treatment

An important principle underlying treatment is that, at present, a drug that can safely eradicate the organism from the host does not exist. All of the available agents, including sulfadiazine, pyrimethamine, and spiramycin, are coccidiostatic, not coccidiocidal—they inhibit multiplication but, in the absence of effective immunity, once tissue drug levels fall, the tachyzoites may be able to proliferate again. By suppressing tachyzoite multiplication (and hence, parasitemia), treatment is effective only in the acute (or recrudescent) stage of infection, and a competent immune response is then required to keep the parasites suppressed, in their quiescent bradyzoite/tissue cyst phase. Furthermore, these agents all have potentially serious problems with long-term toxicity, compounding the difficulties in establishing effective therapeutic drug regimens.

Chemotherapy is indicated only in acutely infected (parasitemic) individuals who have severe disease or are immunocompromised, in pregnant women with convincing evidence of acute infection, in laboratory- and transfusion-acquired infections, and in some cases of ocular disease. In cases of chronic infection, or acute infection with only mild symptoms, it is futile, in view of the refractoriness of bradyzoites to drugs, and potentially dangerous, because of drug toxicity, to administer chemotherapy.

The detailed guidelines for drug therapy that were outlined succinctly a decade ago (3,4) are still valid today. The combination of sulfadiazine and pyrimethamine, with folinic acid supplementation to minimize toxicity, is the mainstay of therapy. While not eradicating infection, it may suppress parasite numbers and thus severe symptoms while an effective host immune response develops. The optimal loading and maintenance doses of each agent and the frequency of administration and duration of treatment have not been defined and vary among authorities (3,4). A variety of newer drugs are currently being tested, but their advantage over present conventional therapy remains to be established by careful clinical trials.

Treatment for 2–4 weeks probably is adequate for immunocompetent patients, at a daily dose of pyrimethamine 1 mg/kg body weight (plus folinic acid 5 mg twice weekly) and sulfadiazine 50–100 mg/kg (divided into 2 daily doses). In the immunocompromised, treatment perhaps should be continued until 4–6 weeks after all evidence of active disease has resolved. Blood white cell and platelet counts should be monitored twice weekly during chemotherapy. AIDS patients with toxoplasmic encephalitis can be expected to show clinical improvement within a week of commencing treatment (3), but unacceptable drug toxicity in most cases precludes extended courses; the infections then usually relapse and chemotherapy needs to be repeated indefinitely. In some of these patients, atovaquone, given at 750 mg 4 times daily, can provide a well-tolerated alternative maintenance therapy (5). Alternative drugs that have been used, with some success, include trimethoprim (instead of pyrimethamine), clindamycin (instead of sulfadiazine), the macrolide antibiotics 5-fluorouracil, minocycline or doxycycline, dapsone, rifabutin, pentamidine, and diclazuril (6), although none is significantly more effective than the standard therapy. Others that show promise from animal experiments include artemisinine and trovafloxacin. Immunocompromised patients might benefit from immunotherapy, including the administration of interleukins, interferon-γ, and tumor necrosis factor–α (albeit at great expense).

In pregnant women who seroconvert, treatment with spiramycin 50 mg/kg twice daily throughout pregnancy, or sulfadiazine plus pyrimethamine/folinic acid after the first trimester, may significantly reduce the risk of fetal infection. Pyrimethamine, because of its potential teratogenicity, is contraindicated in the first trimester. Spiramycin does not cross the placenta effectively; used alone, it may not prevent congenital infection (4). Experimental studies in macaques (7) suggest that pyrimethamine and sulfadiazine, which pass the blood–brain barrier and can accumulate in the CNS, are superior to spiramycin in minimizing the effects of congenital infection, but it is difficult to extrapolate these findings to humans. Adjunct corticosteroid therapy has been advocated.

Early treatment of potentially infected neonates who do not have signs of disease at birth still might protect them from developing sequelae, while in clinically affected neonates treatment may reduce the severity of complications.

In active ocular toxoplasmosis, the lesions may begin to improve within a week of commencing sulfadiazine/pyrimethamine, which should be continued for a month. If inflammation persists, the course may need to be repeated, and high-dose systemic corticosteroids (e.g., prednisolone 50–75 mg/day) may be added for short periods if vision is endangered. However, steroids are not always effective and should never be given without specific anti-*Toxoplasma* cover; sometimes the more severely affected eye is enucleated as a last resort to avert sympathetic ophthalmitis.

5. Prophylaxis

Given the widespread opportunities for exposure to toxoplasmosis, it can be difficult to avoid infection. Obviously, for seropositive individuals, including those who present with acute illness, preventive measures are superfluous. In seronegative pregnant women and immunodeficient individuals, a case could be made for active avoidance of infection by adhering strictly to kitchen and garden hygiene and not handling or eating undercooked meat.

In view of the widespread exhortation for expectant mothers to avoid cats and their feces at all costs, medicolegally prudent physicians might consider advising their pregnant patients not to handle cats or kitty litters. However, it must be kept in mind that fresh cat feces are not infective, as *Toxoplasma* oocysts take several days to sporulate. At any time, only about 1% of cats are passing oocysts, and this state lasts usually for an average of 3 weeks. Further, kittens that are housebound, eat dry or canned food, and use sandboxes, without access to mice, birds, and other reservoirs of toxoplasmosis, should not pose a threat to humans by developing intestinal infection. Fastidious disposal of their feces, into toilets or incinerators, may foster unjustified paranoia in such a pet's owner.

Serological screening of pregnant women has been advocated in populations exposed to a high risk of infection and is compulsory by law in France and Austria (4), but is controversial and of questionable cost effectiveness in most other communities (8). It also carries the risk of needless therapeutic abortions, especially if serology is misinterpreted, and some cases of congenital infection still will be overlooked.

B. Sarcocystosis

1. History of Exposure

Several *Sarcocystis* species are known to infect humans. Those that form tissue cysts in our muscles are acquired as sporocysts ingested from the environment and so are not dealt with here. Human intestinal infections with at least two species, *S. hominis* and *S. suihominis*, are derived, respectively, from undercooked beef and pork, and lead to the release of sporocysts in feces (9).

2. Clinical Manifestations

Little is known about incubation periods or symptomatology, but sarcocystosis is likely to resemble acute food poisoning. Diarrhea, if it occurs, probably is short-lived and/or without specific features.

3. Laboratory Diagnosis

Sporocysts are detected in fecal concentrates, often as coincidental findings. Repeated laboratory examination of feces may be required to confirm the diagnosis.

4. Treatment

If the intestinal phase is short-lived, there may be no need for treatment. Antifolate/sulfonamide combinations may be helpful.

5. Prophylaxis

Sarcocystosis can be prevented by not eating undercooked meat.

C. Trichinellosis

1. History of Exposure

The development of clinical disease from infection with *Trichinella spiralis* and closely related nematodes requires exposure to a significant dose of larvae, the pathogenicity of which may vary according to parasite species and strain (covered in detail in Chapter 10 of this volume). While undercooked pork is the usual source, human trichinellosis has followed the ingestion of meat from a wide range of animals, including warthog, bear, walrus, and horse (10,11). Estimating the time of exposure in relation to the onset of illness can be difficult because the latent interval and severity of symptoms depend on the larval dose ingested. Often, several members of a group who have eaten the same animal become ill concurrently, as was described recently for the first documented epidemic of *T. pseudospiralis* infection (12).

2. Clinical Manifestations

Symptomatology and incubation periods vary, probably according to the larval dose (10–14) and the strain of infecting parasite. In heavy infections, symptoms may develop within hours of exposure and mimic acute food poisoning in response to the invasion of the gastrointestinal mucosa by larvae that then rapidly mature into adult worms. Once a new generation of larvae begins to disseminate hematogenously about a week later, systemic symptoms appear, including fever and muscle (including masseter and diaphragmatic) pain. Severe periorbital edema, subconjunctival hemorrhages, and conjunctivitis may suggest ophthalmological disease. A sustained high temperature may resemble typhoid fever. After 3 weeks, when larvae in tissues other than skeletal muscle begin to perish and degenerate, manifestations of pneumonia, myocarditis, and encephalitis ensue, and may continue for several months, or prove fatal. It may take up to 8 months for systemic symptoms to clear. Trichinellosis acquired from raw walrus in the Canadian Arctic may cause prolonged diarrhea, with less prominent muscular symptoms (15).

3. Laboratory Diagnosis

The infection can be very difficult to identify in its acute intestinal phase as there are no specific findings. Worms are very rarely found in feces, blood, or tissues during early invasion and larval dissemination. Blood eosinophilia develops during the second week, peaks about 4 weeks after exposure, and may be dramatic, with over 20,000 eosinophils/ mm^3. Eosinophil levels correlate with the extent of muscle invasion, but not with the severity of clinical disease, and show little response to corticosteroids or antihelminthics (13). From the second week onward, increasing numbers of larvae deposit in skeletal muscle; they may be detected in a biopsy of the pectoralis, deltoid, biceps, or gastrocnemius muscles, but this is not used as a routine procedure (see Chapter 10, this volume). The biopsy sample can be digested in pepsin and the sediment examined microscopically for larvae, or a portion can be fed to laboratory rats, which then are autopsied 4 weeks later. The specific identity of muscle larvae can be confirmed with molecular techniques (12). Associated with the muscle invasion phase may be significantly elevated serum levels of creatine phosphokinase and lactate dehydrogenase (12). Histology is less sensitive in early or light infections. Serology may be helpful but is not very reliable, particularly in the early phase of the disease (14). A recently developed enzyme-linked immunosorbent assay (ELISA) seems to be sensitive, but its specificity may be unreliable (16). Except in the very early stages of infection, detection of circulating antibodies by indirect immunofluorescence or competitive inhibition assay has been shown to be much more sensitive diagnostically than detection of circulating secretory antigens of *Trichinella* (17) (see also Chapter 15, this volume).

4. Treatment

Views about the chemotherapy of trichinellosis differ which inevitably means that no available drug is totally satisfactory (see also Chapter 10, this volume). According to one authority (14), the only specific agent available is thiabendazole, which destroys adult worms and larvae in the intestine, at a dose of 25 mg/kg body weight twice daily. Clearly, it must be given during the first week of infection. Another view is that thiabendazole is too toxic and that mebendazole, at 200 mg twice daily for 5 days in adults, is preferable because it destroys not only adult worms in the gut but also larvae in muscles (13).

Mebendazole and albendazole (400–800 mg/day for 5–6 days in adults) may have some effect against larvae in tissues, and albendazole seemed to bring about a good response in some patients infected with *T. pseudospiralis* (12). However, the usefulness of any action against the muscle phase is questionable, given that host inflammatory responses to degenerating larvae seem to account for much of the pathogenesis. Anecdotal reports suggest that corticosteroids may be beneficial to patients with severe life-threatening illness; however, steroids also may prolong the intestinal phase of infection (13). Of course, general supportive measures, including the maintenance of fluid and electrolyte balance, may need to be instituted in severe cases. Cardiac function should be assessed and monitored, if necessary, to minimize the risk of arrhythmias and other complications of myocarditis.

5. Prophylaxis

Infection will be avoided by not eating undercooked pork or game, or by first freezing, heating to 77°C (170°F), or curing meat to acceptable standards. Of course, diligent meat inspection and farm management should have eradicated this problem from advanced societies.

D. Gnathostomiasis

1. History of Exposure

Gnathostoma spinigerum and related species of spirurid nematodes are covered in detail in Chapter 11 of this book. Human infections formerly seemed to be restricted predominantly to Southeast Asia, with most cases reported in Thailand (10), although in recent years increasing numbers have been diagnosed in Mexico and neighboring parts of Latin America (18). People become exposed to infection when they eat (or perhaps come in skin contact with) third-stage larvae in undercooked fish, chicken, duck, or pork, or drink water containing copepods that harbor second-stage larvae.

2. Clinical Manifestations

Because the parasite does not develop in the human gut but uses people as paratenic hosts, its larvae may wander extensively through the human body, producing deep abscesses and nodules in subcutaneous tissues, skeletal muscles, and, occasionally, other organs. These lesions occur on the trunk or limbs, may be painful, and usually are transient, lasting up to 4 weeks. Involvement of the spinal cord or orbit has serious consequences (19).

3. Laboratory Diagnosis

The solitary lesion may mimic cutaneous myiasis or other forms of *larva migrans* or a localized pyogenic bacterial infection. Gnathostomiasis should be considered in cases of focal pulmonary, gastrointestinal, neurological, or genitourinary disease that are associated with peripheral eosinophilia (19). The only reliable means of confirming the diagnosis is to identify the worm, which may grow to a length of 12.5 mm in human tissue specimens. Positive serology supports the clinical diagnosis but will not identify the species of parasite (18).

4. Treatment

Effectiveness of chemotherapy has not been established. Preliminary evidence from Thailand and Mexico (18,20) suggests that albendazole, and perhaps praziquantel, may be useful, and sometimes stimulates the larvae to migrate out from the skin, which facilitates species identification. When indicated on clinical grounds, surgery eradicates the lesion and provides a tissue diagnosis.

5. Prophylaxis

In endemic regions (predominantly Southeast Asia), water for drinking should first be filtered or boiled, while undercooked, pickled, or salted fish, poultry, or pork should not be eaten.

E. Taeniasis/Cysticercosis

1. History of Exposure

Taeniasis/cysticercosis infection is covered in detail in Chapter 9 of this volume. With either *Taenia saginata* or *T. solium* infection, there should be a history of eating undercooked beef or pork, respectively, including such processed meat products as smoked sausages or minced meat. About 25% of patients with cysticercosis have concurrent intestinal tapeworm infection (21), indicating that internal autoinfection may be considerable (although this is not accepted universally, being virtually impossible to prove). Otherwise, cysticercosis is acquired through ingesting fecally contaminated food or water, reflecting predisposition through poor sanitation and hygiene. Sharing a house with an infected individual significantly increases the risk of exposure (22).

2. Clinical Manifestations

Most intestinal infections are associated with vague, if any, symptoms (10,23). Larvae of *T. solium* can develop in virtually any human body organ, so the clinical spectrum of cysticercosis is extensive, depending on the number, site, and age of lesions. The most common location is subcutaneous or intramuscular, with lesions presenting as deep, painless lumps (10,21,24). Involvement of the CNS or eye can be serious or lethal. The manifestations of neurocysticercosis (NCC) are protean, again reflecting the size, stage, number, and location of lesions. It is believed to be the leading identifiable cause of seizure disorders in the world (24).

3. Laboratory Diagnosis

Infection with adult worms usually is diagnosed incidentally when feces are examined microscopically for other reasons or when proglottids emerge spontaneously from the anus. Often, cysticercosis also is detected incidentally (e.g., as a result of a skin lump biopsy). Sometimes a plain radiograph reveals calcified nodules in deeper tissues, such as muscles. Peripheral eosinophilia is not a common finding (24). The diagnosis of NCC and the assessment of its activity and response to treatment are established by computed tomography (CT), microscopy of CSF (which typically has reduced glucose levels, increased proteins, and eosinophils), and serology (21). Recently, a sensitive and specific enzyme-linked immunoelectrotransfer blot assay using glycoprotein antigens has been developed for detecting antibodies in either serum or CSF (25). Tests also are being refined for the detection of specific antigens in CSF (26,27).

4. Treatment

For adult-worm eradication, a single dose of either niclosamide (2 gm) or praziquantel (5–10 mg/kg) should be effective (10,21). Paromomycin and quinacrine hydrochloride also have been used to treat taeniasis (24). Claims that albendazole 400 mg/day for 3 days eradicates adults of both species (28) have not been supported by other studies (29). It is important first to identify the worm species, using the morphology of expelled proglottids. If *T. solium* infection is confirmed, it may be advisable to administer an antinauseant before treatment and a purgative 1 h afterward, to minimize the risk of cysticercosis consequent to internal autoinfection (24).

The therapeutic approach to cysticercosis is much more complex. First, it needs to be decided if treatment really is necessary. Asymptomatic skin lesions can be left alone (24). When treatment is indicated, praziquantel and albendazole seem to be the most effective agents available, although cutaneous cysticercosis has responded to metrifonate (30) and diethylcarbamazine (personal communication, Hanoi, Vietnam, 1990, Dr. Kie Tung Lam).

Rational treatment of symptomatic NCC requires precise characterization of the disease, particularly with regard to its activity and the location of lesions (31). Patients with calcified cerebral cysts do not need specific anticysticercal agents, but only symptomatic therapy (e.g., anticonvulsants). If the disease is active, i.e., the patient has changing symptomatology, inflammatory changes in the CSF, and edema around lesions enhanced by contrast medium in CT scans, then specific therapy is indicated.

Albendazole is more effective and cheaper than praziquantel (31–33), and may be given in very short courses; a single daily dose of albendazole 15 mg/kg body weight taken for 3 days is as effective as prolonged treatment (34). It has been found that a single dose (30 mg/kg body weight) of oxfendazole, a close relative of albendazole, destroys cysticerci in pigs (35), suggesting that single-dose treatment of human infection might be effective. Praziquantel is given at 50 mg/kg daily for 2 weeks, or in higher doses for shorter periods (e.g., 100 mg/kg for 4 days) (21). Recently, a 1-day course, comprising 3 doses of 25 mg/kg at 2-h intervals, was found to be just as effective as prolonged treatment, with great cost savings (36). Plasma concentrations of praziquantel can be increased by administering with a high-carbohydrate diet, and also with cimetidine. Its bioavailability can be markedly reduced if given in conjunction with anticonvulsants (e.g., carbamazepine, phenytoin) or corticosteroids (especially dexamethasone).

Corticosteroids may be administered simultaneously to reduce the inflammation that probably accounts for headaches, vomiting, seizures, and other untoward effects (36), while concurrent analgesics and anticonvulsants might also be required. While these side effects are considered reliable predictors of drug efficacy (32), their frequency and severity raise questions about the usefulness of treatment in patients who were not suffering major symptoms (24).

Surgery may be necessary to remove cysts from the ventricles (usually the fourth), spinal cord, and eye (21,24,31), and to decompress obstructive hydrocephalus by inserting shunts. Ocular lesions should not be treated with anticysticercal agents because severe inflammation may be induced.

5. Prophylaxis

Avoidance of raw meat will prevent infection with adult tapeworms, and careful personal hygiene plus treatment of intestinal infection with *T. solium* will minimize the risk of cysticercosis.

F. Sparganosis

1. History of Exposure

Tapeworms of the genus *Spirometra* are widespread through most temperate and tropical regions of the world, and it is their plerocercoid stages that cause sparganosis. Human infection can be acquired by ingesting plerocercoids in the raw flesh of an amphibian, reptile, bird, or mammal; by topically applying such infected tissues to the body, or by swallowing copepods that harbor procercoids in drinking water (10).

2. Clinical Manifestations

Sparganum larvae can invade virtually any tissue, but their most common clinical presentation is as migratory subcutaneous nodules, which may be painful and tender. Invasion of the eye can be very painful and potentially disastrous.

3. Laboratory Diagnosis

Diagnosis never can be made on clinical grounds alone and requires the histological examination of resected lesions in the laboratory. The typical local reaction comprises an eosinophilic granuloma. Peripheral eosinophilia also may occur but usually is not prominent.

4. Treatment

Diagnostic biopsy, usually under local anesthesia, is curative for solitary lesions. Eye involvement is a specialized ophthalmological problem. There does not appear to be any effective treatment for the rare proliferative type of lesion. Limited experience with mebendazole and praziquantel has been disappointing (10). In view of the efficacy of albendazole in cysticercosis (see Sec. II.E.4), perhaps a trial of this drug may be justified in complicated cases, although its use might exacerbate inflammation.

5. Prophylaxis

To avoid infection, water for drinking in endemic areas should be boiled or filtered, poultices of split frogs or other vertebrates should not be applied (regardless of their therapeutic indications), and no type of flesh should be eaten unless well cooked.

G. Pentastomiasis

1. History of Exposure

Larvae of several species of "tongue worms" may invade people who eat raw snakes in India and Asia, or raw viscera of herbivores anywhere in the world (10).

2. Clinical Manifestations

Autopsy studies of fatal cases of pentastomiasis tend to represent heavy infections in vital organs, whereas it is possible that most human infections are subclinical. Larvae or nymphs have been reported from the human intestine, mesentery, liver, spleen, lungs, and eye, and symptomatology reflects the location, activity, and size of the worm (19). In the Middle East, a well-recognized complication of eating *marrara* (a dish comprising raw liver, lungs, trachea, and rumen of goats or sheep, seasoned with chilies and lemon) is *halzoun*, infection of the nasopharynx with nymphs of *Linguatula serrata*. The symptoms can be extremely unpleasant, resulting from acute localized irritation and obstruction by the worm (10,37).

3. Laboratory Diagnosis

The diagnosis usually is made unexpectedly at autopsy or during surgery. In the marrara syndrome, patients often cough or sneeze out the offending parasite.

4. Treatment

Diagnosis and cure usually occur simultaneously. There is no role for specific chemotherapy.

5. Prophylaxis

Pentastomiasis is avoided by not eating raw viscera from domestic or wild animals.

III. FISHBORNE INFECTIONS

A. Capillariasis Philippinensis

1. History of Exposure

People at risk of capillariasis philippinensis are those who have eaten raw, often live, small fish from fresh or brackish water, especially in the northern Philippines or Thailand (38,39).

2.   Clinical Manifestations

Chronic disease, a result of continuing internal autoinfection, can lead to severe and extensive destruction of the small bowel mucosa, giving rise to abdominal pain, vomiting, debilitating diarrhea, malabsorption, weight loss, anemia, and death.

3.   Laboratory Diagnosis

The diagnosis is established microscopically, by finding larvae, adults, or typical eggs in the patient's feces, although repeated examination of concentrated specimens sometimes may be necessary (38,39).

4.   Treatment

Nutritional support and fluid and electrolyte replacement may be life saving while awaiting the effects of antihelminthic therapy. Mebendazole, at a dose of 400 mg daily and continued for 20 days initially or for a further 30 days in relapsed cases, was the first effective treatment used (38), but albendazole, at the same dosage but for 10 days, is now considered the treatment of choice (39), with relapses being rare. Eggs and larvae may be cleared from feces within 4 days, and symptoms generally resolve within a week. However, premature cessation of therapy can be followed by relapse.

5.   Prophylaxis

People who do not eat raw fish do not develop intestinal capillariasis (see also Chapter 11, this volume).

B.   Anisakiasis

1.   History of Exposure

Infection with *Anisakis simplex* and related nematodes follows the ingestion of raw or pickled fish or squid, which is a sine qua non for the diagnosis of anisakiasis.

2.   Clinical Manifestations

Anisakid third-state larvae are capable of invading the human alimentary tract anywhere from the esophagus to the colon, but the vast bulk of diagnosed infections involves the stomach (40), probably because that organ is easily accessible to endoscopy. Primary infections, which often are not symptomatic, may present as low-grade, chronic abdominal pain. In most cases, symptomatic gastrointestinal invasion probably indicates a hypersensitivity response in previously exposed individuals, although primary infection can cause gut inflammation (41). The disease usually is acute in onset, with a wide clinical differential diagnosis.

In gastric anisakiasis, severe, colicky, epigastric pain commences 1–36 h after exposure, associated usually with anorexia, nausea, and vomiting (occasionally hematemesis), but not with fever. Symptoms of intestinal infection, which is diagnosed much less frequently, tend to occur later after exposure and may resemble acute appendicitis or regional ileitis (42). Free fluid may accumulate in the peritoneal cavity secondary to eosinophilic serositis. Less common features include diarrhea and urticaria. Without treatment, the symptoms may settle but persist at a low level for months as a result of the mucosal granulomatous response to the larva. Perforation of the gut wall and penetration by the worm into the peritoneal cavity may lead to an exacerbation of symptoms or, less commonly, acute bacterial peritonitis.

3. Laboratory Diagnosis

In the appropriate setting, a clinical diagnosis can be made confidently. Blood eosinophilia is supportive but is not always present in the early stages of infection. Serology, which is not yet widely available or fully evaluated with regard to specificity (40–43), may prove to be helpful for laboratory diagnosis of anisakiasis (except in the acute stages). However, the definitive diagnosis can be established by gastric endoscopy, which might be both diagnostic as well as therapeutic, by allowing removal of the worm.

Intestinal anisakiasis is much more difficult to diagnose as the site of infection is rarely accessible to endoscopy. The combination of eosinophilic enteritis with positive serology strongly indicates the underlying problem. Upper gastrointestinal contrast radiology, with small bowel follow-through, may present typical findings and sometimes demonstrates the worm (42).

4. Treatment

Endoscopy often allows the detection and removal, by biopsy forceps, of larvae from the stomach, duodenum, or colon, thus effecting a cure (43). It is important to examine the entire gastric mucosa carefully, as up to four worms have been found in one stomach (40); should any larvae remain, the symptoms will persist. Antiallergic therapy, e.g., with corticosteroids, might suppress inflammation and dampen symptoms, but there is no evidence that specific antihelminthic chemotherapy has a role in the management of anisakiasis (41,43). If the presentation is particularly severe and the diagnosis equivocal, laparotomy may be the last resort in treatment.

5. Prophylaxis

People who do not eat raw fish or squid do not develop anisakiasis. Methods of handling fish after catching have almost eradicated the risk from herrings in Europe (41), but in Japan, there is resistance to freezing the mackerel and tuna that are popular in sashimi.

C. Opisthorchiasis/Clonorchiasis

1. History of Exposure

Generally, only heavy, long-standing infections with the liver flukes *Opisthorchis viverrini* and *Clonorchis sinensis* cause clinical problems; that is, there should be a history of potential exposure to the infective stages in people who have lived in areas where the ingestion of raw, freshwater fish is a way of life. These fish are usually harvested from ponds traditionally fertilized with human fecal waste (see also Chapter 11, this volume).

2. Clinical Manifestations

Exposed travelers and most infected people living in endemic areas have only light worm burdens and remain asymptomatic. Acute, heavy infections may be associated with transient, nonspecific complaints (10,19). Chronic, heavy infections present clinically with a variety of hepatobiliary symptoms, usually of a predominantly obstructive nature. Cholangiocarcinoma is a terminal complication of long-standing, high-intensity infections.

3. Laboratory Diagnosis

Laboratory diagnosis is established by finding typical eggs, which may resemble the eggs of some intestinal flukes, in feces (see Sec. III.D).

4. Treatment

The most effective drug for opisthorchiasis/clonorchiasis so far has been praziquantel at a dose of 75 mg/kg body weight, either as a single dose or in three divided doses daily (44,45).

5. Prophylaxis

These infections can be readily avoided by eating only well-cooked fish.

D. Minute Intestinal Fluke Infections

1. History of Exposure

Minute intestinal flukes, which include the heterophyids, echinostomatids, and less common species, are acquired by eating either raw fish or, in some cases, tadpoles, frogs, molluscs, or insects (10,46).

2. Clinical Manifestations

Symptoms, when they occur, are nonspecific and referable to the gastrointestinal tract.

3. Laboratory Diagnosis

The diagnosis is established by detecting eggs on fecal microscopy. Some species can be very difficult, if not impossible, to distinguish from others by egg morphology alone (10,47,48).

4. Treatment

Generally, a single dose of praziquantel, at 15–40 mg/kg, is sufficient to eradicate all flukes from the gut (46).

5. Prophylaxis

Infections with minute intestinal flukes can be prevented by not eating raw intermediate hosts.

E. Diphyllobothriasis

1. History of Exposure

Infection with the large tapeworm *Diphyllobothrium latum* is acquired by consuming plerocercoids in the tissues of carnivorous, freshwater fish from northern regions of Europe, Asia, and America (10,23). These infective larvae may survive in the flesh of fish that have been cooked, salted, pickled, or smoked inadequately.

2. Clinical Manifestations

Most infected people remain asymptomatic, although some experience abdominal discomfort, diarrhea, and even less specific complaints. Megaloblastic anemia, while a recognized complication, develops in less than 2% of infected individuals.

3. Laboratory Diagnosis

The diagnosis is established by finding typical eggs (which are identical with eggs of other *Diphyllobothrium* species occasionally found in humans) or, less commonly, chains of spent proglottids in the feces.

4. Treatment

The worm usually is eradicated effectively by a single dose of praziquantel (5–10 mg/kg body weight) or niclosamide (2 g for adults, less for children) (23).

5. Prophylaxis

Only well-cooked fish should be eaten to prevent diphyllobothriasis.

IV. INVERTEBRATE-BORNE INFECTIONS

A. Cerebrospinal Angiostrongyliasis

1. History of Exposure

The rat lungworm, *Angiostrongylus cantonensis*, already well established in southern Asia, Australasia, and Polynesia, appears to be extending its geographic range, as indicated by recent reports of its discovery in Africa, the Caribbean, and southern United States (10,49,50). In areas where it is particularly common, such as Thailand and Taiwan, human angiostrongyliasis usually results from the consumption of snail intermediate hosts or, in children, from exposure to snail viscera discarded as kitchen waste. Infective larvae also can be encountered in such paratenic hosts as freshwater shrimp and land crabs. In many cases, the source of infection may not be obvious, but careful questioning often provides helpful clues (e.g., in eastern Australia, most adult patients recall having eaten fresh salad vegetables, often home-grown lettuce, 1 or 2 weeks before the onset of symptoms, and cases of human and canine infections are clustered in periods of unseasonal wet weather) (51). Parents or siblings of a young patient may have seen the child ingesting snails or potential paratenic hosts before the onset of illness (51).

2. Clinical Manifestations

Severe headache seems to be an almost universal symptom of this infection, often with low-grade fever (10,52). Other features reflect any combination of focal meningoencephalomyeloradiculitis, perhaps depending on the number of larvae ingested. While most cases are mild and resolve completely within a few weeks, the infection may have permanent or lethal sequelae (50,51,53).

3. Laboratory Diagnosis

The diagnosis is clinical, based on the presentation of meningitic headache with variable focal neurological symptoms and signs (often including transient cranial nerve involvement), supported by laboratory findings of blood and CSF eosinophilia. Serodiagnosis, usually by ELISA, is available in most countries in which the infection is common and is very helpful in supporting clinical suspicions. Focal inflammatory lesions may be detected by cerebral and spinal CT and magnetic resonance (MR) scanning, but this is expensive and not specific.

4. Treatment

There is no convincing evidence that systemic antihelminthic or corticosteroid therapy affects the outcome of the illness (10,50–52); in fact, thiabendazole treatment may exacerbate focal inflammation (53). Management essentially comprises symptomatic support, particularly analgesia, while awaiting the natural resolution of the disease. In Southeast Asia, repeated lumbar puncture, draining up to 10 mL of spinal fluid at a time, has been advocated as an effective means of alleviating headache, with the added bonus of occasionally producing a worm (10,51,52).

5. Prophylaxis

Clearly, the consumption of raw snails is not to be encouraged. To give advice beyond this for prophylaxis is unrealistic because most cases in highly endemic regions reflect established practices. In places where human cases are diagnosed infrequently, the risk of accidentally ingesting a small, infected mollusk concealed within lettuce or cabbage, or perhaps third-stage larvae in mucus deposits, is so remote as not to justify the widespread behavioral modification that would be necessary to prevent infection.

B. Abdominal Angiostrongyliasis

1. History of Exposure

Because *A. costaricensis* is restricted to tropical Latin America, where people do not eat raw slugs, it is thought that infection is acquired in most cases by accidentally ingesting, from the environment, third-stage larvae sequestered in such slug mucus as may occur on vegetables and fallen fruit. This would explain the much higher incidence in children (10,50,54).

2. Clinical Manifestations

The worms inhabit the arteries of the terminal ileum and cecum, so the clinical presentation often resembles that of appendicitis, with right iliac fossa pain, fever, nausea, vomiting, and diarrhea. A mass may be detected on palpation (10,54). There are no specific clinical features.

3. Laboratory Diagnosis

In the laboratory, peripheral blood leukocytosis usually is found, often with a very high eosinophilia. Neither worm eggs nor larvae are passed in human feces, so serology may be particularly useful. However, in many cases, surgery is indicated and provides tissue for diagnosis. Histological examination shows necrotic or granulomatous eosinophilic inflammation of the affected bowel segments, with entrapped worms, eggs, and larvae.

4. Treatment

A variety of antihelminthics, including diethylcarbamazine, thiabendazole, and albendazole, have been used to treat abdominal angiostrongyliasis, but convincing evidence of their effectiveness has not been forthcoming (50). In some cases, the illness resolves spontaneously. Therefore, observation and symptomatic support have been advocated, with surgical intervention in cases where progress is unsatisfactory.

5. Prophylaxis

Because the precise mechanism of human infection is unknown, there are no rational guidelines for individual prophylaxis, apart from meticulous personal hygiene.

C. Paragonimiasis

1. History of Exposure

Infection with lung flukes follows the ingestion of metacercariae encysted within tissues of freshwater or estuarine crustacea, so there should be a history of eating undercooked crabs or crayfish, or their juices, or perhaps even immature worms in wild boar meat (10). While *Paragonimus westermani* is the most common species found in humans, there have been many reports of other, almost identical, trematodes causing human lung disease; these reports range from Africa, through southern and eastern Asia, into the western Pacific, and across to South America (10,47).

2. Clinical Manifestations

The preferred site for development of adult flukes is the lungs, so pulmonary disease is the major problem. Complications include cavitation, bronchiectasis, consolidation, fibrosis, hemoptysis, and secondary infection; paragonimiasis cannot be differentiated either clinically or radiologically from pulmonary tuberculosis.

Immature adults of this species also have a propensity for wandering widely in other tissues, frequently giving rise to bizarre presentations from "ectopic" infections that may occur in almost any body organ. These focal "abscesses" do not have any specific clinical features. The CNS often is affected, resulting in a broad range of symptoms and signs.

3. Laboratory Diagnosis

The diagnosis usually is established microscopically, by finding typical eggs in sputum, feces, pleural effusions, or biopsies. Histology of biopsied material should detect the parasite, often surrounded by eggs released directly into tissues. In light or exclusively ectopic infections, serology may be helpful if a biopsy is not indicated. An ELISA using either secretory or somatic antigens seems to be very sensitive for this infection (while not differentiating among different *Paragonimus* species, of course), but may cross-react with fascioliasis (55), an infection with another fluke that has a propensity for causing ectopic lesions (see Sec. V.A).

4. Treatment

The treatment of choice seems to be praziquantel, given in doses of 25 mg/kg body weight, three times daily after meals, for 3 days. Symptoms should improve in weeks, and radiological abnormalities in the lungs should begin resolving within months (10,47). Complicating problems might require supportive and symptomatic treatment, including physiotherapy and analgesia for respiratory disease, anticonvulsants for neurological disturbances, and antibiotics for secondary infections.

5. Prophylaxis

Prevention of infection requires care in the preparation of crustaceans for food, avoidance of either inadequately cooked or freshly salted or pickled crabs and crayfish, as well as avoidance of foods marinated in their juices.

D. Acanthocephalan Infections

1. History of Exposure

Human infections with "thorny-headed" worms now have been found in all inhabited continents (10,56). Because larvae develop in arthropod intermediate hosts, in all cases there should be a history of either accidental or intentional ingestion of arthropods, usually beetles or cockroaches.

2. Clinical Manifestations

Most cases probably are subclinical and go undetected. The parasite of pigs, *Macracanthorhyncus hirudinaceus*, sometimes causes human intestinal disease, whereas infection with *Moniliformis moniliformis*, which originates from rodents, is less likely to be pathogenic (10,56).

3. Laboratory Diagnosis

While *Moniliformis* matures normally in the human gut to produce eggs in feces, it seems that *Macracanthorhyncus* does not develop as far, so that the finding of eggs of the latter species in human stools probably signifies spurious parasitism (10,56).

4. Treatment

A variety of agents, including thiabendazole, niclosamide, and pyrantel pamoate, have been used with questionable success in the treatment of acanthocephalan infections (56). *Moniliformis* infection in a 12-month-old girl was eradicated with mebendazole 100 mg twice a day for 3 days (57).

5. Prophylaxis

Acanthocephalan infections can be avoided by not eating raw insects, especially beetles and cockroaches.

E. Miscellaneous Infections

Numerous species of insects are eaten in many poorer communities, such as in various parts of Africa and Asia, where they contribute significantly to human protein intake. This exposes the populations to a variety of unusual parasite infections. In more developed countries, some parasites of domestic and wild animals utilize insect hosts, which are ingested accidentally by people. Examples include *Dicrocoelium dendriticum, Hymenolepis diminuta, Dipylidium caninum, Rictularia* species, and other spirurids. As these parasites are mainly of curiosity value, they are not considered further here.

V. PLANTBORNE INFECTIONS

A. Fascioliasis

1. History of Exposure

Fasciola hepatica, the common ovine liver fluke, occurs throughout most temperate regions of the world; *F. gigantica*, its close relative with an almost identical life cycle, is more tropical in its distribution (10). While most infected people have eaten wild watercress, it is possible that in some cases infective metacercariae are acquired through drinking water (10,58,59).

2. Clinical Manifestations

The presentation is related to the intensity of exposure and stage and duration of infection. Most cases probably are subclinical, but when symptoms occur, they usually are referred to the hepatobiliary system. Occasionally, aberrant migration of immature flukes leads to unusual ectopic manifestations (10,58,59).

3. Laboratory Diagnosis

In established infections, fasciolid eggs appear in feces. The eggs of both *F. hepatica* and *F. gigantica* are identical, and may be impossible to distinguish from those of *Fasciolopsis buski* (48). If eggs are absent, such as in the early, tissue-invasive stage or in exclusively ectopic disease, serology may be useful. Blood eosinophilia supports the diagnosis but is not always present. Noninvasive imaging techniques may help to establish the extent of liver damage but do not specifically confirm the diagnosis. Biopsy of ectopic lesions provides a definitive diagnosis. Very occasionally, eggs appear in feces as a result of spurious parasitism, in individuals who have consumed infected sheep liver.

4. Treatment

According to preliminary clinical data and studies in sheep, the most effective and least toxic drug so far used in treatment of fascioliasis is triclabendazole, given in a single dose of 10 mg/kg body weight, sometimes given in two divided doses 6–8 h apart (58,60). Despite claims to the contrary in several texts, praziquantel has proved very disappointing in treating fascioliasis. In ectopic disease, chemotherapy probably has no place; surgery, if indicated, is both diagnostic and curative.

5. Prophylaxis

To avoid fascioliasis, raw leaf vegetables and water should not be consumed if obtained from ponds or streams in fields exposed to sheep and cattle feces.

B. Fasciolopsiasis

1. History of Exposure

Infection with the giant intestinal fluke, *Fasciolopsis buski*, occurs widely but sporadically throughout eastern China, Southeast Asia, and parts of India (10,46,47). People become exposed by eating aquatic plants cultivated in ponds that harbor suitable snail intermediate hosts and are contaminated by pig or human feces.

2. Clinical Manifestations

Light infections generally are subclinical. With high worm burdens, symptoms result from small intestinal irritation and dysfunction.

3. Laboratory Diagnosis

The diagnosis is confirmed microscopically, by demonstrating eggs, which closely resemble those of *Fasciola*, in feces or vomitus.

4. Treatment

Praziquantel, in a single dose of 25 mg/kg body weight, is probably the drug of choice for fasciolopsiasis (60).

5. Prophylaxis

To prevent infection, aquatic plants from endemic areas should be peeled and washed, or immersed briefly in boiling water before consumption.

VI. CONCLUSIONS

Given that humans eat such a diversity of foods, it is not surprising that we should be host to an extensive range of parasites. The variety of parasite life cycles into which we intrude accounts for the spectrum of clinical disturbances that may ensue. Some of these organisms are restricted to more remote localities and do not pose major health threats to communities. Others are more widespread and may be of clinical as well as public health significance; it is this group, particularly those organisms that affect inhabitants of developed countries, that attracts most interest from research groups and pharmaceutical manufacturers.

The management of clinical problems arising from these infections is governed by their socioeconomic context. In deprived third-world populations, even the barest essentials for medical management may be lacking. There is little motivation for clinicians to be familiar with the latest diagnostic and therapeutic technology because it will never be within reach. On the other hand, affluent travelers, on returning home to developed countries after experiencing third-world cultures, will benefit from having the most sophisticated resources available to manage their exotic infections.

It is facile to suggest that instead of proffering individual advice about avoiding many of these parasites we should be making more efforts to control or eradicate them, perhaps by urging the intervention of governments or international aid organizations. The infective organisms are so well established in their habitats, and human behavior and customs so deeply entrenched, that an impossibly herculean effort would be required to have any significant impact on the prevalence of most of these infections. Paradoxically, in some developing communities, improved standards of living have broadened the range of available food, leading to an increased diversity of exotic parasite infections (61).

If only a few individuals suffer, the infections will not be perceived as important or worth eradicating. However, should the parasites assume economic importance, in that they significantly affect animal production, food exports, or public health, research and efforts to develop effective control measures will be stimulated.

It may be a vain hope, but as the general living standards of many third-world populations continue to improve, then reliable access to better quality food, safer facilities in which to store it, more fuel with which to cook it, and improved qualities of water supplies and sanitation, all reinforced by education and a growing sense of communal responsibility, will combine to ensure that many of these parasites eventually will become merely historical curiosities.

REFERENCES

1. Prociv, P., Luke, R., and Quayle, P. (1990). Unidentifiable trematode eggs in faeces of Australian Aborigines from Cape York region, *Med. J. Aust.*, 153:680–682.
2. Dubey, J.P., and Beattie, C.P. (1988). *Toxoplasmosis of Animals and Man*, CRC Press, Boca Raton, FL.
3. McCabe, R.E., and Remington, J.S. (1990). Toxoplasmosis. In: *Tropical and Geographical Medicine*, 2d ed. (K.S. Warren and A.A.F. Mahmoud, eds.), McGraw-Hill, New York, pp. 309–320.
4. Cook, G.C. (1990). *Toxoplasma gondii* infection: a feline zoonosis with potential dangers for the unborn foetus and AIDS sufferers. In: *Parasitic Diseases in Clinical Practice*, Springer-Verlag, London, pp. 167–190.
5. Katlama, C., Mouthon, B., Gourdon, D., Lapierre, D., and Rousseau, F. (1996). Atovaquone as long-term suppressive therapy for toxoplasmic encephalitis in patients with AIDS and multiple drug intolerance, *AIDS*, 10:1107–1112.

6. Fung, H.B., and Kirschenbaum, H.L. (1996). Treatment regimens for patients with toxoplasmic encephalitis, *Clin. Ther.*, 18:1037–1056 (discussion: 1036).

7. van de Ven Schoondermark, E.M., Melchers, W.J., Galama, J.M., Meuwissen, J.H., and Eskes, T.K. (1997). Prenatal diagnosis and treatment of congenital *Toxoplasma gondii* infections: an experimental study in rhesus monkeys, *Eur. J. Obstet. Gynecol. Reprod. Biol.*, 74:183–188.

8. Walpole, I.R., Hodgen, N., and Bower, C. (1991). Congenital toxoplasmosis: a large survey in Western Australia, *Med. J. Aust.*, 154:720–724.

9. Kan, S.P. (1985). A review of sarcocystosis with special reference to human infection in Malaysia, *Trop. Biomed.*, 2:167–175.

10. Beaver, P.C., Jung, R.C., and Cupp, E.W. (1984). *Clinical Parasitology*, 9th ed., Lea and Febiger, Philadelphia.

11. Ljungström, I., Murrell, D., Pozio, E., and Wakelin, D. (1998). Trichinellosis. In: *Zoonoses* (S.R. Palmer, E.J.L. Soulsby, and D.I.H. Simpson, eds.), Oxford University Press, Oxford, pp. 789–802.

12. Jongwutiwes, S., Chantachum, N., Kraivichian, P., Siriyasatien, P., Putaporntip, C., Tamburrini, A., La Rosa, G., Sreesunpasirikul, C., Yingyourd, P., and Pozio, E. (1998). First outbreak of human trichinellosis caused by *Trichinella pseudospiralis, Clin. Infect. Dis.*, 26:111–115.

13. Kociecka, W. (1987). Intestinal trichinellosis, *Clin. Trop. Med. Communi. Dis.*, 2:755–763.

14. Kazura, J.W. (1990). Trichinosis. In: *Tropical and Geographical Medicine*, 2nd ed. (K.S. Warren and A.A.F. Mahmoud, eds.), McGraw-Hill, New York, pp. 442–446.

15. MacLean, J.D., Viallet, J., Law, C., and Staudt, M. (1989). Trichinosis in the Canadian Arctic: report of five outbreaks and a new clinical syndrome, *J. Infect. Dis.*, 160:513–520.

16. Chan, S.W., and Ko, R.C. (1988). Comparison between standard ELISA and dot-ELISA for serodiagnosis of human trichinosis, *Trans. R. Soc. Trop. Med. Hyg.*, 82:892–894.

17. Ivanoska, D., Cuperlovic, K., Gamble, R.H., and Murrell, K.D. (1989). Comparative efficacy of antigen and antibody detection tests for human trichinellosis, *J. Parasitol.*, 75:38–41.

18. Ogata, K., Nawa, Y., Akahane, H., Camacho, S.P.D., Lamothe-Argumedo, R., and Cruz-Reyes, A. (1998). Short report: gnathostomiasis in Mexico. *Am. J. Trop. Med. Hyg.*, 58:316–318.

19. Gutierrez, Y. (1990). *Diagnostic Pathology of Parasitic Infections with Clinical Correlation*, Lea and Febiger, Philadelphia.

20. Kraivichian, P., Kulkumthorn, M., Yingyourd, P., Akarabovorn, P., and Paireepai, C. (1992). Albendazole for the treatment of human gnathostomiasis, *Trans. R. Soc. Trop. Med. Hyg.*, 86: 418–421.

21. Pawlowski, Z.S. (1990). Cestodiases: taeniasis, cysticercosis, diphyllobothriasis, hymenolepiasis and others. In: *Tropical and Geographical Medicine*, 2nd ed. (K.S. Warren and A.A.F. Mahmoud, eds.), McGraw-Hill, New York, pp. 490–504.

22. Camacho, S.D., Ruiz, A.C., Beltran, M.U., and Willms, K. (1990). Serology as an indicator of *Taenia solium* tapeworm infections in a rural community in Mexico, *Trans. R. Soc. Trop. Med. Hyg.*, 84:563–566.

23. Kocieka, W. (1987). Intestinal cestodiases, *Clin. Trop. Med. Commun. Dis.*, 2:677–694.

24. Wortman, P.D. (1991). Subcutaneous cysticercosis, *J. Am. Acad. Dermatol.*, 25:409–414.

25. Tsang, V.C., Brand, J.A., and Boyer, A.E. (1989). An enzyme-linked immunoelectrotransfer blot assay and glycoprotein antigens for diagnosing human cysticercosis (*Taenia solium*), *J. Infect. Dis.*, 159:50–59.

26. Estrada, J.J., Estrada, J.A., and Kuhn, R.E. (1989). Identification *of Taenia solium* antigens in cerebrospinal fluid and larval antigens from patients with neurocysticercosis, *Am. J. Trop. Med. Hyg.*, 41:50–55.

27. Correa, D., Sandoval, M.A., Harrison, L.J.S., Parkhouse, R.M.E., Plancarte, A., Meza-Lucas, A., and Flisser, A. (1989). Human neurocysticercosis: comparison of enzyme immunoassay capture techniques based on monoclonal and polyclonal antibodies for the detection of parasite products in cerebrospinal fluid, *Trans. R. Soc. Trop. Med. Hyg.*, 83:814–816.

28. de Kaminsky, R.G. (1991). Albendazole treatment in human taeniasis, *Trans. R. Soc. Trans. Med. Hyg.*, 85:648–650.

29. Chung, W.C., Fan, P.C., Lin, C.Y., and Wu, C.C. (1991). Poor efficacy of albendazole for the treatment of human taeniasis, *Int. J. Parasitol.*, 21:269–270.

30. Tschen, E.H., Tschen, E.A., and Smith, E.B. (1981). Cutaneous cysticercosis treated with metrifonate, *Arch. Dermatol.*, 117:507–509.

31. Del Brutto, O.H., and Sotelo, J. (1988). Neurocysticercosis: an update, *Rev. Infect. Dis.*, 10: 1075–1087.

32. Sotelo, J., del Brutto, O.Y., Penagos, P., Escobedo, F., Torres, B., Rodriguez-Carbajal, J., and Rubio-Donnadieu, F. (1990). Comparison of therapeutic regimen of anticysticercal drugs for parenchymal brain cysticercosis, *J. Neurol.*, 237:69–72.

33. Cruz, M., Curz, I., and Horton, J. (1991). Albendazole versus praziquantel in the treatment of cerebral cysticercosis: clinical evaluation, *Trans. R. Soc. Trop. Med. Hyg.*, 85:244–247.

34. Alarcon, F., Escalante, L., Duenas, G., Montalvo, M., and Roman, M. (1989). Neurocysticercosis: short course of treatment with albendazole, *Arch. Neurol.*, 46:1231–1236.

35. Gonzalez, A.E., Falcon, N. Gavidia, C., Garcia, H.H., Tsang, V.C., Berna, T., Romero, M., and Gilman, R.H. (1997). Treatment of porcine cysticercosis with oxfendazole: a dose–response trial, *Vet. Rec.*, 141:420–422.

36. Sotelo, J., and Jung, H. (1998). Pharmacokinetic optimisation of the treatment of neurocysticercosis, *Clin. Pharmacokinet.*, 34:503–515.

37. El-Hassan, A.M., Eltoum, I.A., and El-Asha, B.M.A. (1991). The Marrara syndrome: Isolation of *Linguatula serrata* nymphs from a patient and the viscera of goats, *Trans. R. Soc. Trop. Med. Hyg.*, 85:309.

38. Cross, J.H., and Basaca-Sevilla, V. (1987). Intestinal capillariasis, *Clin. Trop. Med. Commun. Dis.*, 2:735–746.

39. Cross, J.H. (1998). Capillariosis. In: *Zoonoses* (S.R. Palmer, E.J.L. Soulsby, and D.I.H. Simpson, eds.), Oxford University Press, Oxford, pp. 759–772.

40. Ishikura, J., and Namika, M., eds. (1989). *Gastric Anisakiasis in Japan: Epidemiology, Diagnosis and Treatment*, Springer-Verlag, Tokyo.

41. Cheng, T.C. (1998). Anisakiosis. In: *Zoonoses* (S.R. Palmer, E.J.L. Soulsby and D.I.H. Simpson, eds.), Oxford University Press, Oxford, pp. 823–840.

42. Ishikura, J., and Kikuchi, K., eds. (1990). *Intestinal Anisakiasis in Japan: Infected Fish, Sero-Immunological Diagnosis, and Prevention*, Springer-Verlag, Tokyo.

43. Bier, J.W., Deardorff, T.L., Jackson, G.J., and Raybourne, R.B. (1987). Human anisakiasis, *Clin. Trop. Med. Commun. Dis.*, 2:723–733.

44. Andrews, P., Thomas, H., Pohlke, R., and Seubet, J. (1983). Praziquantel, *Med. Res. Rev.*, 3: 147–200.

45. Bunnag, D., Pungpark, S., Harinasuta, T., Viravan, C., Vanijanonta, S., Suntharasamai, P., Migasena, S., Charoenlarp, P., Riganti, M., and LooAreesuwan, S. (1984). *Opisthorchis viverrini*: clinical experience with praziquantel in hospital for tropical diseases, *Arzneim.-Forsh/Drug Res.*, 34:1173–1174.

46. Harinasuta, T., Bunnag, D., and Radomyos, P. (1987). Intestinal fluke infections, *Clin. Trop. Med. Commun. Dis.*, 2:695–721.

47. Harinasuta, T., and Bunnag, D. (1990). Liver, lung, and intestinal trematodiasis. In: *Tropical and Geographical Medicine*, 2d ed. (K.S. Warren and A.A.F. Mahmoud, eds.), McGraw-Hill, New York, pp. 473–489.

48. Ash, L.R., and Orihel, T.C. (1997). *Atlas of Human Parasitology*, 4th ed., ASCP Press, Chicago.

49. Gardiner, C.H., Wells, S., Gutter, A.E., Fitzgerald, L., Anderson, D.C., Harris, R.K., and Nichols, D.K. (1990). Eosinophilic meningoencephalitis due to *Angiostrongylus cantonensis* as the cause of death in captive non-human primates, *Am. J. Trop. Med. Hyg.*, 42:70–74.

50. Cross, J.H. (1998). Angiostrongylosis. In: *Zoonoses* (S.R. Palmer, E.J.L. Soulsby, and D.I.H. Simpson, eds.), Oxford University Press, Oxford, pp. 773–781.

51. Prociv, P., and Tiernan, J.R. (1987). Eosinophilic meningitis with permanent sequelae, *Med. J. Aust.*, 147:294–295.

52. Vejjajiva, A. (1990). Eosinophilic meningitis. In: *Tropical and Geographical Medicine*, 2d ed. (K.S. Warren and A.A.F. Mahmoud, eds.), McGraw-Hill, New York, pp. 455–457.

53. Ko, R.C., Chan, S.W., Chan, K.W., Lam, K., Farrington, M., Wong, H.W., and Yuen, P. (1987). Four documented cases of eosinophilic *meningoencephalitis* due to *Angiostrongylus cantonensis* in Hong Kong, *Trans. R. Soc. Trop. Hyg.*, 81:807–810.

54. Morera, P. (1987). Abdominal angiostrongyliasis, *Clin. Trop. Med. Commun. Dis.*, 2:747–753.

55. Maleewong, W., Pariyanonda, S., Wongkham, C., Intapan, P., Daengseegaew, W., and Morakote, N. (1990). Comparison of adult somatic and excretory-secretory antigens in enzyme-linked immunosorbent assay for serodiagnosis of human infection with *Paragonimus heterotremus, Trans. R. Soc. Trop. Med. Hyg.*, 84:840–841.

56. Prociv, P., Walker, J., Crompton, L.J., and Tristram, S.G. (1990). First record of human acanthocephalan infections in Australia, *Med. J. Aust.*, 152:215–216.

57. Goldsmid, J.M., Smith, M.E., and Fleming, F. (1974). Human infections with *Moniliformis* sp. in Rhodesia, *Ann. Trop. Med. Parasitol.*, 68:363–364.

58. Chen, M.G., and Mott, K.E. (1990). Progress in assessment of morbidity due to *Fasciola hepatica* infection, *Trop. Dis. Bull.*, 87:R2–R38.

59. Prociv, P., Walker, J.C., and Whitby, M. (1992). Human ectopic fascioliasis in Australia: first case reports, *Med. J. Aust.*, 156:349–351.

60. WHO Study Group. (1995). Control of foodborne infections. *WHO Tech. Rep. Series*, 849; Geneva.

61. Cross, J.H., and Murrell, K.D. (1991). Report: the 33rd SEAMEO-Tropmed regional seminar on emerging problems in food-borne parasitic zoonoses: impact on agriculture and public health, *S.E. Asian J. Trop. Med. Publ. Health*, 22:4–15.

15

Immunodiagnosis of Infections with Cestodes

Bruno Gottstein
University of Berne, Berne, Switzerland

I. INTRODUCTION

Cestodes (''tapeworms'') have selected humans as hosts basically at two levels. The first includes the human small intestine as the site of residence and development to maturation by several adult stage tapeworm species belonging either to the ''lower'' (Pseudophyllidea) or ''higher'' (Cyclophyllidea) cestodes. The life cycle of these tapeworms usually included one or more intermediate hosts in which the larval stage of the parasite underwent development and differentiation to infectiosity for human final hosts. At another level, humans can function as intermediate hosts for some cestode species. A tapeworm larva will then develop in organ tissues, generally causing much more damage and organ dysfunction than intestinal adult tapeworms.

    Echinococcosis is a group of infectious diseases caused by the larval stage of tapeworms of the genus *Echinococcus*. Two of these parasites are important causative agents of severe diseases in humans. *Echinococcus granulosus* refers to cystic echinococcosis or cystic hydatid disease, whereas infection with *Echinococcus multilocularis* in humans is called alveolar echinococcosis or alveolar hydatid disease. Two other species of the genus *Echinococcus, E. vogeli* and *E. oligarthrus*, mainly are restricted to silvatic animals and occur in some areas of Central and South America. Respective cases of the so-called polycystic echinococcosis, however, are very rare in humans.

Taenia solium can be found as adult tapeworms in the small intestine of humans; unfortunately, the larvae (cysticerci) of this tapeworm also can develop in human tissue and thus can be the causative agent of cysticerocosis. The severity of this disease, especially when the central nervous system is affected (neurocysticercosis), justified an appropriate individual consideration as a separate chapter.

Rarely, larvae from other cestode species may develop in humans, such as those from *Taenia multiceps* and *T. serialis* (coenurosis), *Spirometra* species (sparganosis), and others. Additional cestode species that may infect the human intestine at the adult stage include *Taenia saginata, Hymenolepis nana, H. diminuta*, and *Dipylidium caninum*.

This chapter presents a review of current immunological techniques that can be used to diagnose infections of the cestodes mentioned above. Emphasis is given to antibody detection and demonstration of parasite antigens circulating in the body fluids or in the feces of the hosts. Furthermore, the advent of molecular techniques to synthesize recombinant antigens or to diagnose infections by the demonstration of parasite DNA in diagnostic samples opened new diagnostic principles and thus deserves appropriate consideration.

II. THE PARASITES: BIOLOGY, GEOGRAPHIC DISTRIBUTION, AND THE DISEASES THEY CAUSE

A. *Echinococcus*

1. *Echinococcus granulosus*

Echinococcus granulosus is a small (rarely exceeding 7 mm in length) tapeworm, living in the small intestine of carnivore hosts, predominantly dogs. Ungulates, rarely some other animals and humans are intermediate hosts for *E. granulosus*. In the carnivore, sexual maturity of adult tapeworms is reached within 4–5 weeks. Gravid proglottids, each containing several hundreds of eggs, and eggs liberated from disrupted proglottids are shed with the feces of definitive hosts.

Following ingestion of *Echinococcus* eggs by susceptible intermediate hosts, including humans, a primary larva (the oncosphere) hatches from the egg envelope and penetrates through the intestinal epithelium into the lamina propria. The oncosphere subsequently will be transported passively through blood or lymph vessels to such primary target organs as liver and lungs, and less frequently to other organs. At these sites, the metacestode stage of the parasite will develop, including the terminal formation of protoscolices. Once ingested again by a definitive host, the protoscolices will grow to adult stage and thus complete the parasite's life cycle.

The source of eggs that cause infection of humans is diverse. Taeniid eggs in general are known to stick to the definitive host's fur, predominantly in the perianal area, but also in other areas including nose and tail. Close contact to such *Echinococcus*-harboring animals as pet dogs is considered to be of marked infection risk for the dog owner and contacting persons. Defecation sites of dogs in areas prevalent for *E. granulosus* are at risk, especially for children or any persons contacting the contaminated soil.

Infections with *E. granulosus* occur worldwide. A so-called European form, primarily involving synanthropic hosts in its cycle, has a nearly cosmopolitan distribution (1). This form is related to major public health or economic problems in many rural areas of the world. Another northern form is prevalent in northern parts of the North American continent and Eurasia (1). Endemic areas are related mainly to tundra and taiga, and are delineated by the southern limits of the boreal forest.

In human cystic echinococcosis (hydatidosis), well-delineated spherical primary cysts are formed in target organs, the cysts being filled with the so-called hydatid cyst fluid (2). Cyst localization is most commonly in the liver (approximately 65% of cases), followed by the lungs (25%), and such other organs as kidney, spleen, brain, heart, and bone (2). Cysts cause pathological damages in, or dysfunction of infected organs mainly by the gradual process of space-occupying repression or displacement of vital organs' tissue or vessels. Accidental cyst rupture may be another cause of anaphylactic or acute inflammatory pathology.

Consequently, clinical manifestations primarily are determined by the site, size, and number of the cysts and are markedly variable (2). Accidental rupture of cysts can be followed by a massive release of cyst fluid and dissemination of protoscolices, resulting occasionally in anaphylactic reactions but most frequently in multiple secondary cystic echinococcosis, as each protoscolex has the potential to develop into a secondary cyst within the intermediate host. Successful surgical removal of hydatid cysts is common; therefore, case fatality rates are low, varying between 1% and 4% for cases with first surgical intervention and availability of modern medical facilities (2). The public health importance is reflected mainly by the number of infected persons and their diminished functional capacity, the direct and indirect costs of hospitalization and recovery from surgery, and any residual disability or clinical sequelae (3).

Globally, few data exist on the overall prevalence of human cystic echinococcosis. Regions with good documentation of the prevalence in defined geographic areas include the whole Mediterranean area, the Kenyan Turkana district, large foci in South America, but also many other zones over all continents. North American experiences have been documented by various groups (e.g., Refs. 3–6).

2. Echinococcus multilocularis

Echinococcus multilocularis has a life cycle that is basically very similar to that of *E. granulosus*. *E. multilocularis* occurs as adult tapeworms mainly in red and arctic foxes, but dogs and cats also can be involved incidentally as definitive hosts and may play a significant role in the transmission of parasites to humans due to their close contact as pet animals. Small mammals (microtine and arvicolid rodents, occasionally muskrats and others) are intermediate hosts for *E. multilocularis*. Conversely to *E. granulosus, E. multilocularis* is predominantly maintained by a wildlife cycle. Fox and dog fur contaminated with *E. multilocularis* eggs is considered to be an important health hazard to fox hunters and dog owners and their family members. Outdoor-growing vegetables, forest fruits, and windfall contaminated by fox or dog feces or eggs stripped from the fox fur are considered to be the major source of infections in humans. However, exact data still are lacking and are very difficult to obtain in that retrospective patients' studies cannot provide the required information as the time between infection and occurrence of the first symptoms has been estimated to average 10–15 years (7).

The geographic distribution of *E. multilocularis* seems to be segregated uniquely in the Northern Hemisphere. In North America, the cestode is present in sub-Arctic regions of Alaska and Canada, including St. Lawrence Island (8) and some other islands (1). The parasite has been described in Manitoba and North Dakota (9), and, more recently, in Alberta, Saskatchewan, the states of Illinois, Nebraska (10), Iowa, South Dakota, Montana, Wyoming, and even in South Carolina (11). In Europe, areas with relatively frequent reports of alveolar echinococcosis in humans encompass central and eastern France, Switzerland, Austria, and Germany (12). These European main endemic areas previously have been regarded as isolated foci, but more recent data suggest that they are with conjunction to each other and with neighboring countries (such as The Netherlands, Belgium, Poland, Tschechia) or Asian areas where *E. multilocularis* has been reported. Asian areas with *E. multilocularis* prevalence include the whole zone of tundra from the White Sea eastward to the Bering Strait, covering large parts of the former Soviet Union, Northern China, and portions of other countries (14,15).

The primary localization of *E. multilocularis* metacestodes (larvae) in humans is almost exclusively in the liver (2). Metastasis formation in the lungs, brain, and other organs can occur (2). In contrast to cystic echinococcosis, alveolar echinococcosis is characterized by a hepatic lesion consisting of a dispersed, firm, pale tissue subsegmented by scattered small cysts and vesicles. The diffuse borders commonly are not well delineated from the adjacent liver tissue. A central necrotic cavity often is found in cases of advanced hepatic stages. The lesions may be alloyed with focal, nonperipheral zones of calcification. Microscopically, there is evidence for a vigorous proliferation of fibrous and inflammatory tissue in the periphery of the metacestode, but also regressive changes centrally. In contrast to the infection in rodent hosts, lesions in humans exhibit rarely the formation of protoscolices, brood capsules, and calcareous corpuscles within vesicles.

At diagnosis of human alveolar echinococcosis, usually nonspecific clinical symptoms involve mild upper quadrant and epigastric pain with possible hepatomegaly linked with obstructive jaundice. Occasionally, the initial manifestations may be caused by metastases localized in the lungs or other organs (2,12).

As in cystic echinococcosis, few data exist on the overall prevalence of human alveolar echinococcosis. Cases have been diagnosed in the native Eskimo population at risk in western Alaska, including St. Lawrence Island, at an average annual rate of 28 new cases per 100,000 inhabitants (16). In Switzerland, an annual average morbidity rate of 0.18 case per 100,000 inhabitants was reported (12); however, in various cantons of Switzerland annual morbidity rates are higher (0.2–0.7 cases/100,000 inhabitants). Data were similar for some areas of France, Germany, and Austria (17).

The importance of the disease, however, is not represented by the number of reported cases but rather by the severity of the clinical disease in the individual patient and by a frequently lethal outcome of nontreated cases. For cases without radical surgery, lethality was found to be 92% within 10 years after primary diagnosis (18). In recent times, the rate of lethality significantly decreased to 10–14%, most likely due to marked improvements in diagnosis, surgery, and chemotherapy (19).

3. *Echinococcus vogeli* and *Echinococcus oligarthrus*

Human polycystic echinococcosis has so far been reported from 11 countries, from Nicaragua to Argentina, including 31 case due to *E. vogeli* and three to *E. oligarthrus* (two orbital from Venezuela and Surinam and one cardiac from Brazil). Some yet unproven cases of polycystic echinococcosis have also been described from Nicaragua, Costa Rica, Chile, Argentina, and Uruguay (20). Countries affected by *E. vogeli* are restricted to the geographic range of the bush dog (only definitive host of *E. vogeli*). In 80% of the cases the lesions were in the liver alone or in combination with other organs; the rest were located in the lung or other single sites. The diagnosis of polycystic echinococcus was based on the demonstration of polycystic larval cestode lesions by radiological imaging (X-ray, ultrasound, CT scan) in patients born in tropical sylvatic areas of America where wild carnivores (canids, felids) and rodents (pacas and other species) were present. Serological tests often, but not always, supported the diagnosis (21). The most common clinical presentation was abdominal: hard, round masses in or connected to the liver, hepatomegaly, increased abdominal size, pain, marked weight loss, and fever. Signs of portal hypertension were also present in 25% of cases, all of whom died of the disease or from surgical complications following biliary drainage or partial hepatectomy. Ten percent were in asymptomatic persons. Albendazole treatment resulted in clinical improvement and disappearance or reduction of the size of lesions in some patients observed up to 24 months. As polycystic echinococcosis is still considered to be a very rare disease, the two parasite species are not considered more in detail in this chapter.

B. *Taenia*

Adult taeniid tapeworms live in the small intestine, attached to the mucosa by the scolex, from which a chain of proglottids, called *strobila*, is generated. Proglottids are classified as immature, mature, or, located at the terminating posterior end of the tapeworm, gravid. Shape and structures of the uterus in the gravid proglottids often are employed as main diagnostic criteria for the identification of proglottids passed through host feces. The complex life cycle of several tapeworm species can include humans as intermediate and/or definitive hosts.

1. *Taenia solium*

The adult *Taenia solium* uniquely inhabits the small intestine of humans as definitive hosts. *T. solium* taeniasis is acquired by ingestion of raw or poorly cooked pork meat containing *T. solium* cysticerci (larvae). Development from the cysticercus to the mature adult tapeworm and the occasionally seen clinical pictures are similar to those of *T. saginata*, described below. The size of an adult parasite varies between 2 and 4 m in length, and tapeworms may survive for several to 25 or more years. Usually, a single and occasionally two or several tapeworms can be recovered from one patient. Infection with the adult *T. solium* is relatively common in places with regular pork meat consumption and hygienic or sanitary conditions that enable access of pigs to human defecation places or fecal material. These areas include, for example, Mexico and other Latin American countries, Central and Southeast Asia, southern parts of Africa, and, in Europe, some eastern countries, Yugoslavia, and Portugal (22). Globally, 2.5 million people are estimated to be carriers of adult stage intestinal *T. solium*. In the United States, *T. solium* is seen increasingly among Mexican immigrants, especially in California (23).

After ingestion of *T. solium* eggs, an oncosphere hatches from each egg in the small intestine and penetrates the intestinal wall. This process will occur similarly in the pig and in humans as intermediate hosts. In human tapeworm carriers, eggs liberated in the small intestine from gravid proglottids may become activated; thus, an endogenous autoinfection is assumed to be possible. Transportation of the oncosphere via blood vessels and active travel to various tissues will result in the differentiation of cysticerci, with the most common predilection sites being intermuscular and subcutaneous tissues. Affected organs in which cysticerci almost invariably produce damage and symptoms are the central nervous system (neurocysticercosis) and the eye (ophthalmocysticercosis).

Neurocysticercosis is characterized by immune reactions to cysticerci and marked inflammation around the lesions. Symptomatically, this will result in epileptiform seizures, occasional transient paresis, intermittent obstructive hydrocephalus, dysequilibrium, meningoencephalitis, and visual problems. In some cases, cysticerci may undergo a proliferating form of the so-called racemose type, resembling that of a metastatic tumor. Ophthalmocysticercosis, due to cysticerci located under the conjunctiva or in the vitreous or anterior chamber, may result in permanent eye damage or loss.

Cysticercosis is prevalent mainly in areas where *T. solium* is reported to occur, although it is found increasingly among transient immigrants or travelers around the world. Prognostically, one can assume that cysticercosis and *T. solium* taeniasis will represent an emerging problem in countries exhibiting high immigration movements.

The severity of this disease and its increasing occurrence in nonendemic areas justified an appropriate individual consideration of this cestode as a separate chapter.

2. *Taenia saginata* and Other *Taenia* Species

Taenia saginata is a tapeworm that uniquely includes humans as definitive hosts. Thus infection of humans occurs after ingestion of raw or poorly cooked beef harboring encysted *T. saginata* larvae (cysticerci). After digestion of the meat in the stomach, the larva evaginates and develops to an adult tapeworm in the upper small intestine. Adult worms mature in 5–12 weeks, reaching an average length of 10 m. Gravid terminal proglottids contain approximately 100,000 eggs, which may be released in the intestine upon rupture of proglottids or may be shed within proglottids passed through the feces or crawling actively from the anus of the patient. The life cycle is complete through ingestion of eggs by bovine intermediate hosts. In cattle, oncospheres hatch in the duodenum, penetrate into the lamina propria, and are carried via lymph or blood vessels to the striated muscle, a target site for development into cysticerci.

Taenia saginata taeniasis in humans causes usually no significant pathological alterations. The few symptoms occasionally associated with taeniasis include hunger pains, weight loss, abdominal pains, and diarrhea. The significance of *T. saginata* taeniasis has less to do with the discomfort and embarrassment of intestinal infection than with the veterinary–economical aspect of the problem. Depending on national meat-inspection rules, beef meat containing multiple viable cysticerci normally is confiscated or not released for consumption unless previously deep-frozen. The focal nature of the problem, related to a high egg output by a single tapeworm carrier under low sanitary conditions, may cause severe economical losses in cattle farms, where a high percentage of animals may become infected heavily. *T. saginata* has a cosmopolitan distribution; the estimated worldwide prevalence was estimated to be approximately 3×10^7 cases of human enteral *T. saginata* taeniasis. Comparatively, approximately 0.3–6% of European cattle in abattoirs are found to contain *T. saginata* cysticerci (24).

The so-called Asian *Taenia*, occasionally also called Taiwan *Taenia*, is a recently described species or strain of *Taenia* with a biology and life cycle very similar to those of *T. saginata* and *T. solium* (25,26). Asian *Taenia* has been reported to occur in Indonesia, Korea, Thailand, Madagascar, and Ethiopia. The adult tapeworm lives in the small intestine of humans, with potential symptoms the same as those for *T. saginata* and *T. solium*. Metacestodes exclusively develop in the liver of porcine and bovine intermediate hosts;

susceptibility to experimental infection with Asian *Taenia* eggs also was shown for goats and monkeys.

Several other *Taenia* species, all common tapeworms living in the intestinal tract of wild and domestic carnivores, may be the causative agents of coenurosis or other larval infections, which are found in multiple animals and rarely also in humans. The life cycle of these *Taenia* species are similar to that of *T. solium*. The major species responsible for coenurosis in humans is *T. multiceps*. The clinical manifestation of coenurosis is variable and depends on localization of the larva. The most common site is the brain, followed by subcutaneous or intermuscular tissues and others. Symptoms include intracranial pressure, headache, and others resembling those of neurocysticercosis. Viable cysts are larger than cysticerci, varying between 2 and 10 cm in diameter. Viable cysts may contain hundreds of protoscolices. Coenurosis in humans is a rare disease (approximately 50 cases of human coenurosis have been reported so far) that has been diagnosed mainly in Europe, South and North America, and Africa.

3. Larval Taeniid Cestodes in Immunocompromised Persons

In association with an increasing population of immunosuppressed or immunodeficient persons worldwide, the medical community is facing an increasing number of rare infections usually not affecting immunocompetent human individuals. Among larval cestodes, in the past years several cases of rare or unusual infections have been described, including larval *Taenia crassiceps* in a French AIDS patient (27) and disseminated alveolar hydatid disease in a 6-year-old girl with AIDS (28).

C. Non-Taeniidae

1. *Hymenolepis nana, Hymenolepis diminuta*

A peculiarity of the dwarf tapeworm *Hymenolepis nana* is the fact that it requires no intermediate host in its life cycle. *H. nana* is considered to be the most common tapeworm of the world and usually occurs in children living in tropical or subtropical zones. Among children, parasite egg excretion prevalences of 21% have been reported (29). Ingestion of *H. nana* eggs is followed by egg hatching and liberation of oncospheres, which penetrate the villi of the upper small intestine to develop subsequently into the cysticercoid stage. The cysticercoid migrates from the *lamina propria* back to the lumen of the small intestine. There, attached to the mucosa, it matures within several weeks to very small adult worms (up to about 4 cm in length).

The course of disease often is asymptomatic. Occasionally, such indistinct symptoms as gastrointestinal obstruction and irritability, nausea, and, rarely, diarrhea may be observed. The occurrence of symptoms is assumed to be dependent on the worm burden, which can be generated due to the autoinfection feature of the life cycle.

Some patients occasionally may harbor *H. diminuta*, a biologically closely related tapeworm, which however is found infrequently in humans. Life cycle and morphology are very similar to those of *H. nana*, with the exception that several arthropod species have to serve as intermediate hosts. Infection of humans occurs through accidental ingestion of infected arthropods containing cysticercoids. The worm burden is usually very low so that symptoms do not occur.

2. *Diphyllobothrium* Species

Within the groups of tapeworms belonging to the genus of *Diphyllobothrium, D. latum* is the predominant species with regard to case numbers. Its life cycle includes copepods as first intermediate hosts and freshwater fish as second intermediate hosts. The latter contain plerocercoid larvae. Following ingestion by such definitive hosts as humans, the larvae will mature to adult tapeworms that can reach 10 m or more in length and may contain up to 3000 proglottids. Egg production will start about 35 days after infection.

Diphyllobothrium latum has a worldwide distribution. Major foci include European freshwater lakes (in areas of the former Soviet Union, Finland, Scandinavia, alpine zone, and others) and similar biotopes in Asia and America. Such other *Diphyllobothrium* species (or subspecies of *D. latum*) as *D. ursi* and *D. dalliae* occur much less frequently in North America.

Intestinal infection with *D. latum* often is without symptoms. In some cases, mild gastrointestinal obstruction, rarely diarrhea and abdominal pain, and occasionally also leukocytosis with eosinophilia are present. Anemia occurs in approximately 2% of the patients as shown in studies performed in Finland. The cause is related to a high affinity of the tapeworm surface for vitamin B_{12}, which then is deprived from the host. As the infection is acquired by the consumption of raw freshwater fish, preventive measures mainly would include appropriate cooking of fish meat.

3. *Spirometra* Species and Other Species

Species of the genus *Spirometra* are pseudophyllidian tapeworms that parasitize in the small intestine of carnivores. The tapeworm eggs, shed with the feces of infected carnivores, develop to infective larvae (coracidium) in freshwater. The life cycle is very similar to that of *Diphyllobothrium* species and includes *Cyclops* as a first intermediate host and many vertebrates as second intermediate hosts. A plerocercoid larva, called *sparganum*, develops in the latter intermediate host species as a white ribbon-like organism measuring a few millimeters to several centimeters in length.

The very rare infection of humans with sparganum probably occurs by different modes, the most likely being accidental ingestion of infected *Cyclops* through drinking of water. A second modality of infection is ingestion of raw infected flesh of second intermediate hosts (fish, snakes, amphibians, mammals). Another possibility is by direct passage of the sparganum into human tissue from animal flesh applied to the eyes or the vaginas of patients in the scope of medicinal poultice in some areas of Asia (30).

The clinical manifestations of sparganosis include primarily slowly growing, tender, subcutaneous nodules that occasionally may be migratory. Depending on other locations of spargana, there may be intraocular or orbital tissue damage and abscess formation in the brain, the intestinal wall, or other organs. Living spargana elicit little or no inflammatory reaction, in contrast to dead or dying ones.

Another cestode species rarely recovered from humans is *Dipylidium caninum*, a common tapeworm of dogs and cats that is found throughout the world. Infections of humans have been reported predominantly in children, a fact that can be explained by the life cycle biology of the worm. Egg-containing proglottids are shed from definitive hosts and have to be ingested by juvenile fleas as intermediate hosts. Cysticercoid larvae develop in the flea, which has to be ingested by definitive hosts for subsequent development to the intestinal adult stage of the tapeworm. Ingestion of infected fleas by humans is assumed to occur accidentally during playful interactions of children with pet dogs.

III. NONIMMUNOLOGICAL DIAGNOSTIC METHODS

A. Morphological Diagnosis and Imaging Techniques

Diagnosis of intestinal taeniasis or infection with other tapeworms conventionally is based on the recovery and identification of characteristic eggs and/or proglottids. Proglottids may be passed in chains of a few centimeters to a meter. Depending on the species, diagnosis of mature or gravid proglottids and of single eggs can operate at the species, genus, or family level. An important feature is that morphological speciation regarding *Taenia* and *Echinococcus* is not possible from the eggs because they look identical. Detailed information of the morphological peculiarities in terms of diagnosis have been provided in another chapter of this book and thus are not considered further here.

For infections with the larvae of cestodes, imaging techniques are used primarily to diagnose and characterize the respective diseases. The diagnostic quality will depend largely on the localization and biological status of the parasite and the technique used. Cystic and alveolar echinococcosis of the liver can be diagnosed and differentiated reliably in about 70–95% of cases by ultrasound (US), computed tomography (CT) or magnetic resonance imaging (MRI). Neurocysticercosis and cerebral coenurosis are diagnosed radiologically by MRI and CT, in many cases allowing differentiation of viable cysts from calcified lesions. Cysticercosis, coenurosis, and sparganosis may not allow reliable diagnosis until drainage, surgical removal, or biopsy aspiration of the lesions with subsequent laboratory analysis of the tissue by microscopic, histological, immunological, or molecular biological techniques has been performed.

B. DNA Techniques

A variety of DNA probes have been developed and used by several groups to characterize, identify, or group different *E. granulosus* (31–33) or *E. multilocularis* (34) strains or isolates. Probes and techniques for the molecular identification and discrimination of various *Taenia* species have been generated and evaluated (35–37). Apart from the restricted availability of specific diagnostic DNA probes for in situ hybridization, one major problem basically encountered is the limited sensitivity of hybridization and labeling techniques used. Current hybridization techniques, for example, do not allow identification of single taeniid eggs. The possibility of differentiating single cestode eggs at the species level represents an important goal in parasite diagnosis (32). These technical limitations now essentially can be eliminated by using highly efficient DNA amplification techniques such as the polymerase chain reaction (PCR) (39).

Diagnostic PCR depends on the availability of appropriate target nucleic acid sequences that flank regions of interest, which will help to design synthetic oligonucleotide primers and a suitable DNA isolation technique for the test samples. Based on the development and sequencing of an *E. multilocularis* DNA probe pAL1, appropriate oligonucleotide primers were derived, showing suitability for use in PCR amplification of specific target sequences from diagnostic *Echinococcus* genomic DNA (40). Two designed *E. multilocularis* oligonucleotides BG1 and BG2 defined a 2.6-kilobase-pair (kbp) fragment in the genome of *E. multilocularis*. A PCR study including 14 independent *E. multilocularis* isolates originating from various geographic areas and different other cestodes revealed that the 2.6-kbp PCR product was species specifically amplified only from genomic DNA of all *E. multilocularis* isolates, but not from genomic DNA of other cestode species.

Another primer set BG1 and BG3 defined a 0.3-kbp fragment that resulted in amplification of a genus-specific PCR product from *E. multilocularis, E. granulosus,* and *E. vogeli* genomic DNA only. The diagnostic application of the *E. multilocularis* PCR addressed putatively the identification of fine needle biopsy material obtained from patients with liver lesions of unknown etiology (41), the rapid and easy identification of *E. multilocularis* liver lesions from rodents in the scope of epidemiological studies (42), and, perhaps the most promising and important approach, the demonstration and identification of adult stage parasite tissue, i.e., DNA or eggs in samples derived from feces, small intestines, or anal swabs of definitive carnivore hosts (43,44). Other or similar PCRs have been developed for the reliable discrimination between different *Echinococcus* and *Taenia* species (45).

Generally, genomic DNA libraries are generated for more and more parasite species, and DNA sequencing has become an easy, routine technique available to most laboratories. Consequently, it can be assumed that within the next years species-specific DNA sequences already established for the research laboratory discrimination of different cestode species and strains will be further developed for an appropriate diagnostic PCR tests for clinically or epidemiologically relevant questions (46–48).

IV. IMMUNOLOGICAL TECHNIQUES

A. Cestode Antigens

1. Echinococcosis

For primary serological diagnosis and for support of clinical diagnosis of cystic or alveolar echinococcosis, the selection of a particular immunodiagnostic test involves consideration of the diagnostic operating characteristics of the technique and the purpose for which it will be used. The diagnostic sensitivity and specificity of the tests vary according to the nature and quality of the antigen and the methodological sensitivity of the selected technology. A concise definition of the sera used for the assessment of test parameters is essential, with special attention paid to the definition of pre- or postoperative situations respective to serum sampling time points.

For infection with *E. granulosus*, the use of largely available *E. granulosus* hydatid fluid antigen was reported to exhibit diagnostic sensitivity varying according to the test systems used: overall 70–94% for the indirect hemagglutination (IHA) test and 91–96% for ELISA and hepatic cases and 50–80% for ELISA and nonhepatic cases (50). Both IHA and ELISA perform with a relatively high diagnostic sensitivity. Specificity, however, is usually relatively low (50).

Conversely, one of the most specific conventional immunodiagnostic approaches for cystic hydatid disease (CHD) relies on the demonstration of serum antibodies precipitating an antigen called "antigen 5" in immunoelectrophoresis or similar techniques (51) exhibits low sensitivity. With respect to hepatic CHD, sensitivity has been reported to vary between 50% and 80% (50). Antibodies to "antigen 5" also occurred in serum of human patients with neurocysticercosis (52) and alveolar echinococcosis (AE) (53). Comparative studies showed that 58% of Swiss patients with AE were "arc 5"-positive compared to 74% of patients with CHD (54). Subsequently, the antigenic components of *E. granulosus* responsible for arc 5 phenomena were investigated by various approaches using monoclonal antibodies and/or immunoblotting (55–58). Furthermore, preliminary results had indicated some potential to apply purified antigen 5 in such highly sensitive techniques as ELISA (59). Facon et al. cloned an *E. granulosus* gene encoding for an antigen 5 component and suggested recombinant antigen 5 as an immunodiagnostic reagent (60). Some improvement of the immunodiagnostic properties of the antigen 5 was obtained by synthesizing a peptide stretch of the antigen 5 with appropriate B-cell epitope property (61). However, despite the usefulness of antigen 5 for various immunodiagnostic applications, there is evidence of the lack of high species specificity and problems of diagnostic sensitivity due to the absence of antiantigen 5 antibodies in a proportion of patients. Consequently, multiple attempts to characterize *E. granulosus* antigenic components further and to identify corresponding fractions or molecules with optimal diagnostic characteristics were investigated. Thus, resolution of *E. granulosus* hydatid cyst fluid by sodium dodecyl sulfate/polyacrylamide gel electrophoresis (SDS-PAGE) followed by immunoblotting has allowed the identification of arc-5 subunits, including two subunits with rel. mol. masses estimated by different laboratories between 37–38 kDa and 20–22–24 kDa, respectively (50). Diagnostic assessment of these two antigens by immunoblotting performed in different laboratories have resulted in the publication of discrepant sensitivity and specificity parameters (50). The second major parasite antigen in hydatid cyst fluid is a thermostable lipoprotein called antigen B. The major components of antigen B resolve as three bands of apparently 8/12 kDa, 16 kDa, and 23/24 kDa in western blotting (49,50,58,61,63). The apparent 8/12 kDa and the 23/24 kDa bands were assumed to represent identical antigens, respectively (50). Both antigens proved to exhibit good diagnostic parameters, although a variable range of diagnostic sensitivities and specificities have been documented by different groups (50,62,64–68). The gene encoding for antigen B has been recently cloned and the respective polypeptide has been produced as recombinant antigen for IgG$_4$ detection in predominantly advanced cases of cystic echinococcosis (69). Generally, recombinant DNA technology has been widely used to synthesize *E. granulosus* proteins, many of them exhibiting putative antigenicity. Some recombinant *E. granulosus* antigens have already been described above. Although none have been introduced into routine serodiagnosis of cystic echinococcosis yet, some still demonstrate an interesting potential for routine laboratory application (70).

Nowadays the use of *E. multilocularis* metacestode tissue antigens has been widely established for immunodiagnosis of alveolar echinococcosis. Already using crude *E. multilocularis* antigens provided nonspecific reactions and cross-reactions to a much lesser extent than with heterologous *E. granulosus* antigens. Research thus addressed the question of purifying highly specific antigens from *E. multilocularis*. The first documented attempt was done using affinity chromatographic procedures to immunosorbe cross-reactive antigenic components from a crude *E. multilocularis* metacestode antigen solution (71). The resulting Em2 antigen was used successfully in ELISA to correctly differentiate 95% of human cases with cystic echinococcosis from patients with alveolar echinococcosis (72). In subsequent studies, the antigenic component of the Em2 antigen fraction was purified and characterized by immunochemical means and a monoclonal antibody was raised against the Em2 antigen (73,74). Similar procedures to differentiate both forms of echinococcosis serologically have been undertaken by others (75,76). Thus, lto et al. (77) described the characteristics of two SDS-PAGE resolved antigens Em 18 and Em 16 as specific serological immunoblot markers for alveolar and castic echinococcosis, and Sarciron et al. (78) documented a good immunodiagnostic performance of purified *E. multilocularis* alkaline phosphatase antigen.

Conversely, recombinant *E. multilocularis* antigens have been well developed for multiple routine applications. The recombinant antigen II/3-10 (79) underwent large-scale evaluation by ELISA and showed operating immunodiagnostic characteristics appropriate for immunodiagnosis of alveolar echinococcosis in humans. Another antigen EM10 (80) exhibited sequence homology to the II/3-10 antigen and thus also similar serodiagnostic characteristics, as well as an EM4 antigen cloned by others (81).

2. Taeniasis/Cysticercosis

Conflicting results have been reported with respect to parasite-specific serum antibody detection in patients infected with intestinal *Taenia* species. As the availability of a respective specific immunodiagnostic test still is lacking, potentially corresponding *Taenia* antigens cannot be discussed presently. A considerable number of *T. solium* cysticercus antigens have been described in the past decades for antibody detection in cysticercosis. The techniques used were based primarily on the resolution of mostly (glyco-)proteinic antigens by electrophoresis or other methods and subsequent visualization of antibody binding activity by immunoprecipitation. The cysticercus antigen most frequently recognized in human patients was the so-called antigen B (82). Immunodiagnostic properties of antigen B are discussed below. A large number of various antigen preparations or fractions have been assessed for diagnostic antibody detection in cysticercosis; however, none reached species specificity coupled with high diagnostic sensitivity. A marked improvement in the identification of specific and diagnostically sensitive antigens was obtained by SDS-PAGE resolution of *T. solium* cysticercus antigens followed by immunoblotting techniques (83–85). Especially glycoproteins of *T. solium* cysticerci appeared very suitable as immunodiagnostic antigens (86). These glycoproteins proved their immunodiagnostic applicability for diagnosis of individual patients and for seroepidemiological surveys (87). Purification of identified specific antigenic polypeptides has been attempted by isoelectric focusing (88,89). As for echinococcosis, the generation of recombinant antigens became very attractive for *T. solium* (90), with clones still under study to assess their diagnostic potential (91,92). Cysticercal antigens such as a 150-kDa antigen from *T. solium* have been shown to circulate in the cerebrospinal fluid (CSF) of patients suffering from neurocysticercosis (93). Such circulating antigens demonstrated potential but still preliminary clinical value for diagnosis and posttreatment follow-up of neurocysticercosis.

With regard to adult stage *Taenia* antigens, recent attention was attracted to the use of excretory/secretory (E/S) antigens obtained from in vitro maintained tapeworms as target molecules for diagnostic detection as so-called coproantigens (94–96). Appropriate diagnostic tests have been developed to the stage of practicability (see below), although the corresponding diagnostically relevant antigens have not been identified and characterized yet at the molecular level. The *T. saginata* coproantigens detected in human fecal samples were shown to strongly cross-react with the corresponding antigens in *T. solium* taeniasis (94,96) and thus should allow immunocoprodiagnosis at the genus level. Cestode coproantigens were characterized additionally by extremely stable antigenic properties (97). For instance, native feces from dogs experimentally infected with *E. multilocularis* were left at room temperature for 3 weeks and tested quantitatively positive in identity to freshly tested samples (95). Direct species-specific identification of morphologically completely identical taeniid eggs was developed upon the use of respective monoclonal antibodies for *T. solium* (98) among others (50).

Alternative operational criteria underline the marked potential value of coproantigens in the range of diagnostic methodology for intestinal taeniasis. Antigens of other cestode species not listed above scarcely have been described for use in diagnostic antibody binding tests.

3. Other Cestodes

Using homologous and heterologous hyperimmune sera, Sergeeva was able to demonstrate the antigenetic relationship among various *Diphyllobothrium* species (99). Differences in the pattern of immunoprecipitated antigens enabled discrimination among the species. *Diphyllobothrium* antigens with immunodiagnostic potential have not been demonstrated so far. The potential of developing diagnostic tests based on detection of coproantigens as described for *T. saginata* has not been realized yet for *Diphyllobothrium* but is assumed to be of marked interest.

Diagnostic coproantigens were identified successfully in an experimental rat model for *Hymenolepis diminuta*, and some antigens involved have been characterized at the molecular level (100,101). Also, for *Hymenolepis*, the development of immunocoprological tests for the diagnostic detection of human infections has not reached practicability but surely deserves attention. *Hymenolepis nana* antigens have been used to detect serum antibodies by ELISA in patients infected with *H. nana* (102), and the antigens used have been characterized with regard to cross-reactivity with antibodies from patients with cysticercosis, hydatidosis, and other helminths (85). Similarly, antigens used for antibody binding tests were described for *Spirometra mansoni* (103,104). The approach was focused on the analysis of specific antigenic bands of M_r 36,000 and M_r 29,000 or M_r 36,000 and M_r 31,000, respectively. All showed marked specific antibody binding activity in immunoblotting.

Regarding other cestode species not listed above, experimental approaches to identify antibody-binding antigens generally have not resulted in the identification of parasite antigens that proved suitable for routine diagnostic application.

B. Immunodiagnostic Tests

1. Antibody Detection Methods

a. Echinococcosis: General Comments. The clinical signs and symptoms in hepatic cystic or alveolar echinococcosis resemble those of hepatic carcinoma, cirrhosis, or other liver diseases. Therefore, noninvasive imaging techniques are primarily applied and combined with immunodiagnostic procedures. The role of immunodiagnosis is to confirm clinical findings or to provide diagnostic help with detailed information on parasite or host peculiarities (e.g., species differentiation in unclear cases by imaging techniques, determination of the patient's immune status, etc.). Immunodiagnosis of echinococcosis has been reviewed comprehensively in various articles (2,50). The most important tests and concepts for the serological diagnosis of clinical echinococcosis is described below. Early diagnosis of patients with echinococcosis is considered to be a prerequisite for efficient management and treatment of the disease (105). Consequently, serological screenings have been offered to populations and communities in many areas. However, many studies have clearly demonstrated the limits of using serology alone in epidemiology, especially in areas in which seropositivity is low among patients with cystic echinococcosis (e.g., the Turkana district in Kenya) (106). Recent studies therefore employed what can be considered presently to be the optimal epidemiological tool: US examination for abdominal cystic echinococcosis, if possible combined with immunodiagnosis (107–110).

For *E. multilocularis*, the use of homologous crude or purified antigens has been multiply shown to be superior to the conventional use of *E. granulosus* antigens (2,42,50,76,77,79,111,112). This comes true both for immunodiagnosis of individual patients and for mass screening of populations.

b.   Cystic Echinococcosis.   For serology in individual patients who already exhibit some clinical indications for the infection, diagnostic sensitivity of the antibody detection method is considered as most important operating criterium. The indirect hemagglutination test (IHAT) and the enzyme-linked immunosorbent assay (ELISA) using *E. granulosus* hydatid fluid antigen are diagnostically relatively sensitive for hepatic cases (85–98%). For pulmonary cyst localization the diagnostic sensitivity is markedly lower (50–60%), for multiple organ localization very high (90–100%) (2,50). These tests are usually used for a primary serological screening. Specificity is low with regard to other cestode infection and relatively low for noncestode parasitoses. Specificity might be increased by searching for specific antibody isotypes in ELISA, such as IgG4 (68). In order to increase specificity, primary seropositive sera have usually to be retested in a confirmation test such as antigen 5 precipitation (arc-5 test) or immunoblotting for relatively specific bands such as the 8-kD/12-kD hydatid fluid polypeptide antigen (49,58,63). Although exhibiting some cross-reactivities with serum from patients with alveolar echinococcosis and cysticercosis, these confirmation tests usually allow a reliable serodiagnosis of the infection, especially when specific tests for *E. multilocularis* and cysticercosis are added (49,79).

Serological studies to follow-up patients with cystic echinococcosis postoperatively have been generally considered to be of limited use. After complete surgical resection of hydatid cysts, antibody concentrations significantly decrease or disappear within a year, as do anti-arc-5 antibodies in respective precipitation assays. Much emphasis has been given to detect circulating immune complexes or antigens in patients with cystic hydatid disease (see below). But in all cases, follow-up serology requires an association with instrumental or imaging examination in order to provide an accurate prognostic judgment.

Mass screening programs for human cystic echinococcosis have been carried out using serological tests in many endemic areas so far. In conclusion to these investigations, the most reliable approach consists of the combined application of US screening to the population using portable units as a primary test and serology to confirm image positives. This approach is based on the fact that using only ELISA or a similar test, the probability of obtaining a correct positive result is relatively low (low positive predictive value).

c.   Alveolar Echinococcosis. Complementary to imaging procedures in individual patients, immunodiagnosis represents an important secondary diagnostic tool useful in confirming the nature of the etiological agent (50,79). Serological tests are more reliable in the diagnosis of alveolar than cystic hydatid disease. The use of purified *E. multilocularis* antigens such as the Em2 antigen (79), the Em18 antigen (77), alkalkine phosphatase antigen (78), polysaccharide antigen fraction (113), or recombinant antigens II/3–10 (79) or EM10 (80) exhibit diagnostic sensitivities of 91–100%, with overall specificities of 98–100%. These antigens allow one to discriminate between the alveolar and the cystic form of disease with a reliability of 95%.

Postsurgical cases of alveolar echinococcosis usually are associated with treatment by chemotherapy. As only complete surgical removal of the entire parasite lesion offers a relevant prospect for curative treatment, the assessment of the radicality of the resection is an urgent requirement in the clinical monitoring of patients. Serological tests generally have demonstrated a decrease of parasite-specific serum antibody concentration after surgery such as assessed by various antibody detection systems including Em2 ELISA (112,114,115) and immunoblotting (115,116). Cases of alveolar echinococcosis receiving only chemotherapeutic treatment were very difficult to monitor by conventional serological means (2,12,50,112,114,). In general, a tendency toward a decrease in specific antibody concentrations was observed in chemotherapeutically treated patients with regressive forms of disease, whereas specific antibody concentrations in the sera of patients with nonresectable lesions and/or those who had palliative surgery and a progressive course of disease remained elevated or increasing. To date, the claim by clinicians of a clearly predictive interpretation of serology with regard to progressive or regressive courses of disease has not been affirmed by conventional serology. Parasite-specific antibody isotypes may correlate better with clinical findings than results of classical serological tests, especially with regard to immunoglobulin A (IgA) and immunoglobulin E (IgE) (116, 117).

Also for *E. multilocularis*, mass screening has been offered to many populations living in endemic areas using various different technical approaches (42,118–121). In the frame of one of these studies in Alaska, it was shown by Em2 ELISA that not only asymptomatic cases of human alveolar echinococcosis that were serologically negative by other techniques (118) could be detected, but also unique cases in which the metacestode lesion had died out at an apparently early stage of infection (122). The spontaneous rejection of the infection is expected to imply valuable consequences for future research in the immunology of *E. multilocularis* infection.

d. Taeniasis. Based on the few attempts to use serum antibody detection for serodiagnosis of intestinal taeniasis (123), it can be concluded that this approach deserves no place in the range of diagnostic tools and therefore is not discussed here.

e. Cysticercosis. As for serodiagnosis of infections with most cestode species, there are considerable problems of specificity and diagnostic sensitivity encountered when using crude cysticercus antigens for immunodiagnosis of cysticercosis. Therefore, the suggested strategy for immunodiagnosis of cysticercosis is based on a two-step procedure: (a) initial screening with a diagnostically sensitive (although poorly specific) antigen in ELISA or similar tests and the use of crude or purified (antigen B) antigen, and (b) specificity confirmation using analytical tests such as western blotting with the demonstration of antibody binding activity to species-specific bands (see above).

Crude antigens and antigen B both have been shown to exhibit a diagnostic sensitivity in ELISA of 80% for serum samples and 60–90% for CSF (83,124,125), but specificity was higher for antigen B (126) or GP24 antigen (127). The diagnostic sensitivity of screening tests such as ELISA using crude or semipurified cysticercus antigens was strongly dependent on the biological status, the localization, and the number of cysticercal lesions found in the tissue of patients: (a) false-negative results often include calcified cysts (126); (b) diagnostic sensitivity increases with an increasing number of cysts per patient (128–130); (c) an association with the severity of clinical syndromes was observed (131,132). Documentation of the association of the sensitivity of tests using crude or semipurified cysticercus antigens with severity of disease included patients with parenchymal and/or calcified cysts who were more likely to have a benign clinical course and 75–80% of such patients were seropositive. Conversely, patients with intraventricular cysts, large supratentorial cysts, basal meningitis, multiple granulomatoma, or vasculitis were highly symptomatic and 93% were seropositive, including serum and CSF specimens.

The successful development of such highly specific confirmation tests as immunoblotting offered for the first time the opportunity to perform seroepidemiological surveys with reliable predictive test values. Studies performed in Papua New Guinea (133), Peru (87), and Mexico (134), among others, demonstrated the absolute need for such precise serodiagnostic tools. Diagnostically sensitive but not species-specific prescreening, for instance, resulted in the finding of 36% ELISA-positive persons among a group of 221 refugees originating from Irian Jaya who had immigrated to Papua New Guinea (133). From these ELISA-positive persons, only one subsequently was confirmed by immunoblotting as a patient with cysticercosis. High prevalences of infections with other helminths potentially causing serological cross-reactions may be the origin of so many false-positive reactions in prescreening investigations.

Serological tests used to follow up patients with cysticercosis after treatment with praziquantel or albendazole are the same as those described above for primary serodiagnosis. As a rule, successful treatment of neurocysticercosis (reflected by improvements in clinical and neuroimaging criteria) was followed first by a temporarily short increase with a subsequently very slow decline of parasite-specific antibody concentrations (135). Patients' sera that are negative prior to treatment often become positive a few weeks after initiation of treatment (136); the same observations hold in regard to CSF serology (137). Generally, serology as a follow-up criterion provides little contribution to the assessment of efficacy of treatment and clinical improvement (87). A study focused on the separation of the humoral immune response of cysticercosis patients into antibody classes and subclasses (138). The authors concluded that parasite-specific IgE antibodies, which occasionally are detectable in neurocysticercosis, are low in concentration and probably play no important role in serology. In contrast, elevated concentrations of parasite-specific IgG_4 in serum and CSF were demonstrated to be an important component of the humoral immunity in patients with neurocysticercosis.

f. Other Cestodes.   Antibody detection methods have been developed for various other cestode species. A semipurifed antigen from *Spirometra mansoni* was assessed in ELISA for its value in diagnosing human cases of sparganosis (103). Diagnostic sensitivity and specificity of the ELISA were reported to be high (96% and 97%, respectively), whereas unpurified antigens of *Spirometra* species used in previous studies had shown lower operating characteristics (139–141). Improvement of specificity was achieved when focusing the antibody reactivity to 36- and 31-kDa antigens purified from *S. erinacei* (104).

An ELISA was developed to detect anti–*Hymenolepis nana* IgG in patients infected with *H. nana* (102). The diagnostic sensitivity was 79% and specificity 83%. Specificity was significantly dependent on major cross-reactions related to infections with cysticercosis and hydatidosis. The *H. nana* ELISA also proved suitable for monitoring efficacy of praziquantel treatment of the patients. Antibodies disappeared within 90 days in most patients from whom the parasites had been removed successfully, while antibodies persisted in most patients in which *H. nana* infection was not cleared after treatment.

With regard to other cestode species of public health importance that have not been researched in this regard, we can hope and anticipate that in the future research will provide us with new serological tests to support diagnosis of their respective diseases.

2. Cell-Mediated Immunological Tests

a. Echinococcosis. Skin tests and basophil degranulation tests have been discussed extensively and reviewed earlier (2,50). In addition to these tests, the remarkable developments in basic cellular immunology have attracted the attention of parasite immunologists to parasite-specific host cellular immune responses and implied cytokine interplay at the site of parasitic lesions as well as in their periphery (142,143). In the scope of such studies, the in vitro lymphoproliferative response to *E. granulosus* antigen stimulation was assessed (144,145). One study included 40 patients with cystic echinococcosis (146) and demonstrated the absence of any correlation between serological and lymphoproliferative results. Diagnostic sensitivity of positive test reactions was 75% for both serology and lymphocyte proliferation. The finding of seronegative patients with a positive proliferation assay and, conversely, proliferation-negative but seropositve patients consequently suggested the added application of lymphocyte-specific immunoassays as diagnostic tests.

With regard to *E. multilocularis*, the relevance of cellular immune responses and reactions is suggested by the important granulomatous infiltration surrounding *E. multilocularis* lesions in infected human livers (147). The in vitro determination of lymphocyte proliferation to stimulation with *E. multilocularis* antigens has been proposed as a diagnostic alternative to antibody detection in alveolar echinococcosis (148). The same study addressed parameters of cellular immune response specific to *E. multilocularis* during a 2- to 4-year period of mebendazole treatment. A progressive decrease in the capacity to respond to parasite-specific lymphocyte stimulation was observed in most patients with a regressive course of disease. On the other hand, an increase of stimulation indices usually was shown to be associated with a progression of the liver lesion. Gottstein et al. showed that the in vitro lymphoproliferative response to *E. multilocularis* antigen stimulation was very high in cured patients who had radical surgery or patients with dead lesions; it was significantly lower in patients who had partial or no surgical resection (149). Distinct differences in the parasite-specific humoral and cellular immune status as well as expressed cytokine pattern between self-cured (i.e., demonstrating ''died-out'' lesions at diagnosis) patients and other patient groups with different courses of alveolar echinococcosis were suggested to provide insight into potentially protective immune mechanisms (150–152).

b. Taeniasis/Cysticercosis. Little attention has been given to the potential of cell-mediated immunodiagnostic tests for cysticercosis. This is regrettable insofar as the issue of cellular immune response is probably the most relevant factor in acquired immunoresistance to primary and secondary infections. Such assumptions can be exemplified by other forms of cysticercosis or such related metacestode as bovine or ovine cysticercosis (153, 154).

c. Other Cestodes. Cell-mediated immunity for sparganosis was studied by Moulinier et al. (155), who injected intradermally crude sparganum antigens to demonstrate delayed-type hypersensitivity in a patient with a disseminated proliferative form of sparganosis. Experiences concerning the diagnostic value of cell-mediated immunological tests are minimal and receive no further consideration in this chapter.

3. Antigen Detection Methods

a. Echinococcosis.   Tests for the determination of circulating immune complexes in patients with cystic echinococcosis have recently been carried out (2,50,156). For *E. granulosus*, circulating specific immune complexes (157) and circulating soluble antigens (157,158) have been detected diagnostically in 33–85% of sera from patients with cystic echinococcosis. The determination of circulating parasite antigens proved useful for monitoring the course of disease, including the assessment of the radicality of surgical removal of parasite lesions (2,50,156).

Surprisingly, the diagnostic potential of circulating *E. multilocularis* antigens has been neglected so far. Leikina et al. (159) reported that ruptured lesions with central necrotic areas resulted in the release (leaking) of parasite antigens that could be demonstrated subsequently in the serum by double gel diffusion. It can be assumed that the use of such highly sensitive techniques as sandwich ELISA should enable the detection of antigenic molecules released by active and proliferating metacestode tissue. Such a hypothetical antigen detection may be useful for monitoring therapy because a substantial decrease in circulating antigen, reflecting active metabolism of the parasite, is to be expected, whereas antigens released upon degradation of parasite tissue (damaged by drugs) are anticipated to increase in concentration. Such hypotheses, however, will have to be investigated in the future by identification and characterization of the relevant *Echinococcus* antigens responsible for the different biological status of the metacestodes.

b. Taeniasis/Cysticercosis.   The detection of circulating metabolic antigens (CAg) from *T. solium* cysticerci has been proposed as an alternative to antibody detection in serum or CSF of patients with neurocysticercosis. Using a capture-antibody sandwich ELISA, CAg could be detected in the CSF of 77% of patients with neurocysticercosis and the antigen concentration ranged between 17 and 138 ng/mL (160). Specificity was found to be very low, as cross-reactions now occurred with CSF specimens from patients with various CNS infections. Another group subsequently used a similar test system and reported a diagnostic sensitivity of CAg detection in 72% of CSF samples from neurocysticercosis patients (161).

Antigen detection techniques became especially interesting for diagnosing intestinal taeniasis, as it is almost impossible to diagnose early infection before proglottids and/or eggs start to be excreted. After prepatency, proglottids and eggs are excreted irregularly and the available techniques for fecal examination have a relatively low sensitivity. Therefore, new diagnostic tools based on the immunological detection of coproantigens were developed for assessing the efficacy of chemotherapy and for screening larger populations. The diagnostic detection of cestode antigens (coproantigens) in stool samples by ELISA initially was applied to the diagnosis of rats infected with *Hymenolepis diminuta* (162). For the detection of *Taenia* coproantigens, several different ELISAs were carried out (94–96). Most of these ELISAs used affinity-purified polyclonal antibodies obtained from rabbits hyperimmunized with excretory/secretory antigens derived from *T. saginata*. Investigation of operating characteristics showed very low cross-reactivity with crude antigens from non-*Taenia* helminths. Analysis of diagnostic sensitivity demonstrated approximately 85% diagnostic compared to 62% sensitivity of coprological egg detection. Post-treatment control revealed a high concentration of *T. saginata* coproantigens for 1–4 days after administration of niclosamide or praziquantel, and negative values 9–17 days after treatment. Furthermore, the *Taenia* coproantigens appeared very stable and remained detectable by ELISA even after storage of native feces at 25°C for at least 5 days. Another study was able to show that *Taenia* coproantigens remain detectable even in methiolate-formalin-preserved stool samples (96).

c. Other Cestodes. Although at the experimental level the possibility was shown of detecting tapeworm-specific coproantigens in the feces of rats infected with *Hymenolepis diminuta*, no report of an adequate test has been published that could be applied in routine diagnosis of intestinal non-*Taenia* cestode infection. Nevertheless, we now generally accept the big potential of this technical approach to diagnose reliably intestinal infections with tapeworms, and we believe that in the near future tests will be developed and introduced in laboratory practice.

REFERENCES

1. Rausch, R. L. (1986). Life-cycle patterns and geographic distribution of *Echinococcus* species. In: The Biology of Echinococcus and Hydatid Disease, R.C.A. Thompson, ed., Allen and Unwin, London, pp. 44–80.
2. Gottstein, B., Reichen, J. (1996). Echinococcosis/Hydatidosis. In: Manson's Tropical Diseases, 20th ed., G.C. Cook, ed., Saunders, London, pp. 1486–1508.
3. Schantz, P. M., von Reyn, C. F., Welty, T., and Schultz, M. (1976). Echinococcosis in Arizona and New Mexico. Survey of Hospital Records, 1969–1974. Am. J. Trop. Med. Hyg. 25:312–317.
4. Condie, S. J., Crellin, J. R., Andersen, F. L., and Schantz, P. M. (1981). Participation in a community program to prevent hydatid disease. Pub. Health 95:28–35.
5. Crellin, J. R., Andersen, F. L., Schantz, P. M., and Condie, S. J. (1982). Possible factors influencing distribution and prevalence of *Echinococcus granulosus* in Utah. Am. J. Epidemiol. 116:463–474.
6. Langer, J. C., Rose, D. B., Keystone, J. S., Taylor, B. R., and Langer, B. L. (1983). Diagnosis and management of hydatid disease of the liver. A 15-year North American experience. Ann. Surg. 199:412–417.
7. Gottstein, B., Lengeler, C., Bachmann, P., Hagemann, P., Kocher, P., Brossard, M., Witassek, F., and Eckert, J. (1987). Sero-epidemiological survey for alveolar echinococcosis (by Em2-ELISA) of blood donors in an endemic area of Switzerland. Trans. R. Soc. Trop. Med. Hyg. 81:960–964.
8. Rausch, R. L., and Schiller, E. L. (1954). Studies on the helminth fauna of Alaska. XXIV. *Echinococcus sibiricensis* n. sp., from St. Lawrence Island. J. Parasitol. 40:659–662.
9. Leiby, P. D., and Olsen, O. W. (1964). The cestode *Echinococcus multilocularis* in North Dakota. Science 145:1066.
10. Ballard, N. B., and Vande Vusse, J. (1983). *Echinococcus multilocularis* in Illinois and Nebraska. J. Parasitol. 69:790–791.
11. Strand, S. T., and Kazacos, K. R. (1993). *Echinococcus multilocularis* identified in Indiana, Ohio, and east-central Illinois. J. Parasitol. 79:301–305.
12. Ammann, R. W., Eckert, J. (1996). Cestodes. *Echinococcus*. Gastroenterol. Clin. North Am. 25:655–689.
14. Schantz, P. M. (1986). Hydatid disease (echinococcosis). In: Clinical Medicine, J. A. Spittel, ed., Harper and Row, Philadelphia, pp. 1–12.
15. World Health Organization (1989). Report of the WHO Informal Consultation on Alveolar Echinococcosis. WHO/CDS/VPH/89.85, Geneva.
16. Wilson, J. F., and Rausch, R. L. (1980). Alveolar hydatid disease: a review of clinical features

of 33 indigenous cases of *Echinococcus multilocularis* infection in Alaskan Eskimos. Am J. Trop. Med. Hyg. 29:1340–1355.

17. World Health Organization (1988). Report of the WHO Informal Consultation on *Echinococcus multilocularis* Research. WHO/CDS/VPH/88.78, Geneva.

18. Schicker, H. J. (1976). Die Echinokokkose des Menschen. Stand von Diagnose, Therapie und Prognose bei Echinokokkoseerkrankungen in BadenWOrttemberg in den Jahren 1960–1972. Med. Thesis, University of Tübingen, Germany.

19. Ammann, R., Tschudi, K., von Ziegler, M., Meister, F., Cotting, J., Eckert, J., Witassek, F., and Freiburghaus, A. (1988). Langzeitverlauf bei 60 Patienten mit alveolärer Echinokokkose unter Dauertherapie mit Mebendazol (1976–1985). K/in. Wochenschr. 66:1060–1073.

20. D'Alessandro, A. (1997). Polycystic echinococcosis in tropical America: *Echinococcus vogeli* and *E. oligarthrus.* Acta Trop. 67:43–65.

21. Gottstein, B., D'Alessandro, A., and Rausch, R. L. (1995). Immunodiagnosis of polycystic hydatid disease/polycystic echinococcosis due to *Echinococcus vogeli.* Am. J. Trop. Med. Hyg. 53:558–563.

22. Craig, P. S., Rogan, M. T., and Allan, J. C. (1996). Detection, screening and community epidemiology of taeniid cestode zoonoses: cystic echinococcosis, alveolar echinococcosis and neurocysticercosis. Adv. Parasitol. 38:169–250.

23. Richards, F. O., Schantz, P. M., and Ruiz-Tiben, E. (1985). Cysticercosis in Los Angeles County. JAMA 254:3444–3448.

24. König, M., Busato, A., and Gottstein, B. (1996). Untersuchungen zum Vorkommen der Zystizerkose des Rindes. Swiss Vet 13:5–11.

25. Fan, P. C. (1988). Taiwan Taenia and taeniasis. Parasitol. Today 4:86–88.

26. McManus, D. P., and Bowles, J. (1994). Asian (Taiwan) Taenia: Species or strain? Parasitol. Today 10:273–275.

27. Franccois, A., Favennec, L., Cambon Michot, C., Gueit, I., Biga, N., Tron, F., Brasseur, P., and Hemet, J. (1998). Taenia crassiceps invasive cysticercosis: a new human pathogen in acquired immunodeficiency syndrome? Am. J. Surg. Pathol. 22:488–492.

28. Sailer, M., Soelder, B., Allerberger, F., Zaknun, D., Feichtinger, H., and Gottstein, B. (1997). Alveolar echinococcosis of the liver in a six-year-old girl with acquired immunodeficiency syndrome. J. Pediatr. 130:320–323.

29. Mason, P. R., and Patterson, B. A. (1994). Epidemiology of Hymenolepis nana infections in primary school children in urban and rural communities in Zimbabwe. J. Parasitol. 80:245–250.

30. Zhong, H. L., Shao, L., Lian, D. R., et al. (1983). Ocular sparganosis caused blindness. Chin. Med. J. 96:73–75.

31. Yap, K. W., Thompson, R. C. A., and Pawlowski, I. D. (1988). The development of nonradioactive total genomic probes for strain and egg differentiation in taeniid cestodes. Am. J. Trop. Med. Hyg. 39:472–477.

32. Rishi, A. K., and McManus, D. P. (1987). Genomic cloning of human *Echinococcus granulosus* DNA: isolation of recombinant plasmids and their use as genetic markers in strain characterization. Parasitology 94:369–383.

33. Lymbery, A. J., and Thompson, R. C. A. (1989). Genetic differences between cysts of *Echinococcus granulosus* from the same host. Int. J. Parasitol. 19:961–964.

34. Vogel, M., Moller, N., Gottstein, B., Flury, K., Eckert, J., and Seebeck, T. (1991). *Echinococcus multilocularis*: characterization of a DNA probe. Acta Tropica 48:109–116.

35. Rishi, A. K., and McManus, D. P. (1988). Molecular cloning of Taenia solium genomic DNA and characterization of taeniid cestodes by DNA analysis. Parasitology 97:161–176.

36. Zarlenga, D. S., McManus, D. P., Fan, P. C., and Cross, J. H. (1991). Characterization and detection of a newly described Asian taeniid using cloned ribosomal DNA fragments and sequence amplification by the polymerase chain reaction. Exp. Parasitol. 72:174–183.

37. Harrison, L. J. S., Delgado, J., and Parkhouse, R. M. E. (1990). Differential diagnosis of *Taenia saginata* and *Taenia solium* with DNA probes. Parasitology 100:459–461.

38. McManus, D. P. (1990). Characterisation of taniid cestodes by DNA analysis. Revue Scientifique et Technique de l'Office International des Epizooties 9:489–510.

39. Singh, B. (1997). Molecular methods for diagnosis and epidemiological studies of parasitic infections. Int. J. Parasitol. 27:1135–1145.

40. Gottstein, B., and Mowatt, M. R. (1991). Sequencing and characterization of an *Echinococcus multilocularis* DNA probe and its use in the polymerase chain reaction (PCR). Mol. Biochem. Parasitol. 44:183–194.

41. Diebold-Berger, S., Khan, H., Gottstein, B., Puget, E., Frossard, J. L., and Remadi, S. (1997). Cytologic diagnosis of isolated pancreatic alveolar hydatid disease with immunologic and PCR analyses—a case report. *Acta Cytol.* 41:1381–1386.

42. Gottstein, B., Saucy, F., Wyss, C., Siegenthaler, M., Jacquier, P., Schmitt, M., Brossard, M., and Demierre, G. (1996). Investigations on a Swiss area highly endemic for *Echinococcus multilocularis.* Appl. Parasitol. 37:129–136.

43. Dinkel, A., von Nickisch Roseneck, M., Bilger, B., Merli, M., Lucius, R., and Romig, T. (1998). Detection of *Echinococcus multilocularis* in the definitive host: coprodiagnosis by PCR as an alternative to necropsy. J. Clin. Microbiol. 36:1871–1876.

44. Deplazes, P., and Eckert, J. (1996). Diagnosis of the *Echinococcus multilocularis* infection in final hosts. Appl. Parasitol. 37:245–252.

45. Chapman, A., Vallejo, V., Mossie, K. G., Ortiz, D., Agabian, N., and Flisser, A. (1995). Isolation and characterization of species-specific DNA probes from *Taenia solium* and *Taenia saginata* and their use in an egg detection assay. J. Clin. Microbiol. 33:1283–1288.

46. McManus, D. P., and Bowles, J. (1996). Molecular genetic approaches to parasite identification: their value in diagnostic parasitology and systematics. Int. J. Parasitol. 26:687–704.

47. Gasser, R. B., and Chilton, N. B. (1995). Characterisation of taeniid cestode species by PCR-RFLP of ITS2 ribosomal DNA. Acta Trop. 59:31–40.

48. Weiss, J. B. (1995). DNA probes and PCR for diagnosis of parasitic infections. Clin. Microbiol. Rev. 8:113–130.

49. Porretti, D., Felleisen, E., Grimm, F., Pfister, M., Teuscher, F., Zuercher, C., Reichen, J., and Gottstein, B. (1999). Differential immunodiagnosis between cystic hydatid disease and other cross-reactive pathologies. Am. J. Trop. Med. Hyg. 60:193–198.

50. Lightowlers, M., and Gottstein, B. (1995). Immunodiagnosis of echinococcosis. In: *Echinococcus* and Hydatid Disease, R. C. A. Thompson and A. J. Lymbery, eds., CAB International, Wallingford, UK, pp. 355–410.

51. Capron, A., Vernes, A., and Biguet, J. (1967). Le diagnostic immunoelectrophoretique de Ithydatidose. Journe'es Lyonnaises d'Hydatidologie, SIMEP Editions.

52. Varela-Diaz, V. M., Coltorti, E. A., and D'Alessandro, A. (1978). Immunoelectrophoresis tests showing *Echinococcus granulosus* arc 5 in human cases of Echinococcus vogeli and cysticercosis multiple myeloma. Am. J. Trop. Med. Hyg. 27:554–557.

53. Varela-Diaz, V. M., Eckert, J., Rausch, R. L., Coltorti, E. A., and Hess, U. (1977). Detection of the *Echinococcus granulosus* diagnostic arc 5 in sera from patients with surgically-confirmed E. multilocularis infection. Parasitol. Res. 53:183–188.

54. Gottstein, B., Witassek, F., and Eckert, J. (1986). Neues zur Echinokokkose. Schweiz. Med. Wochenschr. 116:810–817.

55. Di-Felice, G., Pini, C., Afferni, C., and Vicari, G. (1986). Purification and partial characterization of the major antigen of *Echinococcus granulosus* (antigen 5) with monoclonal antibodies. Mol. Biochem. Parasitol. 20:133–142.

56. Lightowlers, M. W., Liu, D., Haralambous, A., and Rickard, M. D. (1989). Subunit composition and specificity of the major cyst fluid antigens of *Echinococcus granulosus.* Mol. Biochem. Parasitol. 37:171–182.

57. Chamekh, M., Facon, B., Dissous, C., Haque, A., and Capron, A. (1990). Use of a monoclonal antibody specific for a protein epitope of *Echinococcus granulosus* antigen 5 in a competitive antibody radioimmunoassay for diagnosis of hydatid disease. J. Immunol. Meth. 134:129–137.

58. Siracusano, A., Ioppolo, S., Notargiacomo, S., Ortona, E., Rigano, R., Teggi, A., DeRosa, F., and Vicari, G. (1991). Detection of antibodies against *Echinococcus granulosus* major antigens and their subunits by immunoblotting. Trans. R. Soc. Trop. Med. Hyg. 85:239–243.

59. Hira, P. R., Bahr, G. M., ~Schweiki, H. M., and Behbehani, K. (1990). An enzyme-linked immunosorbent assay using an arc 5 antigen for the diagnosis of cystic hydatid disease. Ann. Trop. Med. Parasitol. 84:157–162.

60. Facon, B., Chamekh, M., Dissous, C., and Capron, A. (1991). Molecular cloning of an *Echinococcus granulosus* protein expressing an immunogenic epitope of antigen 5. Mol. Biochem. Parasitol. 45:233–240.

61. Chamekh, M., Gras-Masse, H., Bossus, M., Facon, B., Dissous, C., Tartar, A., and Capron, A. (1992). Diagnostic value of a synthetic peptide derived from *Echinococcus granulosus* recombinant protein. J. Clin. Invest. 89:458–464.

62. Leggatt, G. R., Yang, W., McManus, D. P. (1992). Serological evaluation of the 12 kDa subunit of antigen B in *Echinococcus granulosus* cyst fluid by immunoblot analysis. Trans. R. Soc. Trop. Med. Hyg. 86:189–192.

63. Maddison, S. E., Slemenda, S. B., Schantz, P. M., Fried, J. A., Wilson, M., and Tsang, V. C. W. (1989). A specific diagnostic antigen of *Echinococcus granulosus* with an apparent molecular weight of 8 kDa. Am. J. Trop. Med. Hyg. 40:377–383.

64. Verastegui, M., Moro, P., Guevara, A., Rodriguez, T., Miranda, E., Gilman, R. H. (1992). Enzyme-linked immunoelectrotransfer blot test for diagnosis of human hydatid disease. J. Clin. Microbiol. 30:1557–1561.

65. Ioppolo, S., Notargiacomo, S., Profumo, E., Franchi, C., Ortona, E., Rigano, R., and Siracusano, A. (1996). Immunological responses to antigen B from *Echinococcus granulosus* cyst fluid in hydatid patients. Parasite Immunol. 18:571–578.

66. Ayadi, A., Dutoit, E., Sendid, B., and Camus, D. (1995). Specific diagnostic antigens of *Echinococcus granulosus* detected by western blot. Parasite 2:119–123.

67. Leggatt, G. R., and McManus, D. P. (1994). Identification and diagnostic value of a major antibody epitope on the 12 kDa antigen from *Echinococcus granulosus* (hydatid disease) cyst fluid. Parasite Immunol. 16:87–96.

68. Shambesh, M. K., Craig, P. S., Wen, H., Rogan, M. T., and Paolillo, E. (1997). IgG1 and IgG4 serum antibody responses in asymptomatic and clinically expressed cystic echinococcosis patients. Acta Trop. 64:53–63.

69. McVie, A., Ersfeld, K., Rogan, M. T., and Craig, P. S. (1997). Expression and immunological characterisation of *Echinococcus granulosus* recombinant antigen B for IgG4 subclass detection in human cystic echinococcosis. Acta Trop. 67:19–35.

70. Ferreira, H. B., and Zaha, A. (1994). Expression and analysis of the diagnostic value of an *Echinococcus granulosus* antigen gene clone. Int. J. Parasitol. 24:863–870.

71. Gottstein, B., Eckert, J., and Fey, H. (1983). Serological differentiation between *Echinococcus granulosus* and *E. multilocularis* infections in man. Parasitol. Res. 69:347–356.

72. Gottstein, B., Schantz, P. M., Todorov, T., Saimot, A. G., and Jacquier, P. (1986). An international study on the serological differential diagnosis of human cystic and alveolar echinococcosis. WHO Bull. 64:101–105.

73. Gottstein, B. (1985). Purification and characterization of a specific antigen from *Echinococcus multilocularis*. Parasite Immunol. 7:201–212.

74. Deplazes, P., and Gottstein, B. (1991). A monoclonal antibody against *Echinococcus multilocularis* Em2 antigen. Parasitology 103:41–49.

75. Knobloch, J., Lederer, 1., and Mannweiler, E. (1984). Species-specific immunodiagnosis of human echinococcosis with crude antigens. Eur. J. Clin. Microbiol. 3:554–555.

76. Auer, H., Hermentin, K., and Aspock, H. (1988). Demonstration of a specific *Echinococcus multilocularis* antigen in the supernatant of in vitro maintained protoscoleces. Zentralblatt fur Bakteriologie und Hygiene A268:416–423.

77. Ito, A., Wen, H., Craig, P. S., Ma, L., Nakao, M., Horii, T., Pang, X. L., Okamoto, M., Itoh, M., Osawa, Y., Wang, X. G., and Liu, Y. H. (1997). Antibody responses against Em18 and Em16 serodiagnostic markers in alveolar and cystic echinococcosis patients from northwest China. Jpn. J. Med. Sci. Biol. 50:19–26.

78. Sarciron, E. M., Bresson-Hadni, S., Mercier, M., Lawton, P., Duranton, C., Lenys, D., Petavy, A. F., and Vuitton, D. A. (1997). Antibodies against *Echinococcus multilocularis* alkaline phosphatase as markers for the specific diagnosis and the serological monitoring of alveolar echinococcosis. Parasite Immunol. 19:61–68.

79. Gottstein, B., Jacquier, P., Bresson-Hadni, S., and Eckert, J (1993). Improved primary immunodiagnosis of alveolar echinococcosis in humans by an enzyme-linked immunosorbent assay using the Em2^{plus}-antigen. J. Clin. Microbiol. 31:373–376.

80. Frosch, P. M., Frosch, M., Pfister, T., Schaad, V., and Bitter-Suermann, D. (1991). Cloning and characterization of an immunodominant major surface antigen of *Echinococcus multilocularis*. Mol. Biochem. Parasitol. 48:121–130.

81. Hemmings, L., and McManus, D. P. (1989). The isolation by differential antibody screening of *Echinococcus multilocularis* antigen clones with potential for immunodiagnosis. Mol. Biochem. Parasitol. 33:171–182.

82. Flisser, A., Woodhouse, E., and Larralde, C. (1980). Human cysticercosis: antigens, antibodies and non-responders. Clin. Exp. Immunol. 39:27–37.

83. Gottstein, B., Tsang, V. C. W., and Schantz, P. M. (1986). Demonstration of species-specific and cross-reactive components of *Taenia solium* metacestode antigens. Am. J. Trop. Med. Hyg. 35:308–313.

84. Diaz, J. F., Verastegui, M., Gilman, R. H., Tsang, V. C., Pilcher, J. B., Gallo, C., Garcia, H. H., Torres, P., Montenegro, T., and Miranda, E. (1992). Immunodiagnosis of human cysticercosis (*Taenia solium*): a field comparison of an antibody-enzyme-linked immunosorbent assay (ELISA), an antigen-ELISA, and an enzyme-linked immunoelectrotransfer blot (EITB) assay in Peru. The Cysticercosis Working Group in Peru (CWG). Am. J. Trop. Med. Hyg. 46:610–615.

85. Montenegro, T., Gilman, R. H., Castillo, R., Tsang, V., Brandt, J., Guevara, A., Sanabria, H., Verastegui, M., Sterling, C., and Miranda, E. (1994). The diagnostic importance of species specific and cross-reactive components of *Taenia solium*, *Echinococcus granulosus*, and *Hymenolepis nana*. Rev. Inst. Med. Trop. Sao Paulo 36:327–334.

86. Tsang, V. C. W., Brand, J. A., and Boyer, A. E. (1989). An enzyme-linked immunoelectrotransfer blot assay and glycoprotein antigens for diagnosing human cysticercosis (Taenia solium). J. Infect. Dis. 159:50–58.

87. Garcia, H. H., Martinez, M., Gilman, R., Herrera, G., Tsang, V. C. W., Pilcher, J. B., Diaz, F., Verastegui, M., Gallo, C., Porras, M., Alvarado, M., Naranjo, J., and Miranda, E. (1991). Diagnosis of cysticercosis in endemic regions. Lancet 338:549–551.

88. Ko, R. C., Ng, T. F. (1998). Purification of larval *Taenia solium* antigens by isoelectric focusing. Vet. Parasitol. 74:191–202.

89. Rodriguez-Canul, R., Allan, J. C., Fletes, C., Sutisna, I. P., Kapti, I. N., Craig, P. S. (1997). Comparative evaluation of purified *Taenia solium* glycoproteins and crude metacestode extracts by immunoblotting for the serodiagnosis of human *T. solium* cysticercosis. Clin. Diagn. Lab. Immunol. 4:579–582.

90. McManus, D. P., Garcia-Zepeda, E., Reid, A., Rishi, A. K., and Flisser, A. (1989). Human cysticercosis and taeniasis: molecular approach for specific diagnosis and parasite identification. Acta Leidensia 57:81–91.

91. Flisser, A., Plancarte, A., Correa, D., Rodriguez-del-Rosal, E., Feldman, M., Sandoval, M., Torres, A., Meza, A., Parkhouse, R. M. E., Harrison, L. J. S., Wilson, M., Avila, G., Allan, J.,

Craig, P. S., Vallejo, V., Ortiz, D., Garcia, E., and McManus, D. P. (1990). New approaches in the diagnosis of *Taenia solium* cysticercosis and taeniasis. Ann. Parasitol. Hum. Comp 65(Suppl. 1):95–98.

92. Manoutcharian, K., Rosas, G., Hernandez, M., Fragoso, G., Aluja, A., Villalobos, N., Rodarte, L. F., and Sciutto, E. (1996). Cysticercosis: identification and cloning of protective recombinant antigens. J. Parasitol. 82:250–254.

93. Cho, S. Y., Kong, Y., Kim, S. I., Kang, S. Y. (1992). Measurement of 150 kDa protein of *Taenia solium* metacestodes by antibody-sandwich ELISA in cerebrospinal fluid of neurocysticercosis patients. Kisaengchunghak Chapchi 30:299–307.

94. Allan, J. C., Mencos, F., Garcia Noval, J., Sarti, E., Flisser, A., Wang, Y., Liu, D., and Craig, P. S. (1993). Dipstick dot ELISA for the detection of *Taenia* coproantigens in humans. Parasitology 107:79–85.

95. Deplazes, P., Eckert, J., Pawlowski, Z. S., Machowska, L., and Gottstein, B. (1991). An enzyme-linked immunosorbent assay for diagnostic detection of *Taenia saginata* coproantigens in humans. Trans. R. Soc. Trop. Med. Hyg. 85:391–396.

96. Maass, M., Delgado, E., and Knobloch, J. (1991). Detection of *Taenia solium* antigens in methiolate-formalin preserved stool samples. Trop. Med. Parasitol. 42:112–114.

97. Deplazes, P., Gottstein, B., Eckert, J., Jenkins, D. J., Ewald, D., and JimenezPalacios, S. (1992). Detection of *Echinococcus* coproantigens by enzymelinked immunosorbent assay in dogs, dingoes and foxes. Parasitol. Res. 78:303–308.

98. Montenegro, T. C., Miranda, E. A., and Gilman, R. (1996). Production of monoclonal antibodies for the identification of the eggs of *Taenia solium*. Ann. Trop. Med. Parasitol. 90: 145–155.

99. Sergeeva, E. G. (1989). Differentiation of surface antigens of 3 species of diphyllobothriids by serological methods. Meditsinskaya Parazitologiya i Parazitarnye Bolezni 2:61–65 (in Russian).

100. Allan, J. C., and Craig, P. S. (1989). Coproantigens in gut tapeworm infections: Hymenolepis diminuta in rats. Parasitol. Res. 76:68–73.

101. Allan, J. C., and Craig, P. S. (1994). Partial characterization and time course analysis of *Hymenolepis diminuta* coproantigens. J. Helminthol. 68:97–103.

102. Castillo, R. M., Grados, P., Carcamo, C., Miranda, E., Montenegro, T., Guevra, A., and Gilman, R. H. (1991). Effect of treatment on serum antibody to *Hymenolepis nana* detected by enzyme-linked immunosorbent assay. J. Clin. Microbiol. 29:413–414.

103. Cho, S. Y., Kang, S. Y., and Yong, Y. (1990). Purification of antigenic protein of sparganum by immunoaffinity chromatography using a monoclonal antibody. Korean J. Parasitol. 28: 135–142.

104. Morakote, N., Kong, Y. (1993). Antigen specificity of 36 and 31 kDa proteins of *Spirometra erinacei* plerocercoid in tissue invading nematodiasis. Korean J. Parasitol. 31:169–171.

105. Kasai, Y., Koshino, I., Kawanishi, N., Sakamoto, H., Sasaki, E., and Kumagai, M. (1980). Alveolar echinococcosis of the liver. Studies on 60 operated cases. Ann. Surg. 191:145–152.

106. Craig, P. S., Zeyhle, E., and Romig, T. (1986). Hydatid disease: research and control in Turkana, 11. The role of immunological techniques for the diagnosis of hydatid disease. Trans. R. Soc. Trop. Med. Hyg. 80:183–192.

107. Macpherson, C. N. L., Romig, T., Zeyhle, E., Rees, P. H., and Were, J. B. O. (1987). Portable ultrasound scanner versus serology in screening for hydatid cysts in a nomadic population. Lancet 2:259–261.

108. Mlika, N., Larouze, B., Gaudebout, C., Graham, B., Allegue, M., Dazza, M. D., Dridi, M., Gharbi, S., Gaumer, B., Bchir, A., Rousset, J. J, Delattre, M., and Jemmali, M. (1986). Am. J. Trop. Med. Hyg. 35:815–817.

109. Coltorti, E. A., Guarnera, E., Larrieu, E., Santillan, G., and Aquino, A. (1988). Seroepidemiology of human hydatidosis: use of dried blood samples on filter paper. Trans. R. Soc. Trop. Med. Hyg. 82:607–610.

110. Romig, T. (1990). Beobachtungen zur zystischen Echinokokkose des Menschen im Turkana-Gebiet, Kenia. Diss., Hohenheim, Stuttgart, Germany.

111. Sato, H., Mitamura, H., Arai, J., and Kumagai, M. (1983). Serologic diagnosis of human hydatid diseases by enzyme-linked immunosorbent assay (Part 1). Enzyme-linked immunosorbent assay by multilocular *Echinococcus* antigen. Rep. Hokkaido Inst. Pub. Health 33: 8–15.

112. Lanier, A. P., Trujillo, D. E., Schantz, P. M., Wilson, J. F., Gottstein, B., and McMahon, B. J. (1987). Comparison of serologic test for the diagnosis and follow-up of alveolar hydatid disease. Am. J. Trop. Med. Hyg. 37:609–615.

113. Sato, C., and Furuya, K. (1994). Isolation and characterization of a diagnostic polysaccharide antigen from larval *Echinococcus multilocularis*. Jpn. J. Med. Sci. Biol. 47:65–71.

114. Gottstein, B., Tschudi, K., Eckert, J., and Ammann, R. (1989). Em2-ELISA for the follow-up of alvelor echinococcosis after complete surgical resection of liver lesions. Trans. R. Soc. Trop. Med. Hyg. 83:389–393.

115. Ma, L., Ito, A., Liu, Y. H., Wang, X. G., Yao, Y. Q., Yu, D. G., and Chen, Y. T. (1997). Alveolar echinococcosis: Em2plus-ELISA and Em18-western blots for follow-up after treatment with albendazole. Trans. R. Soc. Trop. Med. Hyg. 91:476–478.

116. Wen, H., Bresson-Hadni, S., Vuitton, D. A., Lenys, D., Yang, B. M., Ding, Z. X., and Craig, P. S. (1995). Analysis of immunoglobulin G subclass in the serum antibody responses of alveolar echinococcosis patients after surgical treatment and chemotherapy as an aid to assessing the outcome. Trans. R. Soc. Trop. Med. Hyg. 89:692–697.

117. Gottstein, B., Eckert, J., and Woodtli, W. (1984). Determination of parasite specific immunoglobulins using the ELISA in patients with echinococcosis treated with mebendazole. Parasitol. Res. 70:385–389.

118. Gottstein, B., Schantz, P. M., and Wilson, J. F. (1985). Serologic screening for *Echinococcus multilocularis* infections with ELISA. Lancet 1:1097–1098.

119. Nagano, H., Sato, C., and Furuya, K. (1995). Human alveolar echinococcosis seroprevalence assessed by western blotting in Hokkaido. Jpn. J. Med. Sci. Biol. 48:157–161.

120. Bresson-Hadni, S., Laplante, J. J., Lenys, D., Rohmer, P., Gottstein, B., Jacquier, P., Mercet, P., Meyer, J. P., Miguet, J. P., and Vuitton, D. A. (1994). Seroepidemiologic screening of *Echinococcus multilocularis* infection in a European area endemic for alveolar echinococcosis. Am. J. Trop. Med. Hyg. 51:837–846.

121. Craig, P. S., Deshan, L., MacPherson, C. N., Dazhong, S., Reynolds, D., Barnish, G., Gottstein, B., and Zhirong, W. (1992). A large focus of alveolar echinococcosis in central China. Lancet 340:826–831.

122. Rausch, R. L., Wilson, J. F., Schantz, P. M., and McMahon, B. J. (1987). Spontaneous death of *Echinococcus multilocularis*: cases diagnosed serologically by Em2-ELISA and clinical significance. Am. J. Trop. Med. Hyg. 36:576–585.

123. Flentje, V. B., and Padelt, H. (1981). Wert einer serologischen Diagnostik der Taenia saginata Infestation des Menschen. Angew. Parasitol. 22:65–68.

124. Sloan, L., Schneider, S., Rosenblatt, J. (1993). Evaluation of enzyme-linked immunoassay for serological diagnosis of cysticercosis. J. Clin. Microbiol. 33:3124–3128.

125. Simac, C., Michel, P., Andriantsimahavandy, A., Esterre, P., and Michault, A. (1995). Use of enzyme-linked immunosorbent assay and enzyme-linked immunoelectrotransfer blot for the diagnosis and monitoring of neurocysticercosis. Parasitol. Res. 81:132–136.

126. Espinoza, B., Ruiz-Palacios, G., Tovar, A., Sandoval, M. A., Plancarte, A., and Flisser, A. (1986). Characterization by enzyme-linked immunosorbent assay of the humoral immune response in patients with neurocysticercosis and its application in immunodiagnosis. J. Clin. Microbiol. 24:536–541.

127. Plancarte, A., Fexas, M., and Flisser, A. (1994). Reactivity in ELISA and dot blot of purified GP24, an immunodominant antigen of *Taenia solium*, for the diagnosis of human neurocysticercosis. Int. J. Parasitol. 24:733–738.

128. Gottstein, B., Zini, D., and Schantz, P. M. (1987). Species-specific immunodiagnosis of *Taenia solium* cysticercosis by ELISA and immunoblotting. Trop. Med. Parasitol. 38:299–303.
129. Pammenter, M. D., and Rossouw, E. J. (1987). The value of an antigenic fraction of *Cysticercus cellulosae* in the serodiagnosis of cysticercosis. Ann. Trop. Med. Parasitol. 81:117–123.
130. Wilson, M., Bryan, R. T., Fried, J. A., Ware, D. A., Schantz, P. M., Pilcher, J. B., and Tsang, V. C. W. (1991). Clinical evaluation of the cysticercosis enzyme-linked immunoelectrotransfer blot in patients with neurocysticercosis. J. Infect. Dis. 164:1007–1009.
131. Corona, T., Pascoe, D., Gonzales-Barranco, D., Abad, P., Landa, C., and Estanol, B. (1986). Anticysticercosis antibodies in serum and cerebrospinal fluid in patients with cerebral cysticercosis. J. Neurol. Neurosurg. Psychiatr. 49:1044–1049.
132. Schantz, P. M., Tsang, V. C. W., and Maddison, S. E. (1988). Serodiagnosis of neurocysticercosis. Rev. Infect. Dis. 10:1231–1233.
133. Fritzsche, M., Gottstein, B., Wigglesworth, M. C., and Eckert, J. (1990). Serological survey of human cysticercosis in Irianese refugee camps in Papua New Guinea. Acta Trop. 47:69–77.
134. Larralde, C., Padilla, A., Hernandez, M., Govezensky, T., Sciutto, E., Gutierrez, G., Tapia Conyer, R., Salvatierra, B., and Sepulveda, J. (1995). Seroepidemiology of cysticercosis in Mexico. Salud Publica Mex. 34:197–210.
135. Markwalder, K., Hess, K., Valavanis, A., and Witassek, F. (1984). Cerebral cysticercosis: treatment with praziquantel. Am. J. Trop. Med. Hyg. 33:273–280.
136. Jacquier, P., and Gottstein, B. Immunodiagnostic de la neurocysticercose en Suisse. Schweiz. Med. Wschr. 122:904 (in French).
137. Sotelo, J., Escobedo, F., Rodriguez, J., Torres, B., and Rubio-Donnadieu, F. (1984). Therapy of parenchymal brain cysticercosis with praziquantel. N. Engl. J. Med. 310:1001–1007.
138. Short, J. A., Heiner, D. C., Hsiao, R. L., and Andersen, F. L. (1990). Immunoglobulin E and G4 antibodies in cysticercosis. J. Clin. Microbiol. 28:1635–1639.
139. Kim, H., Kim, S. L., and Choy, S. Y. (1984). Serological diagnosis of human sparganosis by means of micro ELISA. Korean J. Parasitol. 22:222–228.
140. Chang, K. H., Cho, S. Y., Chi, J. G., Kim, W. S., Han, M. C., Kim, C. W., Myung, H., and Choi, K. S. (1987). Cerebral sparganosis: CT characteristics. Radiology. 165:505–510.
141. Ishii, A. (1973). Indirect fluorescent antibody test in human sparganosis. Jpn. J. Parasitol. 22:75–78.
142. De Rycke, P. H., Janssen, D., Osuna, A., and Lazuen, J. (1990). Immunohomeostasis in hydatidosis (Echinococcus granulosus). In: Basic Research in Helminthiases. R. Ehrlich, A. Nieto, and L. Yarzabal, eds., Ediciones Logos, Montevideo, Uruguay, pp. 217–228.
143. Rogan, M. T., and Craig, P. S. (1997). Immunology of Echinococcus granulosus infections. Acta Trop. 67:7–17.
144. Torcal, J., Navarro Zorraquino, M., Lozano, R., Larrad, L., Salinas, J. C., Ferrer, J., Roman, J., and Pastor, C. (1996). Immune response and in vivo production of cytokines in patients with liver hydatidosis. Clin. Exp. Immunol. 106:317–322.
145. Kharebov, A., Nahmias, J., and El On, J. (1997). Cellular and humoral immune responses of hydatidosis patients to Echinococcus granulosus purified antigens. Am. J. Trop. Med. Hyg. 57:619–625.
146. Siracusano, A., Teggi, A., Quintieri, F., Notargiacomo, S., De Rosa, F., and Vicari, G. (1988). Cellular immune response of hydatid patients to Echinococcus granulosus antigens. Clin. Exp. Immunol 72:400–405.
147. Vuitton, D., Lenys, D., Liance, M., Flausse, F., Estavoyer, J. M., and Miguet, J. P. (1985). Specific cell-mediated immunity (CMI) against Echinococcus multilocularis in patients with alveolar echinococcosis. J. Hepatol. 1:149.
148. Bresson-Hadni, S., Vuitton, D. A., Lenys, D., Liance, M., Racadot, E., and Miguet, J. P. (1989). Cellular immune response in *Echinococcus multilocularis* infection in humans. 1. Lymphocyte reactivity to *Echinococcus* antigens in patients with alveolar echinococcosis. Clin. Exp. Immunol. 78:61–66.

149. Gottstein, B., Mesarina, B., Tanner, I., Ammann, R. W., Eckert, J., Wilson, J. F., Lanier, A., and Parkinson, A. (1991). Specific cellular and humoral immune responses in patients with different long-term courses of alveolar echinococcosis (infection with *Echinococcus multilocularis*). Am. J. Trop. Med. Hyg. 45:734–742.

150. Nicod, L., Bresson-Hadni, S., Vuitton, D. A., Emery, I., Gottstein, B., Auer, H., and Lenys, D. (1994). Specific cellular and humoral immune responses induced by different antigen preparations of *Echinococcus multilocularis* metacestodes in patients with alveolar echinococcosis. Parasite 1:261–270.

151. Jenne, L., Kilwinski, J., Scheffold, W., and Kern, P. (1997). IL-5 expressed by CD4(+) lymphocytes from *Echinococcus multilocularis*–infected patients. Clin. Exp. Immunol. 109: 90–97.

152. Godot, V., Harraga, S., Deschaseaux, M., Bresson-Hadni, S., Gottstein, B., Emilie, D., and Vuitton, D. A. (1997). Increased basal production of interleukin-10 by peripheral blood mononuclear cells in human alveolar echinococcosis. Eur. Cytokine Netw. 8:401–408.

153. Lloyd, S. (1987). Cysticercosis. In: Immune Responses in Parasitic Infections, Vol. 2, E. J. L. Soulsby, ed.), CRC Press, Boca Raton, FL, pp. 183–212.

154. Meeusen, E., Barcham, G. J., Gorrell, M. D., Rickard, M. D., and Brandon, M. R. (1990). Cysticercosis: cellular immune responses during primary and secondary infection. Parasite Immunol. 12:403–418.

155. Moulinier, R., Martinez, E., Torres, J., Noya, O., De Noya, B. A., and Reyes, O. (1982). Human proliferative sparganosis in Venezuela: report of a case. Am. J. Trop. Med. Hyg. 31: 358–363.

156. Craig, P. S. (1997). Immunodiagnosis of *Echinococcus granulosus* and a comparison of techniques for diagnosis of canine echinococcosis. In: Compendium on Cystic Echinococcosis in Africa and Middle Eastern Countries with Special Reference to Morocco, F. L. Andersen, H. Ouhelli, and M. Kachani, eds., Brigham Young University, Print Services, Provo, UT, pp. 85–118.

157. Craig, P. S., and Nelson, G. S. (1984). The detection of circulating antigen in human hydatid disease. Ann. Trop. Med. Parasitol. 78:219–227.

158. Gottstein, B. (1984). An immunoassay for the detection of circulating antigens in human echinococcosis. Am. J. Trop. Med. Hyg. 33:1185–1191.

159. Leikina, E. S., Kovrova, E. A., and Krasovskaya, N. N. (1982). Detection of circulating antigens in the sera of patients with unilocular and multilocular hydatidosis or with trichinelliasis. Meditsinskaya Parazitologiya Parazitarnye Bolenzi 51:7–15.

160. Tellez-Giron, E., Ramos, M. C., Dufour, L., Alvarez, P., and Montante, M. (1987). Detection of Cysticercus cellulosae antigens in cerebrospinal fluid by dot–enzyme-linked immunosorbent assay (dot-ELISA) and standard ELISA. Am. J. Trop. Med. Hyg. 37:169–173.

161. Correa, D., Sandoval, M., Harrison, L., Parkhouse, M., Plancarte, A., Meza-Lucas, A., and Flisser, A. (1989). Human neurocysticercosis: comparison of enzyme-immunoassay capture techniques based on monoclonal and polyclonal antibodies for the detection of parasite products in cerebrospinal fluid. Trans. R. Soc. Trop. Med. Hyg. 83:814–816.

162. Allan, J. C., and Craig, P. S. (1989). Coproantigens in gut tapeworm infections: *Hymenolepis diminuta* in rats. Parasitol. Res. 76:68–73.

16

Immunodiagnosis: Nematodes

H. Ray Gamble

U.S. Department of Agriculture, Beltsville, Maryland

I. INTRODUCTION

Immunological diagnosis of human nematodiasis provides an alternative to direct and often invasive methods for the demonstration of infection. It also serves as confirmation of other clinical and laboratory findings. Interpretation of immunological findings must be tempered with an understanding that immune status is not a direct reflection of current parasitological status but rather an indication of acute, chronic, or resolved infections. Considering these limitations, efforts have been made, with varying success, to develop a correlation between serological findings and actual disease state.

Serological methods for diagnosis of nematode infection have paralleled methods developed for diagnosis of other diseases in man and animals. The majority of tests currently used are solid phase, enzyme-linked immunosorbent assay (ELISA) methods based on the use of defined antigens for antibody detection. The definition of specific antigens using monoclonal antibodies and other identification and separation techniques has greatly increased test specificity by eliminating cross-reactions with other infectious agents.

A very limited number of the nematodes infecting humans are transmitted through the ingestion of food products. For meat, *Trichinella spiralis* is the only nematode of importance in causing human disease. However, the potential sources of human trichinellosis continue to expand as pork, bear, other game meats, and, most recently, horsemeat have been implicated in disease transmission. The number of nematodes transmissible to humans from fish and invertebrates are greater in number, although these infections in humans are less common and not as cosmopolitan as trichinellosis. Nematodes transmissible to humans from ingestion of raw or undercooked fish include the gnathostomes, capillarids, and species of the anisakid complex. Angiostrongylid infections are acquired from ingestion of infected snails or slugs.

II. MEATBORNE PARASITES

A. Trichinellosis

Trichinellosis is acquired by ingesting raw or undercooked meat (pork, bear, horse, walrus, etc.) containing infective larvae of the nematode parasite *Trichinella spiralis*. Human infection results in a wide range of symptoms depending on numbers of parasites ingested and the health and susceptibility of the patient. Various signs and symptoms of clinical trichinellosis (abdominal pain, diarrhea, fever, myalgia, malaise, periorbital edema, eosinophilia) allow a presumptive diagnosis (1); however, definitive diagnosis relies on the direct demonstration of parasites by muscle biopsy. Considerable value can be placed on serology results as a noninvasive confirmatory test and this has been reflected in efforts to improve serodiagnostic methods.

Antibody responses to trichinellosis are generally linked to the intensity of infection, i.e., the number of parasites ingested. The major antigens recognized in human infection are those that are secreted by or found on the surface of the infective larval stage. These antigens are initially recognized in the host at the intestinal level during the initial stages of infection, and then again when worms reach host muscle tissue and encyst. Only transient and variable antibody responses occur to antigens from other stages such as adult worms, resident in the intestine, or newborn larvae circulating in the blood.

Various serology tests have been employed for the immunodiagnosis of trichinellosis including indirect immunofluorescence assay (IFA), agglutination or flocculation assays including passive hemagglutination (PHA), latex agglutination (LA) and bentonite flocculation (BF), and enzyme immunoassay (EIA) (2). The performance of these assays, with respect to specificity, is largely dependent on the antigen(s) used in the test. Recent advances in antigen definition and standardization have dramatically improved test results.

The specificity of serodiagnostic tests for trichinellosis is affected by the presence of cross-reacting antibodies in patient serum. Where the probability of other nematode infections is unlikely, less specific antigens such as a crude saline extract of infective larvae might be used successfully. In such cases, IFA, using sections of infective muscle larvae, is a simple and sensitive test (3); however, performance of IFA requires the availability of a fluorescence microscope to analyze results. In cases where other parasites might be coresident in the patient, more specific antigens are required for the diagnosis of trichinellosis; these antigens include preparations of parasite secretions produced by the stichocyte cells. Most cross-reactions in diagnostic tests for trichinellosis have been reported to result from closely related parasites such as *Trichuris* and *Capillaria*. Nevertheless, any intestinal or generalized parasitosis should be taken into consideration in the differential diagnosis of trichinellosis.

Various studies have shown the specificity of crude worm extracts in diagnosis. Chapa-Ruiz et al. (4) reported no cross-reactions in patients coinfected with other parasites including *Ascaris* and *Fasciola*. However, other authors have reported cross-reactions with filarial worms (5), *Trichuris, Ascaris, Dirofilaria, Gnathostoma, Paragonimus, Fasciola* (6–8), and *Capillaria* (8,9). Most commonly, authors have reported a low level of false positives using sera from "negative" controls (2). However, the actual parasitological status of these individuals is generally not known.

To circumvent the problem of false positives due to antibodies directed at antigens shared between parasite species, considerable effort has been directed to improving antigen preparations (10–13). The major component of the host humoral response to *Trichinella* infection is directed to a group of glycoprotein antigens designated as TSL-1 (*Trichinella spiralis* larvae–1) by the International Commission on Trichinellosis' Working Group on Antigen Standardization (14). Present in both crude worm extracts and worm secretions, these antigens are recognized in virtually all of the various types of diagnostic tests. Thus, in IFA, antibody binds to these antigens on the worm surface and in cross-sections of stichocyte cells (3). By selectively recovering enriched preparations of TSL-1 antigens using biochemical separation methods or by recovering worm secretions, cross-reactive components are largely removed from the antigen preparation (7,15–17). Antigen preparations enriched in TSL-1 (designated ES antigens, S3 antigens, or PAW antigens) have shown very limited cross-reactions in extensive diagnostic testing in humans and other species. In at least one case, purified antigens were reported to improve the sensitivity of antibody detection in individuals from an endemic area (17). The TSL-1 antigens have a further advantage in that they are conserved in all subspecies and variants of *Trichinella* (18).

A number of comparisons have been made of test formats for the diagnosis of human trichinellosis, and these have been recently reviewed (2). Although variation exists from laboratory to laboratory, the literature generally supports the fact that the most sensitive tests include ELISA, BF, and IFA (19). The major drawbacks to IFA are the requirement for a fluorescence microscope to determine results and the fact that it employs a crude antigen source (whole larvae). Both BF and ELISA are simple and have the versatility of using either a crude worm extract or more defined antigens.

The BF test has been the standard used by the U.S. Centers for Disease Control for many years. The test is based on the aggregation of antigen-coated bentonite particles in the presence of specific antibodies (20). The bentonite particles are typically coated with an acid-soluble extract of *T. spiralis* muscle larvae, called Melcher's antigen (21). Serial dilutions of serum are mixed with coated bentonite particles on a slide and allowed to react for 15 min. Greater than 50% agglutination of particles is considered a positive result. The results of the BF test are expressed as endpoint titer of flocculation; a titer of 1:5 or higher is considered positive for trichinellosis. One drawback to the BF test is that considerable standardization is required from batch to batch as antigen-coated particles are prepared.

The ELISA test is based on the detection of specific antibodies binding to *T. spiralis* antigens coated onto the plastic wells of microtiter plates. Antigens used in the ELISA include Melcher's antigen, biochemically extracted stichosome antigens, or stichosome products collected from in vitro–cultured *T. spiralis* muscle larvae (5,15,22). ELISA values are compared to known positive and negative controls and results are expressed as endpoint titers, absorbance values, or percentages of reference-positive sera. Commercial ELISA kits using a secretory antigen are available (22,23).

Several studies have compared BF to commercially available ELISA tests (22–24). In general, the ELISA has shown somewhat greater sensitivity with comparable specificity. A low rate of false-positive results has been reported by some authors using ELISA (23,24). However, other studies suggest that false-positive reactions can be eliminated when strict quality control is applied to the production of secretory antigens (25).

Patient antibody response to *Trichinella* antigens is variable and therefore factors affecting this response affect serology results. Animal studies have shown that the temporal appearance of host antibodies is related to the level of infection (25,26), although this has not been thoroughly documented in human studies. Seroconversion can occur in humans from 2 weeks to 2 months or longer after infection. Most studies indicate that positive antibody responses are detected at approximately 3 weeks following exposure (9,27) and may persist for many years following acute disease (28,29). In one study, patients with confirmed trichinellosis were serologically positive at 57 and 120 days post infection, but were antibody-negative at 23 and 700 days after infection (30). Serological findings should be cautiously considered; most important for diagnostic purposes is the finding of rising antibody titers in serial bleedings.

The detection of immunoglobulin G class antibodies is most productive in serological tests for trichinellosis in humans (9). Antibodies of the IgG_1 subclass dominate the IgG response (31) while IgG_4 rises during chronic infection. The appearance of IgM antibodies is essentially coincident with IgG (27) and the level of IgM has been shown to correlate with bentonite flocculation titers (32). Responses of IgA antibodies are transient (31) and while total IgE is often elevated in trichinellosis, detection of parasite-specific IgE is quite variable (32). Successful testing of patient sera has used an anti-IgG class-specific antibody in an ELISA format (22).

Because seroconversion occurs at 3 weeks or later following infection, serology may not be useful during the acute stages of severe trichinellosis. Thus, early diagnosis must rely on signs, symptoms, and laboratory findings suggestive of trichinellosis, a history of eating food consistent with exposure, and possibly biopsy (1). However, it should be recognized that biopsy will only detect infection after newborn larvae have reached and entered muscle tissue; thus biopsy results will not be useful until approximately 2 weeks following infection (1).

An alternative to antibody serology for early detection is a test for circulating antigens (33,34). In one study, antigens were detected in serum from 47% of patients with clinical trichinellosis, but only from 13% of patients with a documented history of exposure but with no clinical signs or symptoms (34). In clinical cases, antigen was detected in some patients prior to seroconversion. In another study, which detected as little as 30 pg of antigen in 5 μL, about one-half of antibody positive samples were also antigen-positive; however, early in infection some patients were antibody-negative but antigen-positive (35). It is possible that with improved sensitivity, antigen detection may provide a noninvasive alternative to biopsy for definitive diagnosis.

III. ZOONOSES TRANSMITTED BY FISH OR INVERTEBRATES

The value of immunodiagnostic methods for these nematode parasites varies relative to the ease of detection by direct methods. For example, diagnosis of intestinal *Capillaria* infection relies entirely on identification of eggs, larvae, or adult worms in stool samples.

There is no good correlation between antibody response and infection, and therefore serology is of no value (36). For other parasite infections (anisakiasis, angiostrongyliasis, gnathostomiasis), serology results can be of some value, particularly as confirmatory tests when clinical signs and symptoms indicate a parasite infection, and direct methods are impractical or ineffective. Two major drawbacks to serology testing need to be overcome for these serology tests to be of clinical value. First, many tests suffer from lack of specificity as evidenced by cross-reactions with other parasite infections. Considerable efforts need to be applied to the area of antigen purification, to eliminate shared parasite antigens and focus on antigens unique to the parasite species being detected. Second, positive serology results are often not obtained during acute disease due to a lag in the time of antibody production in relation to parasite development. Further research on the kinetics of antibody production is needed. Alternative methods such as antigen detection have potential diagnostic value, and further development of these methods might facilitate an early and accurate diagnosis.

A. Gnathostomiasis

Human infection with *Gnathostoma spinigerum* and related species is caused by ingestion of intermediate hosts (fish, frogs, snakes, etc.) containing encysted L3 larvae. Infection results in cutaneous or subcutaneous migratory swelling and, in some cases, eosinophilic myeloencephalitis (37) caused by larvae migrating in the central nervous system (CNS). Definitive diagnosis relies on the recovery of parasites from host tissue, a procedure that is rarely successful. Due to a complex of symptoms that are not readily distinguished from other parasite infections (angiostrongyliasis, cysticercosis, cutaneous larva migrans), a need for confirmatory tests exists.

Most serology tests for gnathostomiasis have used crude extracts of advanced third-stage larvae (38–40) or, in some case, adult worms (41) as the antigen; diagnostic performance of somatic antigens is reported to be superior to metabolic antigens (38). Significant cross-reactions have been reported in serology tests with parasites including *Angiostrongylus*, *Opisthorchis*, *Clonorchis*, *Paragonimus*, *Diphyllobothrium*, *Trichinella*, hookworms, and taeniids (38–40,42,43).

In general, ELISA is superior to other tests (38,39,41,44,45) with sensitivity ranging from 59% to 100% and specificity ranging from 79% to 97% when tested on healthy patients or patients with other parasite infections. Double-diffusion tests have been reported to be more specific (100%) than ELISA (41,42) but have shown considerably lower sensitivity in detecting confirmed cases. Several studies have identified antigens from *Gnathostoma* sp. that appear to offer improved specificity for diagnostic testing. Akao et al. (46) used an L3 extract from *G. hispidum* in immunoblots and identified a 31.5-kDa antigen that reacted specifically with sera from patients with *G. hispidum* and *G. spinigerum*, but did not react with sera from patients with toxocariasis. Tapchaisri et al. (43) found that a 24-kDa protein from extracts of advanced L3 larvae reacted specifically with sera from patients with confirmed gnathostomiasis but was nonreactive with sera from patients with a variety of other parasitic infections. This antigen, which was purified by Nopparatana et al. (47), improved specificity of the ELISA test to 100% while retaining 100% sensitivity.

Rojekittikhun et al. (48,49) purified an approximately 39-kDa antigen from *G. spinigerum* L3 and found that while it lowered overall background, it did not improve overall test sensitivity over use of a crude antigen extract. Tuntipopipat el al. (40) compared excretory-secretory (ES) products with a crude extract of L3. The ES products were effective in reducing most cross-reactions in patients with confirmed angiostrongyliasis, clonorchiasis, and cysticercosis. However, the use of low molecular weight (less than 29 kDa) antigens separated from a crude L3 extract was most effective for diagnosis, with a specificity and sensitivity of 100%. The results of these diagnostic studies suggest potential value in the use of purified antigens in diagnosis of gnathostomiasis and indicate the need for further testing with cases of confirmed and presumptive gnathostomiasis and with sera from patients with other parasite infections.

Little information is available on the kinetics of antibody production or the relative importance of antibody subclasses in cases of gnathostomiasis. In one study, antibody to *G. spinigerum* was detected in cerebrospinal fluid (CSF) from patients who were positive for antibody in serum (50), suggesting that testing CSF could be useful in confirmation of cerebral gnathostomiasis. In the same study, methods for the detection of antigen were not successful, suggesting a lack of parasite antigen in CSF.

B. Angiostrongyliasis

Human infection with *Angiostrongylus cantonensis* results from the ingestion of molluscan intermediate hosts or contaminated vegetation. Infections are characterized by eosinophilic meningitis (EOM) caused by larvae migrating in the CNS. Disease diagnosis is based on clinical signs and symptoms, laboratory findings, and, in some cases, computed tomography (51). Definitive diagnosis relies on the recovery of worms from CSF (52). Because of the difficulty in obtaining such definitive diagnoses, serology is useful as a confirmatory test.

Various tests have been applied to the diagnosis of angiostrongyliasis including indirect hemagglutination, indirect immunofluorescence assays, and ELISA (51,53,54; reviewed in 55). Studies using crude worm extracts as antigen in ELISA formats demonstrate that titers or absorbance values for confirmed cases of angiostrongyliasis are generally higher than values obtained from control sera or serum samples from patients with other parasite infections (56–58). However, these and other studies demonstrate a broad range of cross-reactivity in serology tests when crude extracts of *A. cantonensis* are used as antigens. Cross-reactions occur with sera from patients infected with other parasite species including *Gnathostoma*, *Toxocara*, *Ascaris*, and *Metastrongylus* (57,59,60).

Kliks et al. (60) reported the successful use of a micro-ELISA using crude adult worm extract as antigen and peroxidase-conjugated anti-human IgG as second antibody. These authors suggest that a titer of 1:64 or greater, or a single dilution absorbance value greater that two standard deviations above the mean value of a normal serum pool, in serum or CSF should be considered positive. Furthermore, they suggest that titers of less than 1:64 cannot be considered negative during the first 2 weeks following exposure and that rising titers thereafter indicate active infection. The extent of cross-reactions with other parasites was not discussed.

Improvement in the specific diagnosis of *Angiostrongylus* infection has been achieved in studies using purified antigen preparations. Using techniques of preparative isoelectric focusing, Kum and Ko (59) isolated three antigen fractions from adult *A. cantonensis* with major reactivity at approximate molecular weights of 31.5, 20, and 11 kDa. Each of these antigen preparations eliminated cross-reactions with other parasites and reliably detected antibodies to *A. cantonensis* in serum and CSF from a patient with acute angiostrongyliasis. Kum and Ko (61) increased the specificity of ELISA by passing an extract of adult *A. cantonensis* over affinity columns constructed with rabbit antibodies to *Ascaris suum, Metastrongylus apri*, and *Toxocara canis*. The resulting antigen was devoid of epitopes recognized by sera to these heterologous parasites. Akao et al. (62) used a 29-kDa and a 31-kDa antigen from adult female worms to improve the specificity of diagnosis. Eamsobhana et al. (63) found that this 31-kDa antigen was recognized by 60% of patients with confirmed angiostrongyliasis and did not cross-react with other helminthic infections. In contrast to these authors, Nuamtanong (64) found the 29-kDa antigen to cross-react with sera from patients with gnathostomiasis, toxocariasis, trichinellosis, trichuriasis, hookworm disease, strongyloidiasis, and sparganosis. The 31-kDa antigen cross-reacted with sera from patients with trichinellosis, trichuriasis, and opisthorchiasis. Sensitivity and specificity of diagnosis with the 31-kDa antigen was 69.2% and 82.4%, respectively.

It is fairly clear that detectable antibody does not reliably appear until some 2 weeks following infection in cases of angiostrongyliasis and that most antibodies are directed to antigens from young adult (L5) and adult worms (65). Based on the need for rapid diagnosis, serology testing has limited value. Limited studies have demonstrated that it is possible to detect antigen from *Angiostrongylus* in both CSF and serum of infected patients. Shih and Chen (66) used monoclonal antibodies in a modified antigen capture ELISA to detect a 91-kDa L3 excretory-secretory antigen in 100% of CSF samples and 88% of serum samples from 35 patients with clinical symptoms of EOM. Sera from patients with schistosomiasis, toxocariasis, clonorchiasis, and taeniasis did not cross-react in this test. The time course of appearance and persistence of CSF and circulating antigen was not determined. Chang et al. (67) used monoclonal antibodies to identify a 204-kDa antigen from young adults (L5) of *A. cantonensis*. The antigen epitopes found on this polypeptide were not found in extracts of several other parasites tested. Chye et al. (68) used these monoclonal antibodies in an ELISA to detect circulating antigens. Higher ELISA values were obtained with sera from patients with eosinophilic meningitis and from whom worms had been recovered, as compared with patients with eosinophilic meningitis but not parasitologically confirmed. ELISA values for both groups showed good separation from noninfected control serum samples. Test sensitivity was improved by the use of CSF (97.6%) as compared with serum (81%); the use of either CSF or serum resulted in 100% specificity.

Because a limited number of cases of angiostrongyliasis have been studied in depth by serodiagnosis, a paucity of information is available on the kinetics of antibody production, predominant antibody classes produced, and the general sensitivity of testing procedures. In one study, Yen and Chen (58) found that detection of IgG improves both the sensitivity and specificity of the ELISA; however, considerable overlap existed in the range of values for confirmed positive and negative sera.

C. Anisakiasis

Anisakiasis results from the ingestion of raw or undercooked fish containing larvae of several genera (*Anisakis, Pseudoterranova, Contracaecum*) of nematodes in the family Anisakidae (69). Infection causes epigastric or abdominal pain, nausea, vomiting, and diarrhea. Endoscopy is widely used in Japan for the diagnosis and treatment of gastric anisakiasis. This technique has been applied to duodenal and rectal infections as well (69). Various immunodiagnostic tests have been used for detection of anisakid infections including skin testing, complement fixation, IFA, immunodiffusion, immunoelectrophoresis, ELISA, and the radioallergosorbent test (RAST) (70). Desowitz et al. (71) found that

RAST, which uses an *A. simplex* larval antigen and detects parasite-specific IgE in patient sera, was capable of differentiating infection with *A. simplex* from uninfected controls. However, cross-reactions were obtained with sera from patients with schistosomiasis, ascariasis, visceral larval migrans, and asthma. Sakanari et al. (72) also obtained cross-reactions with sera from a patient with *Ascaris* infection using RAST. These authors found by ELISA that parasite-specific IgE levels were only slightly elevated above normal in *Anisakis*-infected patient sera, whereas parasite-specific IgM and IgG antibody responses were both significantly higher than controls. In other studies, Poggensee et al. (73) and García-Palacios et al. (74) found ELISA for IgG or total antibodies to be more sensitive in detecting patients with anisakiasis than RAST for IgE. However, detection of IgE by RAST resulted in fewer positive reactions in patients with schistosomiasis and ascariasis (73). The replacement of crude antigen with excretory-secretory antigens from third-stage *A. simplex* larvae resulted in lower background values and less cross-reactivity in tests detecting IgG or IgE (73).

Garcia et al. (75) used an immunoblot test measuring IgE to differentiate patients with *A. simplex* allergy from those with high levels of specific IgE but no allergy symptoms. Specific banding patterns, recognizing 30- to 50-kDa antigens, were seen in a predominance of allergic patients sera, whereas those bands were not recognized in sera from nonallergic or control patients.

Yagihashi et al. (76) used a monoclonal antibody (An2) specific to *A. simplex* third-stage larval polypeptides of approximate molecular weights of 40–42 kDa to capture antigen for ELISA testing. These authors obtained good specificity (98–100%) using sera from uninfected patients; sensitivity increased with time after infection, ranging from 46% at 1 day after infection to 100% at 4 weeks post infection. Early detection (1 day post infection) was improved when sera were tested for IgE, while better sensitivity was obtained testing for IgG, IgA, and IgM at later times (4–5 weeks post infection). Iglesias et al. (77) produced several monoclonal antibodies that were specific for *A. simplex* antigens and had limited or no cross-reactivity with *A. suum, T. canis,* or *Hysterothylacium aduncum*. Two of these monoclonals, UA2 recognizing 48- and 67-kDa antigens, and UA3 recognizing 139- and 154-kDa antigens, performed well in a sandwich ELISA for detection of specific *A. simplex* immunoglobulins in human sera. Antibody detection was more sensitive as compared with the same test using the An2 monoclonal antibody.

Considerable work remains before a reliable diagnostic test is available for anisakiasis, including identifying and validating specific antigens and a more in-depth analysis of the level and kinetics of an antibody response. Based on the need for early diagnosis, it is likely that serology will have limited value.

IV. CONCLUSIONS

The value of serological testing for detection of foodborne parasite infection in humans depends on the difficulty of detection by other methods (such as direct demonstration of the parasite) and the reliability (sensitivity and specificity) of the test. Parasites that are tissue dwellers are often difficult to detect directly; therefore, diagnosis sometimes requires alternative methods. When use of early direct detection methods is not feasible, patient seroconversion can be a valuable diagnostic tool. Considerable improvements are needed for serodiagnostic tests for several parasites. Improvements include the identification of specific antigens to eliminate cross-reactivity and optimization of antibody detection systems, with respect to antibody class specificity, temporal expression of antibody, and signal detection. Further efforts to improve early diagnosis should also consider improvements in methods for the demonstration of parasite antigens in patient body fluids as well as other methods such as ultrasound or nuclear magnetic resonance for direct detection of parasites.

REFERENCES

1. H. R. Gamble and K. D. Murrell. Trichinellosis. In: Laboratory Diagnosis of Infectious Disease: Principles and Practice (W. Balows, ed.), Springer-Verlag, New York, pp. 1018–1024 (1988).
2. M. R. Chapa-Ruiz, M. R. Salinas-Tobon, R. Martinez Maranon, R. Cedillo, and E. Garcia-Latorre. Diagnosis of human trichinosis by indirect enzyme linked immunosorbent assay (ELISA). Rev. Latamer. Microbiol., 31: 133–136 (1989).
3. S. Romand, P. Bouree, and D. Limonne. Evaluation du serodiagnostic de trichinellose par un test ELISA. Bull. Soc. Franciase Parasitol., 12: 11–16 (1994).
4. E. J. Ruitenberg, P. A. Steerenberg, B. J. M. Brosi, and J. Buys. The reliability of enzyme-linked immunosorbent assay (ELISA) for the serodiagnosis of *Trichinella spiralis* infection in conventionally raised pigs. J. Immunol. Meth., 10: 67–83 (1976).
5. M. Kobayashi, M. Niimura, M. Yokogawa, and T. Yamaguchi. Studies of immunoserological tests for acute human trichinellosis. Jpn. J. Parasitol., 36: 248–253 (1987).
6. N. Morakote, C. Khamboonruang, V. Siriprasert, S. Suphawitayanukul, S. Marcanantachoti, and W. Thamasonthi. The value of enzyme-linked immunosorbent assay (ELISA) for diagnosis of human trichinosis. Trop. Med. Parasitol., 42: 172–174 (1991).
7. M. Niimura, M. Kobayashi, and S. Kojima. Immunodiagnostic value of an α-stichocyte-derived antigen isolated by affinity chromatography for trichinosis. Jpn. J. Parasitol., 41: 287–293 (1992).
8. P. Mahannop, P. Setasuban, N. Morakote, P. Tapchaisri, and W. Chaicumpa. Immunodiagnosis of human trichinellosis and identification of specific antigen for *Trichinella spiralis*. Int. J. Parasitol., 25: 87–94 (1995).
9. R. C. Ko. Application of serological techniques for the diagnosis of trichinellosis. In: Current Concepts in Parasitology (R. C. Ko, ed.), Hong Kong University Press, Hong Kong, pp. 81–100 (1989).
10. H. R. Gamble, W. R. Anderson, C. E. Graham, and K. D. Murrell. Serodiagnosis of swine trichinosis using an excretory secretory antigen. Vet. Parasitol., 13: 349–361 (1983).
11. H. R. Gamble and C. E. Graham. A monoclonal antibody-purified antigen for the immunodiagnosis of trichinosis. Am. J. Vet. Res., 45: 67–74 (1984).
12. D. D. Despommier and A. Lacetti. *Trichinella spiralis*: partial characterization of antigens isolated by immuno-affinity chromatography from the large particle fraction of the muscle larvae. J. Parasitol., 67: 332–339 (1981).
13. S. W. Chan and R. C. Ko. Specificity of affinity-purified *Trichinella spiralis* antigens. Vet. Parasitol., 41: 109–120 (1992).
14. J. A. Appleton, R. G. Bell, W. Homan, and F. van Knapen. Consensus on *Trichinella spiralis* antigens and antibodies. Parasitol. Today, 7: 190–192 (1991).
15. G. L. Seawright, W. J. Zimmerman, R. A. Isenstein, and D. D. Despommier. Enzyme immunoassay for swine trichinellosis using antigens purified by immunoaffinity chromatography. Am. J. Trop. Med. Hyg., 32: 1275–1284 (1983).
16. H. R. Gamble, D. Rapic, A. Marinculic, and K. D. Murrell. Influence of cultivation conditions on specificity of excretory-secretory antigens for the immunodiagnosis of trichinellosis. Vet. Parasitol., 30: 131–137 (1988).

17. M. R. Chapa-Ruiz, M. R. Salinas-Tobón, D. J. Aguilar-Álvarez, and R. Martínez-Marañón. Recognition of *Trichinella spiralis* muscle larvae antigens by sera from human infected with this parasite and its potential use in diagnosis. Rev. Lat. Am. Microbiol., 34: 95–99 (1992).

18. H. R. Gamble and K. D. Murrell. Conservation of diagnostic antigen epitopes among biologically diverse isolates of *Trichinella spiralis*. J. Parasitol., 72: 921–925 (1986).

19. J. Stumpf, K. Undeutsch, and H. Kandgraf. Results of the clinical and serological diagnosis of an epidemic of *Trichinella spiralis*. In: Trichinellosis (C. W. Kim, E. J. Ruitenberg, and J. S. Teppema, eds.), Reedbooks, Ltd., Windsor, Berks, pp. 279–282 (1981).

20. I. G. Kagan and L. Norman. The serology of trichinosis. In: Trichinosis in Man and Animals (S. E. Gould, ed.), Charles C Thomas, Springfield, pp. 222–268 (1970).

21. L. R. Melcher. An antigenic analysis of *Trichinella spiralis*. J. Infect. Dis., 73: 31–39 (1943).

22. D. G. Oliver, P. Singh, L. S. Jang, A. Boron-Kaczmarska, N. Skovgaard, J. M. Flink, and E. Ingerslev. Enzyme linked immunoassay for detection of trichinellosis in humans. In: Trichinellosis (C. W. Tanner, A. R. Martinez-Fernandez, and F. Bolas-Fernandez, eds.), Consejo Superior de Investigaciones Cientificas Press, Madrid, pp. 250–253 (1989).

23. M. Wilson, D. A. Ware, and J. B. McAuley. Comparison of the bentonite flocculation test with the LMD ELISA kit for detection of antibodies to *Trichinella spiralis*. Am. J. Trop. Med. Hyg., 45 (3 Suppl.):160 (1991).

24. C. Boodram and I. Kagan. Evaluation of LMD Laboratories' 20 minute microtiter EIA for the detection of antibodies to *Trichinella spiralis*. Proceedings 91st Meeting of the American Society for Microbiology, p. 397 (1991).

25. H. R. Gamble, D. Rapic, A. Marinculic, and K. D. Murrell. Factors influencing the efficacy of excretory-secretory antigens in the serodiagnosis of swine trichinellosis. In, Trichinellosis (C. W. Tanner, A. R. Martinez-Fernandez, and F. Bolas-Fernandez, eds.), Consejo Superior de Investigaciones Cientificas Press, Madrid, pp. 202–209 (1989).

26. F. van Knapen, J. H. Franchimont, E. J. Ruitenberg, P. Andre, B. Baldelli, T. E. Gibson, C. Gottal, S. A. Hendriksen, G. Kohler, O. Roneus, N. Skovgaard, C. Soule, K. L. Strickland, and S. M. Taylor. Comparison of four methods for the early detection of *Trichinella spiralis* infections in pigs. Vet. Parasitol., 9:117–123 (1981).

27. H. Feldmeier, H. Fischer, and G. Blaummeiser. Kinetics of humoral response during the acute and the convalescent phase of human trichinosis. Zentralbl. Bakteriol. Mikrobiol. Hyg. A, 264: 221–234 (1987).

28. H. Feldmeier, U. Bienzle, R. Jansen-Rosseck, P. G. Kremsner, H. Weiland, G. Dobos, S. Schroeder, D. Fengler-Dopp, and H. H. Peter. Sequelae after infection with *Trichinella spiralis*: A prospective cohort study. Wein. Klin. Wochenschr., 103/4: 111–116 (1991).

29. W. Froscher, F. Gullotta, M. Saathoff, and W. Tackmann. Chronic trichinosis: Clinical, bioptic, serological and electromyographic studies. Eur. Neurol., 28: 221–226 (1988).

30. P. Mahannop, W. Chaicumpa, P. Setasuban, N. Morakote, and P. Tapchaisri. Immunodiagnosis of human trichinellosis using excretory-secretory (ES) antigen. J. Helminthol. 66: 297–304 (1992).

31. I. Ljungstrom, L. Hammarstrom, W. Kociecka, and C. I. Smith. The response of IgG subclasses and IgE during early and late stages of human *Trichinella spiralis* infection. In: Trichinellosis (C. W. Tanner, A. R. Martinez-Fernandez, and F. Bolas-Fernandez, eds.), Consejo Superior de Investigaciones Cientificas Press, Madrid, pp. 210–217 (1989).

32. R. Patterson, M. Roberts, G. Slonka, and J. McAninch. Studies of immunoglobulins, bentonite flocculation and IgE, IgG and IgM antibodies in serum from patients with trichinosis. Am. J. Med., 58: 787–793 (1975).

33. E. Candolfi, Ph. Frache, M. Liance, R. Houin, and T. Kien. Detection of circulating antigen in trichinellosis by immuno-enzymology. Comparative results in mice, rats and humans. In: Trichinellosis (C. W. Tanner, A. R. Martinez-Fernandez, and F. Bolas-Fernandez, eds.), Consejo Superior de Investigaciones Cientificas Press, Madrid, pp. 194–201 (1989).

34. D. Ivanoska, K. Cuperlovic, H. R. Gamble, and K. D. Murrell. Comparative efficacy of antigen and antibody detection tests for human trichinellosis. J. Parasitol., 75: 38–41 (1989).

35. T. H. Dzbeński, E. Bitkowska, and W. Plonka. Detection of circulating parasitic antigen in acute infections with *Trichinella spiralis*: diagnostic significance of findings. Zbl. Bakt. 281: 519–525 (1994).

36. J. H. Cross. Capillariasis. In: Laboratory Diagnosis of Infectious Disease: Principles and Practice (W. Balows, ed), Springer-Verlag, New York, pp. 793–800 (1989).

37. N. Jaroonvesama. Differential diagnosis of eosinophilic meningitis. Parasitol. Today, 4: 262–266 (1988).

38. W. Maleewong, N. Morakote, W. Thamasonthi, K. Charuchinda, S. Tesana, and C. Khamboonruang. Serodiagnosis of human gnathostomiasis. Southeast Asian J. Trop. Med. Publ. Health, 19: 201–205 (1988).

39. P. Suntharasamai, V. Desakorn, S. Migasena, D. Bunnag, and T. Harinasuta. Southeast Asian J. Trop. Med. Publ. Health, 16: 274–279 (1985).

40. S. Tuntipopipat, R. Chawengkirttikul, and S. Sirisinha. A simplified method for the fractionation of *Gnathostoma*-specific antigens for the serodiagnosis of human gnathostomiasis. J. Helminthol., 67: 297–304 (1993).

41. M. T. Anantaphruti. ELISA for diagnosis of gnathostomiasis using antigens from *Gnathostoma doloresi* and *G. spinigerum*. Southeast Asian J. Trop. Med. Publ. Health, 20: 297–304 (1989).

42. I. Tada, T. Araki, H. Matsuda, K. Araki, H. Akahane, and T. Mimori. A study on immunodiagnosis of gnathostomiasis by ELISA abd double diffusion with special reference to the antigenicity of *Gnathostoma doloresi*. Southeast Asian J. Trop. Med. Publ. Health, 18: 444–448 (1987).

43. P. Tapchaisri, C. Nopparatana, W. Chaicumpa, and P. Setasuban. Specific antigen of *Gnathostoma spinigerum* for immunodiagnosis of human gnathostomiasis. Int. J. Parasitol., 21: 315–319 (1991).

44. A. Dharmkrong-At, S. Migasena, P. Suntharasamai, D. Bunnag, R. Priwan, and S. Sirisinha. Enzyme-linked immunosorbent assay for detection of antibody to *Gnathostoma* antigen in patients with intermittent cutaneous migratory swelling. J. Clin. Immunol., 23: 847–851 (1986).

45. T. Mimori, I. Tada, M. Kawabata, W. L. Ollague, G. H. Calero, and Y. F. deChong. Immunodiagnosis of human gnathostomiasis in Ecuador by skin test and ELISA using *Gnathostoma doloresi* antigen. Jpn. J. Trop. Med. Hyg., 15: 191–196 (1987).

46. N. Akao, T. Ohyama, K. Kondo, and Y. Takakura. 1989. Immunoblot analysis of human gnathostomiasis. Ann. Trop. Med. Parasitol., 83: 635–637 (1989).

47. C. Nopparatana, P. Setasuban, W. Chaicumpa, and P. Tapchaisri. Purification of *Gnathostoma spinigerum* specific antigen and immunodiagnosis of human gnathostomiasis. Int. J. Parasitol., 21: 677–687 (1991).

48. W. Rojekittikhun, T. Yamashita, A. Saito, T. Watanabe, T. Azuma, and F. Sendo. Purification of *Gnathostoma spinigerum* larval antigens by monoclonal antibody affinity chromatography. Southeast Asian J. Trop. Med. Publ. Health, 24: 680–684 (1993).

49. W. Rojekittikhun, A. Saito, and F. Sendo. Crude and monoclonal antibody affinity-purified *Gnathostoma spinigerum* larval antigens for the immunodiagnosis of human gnathostomiasis. Southeast Asian J. Trop. Med. Publ. Health, 26: 439–442 (1995).

50. S. Tuntipopipat, R. Chawengkiattikul, R. Witoonpanich, S. Chiemchanya, and S. Sirisinha. Antigens, antibodies and immune complexes in cerebrospinal fluid of patients with cerebral gnathostomiasis. Southeast Asian. J. Trop. Med. Publ. Health, 20: 439–446 (1989).

51. R. C. Ko, M. C. Chiu, W. Kum, and S. H. Chan. First report of human angiostrongyliasis in Hong Kong diagnosed by computerized axial topography (CAT) and enzyme linked immunosorbent assay. Trans. R. Soc. Trop. Med. Hyg., 78: 354–355 (1984).

52. R. R. Chen. Angiostrongyliasis and eosinophilic meningitis on Taiwan: a review. In: Studies

on Angiostrongyliasis in Eastern Asia and Australia. (J. H. Cross, ed.), NAMRU-2 Special Publication (NAMRU-2-SP-44), Taipei, Taiwan (1979).

53. J. H. Cross. Clinical manifestations and laboratory diagnosis of eosinophilic meningitis syndrome associated with angiostrongyliasis. Southeast Asian J. Trop. Med. Publ. Health, 2: 161–170 (1978).

54. C. Y. Lin and S. N. Chen. Clinical observations and immunodiagnosis of angiostrongyliasis in North Taiwan. Med. J. Osaka. Univ., 31: 1–6 (1980).

55. R. C. Ko. Application of serological techniques for the diagnosis of angiostrongyliasis. In: Current Concepts in Parasitology (R. C. Ko, ed.), Hong Kong University Press, Hong Kong, pp. 101–110 (1987).

56. N. Jaroonvesama, K. Charoenlarp, P. Buranasin, G. G. Zaraspe, and J. H. Cross. ELISA testing in cases of clinical angiostrongyliasis in Thailand. Southeast Asian J. Trop. Med. Publ. Health, 16: 110–112 (1985).

57. S. N. Chen. Enzyme-linked immunosorbent assay (ELISA) for the detection of antibodies to *Angiostrongylus cantonensis*. Trans. R. Soc. Trop. Med. Hyg., 80: 398–405 (1986).

58. C. M. Yen and E. R. Chen. Detection of antibodies to *Angiostrongylus cantonensis* in serum and cerebrospinal fluid of patients with eosinophilic meningitis. Int. J. Parasitol., 21: 17–21 (1991).

59. W. W. S. Kum and R. C. Ko Isolation of specific antigens from *Angiostrongylus cantonensis*. 1. Preparative flatbed isoelectric focusing. Z. Parasitenkd., 71: 789–800 (1985).

60. M. M. Kliks, W. K. K. Lau, and N. E. Palumbo. Neurologic angiostrongyliasis: parasitic eosinophilic menigoencephalitis. In: Laboratory Diagnosis of Infectious Disease: Principles and Practice (W. Balows, ed.), Springer-Verlag, New York, pp. 754–767 (1989).

61. W. W. S. Kum and R. C. Ko. Isolation of specific antigens from *Angiostrongylus cantonensis*. 2. By affinity chromatography. Z. Parasitenked., 72: 511–516 (1986).

62. N. Akao, K. Kondo, T. Ohyama, E. Chen, and M. Sano. Antigens of adult female worm of *Angiostrongylus cantonensis* recognized by infected humans. Jpn. J. Parasitol., 41: 225–230 (1992).

63. P. Eamsobhana, J. W. Mak, and H. S. Yong. Development of specific immunodiagnosis for human parastrongyliasis. Workshop on parasitological diagnosis by immunology. 24–25 August, 1995, Bangkok, Thailand.

64. S. Nuamtanong. The evaluation of the 29 and 31 kDa antigens in female *Angiostrongylus cantoensis* for serodiagnosis of human angiostrongyliasis. Southeast Asian J. Trop. Med. Publ. Health, 27: 291–295 (1996).

65. J. H. Chang, C. M. Yen, E. R. Chen, L. Y. Chung, J. J. Wang, S. M. Chye, and L. C. Wang. Detection of antibodies to surface antigens of *Angiostrongylus cantonensis* by ELISA. Ann. Trop. Med. Parasitol., 89: 569–572 (1995).

66. H. H. Shih and S. N. Chen. Immunodiagnosis of angiostrongyliasis with monoclonal antibodies recognizing a circulating antigen of mol. wt. 91,000 from *Angiostrongylus cantonensis*. Int. J. Parasitol., 21: 171–177 (1991).

67. J. H. Chang, C. M. Yen, and E. R. Chen. Characterization of monoclonal antibodies to young-adult worms of *Angiostrongylus cantonensis*. Hybridoma, 9:465–471 (1990).

68. S-M. Chye, C-H. Yen, and E-R. Chen. Detection of circulating antigen by monoclonal antibodies for immunodiagnosis of angiostrongyliasis. Am. J. Trop. Med. Hyg., 56: 408–412 (1997).

69. J. W. Bier. Anisakiasis. In: Laboratory Diagnosis of Infectious Disease: Principles and Practice (W. Balows, ed.), Springer-Verlag, New York, pp. 768–774 (1989).

70. J. A. Sakanari and J. H. McKerrow. Anisakiasis. Clin. Microbiol. Rev., 2: 278–284 (1989).

71. R. S. Desowitz, R. B. Raybourne, H. Ishikura, and M. M. Kliks. The radioallergosorbent test (RAST) for the serological diagnosis of human anisakiasis. Trans. R. Soc. Trop. Med. Hyg., 79: 256–259 (1985).

72. J. A. Sakanari, H. M. Loinaz, T. L. Deardorff, R. B. Raybourne, J. H. McKerrow, and J. G.

Frierson. Intestinal anisakiasis: a case diagnosed by morphologic and immunologic methods. Am. J. Clin. Pathol., 90: 107–113 (1988).

73. U. Poggensee, G. Schommer, R. Jansen-Rosseck, and H. Feldmeier. Immunodiagnosis of human anisakiasis by use of larval excretory-secretory antigen. Zbl. Bakt. Hyg., 270: 503–510 (1989).

74. L. García-Palacios, M. L. González, M. I. Esteban, E. Mirabent, M. J. Perteguer, and C. Cuéllar. Enzyme-linked immunosorbent assay, immunoblot analysis and RAST fluoroimmunoassay analysis of serum responses against crude larval antigens of *Anisakis simplex* in a Spanish random population. J. Helminthol., 70: 281–289 (1996).

75. M. Garcia, I. Moneo, M. T. Audicana, M. D. del Pozo, D. Muñoz, E. Fernández, J. Díez, M. A. Etxenzguisa, I. J. Ansotegui, and L. Fernández de Corres. The use of IgE immunoblotting as a diagnostic tool in *Anisakis simplex* allergy. J. Allergy Clin. Immunol., 99: 497–501 (1997).

76. A. Yagihashi, N. Sato, S. Takahashi, H. Ishikura, and K. Kikuchi. A serodiagnostic assay by microenzyme-linked immunosorbent assay for human anisakiasis using a monoclonal antibody specific for *Anisakis* larvae antigen. J. Infect. Dis., 161: 995–998 (1991).

77. R. Iglesias, J. Leiro, M. T. Santamarina, M. L. Sanmartín, and F. M. Ubeira. Monoclonal antibodies against diagnostic *Anisakis simplex* antigens. Parasitol. Res., 83: 755–761 (1997).

17
Diagnosis of Toxoplasmosis

Alan M. Johnson
University of Technology, Sydney, New South Wales, Australia

J. P. Dubey
U.S. Department of Agriculture, Beltsville, Maryland

The diagnosis of toxoplasmosis usually is based on one or more of the following: clinical signs, parasite isolation, antibody detection, cell-mediated responses, and antigen detection. However, more recently, the detection of *Toxoplasma gondii* DNA by polymerase chain reaction (PCR) for diagnosis has become widespread. No single procedure gives absolute proof of infection rather than just infestation (i.e., chronic subclinical presence of the tissue cysts over many years); hence, ideally, a combination of two or more tests should be used.

I. CLINICAL SIGNS

A. Humans

The clinical diagnosis of toxoplasmosis in humans is complicated due to the vast range and variation in degree of severity of symptoms. Acquired toxoplasmosis in the immuno-competent host usually is asymptomatic but may present as lymphadenopathy, commonly of the cervical nodes (1). In severe cases, the symptoms may include fever, sore throat, headache, myalgia, malaise, spleen or liver involvement, and a misdiagnosis of infectious mononucleosis or common cold may be made. In rare cases, myocarditis, pericarditis, hepatitis, polymyositis, encephalitis, or meningoencephalitis may be present (2–6). However, these symptoms are nonspecific and may not be recognized as being caused by toxoplasmosis.

The most characteristic feature of acquired toxoplasmosis in the immunoincompetent host is central nervous system (CNS) involvement. Until recently, these neurological symptoms often were attributed to the effects of immunosuppressive therapy for malignant disease, organ transplantation, or collagen vascular disease, and the diagnosis of toxoplasmosis often was not made until the patient was moribund or had died and an autopsy was performed (4,5). However, the recent rise in the number of cases of acquired immunodeficiency syndrome (AIDS), and the role of CNS toxoplasmosis and other toxoplasmic lesions (7) in the death of a significant number of AIDS patients, has heightened the awareness of toxoplasmosis in immunocompromised patients.

Ocular complications of toxoplasmosis occur in about 2–3% of cases that present with other symptoms and are more likely to occur when the CNS is affected (8). The most characteristic ocular lesion in both acquired and congenital toxoplasmosis is a focal retinochoroiditis. When active, this may cause blurred vision, pain, photophobia, or epiphoria (9). The diagnosis of ocular toxoplasmosis can be made by finding a morphologically acceptable lesion in the fundus and, like the clinical diagnosis associated with the other forms of toxoplasmosis, should be accompanied by demonstration of *T. gondii* parasites, antibody, or antigen detection (10–12). One positive antibody titer at any dilution supports the diagnosis; a rising antibody titer is not required (13). Most cases of toxoplasmic chorioretinitis were considered to be the result of congenital infection (8,9). However, recently 19% of patients developed chorioretinitis following the largest outbreak of toxoplasmosis linked to drinking water contaminated with oocysts (14) and ocular toxoplasmosis caused by acquired infection is being reported increasingly (15). Silveira et al. (16) had previously reported many cases of acquired ocular toxoplasmosis in siblings in Brazil.

Congenital infection in pregnant women may result in abortion, fetal abnormalities, or perinatal death. It is less common in infants born to mothers infected during the first trimester but, if contracted then, the disease is most severe. Conversely, congenital infection is more common in infants born to mothers infected during the third trimester, but the infection is less severe and may be asymptomatic (17). The symptoms shown by overtly infected neonates may include hydrocephaly or microcephaly, retinochoroiditis, cerebral calcification, fever, hepatomegaly, splenomegaly, jaundice, or rash (9). However, many of these symptoms are manifested by such other neonatal infections as rubella, cytomegalovirus, *Herpes simplex*, or syphilis, and consequently, once again a definitive diagnosis of congenital toxoplasmosis should also involve at least one of the other diagnostic techniques.

Congenital toxoplasmosis in humans has been comprehensively reviewed up to 1982 (18). Recent studies indicate that, of congenitally infected neonates asymptomatic at birth, as many as 75% may develop such adverse sequelae as mental retardation and/or hearing defects later in life (19). As many as 90% may suffer ocular problems as they grow older (20).

B. Animals

Along with the loss and reduction in quality of human life, toxoplasmosis causes great financial loss in the agricultural industry. In New Zealand, Britain, the United States, and Australia, toxoplasmosis is one of the more common causes of abortion and stillbirth in sheep and goats. Toxoplasmosis is estimated as having caused up to 40% of outbreaks of abortion and neonatal death in sheep in Tasmania (21) and an annual loss of as many as 100,000 lambs in Britain (22). Toxoplasmosis also has been identified as a cause of abortion in Angora goats in both Tasmania and New South Wales (23). Dubey and Beattie have reviewed the clinical signs of toxoplasmosis extensively in a wide range of domestic animals (24).

II. PARASITE ISOLATION

Culture of the parasite from an appropriate clinical specimen such as blood, cerebrospinal fluid, or placental products gives convincing proof that *T. gondii* is responsible for the associated clinical symptoms (25). However, isolation of the parasite from an inappropriate specimen such as muscle, lung, or brain may reflect only the presence of dormant tissue cysts (9,10).

Mouse inoculation, tissue culture, and egg inoculation all have been used for the isolation of the parasite; however, the first is probably the easiest, most commonly used, and most sensitive method (26). Virulent strains usually kill the mice within 4–14 days; less virulent strains form tissue cysts that can be easily identified microscopically in squash preparations of the brains of the mice about 8 weeks after inoculation. The main advantages of this method of diagnosis are that a greater volume of material may be examined and it enables the parasite to proliferate, thereby resulting in an increased level of detection. Its main disadvantage is the time required to make the diagnosis. Dubey (27) has recently updated knowledge on the life cycle of *T. gondii*.

III. ANTIBODY DETECTION

Many serological tests have been used for the detection of immunoglobulin G (IgG) *T. gondii* antibodies: Sabin-Feldman dye test (DT), indirect hemagglutination (IHA), complement fixation (CF), modified agglutination test (MAT), latex agglutination (LA), indirect fluorescent antibody (IFA), and enzyme-linked immunosorbent assay (ELISA). Of these, IFA, ELISA, and MAT have been modified to detect immunoglobulin M (IgM) antibodies.

A. Dye Test

The DT remains the definitive test for human toxoplasmosis but, because of technical difficulties, it is performed in only a very few laboratories in the world. More is known of the DT antibodies in humans than of any other type of antibody. This is essentially a complement-mediated neutralizing type of antigen–antibody reaction. In this test, live tachyzoites are incubated with accessory factor (complement from human serum) and the test serum at 37°C for 1 h and a dye (methylene blue) is added. Tachyzoites unaffected by antibody are stained uniformly with methylene blue. Specific antibody induces complement-mediated cytolysis of tachyzoites and the cytoplasm leaks out (28,29). As a result, tachyzoites affected by antibody do not incorporate methylene blue and appear as "ghosts." The titer reported is the dilution at which 50% of tachyzoites remain unstained.

It is important that the reports of serological testing should be consistent and comparable. For this reason, the World Health Organization (WHO) Collaborating Center for Research and Reference on Toxoplasmosis, established in the Statens Seruminstitut, Copenhagen, Denmark (30), supplies standard anti-*Toxoplasma* sera. Results are expressed in international units (IU). One ampule contains 2000 IU, corresponding to 2 mL serum, the titer of which is 1000 IU. The DT is highly specific and sensitive with no evidence of false results in humans. However, it is expensive, requires a high degree of technical expertise, and potentially is unsafe because of the use of live virulent organisms. It measures complement-fixing antibodies. Most hosts develop DT antibodies within 4 weeks and titers may remain stable for months or years. The test has been standardized and adopted to microtiter plates (31–33). Although DT measures antibodies specific to *T. gondii* in many hosts, ruminant sera have substances that give false results unless the sera are inactivated at 60°C for 30 min (34).

B. Indirect Hemagglutination Test

In the IHA test, soluble antigen from tachyzoites is coated on tanned red blood cells that then are agglutinated by immune serum (35). Full technical details are given in a serology manual prepared by Palmer et al. (36), and many commercial kits are available. Although the IHA test is simple in principle, it has many variables (37). Furthermore, IHA detects antibodies later in the disease than the DT and titers remain elevated for a long period, so that acute infections are likely to be missed by this test. The IHA test also frequently is negative in congenital infections (18). In animals, titers lower than 1:128 may be nonspecific (38).

C. Complement Fixation Test

It generally is believed that CF antibodies appear later than DT antibodies and the test is positive during acute infection, but this depends largely on the antigenic preparation. This is not a test of choice because of complex procedures and lack of standardization of the antigens and reagents.

D. Modified Agglutination Test

In the MAT, sera are treated with 2-mercaptoethanol to remove cross-reacting IgM-like substances and to react with whole, killed tachyzoites. The test is simple, easy to perform, and kits are available commercially. Titers in the MAT parallel those in the DT (39) in both human sera and animal sera (38,40,41). Technical details are given by Desmonts and Remington (39) and Dubey and Desmonts (37). The reliability of the MAT for the diagnosis of latent toxoplasmosis has been evaluated extensively in pigs and cats (43–46).

MAT antibody titers may be detectable as early as 3 days postinfection and titers may peak within a month of infection. MAT antibody titers may persist for many years. Although MAT may not be useful in detecting recent infections, a modification of the original MAT using acetone-fixed (instead of formalin-fixed) tachyzoites has proved useful in detecting recent acute infection, particularly in AIDS patients. The acetone-fixed tachyzoites react only to IgG antibodies produced during acute infection, whereas formalin-fixed tachyzoites react to IgG antibodies during both acute and chronic infection. The MAT has proved very useful in the diagnosis of toxoplasmosis in animals.

Desmonts et al. developed an ingenious modification of IgM-ELISA to eliminate the necessity for an enzyme conjugate and combined it with the immunosorbent agglutination assay (ISAGA) agglutination test (IgM-ISAGA) (47). In this test, wells of microtiter plates are coated with anti–human IgM antibodies. After washing, test sera are added to the wells and incubated overnight at 37°C to allow the binding of IgM. The plates then are washed and a suspension of whole tachyzoites (as in the direct agglutination test of Desmonts and Remington discussed in Ref. 39) is added to the wells. The pattern of agglutination is read as for the MAT. If the patient's serum has specific IgM, then it binds to the antispecies IgM and then will agglutinate a fixed parasite antigen. If the patient's serum is negative for IgM, then *T. gondii* settles at the bottom of the well and gives a negative reaction (47,48).

This test is simpler, more rapid, and easier to perform than the IgM-ELISA. However, IgM-ISAGA requires large numbers of tachyzoites. To overcome this, Remington et al. modified the IgM-ISAGA by substituting whole tachyzoites with latex beads coated with soluble *T. gondii* antigen (49). However, this test has not been adopted by other laboratories as the type of latex beads used is apparently critical.

E. Latex Agglutination Test

In the LA test, soluble antigen is coated on latex particles and the pattern of agglutination is observed when the serum to be tested is added (50). The LA test is available commercially. It is easy to perform and does not require special training or equipment. Titers in the LA test compared favorably with those for the DT and the test was judged to be useful for screening human sera (51,52). Its sensitivity for livestock sera needs too be improved (38,41).

F. Indirect Fluorescent Antibody Test

In the conventional IFA test, whole killed tachyzoites are incubated with serum and antibody detection is enhanced by adding fluorescent-labeled antispecies IgG (or whole immunoglobulin) and viewing with a fluorescence microscope. IFA titers generally correspond with those of the DT, and the test has been standardized in some laboratories (53,54). Its disadvantages are the need for a fluorescence microscope, for species-specific conjugates, and cross-reaction with rheumatoid factor (RF) and antinuclear antibodies (ANAs) (55).

A modification of the IFA (IgM-IFA) was developed to detect IgM antibodies in congenitally infected children because the heavier IgM antibodies do not cross the placenta, whereas the lighter IgG antibodies can cross the human placenta (56). However, the test detects only 75% of congenital infections because

1. False-positive results due to cross-reaction with ANA occur.
2. There is a lack of specificity of reagents due to contamination of IgM conjugates with IgG.
3. There is delayed synthesis in the child.
4. There is a suppressive effect of passively transferred IgG.

Some of these problems can be overcome by separation of IgM and IgG from RF and ANAs by filtration (57).

G. Enzyme-Linked Immunosorbent Assay

Since the first report of the use of the ELISA to measure antibodies to *T. gondii* (58), hundreds of publications have described improvements or modifications of the test for human antibodies. The ELISA has become the almost exclusive test of choice for the measurement of human antibodies to *T. gondii*. In addition, ELISA has been used to measure antibodies to *T. gondii* in sheep (59), swine (60,61), and Australian marsupials (62).

The first major advance in the use of ELISA to measure human antibodies was the development of the antibody class capture system for parasite-specific IgM (63–65). Then, as our understanding of the antigenic structure of the parasite has increased, so too has the sophistication of the antigenic fractions and immunological reagents used in the ELISA for toxoplasmosis. Such various simple antigenic fractions as whole tachyzoites (66,67) and soluble antigen (67,68) have been used, and the time required to complete at least one version of the test has been reduced to 2 h (69). More complex purified antigen fractions of the cell surface protein termed P30 have been used in ELISA to increase specificity (70,71), and monoclonal antibodies (MAb) to P30 have been used as conjugates to increase specificity and sensitivity for measurement of both IgG and IgM (72–75). However, tests that use MAb may be subject to problems (76–79), so this has led to the use of more sophisticated tests for the detection of antibody to *T. gondii*.

Several groups have used ELISA or nonenzymatic assays to detect parasite-specific immunoglobulin A (IgA), and they found them to be as good as, or more useful than, the detection of specific IgM for the diagnosis of acute (80–83) or congenital (83) toxoplasmosis. Decoster et al. combined the best elements of the ELISA technologies used for toxoplasmosis and developed an ELISA that uses P30 as the antigenic fraction to detect parasite-specific IgA (84). The presence of these antibodies in serum was a clear marker of acute rather than chronic toxoplasmosis, and natural IgM antibodies, RF, or ANAs did not interfere with the test. More importantly, *T. gondii*-specific IgA antibodies were found to be a better marker of congenital toxoplasmosis than was parasite-specific IgA. In addition, five of 18 uninfected neonates gave a false-positive IgM test but were negative for IgA.

The continued development of such tests seems sure to increase the specificity and sensitivity of *T. gondii* antibody detection, but there are still problems with using antibody detection as an indicator of infection. The antibody titers take time to rise and, especially in immunosuppressed patients, false-positive or false-negative results may occur (85,86). Very few AIDS patients possess parasite-specific IgM or show rises in specific IgG (87). Suzuki et al. used the MAT with formalin- or acetone-fixed tachyzoites to overcome these problems in AIDS patients (88). Measurement of anti-*T. gondii* titers in the cerebrospinal fluid may be of benefit if serum antibody titers are not diagnostic in cases of suspected CNS toxoplasmosis (89). In addition, detection of the whole parasite or its antigen can also be of benefit in the diagnosis of toxoplasmosis.

A significant advance with respect to the serological diagnosis of toxoplasmosis by ELISA was the cloning of two *T. gondii* gene fragments, termed H4 and H11 (90). The fusion proteins encoded by these gene fragments can be used as antigens in ELISA to detect antibodies to *T. gondii* in the sera of naturally infected humans, sheep, and cats (91–93) and experimentally infected mice (94). The use of these fusion proteins will assist the standardization of ELISA for toxoplasmosis greatly.

IV. CELL-MEDIATED RESPONSES

A. Skin Tests

Some skin tests using impure preparations from mouse ascites have been used for surveys (95). Generally, they are of no value in acute infection. To overcome this, Rougier and Ambroise-Thomas introduced a purified secretory/excretory antigen obtained from tissue culture (96). They found that results with it correlated well with results in the IFA test. This test could be useful as the initial screening test for *T. gondii* infection in pregnant women, as it is simple and inexpensive.

B. Antigen-Specific Lymphocyte Transformation

Cultures of immune peripheral lymphocytes in the presence of soluble *T. gondii* antigen show an increase in uptake of tritiated thymidine compared with nonimmune individuals; this test may be useful in the diagnosis of congenital infections (97–99). The method using the secretory antigen appears to be highly specific (98), although it seems too complicated for large-scale routine use.

V. ANTIGEN DETECTION

The detection of *T. gondii* antigen can be subdivided usefully into those tests detecting circulating soluble antigen (CSA) in the body fluids (serum, urine, cerebrospinal fluid, aqueous humor) of experimentally infected animals or naturally infected humans, and those immunohistochemical tests identifying the intact organism in solid tissue.

A. Circulating Soluble Antigen

In 1975, Raizman and Neva first detected CSA, using immunoprecipitation, in animals experimentally infected with *T. gondii* (100). Shortly thereafter, Van Knapen and Panggabean developed an ELISA to detect CSA in the acute stages of murine and human toxoplasmosis (101); the majority of tests used since then have been based on ELISA technology.

In general, the use of ELISA to detect CSA in toxoplasmosis suggests that it may be able to discriminate, after sonication and emulsion of the tissue, between chronic brain infection and recent infection of the brain by tachyzoites (102). It can be used to detect

CSA in aqueous humor (103) or urine (104), as well as serum (105–107), and is best based on a polyclonal antibody rather than an MAb (108).

The ELISA detection of CSA in immunocompetent patients suggests that CSA is found only during a short period of the active phase of the infection and that it is cleared before the loss of the specific IgM antibody rise, probably as a result of the formation of immune complexes between the CSA and the large amounts of immunoglobulin (107). Van Knapen et al. developed an ELISA to detect these circulating immune complexes, but it gave positive results in 11% of 121 healthy blood donors, suggesting that exacerbations of chronic infestations or reinfections occurred not infrequently without causing symptoms (109). In either situation, the results mean that the interpretation of a test to detect these circulating immune complexes should be made with great care.

Hassl et al. developed an ELISA to detect CSA in AIDS patients. It was positive in 16% of 125 patients (110). The results with negative controls were not stated, but 5 of the 20 AIDS patients with CSA had either very low or no *T. gondii* antibody titers, consistent with a high false-positive rate.

The sensitivity of ELISA for detection of CSA appears to be in the range 250 ng protein/mL (105) to 0.3 ng protein/mL (111). Hughes found that CSA was composed of two components with apparent intracellular origin and released from the tachyzoite by immune lysis (112,113). Ise et al. found that CSA was composed of four fractions (greater than 400, 220, 130, and 45 kDa), with the 220- and 130-kDa proteins being predominant (106). Hassl et al. found that CSA consisted of at least two antigens with apparent molecular masses of 57 kDa and 27 kDa, and suggested that they may be fragments of the *T. gondii* nucleoside triphosphate hydrolase (110).

Brooks et al. developed an assay to detect CSA by dot immunobinding to nitrocellulose membrane (111). This test had a sensitivity of detection of 40–123 pg of CSA suspended in 4 pL of serum or cerebrospinal fluid. CSA was detected in the cerebrospinal fluid of four of six congenitally infected infants. Suzuki and Kobayashi developed an agglutination test based on latex particles coated with anti-*T. gondii* antibodies to detect CSA in the serum of experimentally infected athymic nude mice (112). This system, designed to mimic toxoplasmosis in such severely immunocompromised hosts as AIDS patients, had a level of sensitivity of greater than 78 ng of protein/mL.

B. Immunohistochemistry

Toxoplasmosis has been diagnosed on histological criteria that were based in part on serological verification that did not involve rising antibody titers or demonstration of parasite-specific IgM (113,114). In fact, the histological findings of some of these publications were nonspecific and found to a greater or lesser degree in such other conditions as Hodgkin's disease, mycobacterial or *Histoplasma* infection, syphilis, or leishmaniasis (115–120). However, these criteria are still often used to diagnose toxoplasmosis and it therefore is possible, and indeed probable, that toxoplasmosis is overdiagnosed when diagnosis is based on histological grounds alone. The minimum requirement for diagnosis should be a positive IgM-specific antibody determination, a rising IgG antibody titer, or, if tissues are available, the demonstration of tachyzoites by immunohistochemistry.

The diagnosis of toxoplasmosis has been confirmed by direct immunohistological detection of the parasite in tissue. These studies used fluorescent and/or enzyme labels attached to human or rabbit polyclonal antisera (121–126). There are, however, problems with the use of such antisera for immunohistological diagnosis of toxoplasmosis. Polyclonal antisera may contain antibodies to other parasites (reviewed in Ref. 127), and different animals even of the same species produce different antibodies to the same antigen (76).

The use of MAb to specific parasite antigens offers a way to increase both the sensitivity (due to their extremely high titers) and the specificity of the immunohistological diagnosis of toxoplasmosis. MAb have been found to be at least as good as, and in most cases superior to, polyclonal antisera for immunohistochemical identification of a wide range of antigens in paraffin sections and/or cryostat sections (128,129). Immunohistochemical staining of tachyzoites using MAb should assist in the detection of tachyzoites by producing a stain on a clear background only when parasites are present. Using a number of different MAb, we compared several immunohistochemical staining procedures, evaluating them with respect to efficiency of staining and robustness of use (130).

The MAb studied react with several different proteins of the *T. gondii* tachyzoite (131–133). FMC 18 and FMC 19 were unsuitable for staining formalin-fixed, paraffin-embedded tissue despite their strong reaction with isolated formalin-killed tachyzoites by immunofluorescence (133). Presumably, the antigens are denatured by wax embedding or possibly are extracted into the solvents. FMC 20, FMC 22, and FMC 23 produced staining. FMC 20 stained the cell membrane of the tachyzoite and not the cell cytoplasm, despite reacting biochemically with soluble rather than membrane-bound antigen. This discrepancy has been noted previously in an electron microscopic study (132). FMC 22 proved to be the most robust and intense reagent for detecting *T. gondii* tachyzoites. Its use alone gives intense staining. In no instances have the other antibodies provided positive staining in material negative for FMC 22. Combination of different antibodies is not warranted since extra antibodies could contribute background staining that might make detection of parasites more difficult.

Of the detection systems evaluated (peroxidase-conjugated antimouse immunoglobulin, monoclonal alkaline phosphatase-antialkaline phosphatase, avidin peroxidase ABC, streptavidin peroxidase, silver intensification of protein-A gold), the streptavidin and avidin peroxidase systems gave the best staining for routine use. There was virtually no background in normal tissue and the tachyzoites were stained strongly. The alkaline phosphatase methods suffered from higher backgrounds and lack of optical resolution because of the aqueous mountant. The gold method gave the strongest staining; however, occasional overstaining and consequent unacceptable high background is a disadvantage.

For routine use we adopted the streptavidin peroxidase method incorporating imidazole intensification into the diaminobenzoic acid reaction (134). We have applied this method successfully to a variety of clinical specimens as well as specimens from several mammalian and avian species. The immunoperoxidase method should find wide use in the examination of brain biopsies since it has many advantages over the immunofluorescence method recently described by Sun et al. (135). The peroxidase method of immunohistochemistry is more sensitive than direct immunofluorescence; the method is permanent, unlike fluorescence, which fades, and the technique requires no special fluorescence microscope.

VI. DNA DETECTION

The DNA method of detection and diagnosis is based on the PCR (136). Two primers are added to the solution being tested for the presence of *T. gondii*: DNA together with dNTPs and DNA polymerase. Under the correct conditions of thermal cycling, the DNA sequence between the priming sites is amplified millions of times. This DNA then can be visualized on an agarose gel and used to confirm the presence of *T. gondii* in the test solution. DNA from as little as one tachyzoite can be detected. Various primers have been used to amplify *T. gondii* DNA; these are based on the B1 gene (137,138), P30 gene (139), and ribosomal RNA gene (140). At present, the combination obtained by using primers for the single-copy P30 gene and the multicopy ribosomal RNA gene appears ideal for the diagnosis of toxoplasmosis (141). However, given the exquisite sensitivity of the PCR, it will be essential to differentiate between *T. gondii* disease and the presence of dormant tissue cysts. The detection of *T. gondii* by PCR was reviewed recently (142).

ABBREVIATIONS

AIDS acquired immunodeficiency syndrome
ANA antinuclear antibody
CF complement fixation
CNS central nervous system
CSA circulating soluble antigen
DT dye test
ELISA enzyme-linked immunosorbent assay
IFA indirect fluorescent antibody
IgA immunoglobin A
IgG immunoglobin G
IgM immunoglobin M
IHA indirect hemagglutination
ISAGA immunosorbent agglutination assay
IU international unit
LA latex agglutination
MAb monoclonal antibody
MAT modified agglutination test
RF rheumatoid factor
WHO World Health Organization

REFERENCES

1. Johnson, A. M., Gu, Q. M., and Roberts, H. (1987). Antibody patterns in the serological diagnosis of acute lymphadenopathic toxoplasmosis, Aust. N. Z. J. Med., 17:430.
2. Vischer, T. L., and Bernheim, C. (1967). Two cases of hepatitis due to *Toxoplasma gondii*, Lancet, 2:919.
3. Theologides, A., and Kennedy, B. J. (1969). Toxoplasmic myocarditis and pericarditis, Am. J. Med., 47:169.
4. Townsend, T. J., Wolinsky, J. S., Baringer, J. R., and Johnson, P. C. (1975). Acquired toxoplasmosis: a neglected cause of treatable nervous system disease, Arch. Neurol., 32:335.
5. Gleason, T. H., and Hamlin, W. B. (1974). Disseminated toxoplasmosis in the compromised host, Arch. Intern. Med., 134:1059.
6. Samuels, B. S., and Rietschel, R. L. (1976). Polymyositis and toxoplasmosis, J. Am. Med. Assoc., 235:60.

7. Bertoli, F., Espino, M., Arosemena, J. R., Fishback, J. L., and Frenkel, J. K. (1995). A spectrum in the pathologu of toxoplasmosis in patients with acquired immunodeficiency syndrome, Arch. Pathol. Lab. Med., 119:214.

8. Perkins, E. S. (1973). Ocular toxoplasmosis, Br. J. Ophthalmol., 57:1.

9. Anderson, S. E., and Remington, J. S. (1975). The diagnosis of toxoplasmosis, South. Med. J., 681:433.

10. Desmonts, G. (1966). Definitive serological diagnosis of ocular toxoplasmosis, Arch. Ophthalmol., 76:839.

11. O'Connor, G. R. (1974). Manifestations and management of ocular toxoplasmosis, Bull. N. Y. Acad. Med., 50:192.

12. Nussenblatt, R. B., and Belfort, R. (1994). Ocular toxoplasmosis, an old disease revisited. JAMA, 271:304.

13. Dutton, G. N. (1984). The diagnosis of toxoplasmosis, Br. Med. J., 289:1078.

14. Bowie, W. R., King, A. S., Werker, D. H., Isaac-Renton, J. L., Bell, A., Eng, S. B., and Marion, S. A. (1997). Outbreak of toxoplasmosis associated with municipal drinking water. Lancet 350:173.

15. Rothova, A. (1993). Ocular involvement in toxoplasmosis. Br. J. Ophthalmol. 77:371.

16. Silveira, C., Belfort, R., Nussenblatt, R., Farah, M., Takahashi, W., Imamura, P., and Burnier, M. (1989). Unilateral pigmentary retinopathy associated with ocular toxoplasmosis, Am. J. Ophthalmol., 107:682.

17. Desmonts, G., and Couvreur, J. (1974). Toxoplasmosis in pregnancy and its transmission to the fetus, Bull. N. Y. Acad. Med., 50:146.

18. Remington, J. S., and Desmonts, G. (1983). Toxoplasmosis. In: Infectious Diseases of the Fetus and Newborn Infant, J. S. Remington and J. O. Klein, eds., W. B. Saunders, Philadelphia.

19. Wilson, C. B., Remington, J. S., Stagno, S., and Reynolds, D. W. (1980). Development of adverse sequelae in children born with subclinical congenital Toxoplasma infection, Pediatrics, 66:767.

20. Koppe, T. G., Loewer-Sieger, D. H., and de Roever-Bonnet, H. (1986). Results of 20-year follow-up of congenital toxoplasmosis, Lancet, 1:254.

21. Munday, B. L. (1970). The Epidemiology of Toxoplasmosis with Particular Reference to the Tasmanian Environment, M.V.Sc. thesis, University of Melbourne.

22. Beverley, J. K. A. (1976). Toxoplasmosis in animals, Vet. Rec., 99:123.

23. Munday, B. L., and Mason, R. W. (1979). Toxoplasmosis as a cause of perinatal death in goats, Aust. Vet. J., 55:485.

24. Dubey, J. P., and Beattie, C. P. (1988). Toxoplasmosis of Animals and Man, CRC Press, Boca Raton, FL.

25. Remington, J. S., and Cavanaugh, E. N. (1965). Isolation of the encysted form of *Toxoplasma gondii* from human skeletal muscle and brain, N. Engl. J. Med., 273:1308.

26. Abbas, A. M. A. (1967). Comparative study of methods used for the isolation of *Toxoplasma gondii*, Bull. WHO, 36:344.

27. Dubey, J. P. (1998). Advances in the life cycle of *Toxoplasma gondii*. Int. J. Parasitol. 28: 1019.

28. Endo, T., and Kobayashi, A. (1976). *Toxoplasma gondii*: electron microscopic study on the dye test reaction, Exp. Parasitol., 40:170.

29. Schreiber, R. D., and Feldman, H. A. (1980). Identification of the activator system for antibody to *Toxoplasma* as the classical complement pathway, J. Infect. Dis., 141:3566.

30. Lyng, J., and Siim, J. C. (1982). The WHO international standard for anti-*Toxoplasma* human serum, Lyon Med., 248:107.

31. Feldman, H. A., and Lamb, G. A. (1966). A micromodification of the *Toxoplasma* dye test, J. Parasitol., 52:415.

32. Frenkel, J. K., and Jacobs, L. (1958). Ocular toxoplasmosis. Pathogenesis, diagnosis and treatment, Arch. Ophthalmol., 59:260.

33. Waldeland, H. (1976). Toxoplasmosis in sheep. The reliability of a microtiter system in Sabin and Feldman's dye test, Acta Vet. Scand., 17:426.

34. Berger, J. (1966). Serologische Untersuchungen uber Toxoplasma-Infektionen bei Tieren im Einzugsbereich des Staatlichen Veterinaruntersuchung-samtes Frankfurt/Main, Dtsch. Tieraerztl. Wochenschr., 73:261.

35. Jacobs, L., and Lunde, M. N. (1957). A hemagglutination test for toxoplasmosis, J. Parasitol., 43:308.

36. Palmer, D. R., Cavallaro, J. J., Walls, K., Sulzer, A., and Wilson, M. (1976). Serology of toxoplasmosis. Immunology Series No. 1, Procedural Guide, Centers for Disease Control, Public Health Service, U.S. Department of Health, Education, and Welfare, Atlanta, GA.

37. Caruana, L. B. (1980). A study of variation in the indirect hemagglutination antibody test for toxoplasmosis, Am. J. Med. Technol., 46(30):386.

38. Dubey, J. P., Desmonts, G., McDonald, C., and Walls, K. W. (1985). Serologic evaluation of cattle inoculated with *Toxoplasma gondii*: comparison of Sabin-Feldman dye test and other agglutination tests, Am. J. Vet. Res., 46:1085.

39. Desmonts, G., and Remington, J. S. (1980). Direct agglutination test for diagnosis of *Toxoplasma* infection: method for increasing sensitivity and specificity, J. Clin. Microbiol., 11:562.

40. Dubey, J. P. (1985). Serologic prevalence of toxoplasmosis in cattle, sheep, goats, pigs, bison, and elk in Montana, J. Am. Vet. Med. Assoc., 186:969.

41. Dubey, J. P., Desmonts, G., Antunes, F., and McDonald, C. (1985). Serologic diagnosis of toxoplasmosis in experimentally infected pregnant goats and transplacentally infected kids, Am. J. Vet. Res., 46:1137.

42. Dubey, J. P., and Desmonts, G. (1987). Serological responses of equids fed *Toxoplasma gondii* oocysts, Equine Vet. J., 19:337.

43. Dubey, J. P., Thulliez, P., Weigel, R. M., Andrews, C. D., Lind, P., and Powell, E. C. (1995). Sensitivity and specificity of various serologic tests for detection of *Toxoplasma gondii* infection in naturally infected sows. Am. J. Vet. Res. 56:1030.

44. Dubey, J. P., Lappin, M. R., and Thulliez, P. (1995). Long-term antibody responses of cats fed *Toxoplasma gondii* tissue cysts. J. Parasitol. 81:887.

45. Dubey, J. P., Thulliez, P., and Powell, C. (1995). *Toxoplasma gondii* in Iowa sows: comparison of antibody titers to isolation of *T. gondii* by bioassays in mice and cats. J. Parasitol. 81:48.

46. Dubey, J. P. (1997). Toxoplasmosis. In: Topley and Wilson's Microbiology and Microbial Infections., F. E. G. Cox, ed., Arnold, London, pp. 303–318.

47. Desmonts, G., Naot, Y., and Remington, J. S. (1981). Immunoglobulin M-immunosorbent agglutination assay for diagnosis of infectious diseases: diagnosis of acute congenital and acquired *Toxoplasma* infections, J. Clin. Microbiol., 14:486.

48. Desmonts, G., and Thulliez, Ph. (1985). The *Toxoplasma* agglutination antigen as a tool for routine screening and diagnosis of *Toxoplasma* infection in the mother and infant, Dev. Biol. Stand., 62:31.

49. Remington, J. S., Eimstad, W. M., and Araujo, F. G. (1983). Detection of immunoglobulin M antibodies with antigen-tagged latex particules in an immunosorbent assay. J. Clin. Microbiol., 17:939.

50. Ohshima, S., Tsubota, N., and Hiraoka, K. (1981). Latex agglutination microtiter test for diagnosis of *Toxoplasma* infection in animals, Zentralbl. Bakteriol. Parasitenkd. Infektionskr. Hyg. Abt. 1 Orig., Reihe A, 250:376.

51. Balfour, A. H., Fleck, D. G., Hughes, H. P. A., and Sharp, D. (1982). Comparative study of three tests (dye test. indirect haemagglutination test, latex agglutination test) for the detection of antibodies to *Toxoplasma gondii* in human sera, J. Clin. Pathol., 35:228.

52. Walls, K. W., and Remington, J. S. (1983). Evaluation of a commercial latex agglutination method for toxoplasmosis, Diagn. Microbiol. Infect. Dis., 1:265.

53. Carmichael, G. A. (1975). The application of the indirect fluorescent antibody technique for the detection of toxoplasmosis, Can. J. Med. Technol., 37:168.

54. Niel, C., Desmont, G., and Gentilini, M. (1973). On the quantitative immune-fluorescence test as applied to the serological diagnosis of toxoplasmosis: introduction of international units in the expression of positivity, Pathol. Biol., 12:157.

55. Araujo, F. G., Barnett, E. V., Gentry, L. O., and Remington, J. S. (1971). False-positive anti-*Toxoplasma* fluorescent-antibody tests in patients with antinuclear antibodies, Appl. Microbiol., 22:270.

56. Remington, J. S. (1969). The present status of the IgM fluorescent antibody technique in the diagnosis of congenital toxoplasmosis, J. Pediatr., 75:1116.

57. Sulzer, A. J., Franco, E. L., Takafuji, E., Benenson, M., Walls, K. W., and Greenup, R. L. (1986). An oocyst-transmitted outbreak of toxoplasmosis: patterns of immunoglobulin G and M over one year, Am. J. Trop. Med. Hyg., 35:290.

58. Voller, A., Bidwell, D. E., Bartlett, A., Fleck, D. G., Perkins, M., and Oladehin, B. A. (1976). Microplate enzyme-immunoassay for *Toxoplasma* antibody, J. Clin. Pathol., 29:150.

59. O'Donoghue, P. J., Riley, M. J., and Clarke, J. F. (1987). Serological survey for *Toxoplasma* infections in sheep, Aust. Vet. J., 64:40.

60. Waltman, W. D., Dreesen, D. W., Prickett, M. D., Blue, J. L., and Oliver, D. G. (1984). Enzyme-linked immunosorbent assay for the detection of toxoplasmosis in swine: interpreting assay results and comparing with other serological tests, Am. J. Vet. Res., 45:1719.

61. Takahashi, J., and Konishi, E. (1986). Quantitation of antibodies to *Toxoplasma gondii* in swine sera by enzyme-linked immunosorbent assay, J. Immunoassay, 7:257.

62. Johnson, A. M., Roberts, H., and Munday, B. L. (1988). Prevalence of *Toxoplasma gondii* in wild macropods, Aust. Vet. J., 65:199.

63. Naot, Y., and Remington, J. S. (1980). An enzyme-linked immunosorbent assay for detection of IgM antibodies to Toxoplasma gondii: use for diagnosis of acute acquired toxoplasmosis, J. Infect. Dis., 142:757.

64. Duermeyer, W., Wielaard, F., Gruithuijsen, H., and Swinkels, J. (1980). Enzyme-linked immunosorbent assay for detection of immunoglobulin M antibodies against *Toxoplasma gondii*, J. Clin. Microbiol., 12:805.

65. Franco, E. L., Walls, K. W., and Sulzer, A. J. (1981). Reverse enzyme immunoassay for detection of specific anti-*Toxoplasma* immunoglobulin M antibodies, J. Clin. Microbiol., 13:859.

66. Picher, O., Aspock, H., Auer, H., and Hermentin, K. (1982). Enzyme-linked immunosorbent assay with whole trophozoites of *Toxoplasma gondii*, Zbl. Bakt. Hyg., 1. Abt. Orig. A., 253:397.

67. Verhofstede, C., Sabbe, L., and Van Renterghem, L. (1987). Ability of enzyme-linked immunosorbent assays to detect early immunoglobulin G antibodies to *Toxoplasma gondii*, Eur. J. Clin. Microbiol., 6:147.

68. Hughes, H. P. A., Van Knapen, F., AtWnson, H. J., Balfour, A. H., and Lee, D. L. (1982). A new soluble antigen preparation of *Toxoplasma gondii* and its use in serological diagnosis, Clin. Exp. Immunol., 49:239.

69. Tomasi, J. P., Schlit, A. F., and Stadtsbaeder, S. (1986). Rapid double-sandwich enzyme-linked immunosorbent assay for detection of human immunoglobulin M anti-*Toxoplasma gondii* antibodies, J. Clin. Microbiol., 24:849.

70. Santoro, F., Afchain, D., Pierce, R., Cesbron, J. Y., Ovlaque, G., and Capron, A. (1985). Serodiagnosis of Toxoplasma infection using a purified parasite protein (P30), Clin. Exp. Immunol., 62:262.

71. Lindenschmidt, E. G. (1986). Demonstration of immunoglobulin M class antibodies to *Toxoplasma gondii* antigenic component P35000 by enzyme-linked antigen immunosorbent assay, J. Clin. Microbiol., 24:1045.

72. Deletoille, P., Ovlaque, G., and Slizewicz, B. (1988). Advantage of using 35,000-molecular-

weight protein for testing of *Toxoplasma gondii* immunoglobulin M, J. Clin. Microbiol., 26: 796.

73. Cesbron, J. Y., Capron, A., Ovlaque, G., and Santoro, F. (1986). Use of a monoclonal antibody in a double-sandwich ELISA for detection of IgM antibodies to *Toxoplasma gondii* major surface protein (P30), J. Immunol. Meth., 83:151.

74. Cesbron, J. Y., Caron, A., Santoro, F., Wattre, P., Ovlaque, G., Pierce, R. J., Delagneau, J. P., and Capron, A. (1986). A new ELISA method for the diagnosis of toxoplasmosis: serum IgM assay by immunocapture with a monoclonal antibody against *Toxoplasma gondii*, Presse Med., 15:737.

75. Payne, R. A., Joynson, D. H. M., Balfour, A. H., Harford, J. P., Fleck, D. G., Mythen, M., and Saunders, R. J. (1987). Public health laboratory service enzyme linked immunosorbent assay for detecting *Toxoplasma*-specific IgM antibody, J. Clin. Pathol., 40:276.

76. Crowle, A. S. (1973). Immunodiffusion, Academic Press, New York.

77. Zola, H. (1985). Speaking personally: Monoclonal antibodies as diagnostic reagents, Pathology, 17:53.

78. Yolken, R. H. (1988). Nucleic acids or immunglobulins: which are the molecular probes of the future? Mol. Cell. Probes, 2:87.

79. Payne, W. J., Marshall, D. L., Shockley, R. K., and Martin, W. J. (1988). Clinical laboratory applications of monoclonal antibodies, Clin. Microbiol. Rev., 1:313.

80. Van Loon, A. M., Van der Logt, J. T. M., Hessen, F. W. A., and Van der Veen, J. (1983). Enzyme linked immunosorbent assay that uses labeled antigen for detection of immunoglobulin M and A antibodies in toxoplasmosis: comparison with indirect immunofluorescence and double sandwich enzyme linked immunosorbent assay, J. Clin. Microbiol., 17:997.

81. Turunen, H., Vuorio, K. A., and Leinikki, P. O. (1983). Determination of IgG, IgM and IgA antibody responses in human toxoplasmosis by enzyme-linked immunosorbent assay (ELISA), Scand. J. Infect. Dis., 15:307.

82. Favre, G., Bessieres, M. H., and Seguela, J. P. (1984). Dosage des IgA seriques specifiques de la toxoplasmose par une methode ELISA: application a 120 cas, Bull. Soc. Fr. Parasitol., 3:139.

83. Pinon, J. M., Thoannes, H., Pouletty, P. H., Poirriez, J., Damiens, J., and Pellefier, P. (1986). Detection of IgA specific for toxoplasmosis in serum and cerebrospinal fluid using a non-enzymatic IgA-capture assay, Diag. Immunol., 4:223.

84. Decoster, A., Darcy, F., Caron, A., and Capron, A. (1988). IgA antibodies against P30 as markers of congenital and acute toxoplasmosis, Lancet, 2:1104.

85. Naot, Y., Luft, B. J., and Remington, J. S. (1981). False-positive serological tests in heart transplant recipients, Lancet, 2:590.

86. McCabe, R. E., Gibbons, D., Brooks, R. G., Luft, B. J., and Remington, J. S. (1983). Agglutination test for diagnosis of toxoplasmosis in AIDS, Lancet, 2:680.

87. Luft, B. J., Brooks, R. G., Conley, F. K., McCabe, R. E., and Remington, T. S. (1984). Toxoplasmic encephalitis in patients with acquired immune deficiency syndrome, J. Am. Med. Assoc., 252:913.

88. Suzuki, Y., Israelski, D. M., Dannemann, B. R., Stepick-Biek, P., Thulliez, P., and Remington, J. S. (1988). Diagnosis of toxoplasmic encephalitis in patients with acquired immunodeficiency syndrome by using a new serological method, J. Clin. Microbiol., 26:2541.

89. Wong, B., Gold, J., Brown, A. E., Lange, M., Fried, R., Grieco, M., Mildvan, D., Giron, J., Tapper, M. L., Lerner, C. W., and Armstrong, D. (1984). Central-nervous system toxoplasmosis in homosexual men and parenteral drug abusers, Ann. Intern. Med., 100:36.

90. Johnson, A. M., and Illana, S. (1991). Cloning of *Toxoplasma gondii* gene fragments encoding diagnostic antigens, Gene., 99:127.

91. Tenter, A. M., and Johnson, A. M. (1991). Recognition of recombinant *Toxoplasma gondii* antigens by human sera in an ELISA, Parasitol. Res., 77:197.

92. Johnson, A. M., Roberts, H., and Tenter, A. M. (1992). Evaluation of a recombinant antigen

ELISA for the diagnosis of acute toxoplasmosis and comparison with traditional antigen ELISAs, J. Med. Microbiol., 37:409.

93. Tenter, A. M., Vietmeyer, C., and Johnson, A. M. (1992). Development of ELISAs based on recombinant antigens for the detection of *Toxoplasma gondii*–specific antibodies in sheep and cats, Vet. Parasitol., 48:189.

94. Parker, S. J., Smith, F. M., and Johnson, A. M. (1991). Murine immune responses to recombinant *Toxoplasma gondii* antigens, J. Parasitol., 77:402.

95. Frenkel, J. K. (1948). Delayed hypersensitivity to *Toxoplasma* antigens (toxoplasmosis), Proc. Soc. Exp. Biol. Med., 68:634.

96. Rougier, D., and Ambroise-Thomas, P. (1985). Detection of toxoplasmic immunity by multi-puncture skin test with excretory-secretory antigen, Lancet, 2:121.

97. Hughes, H. P. A., and Van Knapen, F. (1983). Characterization of a secretory antigen from *Toxoplasma gondii* and its role in circulating antigen production, Int. J. Parasitol. 12:433.

98. Hughes, H. P. A. (1985). Toxoplasmosis: the need for improved diagnostic techniques and accurate risk assessment, Curr. Top. Microbiol. Immunol., 120:105.

99. Krahenbuhl, J. L., Gaines, J. D., and Remington, J. S. (1972). Lymphocyte transformation in human toxoplasmosis, J. Infect. Dis., 125:283.

100. Raizman, R. E., and Neva, F. A. (1975). Detection of circulating antigen in acute experimental infections with *Toxoplasma gondii*, J. Infect. Dis., 132:44.

101. Van Knapen, F., and Panggabean, S. O. (1977). Detection of circulating antigen during acute infections with *Toxoplasma gondii* by enzyme-linked immunosorbent assay, J. Clin. Microbiol., 6:545.

102. Van Knapen, F., and Panggabean, S. O. (1982). Detection of *Toxoplasma* antigen in tissues by means of enzyme-linked immunosorbent assay (ELISA), Am. J. Clin. Pathol., 77:755.

103. Rollins, D. F., Tabbara, K. F., O'Connor, G. R., Araujo, F. G., and Remington, J. S. (1983). Detection of toxoplasmal antigen and antibody in ocular fluids in experimental ocular toxoplasmosis, Arch. Ophthalmol., 101:455.

104. Turunen, H. J. (1983). Detection of soluble antigens of *Toxoplasma gondii* by a four-layer modification of an enzyme immunoassay, J. Clin. Microbiol., 17:768.

105. Ise, Y., Lida, T., Sato, K., Suzuki, T., Shimada, K., and Nishioka, K. (1985). Detection of circulating antigens in sera of rabbits infected with *Toxoplasma gondii*, Infect. Immun., 48: 269.

106. Lindenschmidt, E. G. (1985). Enzyme-linked immunosorbent assay for detection of soluble *Toxoplasma gondii* antigen in acute-phase toxoplasmosis, Eur. J. Clin. Microbiol., 4:488.

107. Candolfi, E., Derouin, F., and Kien, T. (1987). Detection of circulating antigens in immunocompromised patients during reactivation of chronic toxoplasmosis, Eur. J. Clin. Microbiol., 6:44.

108. Aranjo, F. G., Handman, E., and Remington, J. S. (1980). Use of monoclonal antibodies to detect antigens of *Toxoplasma gondii* in serum and other body fluids, Infect. Immun., 30: 12.

109. Van Knapen, F., Panggabean, S. O., and Van Leusden, J. (1985). Demonstration of *Toxoplasma* antigen containing complexes in active toxoplasmosis, J. Clin. Microbiol., 22:645.

110. Hassl, A., Aspock, H., and Flamm, H. (1985). Circulating antigen of *Toxoplasma gondii* in patients with AIDS: significance of detection and structural properties, Zbl. Bakt. Hyg., A270:302.

111. Brooks, R. G., Sharma, S. D., and Remington, J. S. (1985). Detection of *Toxoplasma gondii* antigens by dot-immunobinding technique, J. Clin. Microbiol., 21:113.

112. Suzuki, Y., and Kobayashi, A. (1987). Presence of high concentrations of circulating *Toxoplasma* antigens during acute *Toxoplasma* infection in athymic nude mice, Infect. Immun., 55:1017.

113. Stansfeld, A. G. (1961). The histological diagnosis of toxoplasmic lymphadenitis, J. Clin. Pathol., 14:565.

114. Miettinen, M. (1981). Histological differential diagnosis between lymph node toxoplasmosis and other benign lymph node hyperplasias, Histopathology, 5:205.

115. Harstock, R. J., Halling, L. W., and King, F. M. (1970). Luetic lymphadenitis: a clinical and histologic study of 20 cases, Am. J. Clin. Pathol., 53:304.

116. Turner, P. R., and Wright, D. J. M. (1973). Lymphadenopathy in early syphilis, J. Pathol., 110:305.

117. Schnitzer, B. (1981). Reactive lymphoid hyperplasia. In: Surgical Pathology of the Lymph Nodes and Related Organs, E. S. Jaffe, ed., W. B. Saunders, Philadelphia.

118. Miettinen, M., and Franssila, K. (1982). Malignant lymphoma simulating lymph node toxoplasmosis, Histopathology, 6:129.

119. Daneshob, K. (1978). Localized lymphadenitis due to Leishmania simulating toxoplasmosis: Value of electron microscopy for differentiation, Am. J. Clin. Pathol., 69:462.

120. Mans, R. A., and Sanchez, G. (1979). Histoplasma capsulatum lymphadenitis simulating glandular toxoplasmosis, Lab. Med., 10:156.

121. Calderon, C., Atias, A., Saavedra, P., and Diaz, S. (1973). Identification of *Toxoplasma gondii* in lymph-nodes by immunofluorescence, Lancet, 2:1264.

122. Matossian, R. M., Nassar, V. H., and Basmadji, A. (1977). Direct immuno-fluorescence in the diagnosis of toxoplasmic lymphadenitis, J. Clin. Pathol., 30:847.

123. Andres, T. L., Dorman, S. A., Winn, W. C., Trainer, T. D., and Perl, D. P. (1981). Immunahistochemical demonstration of *Toxoplasma gondii*, Am. J. Clin. Pathol., 75:431.

124. Conley, F. K., Jenkins, K. A., and Remington, J. S. (1981). *Toxoplasma gondii* infection of the central nervous system, Hum. Pathol., 12:690.

125. Tabei, S. Z. (1982). Immunohistologic demonstration of *Toxoplasma gondii*, N. Engl. J. Med., (November 25):1404.

126. Dutton, G. N., Hay, J., and Ralston, J. (1984). The immunocytochemical demonstration of *Toxoplasma* within the eyes of congenitally infected mice, Ann. Trop. Med. Parasitol. 78:431.

127. Frenkel, J. K., and Piekarski, G. (1978). The demonstration of *Toxoplasma* and other organisms by immunofluorescence: A pitfall, J. Infect. Dis., 138:265.

128. Lindgren, J., Wahlstrum, T., Bang, B., Hurme, M., and Makela, O. (1982). Immunoperoxidase staining of carcinoembryonic antigen with monoclonal antibodies in adenocarcinoma of the colon, Histochemistry, 74:223.

129. Naritoku, W. Y., and Taylor, C. R. (1961). A comparative study of the use of monoclonal antibodies using three different immunohistochemical methods: an evaluation of monoclonal and polyclonal antibodies against human prostatic acid phosphatase, J. Clin. Pathol., 14:565.

130. Hodgson, A., Johnson, A. M., Skinner, J. S., and Zola, H. Unpublished data, 1984.

131. Johnson, A. M., McDonald, P. J., and Neoh, S. H. (1983). Monoclonal antibodies to *Toxoplasma* cell membrane surface antigens protect mice from toxoplasmosis, J. Protozool., 30:351.

132. Johnson, A. M., Haynes, W. D., Leppard, P. J., McDonald, P. J., and Neoh, S. H. (1983). Ultrastructural and biochemical studies on the immunohistochemistry of *Toxoplasma gondii* antigens using monoclonal antibodies Histochemistry, 77:209.

133. Johnson, A. M., McNamara, P. J., Neoh, S. H., McDonald, P. J., and Zola, H. (1981). Hybridomas secreting monoclonal antibody to *Toxoplasma gondii*, Aust. J. Exp. Biol. Med. Sci., 59:303.

134. Straus, W. (1982). Imidazole increases the sensitivity of the cytochemical reaction for peroxidase with diaminobenzidine at a neutral pH., J. Histochem. Cytochem., 30:491.

135. Sun, T., Greenspan, J., Tenenbaum, M., Farmer, P., Jones, T., Kaplan, M. (1986). Diagnosis of cerebral toxoplasmosis using fluorescein-labeled anti-*Toxoplasma* monoclonal antibodies, Am. J. Surg. Pathol., 10:312.

136. Mullis, K. B., and Faloona, F. A. (1987). Specific synthesis of DNA in vitro via a polymerase-catalyzed chain reaction, Meth. Enzymol., 155:335.

137. Van de yen, E., Melchers, W., Galama, J., Camps, W., and Meuwissen, J. (1991). Identification of *Toxoplasma gondii* infections by B1 gene amplification, J. Clin. Microbiol., 29:2120.

138. Burg, J. L., Grover, C. M., Pouletty, P., and Boothroyd, J. C. (1989). Direct and sensitive detection of a pathogenic protozoan, *Toxoplasma gondii* by polymerase chain reaction, J. Clin. Microbiol., 27:1787.

139. Sarra, D. V., and Holliman, E. E. (1990). Diagnosis of toxoplasmosis using DNA probes, J. Clin. Pathol., 43:260.

140. Cazenave, J., Forestier, F., Bessieres, M. H., Broussin, B., and Begueret, J. (1992). Contribution of a new PCR assay to the prenatal diagnosis of congenital toxoplasmosis, Prenat. Diag., 12:119.

141. Cazenave, J., Cheyrou, A., Blouin, A., Johnson, A. M., and Begueret, J. (1991). Use of polymerase chain reaction to detect *Toxoplasma*, J. Clin. Pathol., 44:1037.

142. Ellis, J. T. (1998). Polymerase chain reaction approaches for the detection of *Neospora caninum* and *Toxoplasma gondii*. Int. J. Parasitol., 28:1053.

18

Seafood Parasites: Prevention, Inspection, and HACCP

Ann M. Adams
U.S. Food and Drug Administration, Bothell, Washington

Debra D. DeVlieger
U.S. Food and Drug Administration, Washington, D.C.

I. INTRODUCTION

Numerous species of parasites naturally infect fishes and, although many are lost prior to the presentation of a seafood item for consumption, parasites cannot be entirely eliminated. In general, parasitic organisms are very rarely introduced to a seafood product after capture and during processing. Rather, fishes and other aquatic organisms become infected during their normal life activities. Some of these parasites can pose a hazard to human health, whereas others may only affect the aesthetics of the product. Measures can be taken to decrease the incidence of parasites in seafood and to negate the infectivity of those parasites remaining.

    This chapter will consider practices commonly occurring in the seafood industry and to discuss those that can mitigate hazards posed by the presence of parasites. These practices will be contrasted with other programs within the seafood industry to clarify the goals of each system. The effect of these practices and programs on particular parasites, primarily the helminths, will also be addressed. In this chapter, seafood refers to fresh- or saltwater finfish, crustaceans, and other aquatic animals (excluding birds and mammals) intended for human consumption, although finfish will be of primary concern.

II. PREREQUISITE PROGRAMS

Prerequisite programs are defined as those steps or procedures that control the in-plant environmental conditions that provide a foundation for safe food production. Examples of such programs include Good Manufacturing Practices (GMPs), Sanitation and Sanitation Standard Operating Procedures (SSOPs), training, recall programs, preventive maintenance, and product identification programs and coding. Although these programs have little if any effect on the presence and infectivity of parasites within seafood, they are important to mention in this chapter because they are instrumental in the development of a Hazard Analysis and Critical Control Point (HACCP) plan. With programs such as SSOPs and GMPs in place, HACCP plans are more effective because industry can concentrate on the hazards associated with the food or processing rather than on the processing environment (1,2).

In the United States, the Food and Drug Administration (FDA) is responsible for the enforcement of regulations promulgated under the Federal Food, Drug and Cosmetic Act. Under this Act, Sections 402(a)(1) through (a)(7) define what constitutes an adulterated food. Specifically, 402(a)(4) states that a food is adulterated if it has been prepared, packed, or held under insanitary conditions whereby it may have become contaminated with filth, or whereby it may have been rendered injurious to health.

The current GMPs describe the conditions and practices that must be followed to avoid producing adulterated food products. Title 21 of the Code of Federal Regulations, Part 110 (21 CFR Part 110), outlines the basis for determining whether the facilities, methods, practices, and controls used to process food products are safe and whether the products have been processed under sanitary conditions. In general, GMPs cover major blocks or units dealing with sanitation. These units are personnel; buildings and facilities; plants and grounds; sanitary operations; sanitary facilities and controls; equipment; processes and controls; raw materials; manufacturing operations; and warehousing and distribution.

As stated previously, GMPs do not specifically address the presence and infectivity of parasites. However, Subpart E of 21 CFR Part 110.80(a)(2) is interpreted to be applicable to the presence of parasites as a filth issue. The viability or infectivity of the parasites, which may pose a food safety hazard, is not considered in this portion of the regulation. Specifically, this section states in part: "Raw materials . . . susceptible to contamination with pests, undesirable microorganisms, or extraneous material shall comply with all applicable Food and Drug Administration regulations, guidelines, and defect action levels for natural or unavoidable defects if a manufacturer wishes to use the material in manufacturing food" (3). The FDA's Compliance Policy Guide Section 540.590 establishes defect action levels that relate to parasites as a filth issue for freshwater and rose fish. Parasites of marine fishes are covered under GMP levels established by FDA's Office of Seafood.

The FDA also provides sanitation requirements specific to seafood processors in 21 CFR Part 123, or what is commonly referred to as the Seafood HACCP regulation. In the preamble of this regulation, the FDA concluded that the requirements in Part 110 alone did not prove adequate for the seafood industry. Although the FDA had been enforcing the sanitation standards found in Part 110 for many years, the seafood industry had not succeeded in developing a culture in which processors have assumed an operative role in controlling sanitation in their plants. The regulation encourages, but does not mandate, that seafood processors develop SSOPs. If an SSOP is developed, it should describe how the processor will ensure that certain key sanitation conditions and practices will be met.

It should also describe how the plant operations will be monitored to ensure that the conditions and practices will be met. Moreover, the FDA pointed out that Part 110 was silent with regard to monitoring by the processor and that monitoring was important in assuring that sanitation controls were being followed (4). Hence, Part 123 outlines eight key areas of sanitation that must be monitored: safety of water; condition and cleanliness of food contact surfaces; prevention of cross-contamination; maintenance of handwashing, hand sanitizing, and toilet facilities; protection from adulterants; labeling, storage, and use of toxic compounds; employee health conditions; and exclusion of pests.

Many of these areas can be of concern in regard to parasites, in particular the quality of water; cross-contamination; employee hygiene and health; and transportation issues. In aquaculture, the source of water and possible routes of contamination should be considered. Culturing of freshwater fishes in Asian countries often involved the location of latrines above the surface of ponds as a means of fertilization (5). This practice contributed to the infection of snails and fishes with trematode larvae of *Clonorchis sinensis, Opisthorchis viverrini*, and *O. felineus* (6). Consumption of the fishes in raw or undercooked dishes resulted in human infections, thereby completing the life cycles of the parasites. The removal of latrines from fish-rearing ponds and the use of other forms of fertilizer, such as the addition of rice bran as practiced in Indonesia (5, p. 45), can contribute to the decrease in the incidence of trematode infections.

The quality of water used in processing and washing of food products can also determine the potential of pathogen contamination. In particular, the source of water, whether municipal, well, marine, etc., should be safe for use. The water should be free of contaminants, both chemical and biological. Numerous municipal water sources have become contaminated with oocysts of *Cryptosporidium parvum*, resulting in several outbreaks of cryptosporidiosis. The most notable outbreak occurred in Milwaukee, resulting in over 400,000 cases (7). If such water is used during the processing of a seafood product, contamination might be a possibility.

The reported presence of *C. parvum* is oysters from the Chesapeake Bay, Maryland (8) indicates that the protozoan can tolerate the salinity levels in estuarine areas. Although no outbreaks of *C. parvum* have been reported from the consumption of seafood (8), the protozoan has been shown to retain its infectivity after passage through oysters (9). Experimental work has demonstrated that freshwater benthic clams can be infected by both *C. parvum* and *Cyclospora cayetanensis* (10,11) and that rainbow trout can be infected by the former protozoan (12). Oysters can also be infected experimentally with *Giardia lamblia* (13), although Fayer et al. (8) did not recover this protozoan from oysters in the Chesapeake Bay.

Seafood processors that utilize water from the marine environment for storage or actual processing may wish to evaluate the potential for the presence of the protozoans in the environment. The protozoans would be absent from offshore fishing areas, but may be present in areas that are prone to runoff from pastures or populated areas that may have failed septic systems. In addition, companies that process oysters should perhaps consider the potential of cross-contamination between infected and uninfected oysters during harvesting, processing, storage, and shipping of their product.

Employee hygiene and health, in addition to the sanitation procedures of a processor, can have a direct impact on the possibility of contamination. Although the protozoans mentioned above are generally transmitted by contaminated water, foodborne outbreaks of *Giardia* and *Cryptosporidium* have been traced to improper sanitation by food preparers. In the first known outbreak for *Giardia*, a woman had drained and transferred canned salmon to containers after diapering her grandson. The child was later found to be excreting numerous cysts (14). In contrast, foodborne outbreaks of cryptosporidiosis generally have involved agricultural products that were likely exposed to manure or infected farm animals, such as cold-pressed apple cider (15,16) and green onions (17). However, transmission of *Cryptosporidium* by improper hygiene can occur as demonstrated by an outbreak in Minnesota in which chicken salad was implicated. The hostess who prepared the dish also operated a day care facility in her home and had changed diapers during the period in question (18).

Except for the first case of foodborne giardiasis (above), no outbreaks of protozoal disease have been reported from the consumption of seafood, but processors and employees need to be cognizant of the possibility of contamination through poor hygiene and sanitation, particularly for those seafood products that are eaten raw or undercooked. In a study of sushi and sashimi from restaurants, no fecal coliforms were detected, but low levels of *Staphylococcus aureus* were found in samples from six restaurants (19). Fecal coliforms are often used as indicators of fecal contamination, although the lack of these bacteria reflected acceptable personal hygiene in regard to the use of toilet facilities. However, the presence of *S. aureus* indicated poor handwashing during food preparation and the possibility of contamination of foods by the handlers. Lastly, employees should be instructed as to the possibility of infections through insanitary handling of a product. A worker who never ate raw or cold smoked fish became infected with the trematode *Nanophyetus salmincola* while working with heavily infected juvenile coho salmon (20). The employee did not wear protective gloves and probably became infected when he failed to wash his hands properly before eating or smoking.

Contamination of products can also occur during the transportation of the product from different points during production up to delivery to the consumer. This is particularly a concern for pathogenic bacteria rather than for seafood parasites. A greater problem for seafood is the introduction of parasites to a population outside of the natural geographic region of the organism. Knowledge of a parasite and the means of preventing infection may be absent from the new market area. For example, the trematode *N. salmincola* can be found in anadromous and freshwater fishes from rivers and streams in western Washington, Oregon, and northern California (21). Over 20 cases of nanophyetiasis have been reported from Oregon (22,23), but the most noteworthy infections occurred in New Orleans (Dr. Maurice D. Little, personal communication). Fresh salmon from the Pacific Northwest was shipped to Louisiana and sold to restaurants. Two infections resulted from the consumption of sushi made from the salmon. Transportation and sale of fresh seafood will increase in the future and continues to pose public safety issues. Although seafood products from Asia are mostly shipped frozen to the United States at this time, if fresh product is shipped in the future, the possibility of the presence of infective larvae of the trematodes *Clonorchis sinensis*, *Opisthorchis viverrini*, *O. felineus*, and *Paragonimus westermani* need to be considered.

III. PREVENTIVE PRACTICES

Preventive practices may decrease the incidence of parasites or lower the possibility of infection, but they are not a means of avoiding the actual production of an adulterated food. Preventive practices do not entirely eliminate the risk of infection. Such programs are not required by regulations, but individual companies or industries may require these activities among their suppliers and employees. Two preventive programs are considered here: Good Agricultural Practices (GAPs) and Good Fishery Practices (GFPs).

A. Good Agricultural Practices

Although in general, agricultural practices will not pertain to production and supply of seafood, GAPs are included here for several reasons. In many respects, GAPs are considered the equivalent of agricultural GMPs, and so may be contrasted with the latter program. Also, under some conditions, principles of GAPs may be applicable to the seafood industry. Finally, in many respects, GAPs are simply those concepts taught in introductory parasitology courses that consider the disruption of life cycles and of the transmission of parasites. The concepts themselves are useful in the development of many programs addressing foodborne pathogens.

The development of GAPs is relatively recent, resulting from the recognition of the increasing risk of microbial contamination in fresh produce. In 1997, the Food Safety Initiative was announced as a means of improving the safety of the food supply. One result of the Initiative was the issuance by the FDA and the U.S. Department of Agriculture of a guidance for fruits and vegetables in final form in October 1998 (24). This guidance, regulations, and other publications by the FDA are also available on the agency's web page (http://www.fda.gov). The guidance does not consist of new regulations, and therefore does not have the force of law and is not subject to enforcement. Rather, the use of principles within the guidance is voluntary and meant to assist in a framework to minimize the risk of microbial contamination of fresh produce. This contamination can consist of bacteria, parasites, or viruses, although only parasites will be considered here. The guide focuses on risk reduction, not risk elimination, during the growing, harvesting, washing, sorting, packing, and transporting of fresh produce. Basic principles are applied to the use and source of water and organic fertilizers, employee health and sanitation, field and facility sanitation, and transportation of products. Parasites that can be acquired from contaminated produce and are covered by this guide include the nematode *Ascaris lumbricoides*, and the protozoans *C. parvum, G. lamblia*, and *Cyclospora cayetanensis*.

The quality of water used in any aspect of food production, whether in agriculture, in the seafood industry, or in other food production, is of primary importance in food safety. The source of water, whether from municipalities, wells, rivers, ponds, etc., should be considered when evaluating the possibility of exposure to pathogens. The major source of microbial contamination is the exposure of water to human and animal feces. This can be from improper sewage treatment or disposal, by water runoff from pastures and other areas where manure may be located, or by direct contamination from animals with access to water sources. In agriculture, exposure of irrigation water to edible portions of produce should be minimized. If the water has been previously contaminated with pathogens, human infections may result from the consumption of the raw produce.

In processing, the guidance document for GAPs recommends that the quality of water be consistent with requirements for drinking water. The guide also suggests treatment of water with antimicrobial chemicals (chlorine, trisodium phosphate, or organic acids), ozone, ultraviolet radiation, or irradiation. The effectiveness of such treatments depends in part on the target organism(s). For example, although chlorination tends to be very effective against bacteria, the protozoans *C. parvum, G. lamblia*, and *Cyclospora cayetanensis* are highly resistant to the levels used for drinking water (25,26).

B. Good Fisheries Practices

GFPs include measures that can be taken during harvest or processing to reduce the abundance or likelihood of parasites within seafood products. Such actions do not necessarily remove all parasites and so do not eliminate the possibility of infection. Many factors can be considered when evaluating the probability of parasites within seafood. The type and size of fish harvested should be considered, as well as the feeding habits and environment of the species selected. Groundfish, such as arrowtooth flounder (*Atheresthes stomias*) and many species of sole, can accumulate large numbers of anisakid nematodes (*Anisakis* spp. and *Pseudoterranova decipiens*). Anadromous fishes, in particular the salmons, can acquire both freshwater and marine parasites during their life spans. Although external and intestinal parasites may be lost during the transition from the different environments, parasites residing within the tissues, such as larval anisakids and diphyllobothriids, are retained. Because these parasites can accumulate over time, the size and age of fish can often be correlated with the numbers of parasites present. For example, in one study, 83% of cod harvested before they reached 60 cm in length had not yet acquired heavy infections of anisakids as compared to larger fish (27). In addition, the prey selected by fish may change as they mature, exposing the fish to different and greater numbers of parasites. Rainbow trout, *Oncorhynchus mykiss*, did not become infected with diphyllobothriid tapeworm larvae until they had attained lengths of approximately 150 mm and were able to prey on sticklebacks (28). To minimize the extent of processing necessary to provide a marketable product, some companies specifically purchase smaller fish of species known to have parasite problems.

Particular stocks of fish or geographic locations may be known to have excessive numbers of parasites. "Problem areas" may include sites frequented by marine mammals, the definitive hosts for many parasites of concern to the seafood industry. Fishes from inland and coastal areas, haul-out locations, and rookeries often sustain appreciably heavier infections of parasites in their flesh. Under optimal circumstances these areas would be avoided by fishing vessels or any fish harvested would be heavily processed. Processing may consist of candling (discussed below) and removal of parasites, or the fish may be made into minced products or analog (surimi, imitation crab, etc.). To decrease the degree of processing required, a firm may pay a premium for fish caught outside of areas frequented by marine mammals.

At the present time, many harvests are regulated on the basis of the quantity of catch, often resulting in short or restricted fishing seasons. Such time constraints limit the flexibility that fishermen can exercise in regard to harvest location. The fishermen may need to fish closer to shore and to their off-loading site to fully meet their quota. These areas can include the above-mentioned areas frequented by marine mammals. In these circumstances, harvesting lower quality or heavily infected fish is often preferable for the harvester than having a decreased catch. In the year 2000, new restraints will be placed on the harvest of fish in the Bering Sea because of the decreasing populations of some marine mammals. Fishing seasons will be replaced by quotas for the harvesters, eliminating the time constraints. Fishermen will have the ability to harvest fish in areas farther from shore and away from rookeries and haul-outs. The harvest can be more selective and processors may receive fishes with lower numbers of parasites.

The method of capture can affect the ability to inspect for and remove parasites during processing. Fish caught with long lines rather than with nets tend to be fresher and of better quality. Such fish are bled immediately after death and then chilled or frozen.

The resulting product has a whiter flesh and is easier to candle; encysted nematodes or tapeworms are more easily seen and removed.

The majority of parasites present in and on fishes are never seen by the consumer and are lost during capture and processing. Parasites, both larval and adult, can be found throughout the bodies of fishes. Parasitic copepods and monogenetic trematodes may be found on the skin; larval trematodes can encyst under the skin, in the eyes, brain, viscera, and flesh; nematodes and tapeworms may be in the viscera or flesh. The act of capture and the resulting handling dislodges many of the parasites on the surface of the fish. After capture, whole fish may be stored chilled or frozen, or may undergo some preliminary processing. Larval helminths accumulate primarily in the viscera and secondarily in the flesh (29). Freezing or rapid chilling reduces the possibility of larvae migrating from the viscera into the flesh after the death of the host. Proper cold storage of the catch results in a higher quality product, a longer shelf life, and lower numbers of parasite larvae in the edible flesh, particularly in species that appear more susceptible to postmortem migration of parasites (e.g., salmon, herring, arrowtooth flounder). Freezing the fish at appropriate temperatures and durations will kill the parasites and eliminate the risk of infection (see below). Some fishes are headed and gutted soon after capture. During gutting, the entire viscera and any attendant parasites are removed. Parasites present in the flesh prior to capture will not be affected, but further migration of larval helminths into the flesh can be prevented.

Processors may also place restrictions on the harvesters for those fishes known to host parasites. For those species in which migration of larval parasites can be a problem, the processor may request that fishermen either reduce the time spent at sea or refrain from harvesting the fishes until the last few days of fishing. The catch may also be sampled by the processor to determine the overall levels of infection and the extent of migration. For the latter, a portion of the catch may be candled. If larvae are found above the lateral line in fillets, the processor may refuse the catch. If the processor owns the fishing vessels or otherwise has control over the fleet, the company may continuously monitor the quality of the catches, including the degree of parasitism. If candling reveals particularly high numbers of parasites, the processor may direct the vessels away from the implicated areas and fish stocks.

As mentioned previously, a processor may perform several steps in the handling of a catch to prepare the product for market. These steps include heading, gutting, filleting, and skinning. The resulting fillets are often trimmed, with or without candling. During candling, a fish fillet is placed on a light table to enable the detection of parasites (30). This process does not necessarily reveal all of the parasites present, nor is it appropriate for all types of fish. Factors that may interfere with detection include the thickness of the fillet; the presence of skin on the fillet; the oil content of the flesh; pigmentation; and the level of experience of the operator (31). When slices of salmon averaging 0.7 cm in thickness were candled, no anisakid larvae were detected, although 10% of the slices were found to contain the parasites per artificial digestion (19). Both the pigmentation and oil content were suggested as factors in the poor results from candling. In a comparison of white-fleshed fillets from rockfish, sole, arrowtooth flounder, and true cod, candling detected infections in 53–79% of the fillets and 43–76% of the anisakids present (31). After detection by candling, the parasites may be removed with forceps or the infected portion of the fillet may be trimmed away. Generally, the larval anisakids accumulate in the belly flap region and below the lateral line. The belly flap may be trimmed automatically without candling, or the fillets may be candled and then trimmed if found to be infected.

IV. HACCP

Inasmuch as parasites can be reduced through means of GMPs or GFPs or other prerequisite programs, they are not eliminated nor is the hazard of parasitic infection negated with these programs. The FDA has implemented a regulation (21 CFR Part 123) that applies to all fish and fishery products and is based on a Hazard Analysis and Critical Control Point (HACCP) system that specifically addresses a food safety hazard, and not just a filth or aesthetic issue that is inherent in the presence of parasites. HACCP is a preventive system of hazard control, rather than a reactive one, thereby ensuring a safer food product. HACCP systems are designed to identify hazards, establish controls for the identified hazards, and monitor the control measures. In this system, hazards are identified as microbiological, chemical, or physical contaminants that can occur from the fish species and/ or the processing method.

HACCP systems are not new and have been in existence since the 1960s when the Pillsbury Company pioneered the application of HACCP to food products being produced for the U.S. space program (32). Early on, Pillsbury decided that existing quality assurance programs and end-product testing would not provide the needed assurances for the program; the only way to ensure safety would be to develop a preventive system that controlled hazards during food production. The FDA first required HACCP-type controls in 1973 when they published 21 CFR Part 113, the Low Acid Canned Food Regulations, to control low-acid canned foods against *Clostridium botulinum*. In 1985, the U.S. National Academy of Sciences assessed the effectiveness of food regulations and concluded that the HACCP approach should be adopted by all regulatory agencies. This recommendation led to the formation of the National Advisory Committee on Microbiological Criteria for Foods (NACMCF) in 1988 and the standardization of the HACCP principles that are currently used in the food industry and by regulatory authorities today (1).

FDA patterned their seafood HACCP regulation, 21 CFR Part 123, after the seven principles outlined in the work done by NACMCF. The principles are:

1. *Conduct a hazard analysis (HA).* Prepare a list of processing steps and determine where significant hazards are reasonably likely to occur, and describe the preventive measure or control of the hazard. As previously stated, these hazards are biological, chemical, or physical contaminants that may cause a food to be unsafe.
2. *Identify the critical control points (CCPs).* The regulation defines a critical control point as a step or procedure at which control can be applied and a food safety hazard can be prevented, eliminated, or reduced to an acceptable level. A CCP will always be a specific point in the process flow where the application of a preventive measure will effectively control the identified hazard.
3. *Establish critical limits (CLs)* for the preventive measures that were identified at the CCPs. A critical limit is defined as the maximum or minimum value to which a biological, chemical, or physical parameter must be controlled in order to prevent, eliminate, or reduce the hazard to an acceptable level.
4. *Establish monitoring procedures* at each identified critical control point to assess whether the CCP is under control and to produce a record for future use in verifying that the controls were in place and the established critical limits were not exceeded without taking appropriate corrective action. These proce-

dures should describe what is being monitored, the frequency of the monitoring, and who is performing the monitoring.

5. *Outline corrective action plans* to be followed in response to a deviation from a critical limit at a critical control point.
6. *List verification procedures* and their frequency in order to verify that the HACCP system is working.
7. *Establish a record-keeping system* to document the HACCP system, including monitoring of critical limits and the application of appropriate corrective actions when the limits were exceeded.

To apply these seven principles as they relate to parasites, the potential hazard must first be recognized; the significance of the hazard is determined; and then the preventive measures for controlling the parasite hazard are described.

Some parasites, when present in a product consumed raw or undercooked, can contribute to a food safety hazard. Of most concern in seafood are the nematodes or roundworms (*Anisakis* spp., *P. decipiens*, *Eustrongylides* spp., and *Gnathostoma* spp.), cestodes or tapeworms (*Diphyllobothrium* spp.), and trematodes or flukes (*C. sinensis*, *Opisthorchis* spp., *Heterophyes* spp., *Metagonimus* spp., *N. salmincola*, and *Paragonimus* spp.). Several products fall into the category of "consumed raw or undercooked" and have been implicated in parasitic infections in humans. Some of these are ceviche (fish and spices marinated in lime juice); lomi lomi (salmon marinated in lemon juice, onions, and tomato); salmon roe; sashimi (slices of raw fish); sushi (pieces of raw fish with rice and other ingredients); drunken crabs (crabs marinated in wine and peppers); cold smoked fish; and rare (undercooked) grilled fish. To determine whether the parasite hazard is significant, the intended use of the product must be determined. If the fish is to be cooked by the consumer prior to consumption, the hazard is not considered significant. Also, if a primary processor has knowledge that a secondary processor of the same product uses a process that kills parasites, the hazard would not be considered significant.

A parasite hazard is deemed significant if it is reasonably likely that parasites will be introduced at any of the processing steps. Generally, only the receiving step is considered for parasite hazards because they are naturally present and will be in the raw material. The Fish and Fishery Products Hazards and Controls Guide (the Guide) has listed those fish species for which the FDA has information that a potential parasite hazard exists (33). Typically, if the fish species is listed in the guide as having a possible parasite hazard and the product will be consumed raw or undercooked, the receiving step would be identified as where the hazard exists.

An exemption is provided in the HACCP regulation for fishes produced by acceptable aquaculture practices in regard to possible parasite hazards. If the aquaculture operation is conducted such that the fish are not exposed to parasite infections, the hazard in the resulting product is not considered significant. In aquaculture, fish can be indoctrinated into recognizing only pelleted food as acceptable prey. Although crustaceans and smaller fishes may be present within the net pens, indoctrinated fish do not consume them. Under these conditions, the life cycles of the nematodes and tapeworms are interrupted and the aquacultured fish remain free of these parasites. However, if the fish are kept at any time in pens that contain intermediate hosts for trematodes, the possibility exists for the fish to become infected from cercaria released into the water from the snails. This scenario is of concern in freshwater ponds in Asia where trematodes such as *C. sinensis* may be present. In the United States, hatcheries in the Pacific Northwest may be inhabited by the snail *Juga plicifera*, the intermediate host for *N. salmincola*. Salmon from these hatcheries may be infected with the trematode prior to their arrival at the net pens. These infections can be avoided if hatcheries use well water and prevent the establishment of the snail host. In a study of 237 Atlantic salmon, *Salmo salar*, from two net-pen sites in the Puget Sound (Washington State), the viscera and edible flesh of all fish were free of anisakids, diphyllobothriids, and metacercaria of *N. salmincola* (34). In contrast, aquacultured fish that are fed processing waste and by-catch may become infected with parasites and pose a hazard, even if the wild stock is generally uninfected. Processors using fish from such operations must consider parasites as a significant hazard.

Treatments such as irradiation, salting, and brining can have lethal effects for parasites, but these steps are not generally accepted as preventive measures because of unacceptable or inconsistent results. These processes are mentioned here to explain and stress the reasons for which these do not meet the requirements as preventive measures.

The food industry continues to search for new technology that will provide for safety of food products. Irradiation is one such method which has been proposed as an additional step prior to marketing of many agricultural products (35). Gamma irradiation is the most common form used in commercial food processing. It is an effective means of destroying many types of microorganisms; is not destructive to some heat-sensitive foods, may extend product shelf life, and can penetrate dense materials. However, irradiation has limitations that prevent its use in some situations. First and foremost, irradiation is not currently approved for use in fish processing and therefore cannot be considered a preventive measure in a HACCP program. The method is expensive, can cause changes in the food product, must be approved as a food additive, and raises the question of safety and environmental concerns in γ-ray facilities (36). In addition, many consumers do not understand the irradiation process and public perception of irradiation has resulted in a requirement that irradiated foods bear the universal radiation label. The amount of irradiation necessary to eliminate a hazard varies according to the targeted organism, particularly for parasites. For example, tissue cysts of the protozoan *Toxoplasma gondii* were killed after irradiation at 0.4 kGy (37); oocysts of *T. gondii*, which are the environmental stages, were killed at 0.25 kGy (38). To eliminate the risk of trichinosis, pork can be irradiated at 0.15–0.30 kGy (39). The larval worms are not killed at these low levels; rather, they are rendered sexually sterile. However, irradiation is inappropriate for anisakid nematodes; the larvae present in the seafood product cause anisakiasis directly. Therefore, the level of irradiation must be high enough to kill the nematodes (0.5 Mrad or 10 kGy), rather than rendering the parasites sexually sterile (40). Reproduction for anisakids does not occur in the human host, regardless of the sexual status of the larvae. At these levels of irradiation, the fish acquires an unpalatable texture and taste.

In addition to fresh and frozen seafood, the consumer can purchase products that have been prepared by brining, pickling, marinating, or salting. These methods may also be used by the consumer to prepare similar dishes at home. The primary components of pickling solutions, brines, and marinades are salt and acid. Because anisakid nematodes normally reside in the stomachs of marine mammals, the parasites are highly resistant to the acid levels present in the solutions. Any lethality of parasites from the use of these methods is caused by the presence and concentration of salt. Direct exposure of the worms to dry salt results in the death of anisakids within 10 min (41), although the progression of the salt through the flesh of fish products will take much longer. As the salt concentration decreases, the time required to achieve death of the nematodes increases. In a saturated salt solution of 22%, the worms are dead in 10 days; a 15% salt and 7% acid solution kills 97% of the parasites after 30 days. Most pickling solutions contain approximately 6% salt and 4% acid and more than 70 days is required for the worms to be killed (41). Prior to processing with a salt solution, freezing of fish is recommended to ensure consistent control of the parasite hazard.

Even though the significant hazard is identified at the receiving step, the preventive measure or control is not applied at that time. Accepted preventive measures for parasites usually include cooking, retorting, pasteurizing, or freezing, and the critical control point (CCP) is identified at one of these points or steps in the process. After identifying the CCP, the critical limits (CLs) must be set. The CLs should be established at the point where if they are not met the safety of the product would be questionable. If a CL is too strict, corrective actions may be taken even though no safety issue exists. In contrast, if the CL is too loose, an unsafe product may result.

For seafood, there are basically two ways to control parasites in a HACCP plan: heating and freezing. If parasites are controlled by heating at a cook (including hot smoking), pasteurization, or thermal processing (retort) step, the time/temperature parameters at that step will be based on the destruction of bacterial pathogens. Processes used to heat raw fish sufficiently to kill bacterial pathogens are also sufficient to kill parasites. For example, in canned salmon or pasteurized crab meat, the identification of parasites as a significant hazard is not necessary because the cook step designed to control the target organism, *Clostridium botulinum*, is exceptionally lethal to parasites. Even if these products were significantly underprocessed, the safety of the product relative to parasites would not be jeopardized. The 1997 Food Code provides minimum cooking times and temperatures for consumers and retail operations to kill both parasites and bacterial pathogens. For conventional cooking, including baking, the internal temperature of the thickest part of the seafood product should reach a minimum of 63°C (145°F) for 15 s or longer to kill parasites. Time/temperature parameters to kill bacteria in raw fish encompass those for parasites; fish should be heated to an internal temperature of 68°C (155°F) for 15 s, or 66°C (150°F) for 1 min, or 63°C (145°F) for 3 min for a 5 log decimal reduction in microorganisms (43). Cooking in a microwave oven requires a higher temperature to kill anisakids due to the uneven heating intrinsic to the process; a temperature of 77°C (170°F) in the thickest part of the product is recommended (44). Turning the food during cooking, covering the food, and/or adding liquid will decrease the uneven heating and can lower the required temperature to kill parasites.

As mentioned previously, if a likelihood exists that a seafood product will be consumed raw or undercooked, the processor is required to institute a control plan for parasites. The hazard is generally controlled by freezing at one of several steps. For example, fish frozen during harvesting need not be refrozen by the processor as a control step. During the production of Nova Scotia lox, salmon is frozen after the cold-smoking process to cut thin slices of the fish, which is characteristic of the product. Cold-smoked products are essentially raw because temperatures necessary to kill parasites and other pathogens are not reached during the smoking process (45); therefore, a CCP for parasites is necessary. Similar to heating, freezing is time/temperature-dependent. The colder the temperature, the less time is required to kill the parasites. Guidance on setting the CL for controlling parasites during freezing has been published in the FDA's Food Code, and Fish and Fishery Products Hazards and Controls Guide. Minimum freezing guidelines are −20°C

($-4°F$) or below (internal or external) for 7 days or $-35°C$ ($-31°F$) or below (internal) for 15 h (33,34). Other time/temperature parameters can be used if the processor can show that any parasite hazard will be controlled.

In order to assure that CLs are consistently met, monitoring procedures need to be developed. The program must describe what will be monitored, how it will be monitored, the frequency of monitoring, and who will perform the monitoring. Whether an external or internal temperature is to be measured will determine what is monitored. For external temperatures, the freezer temperature and the length of freezing is monitored; for internal temperatures, the internal fish temperature and the length of time the fish is held at the desired internal temperature is monitored. Monitoring will be performed using some sort of temperature indication device, e.g., a dial thermometer, digital time/temperature data logger, etc., for both external and internal temperature measurements. The temperature indication device is monitored visually. The frequency at which these temperatures are monitored will differ in regard to the chosen temperature indicating device. For instance, if a recording device or digital time/temperature data logger is used, the monitoring will be continuous, with perhaps a visual check twice a day (but not less than once a day) during the freezing cycle. For other temperature indication devices, the temperature (either external or internal) should be checked at least once a day and the time should be checked at the start and end of each freezing cycle. Monitoring of the freezing cycle can be accomplished by an employee responsible for the freezer, a quality assurance technician, a production supervisor, or anyone who has a thorough understanding of the CLs and the monitoring device.

Deviation from a CL is always possible. Therefore, predetermination of corrective actions is prudent and necessary to regain control over the process step and to ensure a safe product. To do this, corrective actions should include repairing the freezer unit or moving the product to a functioning freezer, and then refreezing the product using the same values that were established to ensure no deviations from the CL.

A record keeping system is important to demonstrate that the monitoring procedures are being followed and that the values recorded did not deviate from CLs without corrective action being taken. Records must contain the actual values and observations from the monitoring procedures. Records of the freezing times and temperatures would include, but not be limited to, temperature recording charts or freezer logs.

Verification of the HACCP system is important to assure that the HACCP plan is adequately addressing the parasite hazard and that the plan is consistently being followed. Reviewing records for monitoring and corrective action is paramount to determining whether the hazard is being controlled consistently. The verification program must also guarantee the accuracy of the temperature indication devices. As such, calibration of these devices should be frequent enough to assure their performance.

V. INSPECTION AND ANALYSIS

HACCP is only one element of a seafood inspection and not the entire agenda. As previously discussed, there are other concerns such as sanitation, economic fraud, decomposition, filth, and GMP violations. Furthermore, each inspection is unique as products and processes differ. Therefore, even though a standardized approach to seafood inspection exists, that approach might need to change based on what is encountered during the inspection (46).

However, the FDA's Seafood HACCP Regulators training manual refers to seven basic components of a HACCP inspection (47). They are:

1. *Conducting the initial interview.* During the initial interview, administrative issues such as showing credentials and issuing a notice of inspection occur. More importantly, the inspector selects the product that will be covered during the inspection. The product should be one in production on the day of the inspection, unless a specific product was targeted. The adequacy of the HACCP plan and whether or not it is being implemented properly can be evaluated in this way. This evaluation cannot be done by reviewing the monitoring records alone. If the firm is processing more than one product and if only one product can be inspected because of time constraints, the product that has the highest potential for a safety hazard or a noncompliant product from a previous inspection should be chosen. In considering a parasite hazard, the emphasis for inspection would be placed on a product to be consumed raw, such as sushi or sashimi.

2. *Develop a HACCP plan.* After the product is chosen, a hazard analysis independent of the processor is developed by the inspector. This is typically done during a walk-through of the plant to observe the processing from the receipt of the raw material to the shipment of the final product.

3. *Compare HACCP plans.* The inspector and the processor should compare their HACCP plans. If the hazard analyses differ, both the inspector and the processor must be prepared to discuss the justifications and bases for their decisions. The Guide contains tables with both species-specific and processing-specific hazards that might be associated with seafood (33). The FDA inspectors use this guide as a reference document when conducting their hazard analysis and for decisions of what hazards need to be considered.

4. *Evaluate the processor's plan.* After agreements have been reached on what hazards are significant and what preventive measures are in place to control the hazards, the processor's HACCP plan must be evaluated. The inspector will determine the adequacy of the critical limits, the monitoring procedures, any corrective actions to be taken, and the verification activities.

5. *Implementation of the HACCP plan.* The inspector must then determine if the processor is implementing the HACCP plan. Are they really doing what they say they are doing? This evaluation is best accomplished by observation and communication with plant employees to determine what occurs during production and evaluating whether the monitoring, corrective actions, and verification procedures are being followed.

6. *Reviewing records.* One of the most important components of an inspection is reviewing HACCP records. Records play a vital role because an inspector can determine whether the HACCP plan is being properly and consistently implemented. Typically, record review is done near the end of an inspection because records are easier to review and more meaningful when you understand their relationship to the processing operation and you have seen how they are prepared, and because it usually takes time for the processor to assemble the records that are requested by the inspector for review.

7. *Documentation of violations.* Lastly, the inspector documents any violations of the requirements of the regulation that are encountered during the inspection. These observations are listed and discussed with the top level managers at the

conclusion of an inspection. Voluntary correction of deficient items is the goal of the inspection.

Sampling has always been an integral part of inspection and will remain as such. The FDA has historically collected samples for compliance purposes or for what is referred to as ''for cause.'' These samples are usually collected to support an observed deficiency or legal action. However, sampling is not a very effective way of identifying problems such as low-level microbiological or physical contamination. Also, large sample sizes of perhaps 10% or more may be needed to determine if the product is contaminated. For collections that large, the time and cost of analysis can be cost-prohibitive.

With the Seafood HACCP regulation in place, inspectors will not necessarily collect compliance samples. Historically, a product was deemed adulterated after collection and analysis. In contrast, the FDA's Seafood HACCP regulation was written based on various sections of the Federal Food, Drug and Cosmetic Act that define what constitutes an adulterated product. Therefore, a product can be determined to be adulterated after observation and documentation of a deficient HACCP program showing failure to control a hazard that is reasonably likely to occur in a seafood product. A physical sample is not needed to support the charge. However, situations will exist where inspection leads to the collection of a sample. The FDA continues to collect samples to fulfill requirements of compliance programs for domestic and imported products, or of special assignments such as surveys. In addition, if a processor's program is in substantial compliance with the regulation, samples may also be collected to ''verify'' that the HACCP plan is controlling the significant hazards.

Analysis of parasite samples collected for compliance or special programs usually entails nondestructive methods and follows the procedures contained within the *FDA Bacteriological Analytical Manual* (48). These methods reflect what is available to industry as a means to control the quality of their products and the incidence of parasites within those products. In most analyses, the candling technique is used to detect and enumerate parasites in samples of finfish. If a verification sample was collected to determine adequate control of parasite hazards, the analysis would need to isolate the parasites and evaluate the viability of the parasites. The actual number of parasites present in the product would not be of concern; rather, determination of whether all of the parasites present had been rendered noninfectious by the control measure would be important. Artificial digestion of the seafood sample with a pepsin solution (49) would most likely be the method of choice for the analysis of verification samples for parasite hazards.

REFERENCES

1. Ward, D, and Hart, K. (1995). HACCP: Hazard analysis and critical control point training curriculum. National Seafood HACCP Alliance for Training and Education, University of North Carolina Sea Grant Program, Raleigh, North Carolina, 236 pp.
2. DeVlieger, D. D. (1997). Module 5: Prerequisite programs. In: Food Microbiological Control, Division of Human Resource Development Training Branch, Food and Drug Administration, U.S. Department of Health and Human Services, Washington, D.C., pp. V-1–V-15.
3. Food and Drug Administration (1998). Code of Federal Regulations 21 Part 110. U.S. Department of Health and Human Services, Washington, D.C., pp. 214–223.
4. Food and Drug Administration (1995). 21 CFR Part 123 and 1240: Procedures for the Safe

and Sanitary Processing and Importing of Fish and Fishery Products; Final. U.S. Department of Health and Human Services, Washington, D.C., pp. 65096–65202.

5. World Health Organization (1995). Control of foodborne trematode infections. WHO Technical Report Series, No. 849. Geneva.

6. Cross, J. H. (1994). Fish and invertebrate-borne helminths. In: Foodborne Disease Handbook, Vol. 2, Diseases Caused by Viruses, Parasites, and Fungi (Y. H. Hui, J. R. Gorham, K. D. Murrell, and D. O. Cliver, eds.), Marcel Dekker, New York, pp. 279–329.

7. MacKenzie, W. R., Hoxie, N. J., Proctor, M. E., Gradus, M. S., Blair, K. A., Peterson, D. E., Kazmierczak, J. J., Addiss, D. G., Fox, K. R., Rose, J. B., and Davis, J. P. (1994). A massive outbreak in Milwaukee of *Cryptosporidium* infection transmitted through the public water supply. N. Engl. J. Med., 331:161–167.

8. Fayer, R., Graczyk, T. K., Lewis, E. J., Trout, J. M., and Farley, C. A. (1998). Survival of infectious *Cryptosporidium parvum* oocysts in seawater and eastern oysters (*Crassostrea virginica*) in the Chesapeake Bay. Appl. Environ. Microbiol., 64:1070–1074.

9. Fayer, R., Farley, C. A., Lewis, E. J., Trout, J. M., and Graczyk, T. K. (1997). Potential role of the eastern oyster, *Crassostrea virginica*, in the epidemiology of *Cryptosporidium parvum*. Appl. Environ. Microbiol., 63:2086–2088.

10. Graczyk, T. K., Fayer, R., Cranfield, M. R., and Conn, D. B. (1998). Recovery of waterborne *Cryptosporidium parvum* oocysts by freshwater benthic clams (*Corbicula fluminea*). Appl. Environ. Microbiol., 64:427–430.

11. Graczyk, T. K., Ortega, Y. R., and Conn, D. B. (1998). Recovery of waterborne oocysts of *Cyclospora* cayetanensis by Asian freshwater clams (*Corbicula fluminea*). Am. J. Trop. Med. Hyg., 59:928–932.

12. Freire-Santos, F., Vergara-Castiblanco, C. A., Tojo-Rodriguez, J. L., Santamarina-Fernandez, T., and Ares-Mazas, E. (1998). *Cryptosporidium parvum*: an attempt at experimental infection in rainbow trout *Oncorhynchus mykiss*. J. Parasitol., 84:935–938.

13. Graczyk, T. K., Farley, C. S., Fayer, R., Lewis, E. J., and Trout, J. M. (1998). Detection of *Cryptosporidium* oocysts and *Giardia* cysts in the tissues of eastern oysters (*Crassostrea virginica*) carrying principal oyster infectious diseases. J. Parasitol., 84:1039–1042.

14. Osterholm, M. T., Forfang, J. C., Ristinen, T. L., Dean, A. G., Washburn, J. W., Godes, J. R., Rude, R. A., and McCullough, J. G. (1981). An outbreak of foodborne giardiasis. N. Engl. J. Med., 304:24–28.

15. Centers for Disease Control (1997). Outbreaks of *Escherichia coli* O157:H7 infection and cryptosporidiosis associated with drinking unpasteurized apples cider—Connecticut and New York, October 1996. MMWR, 46:4–8.

16. Millard, P. S., Gensheimer, K. F., Addiss, D. G., Sosin, D. M., Beckett, G. A., Houck-Jankoski, A., and Hudson, A. (1994). An outbreak of cryptosporidiosis from fresh-pressed apple cider. JAMA, 272:1592–1596.

17. Centers for Disease Control (1997). Foodborne outbreak of cryptosporidiosis—Spokane, Washington, 1997. MMWR, 47:565–567.

18. Centers for Disease Control (1996). Foodborne outbreak of diarrheal illness associated with *Cryptosporidium parvum*—Minnesota, 1995. MMWR, 45:783–784.

19. Adams, A. M., Leja, L. L., Jinneman, K., Beeh, J., Yuen, G. A., and Wekell, M. M. (1994). Anisakid parasites, *Staphylococcus aureus* and *Bacillus cereus* in sushi and sashimi from Seattle area restaurants. J. Food Protect., 57:311–317.

20. Harrell, L. W., and Deardorff, T. L. (1990). Human nanophyetiasis: transmission by handling naturally infected coho salmon (*Oncorhynchus kisutch*). J. Infect. Dis., 161:146–148.

21. Millemann, R. E., and Knapp, S. E. (1970). Biology of *Nanophyetus salmincola* and "salmon poisoning" disease. In Advances in Parasitology, Vol. 8: Dawes, B., ed., Academic Press, New York, pp. 1–41.

22. Eastburn, R. L., Fritsche, T. R., and Terhune, C. A., Jr. (1987). Human intestinal infection with *Nanophyetus salmincola* from salmonid fishes. Am. J. Trop. Med. Hyg., 36:586–591.

23. Fritsche, T. R., Eastburn, R. L., Wiggins, L. H., and Terhune, C. A., Jr. (1989). Praziquantel for treatment of human *Nanophyetus salmincola* (*Troglotrema salmincola*) infection. J. Infect. Dis., 160:896–899.

24. Food and Drug Administration (1998). Guidance for industry: guide to minimize microbial food safety hazards for fresh fruits and vegetables. U.S. Department of Health and Human Services, Washington, D.C.

25. Jarroll, E. L., Bingham, A. K., and Meyer, E. A. (1981). Effect of Chlorine on *Giardia lamblia* cyst viability. Appl. Environ. Microbiol., 41:483–487.

26. Korich, D. G., Mead, J. R., Madore, M. S., Sinclair, N. A., and Sterling, C. R. (1990). Effects of ozone, chlorine dioxide, chlorine, and monochloramine on *Cryptosporidium parvum* oocyst viability. Appl. Environ. Microbiol., 56:1423–1428.

27. Young, P. C. (1972). The relationship between the presence of larval anisakine nematodes in cod and marine mammals in British home waters. J. Appl. Ecol., 9:459–485.

28. Adams, A. M., and Rausch, R. L. (1997). Diphyllobothriasis. In: Pathology of Infectious Diseases, Vol. 2: Connor, D. H., Chandler, F. W., Schwartz, D. A., Manz, H. J., and Lack, E. E., eds., Appleton and Lange, Stamford, CT, pp. 1377–1389.

29. Meyers, B. J. (1979). Anisakine nematodes in fresh commercial fish from waters along the Washington, Oregon and California coasts. J. Food Protect., 42:380–384.

30. Helrich, K. (ed.) (1990). Parasites in fish muscle (method 985.12). Official Methods of Analysis of the Association of Official Analytical Chemists, Vol. 2. Food Composition: additives; natural contaminants, 15th Ed., Association of Official Chemists, Inc., Arlington, Virginia, pp. 882–883.

31. Adams, A. M., Murrell, K. D., and Cross J. H. (1997). Parasites of fish and risks to public health. Rev. Sci. Tech. Off. Int. Epiz., 16:652–660.

32. Bauman, H. (1990). HACCP: Concept, development, and application. Food Technol., May 1990:156–158.

33. Food and Drug Administration (1998). Fish and Fishery Products Hazards and Controls Guide: Second Edition. Center of Food Safety and Applied Nutrition, Office of Seafood, U.S. Food and Drug Administration, U.S. Department of Health and Human Services, Washington, D.C.

34. Deardorff, T. L., and Kent, M. L. (1989). Prevalence of larval *Anisakis* simplex in penreared and wild-caught salmon (*Salmonidae*) from Puget Sound, Washington. J. Wildl. Dis., 25:416–419.

35. Thayer, D. W., Josephson, E. S., Brynjolfsson, A., and Giddings, G. G. (1996). Radiation pasteurization of food. Council of Agricultural Science and Technology, Ames, Iowa, Issue paper No. 7.

36. Ellison, M. (1998). Module 8: Emerging technology for microbiological control. In: Food Microbiological Control, Second Edition, Division of Human Resource Development Training Branch, Food and Drug Administration, U.S. Department of Health and Human Services, Washington, D.C., pp. 121–124.

37. Dubey, J. P., and Thayer, D. W. (1994). Killing of different strains of *Toxoplasma gondii* tissue cysts by irradiation under defined conditions. J. Parasitol., 80:764–767.

38. Dubey, J. P., Jenkins, M. C., Thayer, D. W., Kwok, O. C. H., and Shen, S. K. (1996). Killing of *Toxoplasma gondii* oocysts by irradiation and protective immunity induced by vaccination with irradiated oocysts. J. Parasitol., 82:724–727.

39. Engel, R. E., Post, A. R., and Post, R. C. (1988). Implementation of irradiation of pork for trichina control. Food Technol., 42:71–75.

40. Farkas, J. (1987). Decontamination, including parasite control, of dried, chilled and frozen foods by irradiation. Acta Alimentaria, 16:351–384.

41. Khalil, L. F. (1969). Larval nematodes in the herring (*Clupea harengus*) from British coastal waters and adjacent territories. J. Mar. Biol. Assoc. U.K., 49:641–659.

42. Smith, J. W., and Wootten, R. (1978). *Anisakis* and anisakiasis. In: Advances in Parasitology,

Vol. 16: Lumsden, W. H. R., Muller, R., and Baker, J. R., eds., Academic Press, New York, pp. 93–163.

43. Food and Drug Administration (1997). 1997 Food code. U.S. Department of Health and Human Services, Washington, D.C.

44. Adams, A. M., Miller, K. S., Wekell, M. M., and Dong, F. M. (1999). Survival of *Anisakis simplex* in microwave-processed arrowtooth flounder (*Atheresthes stomias*). J. Food Prot., 62: 403–409.

45. Gardiner, M. A. (1990). Survival of *Anisakis* in cold smoked salmon. Can. Inst. Food Sci. Technol. J., 23:143–144.

46. Archer, D. L. (1990). The need for flexibility in HACCP. Food Technol., May 1990: 174–178.

47. Food and Drug Administration (1997). Seafood HACCP Regulator Training Program. Center for Food Safety and Applied Nutrition, Office of Seafood, U.S. Food and Drug Administration, U.S. Department of Health and Human Services, Washington, D.C., pp. I-1–XIV-4.

48. Bier, J. W., Jackson, G. J., Adams, A. M., and Rude, R. A. (1995). Chapter 19: Parasitic animals in foods. In: FDA Bacteriological Analytical Manual, 8th ed. AOAC International, Gaithersburg, MD, pp. 19.01–19.09.

49. VanVelzen, W. T. (1990). Extraction of nematodes from fish and other seafood products. Laboratory Information Bulletin, U.S. Food and Drug Administration, U.S. Department of Health and Human Services, Washington, D.C., 12 February 1990.

19

Foodservice Operations: HACCP Principles

O. Peter Snyder, Jr.
Hospitality Institute of Technology and Management, St. Paul, Minnesota

I. INTRODUCTION

As more retail food operations across the United States and throughout the world compete to feed consumers, it becomes essential that uniform hazard analysis and control guidelines for producing, buying, and selling food products be developed and used. These guidelines must be based on science and validated in actual operation. At this time, consumers in the United States are doing less food preparation themselves and are relying on retail food outlets for ready-prepared items. Retail food operations (as referred to in this chapter) include food markets where food is sold to be prepared in the home; food preparation and foodservice establishments that include restaurants, institutional foodservice units, street vending operations, hotel and lodging operations, military commissaries, and even the home, which is actually a miniature foodservice unit.

Food science and technology have improved the understanding of the potential microbiological, chemical, and physical hazards in foods. This knowledge can be used to determine the criteria necessary to assure that food products and commodities meet consumer safety expectations with an acceptable risk at the raw material level, the distributor level, and the consumer level. International trade and tourism will be enhanced throughout the world when there is a clearer understanding between the producer or seller and the buyer of food concerning the potential hazards in food and the level of risk associated with consuming that food.

Beginning with *Codex Alimentarius* (13) and the International Commission on Microbiological Specifications for Food, and continuing with the National Advisory Committee on Microbiological Criteria for Foods (62,64), there has been a movement for many years toward more complete safety specifications for foods in local and international trade. The result is the current emphasis on Hazard Analysis and Critical Control Points (HACCP) in food production facilities and retail food operations. With this emphasis, however, people have lost sight of the fact that HACCP is only a part of a company's food production quality management program. A company cannot accomplish process hazard control until it has process quality control. Hazards and critical control points can be easily identified. However, it is a separate issue to actually operate in such a way that there is a very low chance of process deviation because of correct employee task performance and low risk of a hazardous food product(s) being produced.

II. GOVERNMENT MICROBIOLOGICAL STANDARDS FOR RAW AND PASTEURIZED FOOD

The Code of Federal Regulations [9 CFR §301.2 C] (12) has been interpreted by the U.S. Department of Agriculture and the Food and Drug Administration to mean that if a sample of a processed food is found to be contaminated with *E. coli* O157:H7, *Salmonella* spp., or *Listeria monocytogenes*, the food is deemed unfit for human consumption. Sample size is variable, but the presence of any one of these three pathogens must be negative. Measurement sensitivity is from 1 per gram to 1 per 25 grams, depending on the laboratory and procedure.

In retail food operations, it can be expected that *L. monocytogenes* will be found on fresh produce, meat, fish, and poultry, as well as on floors and in floor drains. It has been estimated by Farber (21) that raw food (e.g., cole slaw) can contain 100–10,000 CFU/g of *L. monocytogenes*.

III. FOODBORNE ILLNESS IN THE UNITED STATES: ESTIMATES AND COST

While the United States has the capability of producing the "safest" food in the world, there are still an enormous number of illnesses and deaths that occur each year because of inadequate hazard control when food is prepared. It is *estimated* that 6.5–81 million Americans become ill each year due to foodborne illness with from 525 to more than 9000 associated deaths (6,26,74,84). It has also been estimated that the cost of foodborne illness in the United States ranges from $4 to $23 billion a year in terms of medical expenses, lost wages, insurance costs, and liability (26,73,74,84). Statistics clearly show that microorganisms consumed in food and water are a significant cause of illness (see Table 1). In most instances, the illnesses and deaths that result from these sources can be prevented.

Most American consumers assume that if the food looks fresh and appetizing, it is safe. They do not realize that the visual appearance, smell, and taste of food are not indications of safety. Pathogens and toxic compounds can be present in food products despite the food's attractive appearance, smell, and taste.

The sources of pathogens responsible for foodborne illness and disease are everywhere. Because of environmental and animal contamination, food and food products will always be contaminated with low levels of pathogens. At low levels, the presence of pathogenic microorganisms in food causes no problems, and people may even develop immunity to their presence in food (34,39,41,66,72,82,90).

A noted microbiologist (43) has given the following hypothesis: "With advances and improvements in cleaning/sanitizing compounds, and with a better understanding of how to control biofilms on food surfaces and equipment, our fresh meats are so *clean* now that they do not contain enough harmless background organisms to prevent the proliferation of bacterial pathogens." Bacterial interference occurs due to production and excretion of substances resulting from the growth of some microorganisms that are inhibitory or lethal to other organisms. When different types of microorganisms grow in a substance or food, there is competition for nutrients and competition for attachment sites. Thus the environment may become unfavorable or undesirable for the growth of other organisms. For example, lactic acid bacteria in fermented dairy and other food products produce bacteriocins that inhibit the growth of pathogenic bacteria, and in this way are responsible for the long history of safety for these products. At illness thresholds, however, pathogens in food can make people ill and cause death. Pathogens in food can only be controlled when food producers, food retailers, and consumers know the potential hazards in food, and handle and prepare food by methods designed to assure safety.

How do pathogens in food reach high levels? Pathogen contamination and cell population increases may occur anywhere in the food chain, e.g., from the time the animal, fish, vegetable, or fruit was grown to the time it was consumed. Everyone throughout the food chain must have knowledge of the causes of foodborne disease and illness, and must establish a program that assures safety before they produce and sell food. If this is not the case, incomplete hazard control processes are implemented.

Table 1 Foodborne Illness Annual Estimated Cases and Deaths in the United States

Cause	Cases	Deaths
BACTERIA		
Spore-forming Bacteria		
Bacillus cereus	84,000	0
Clostridium perfringens	650,000	6–7
Clostridium botulinum (adults)	100	2–3
Infant botulism	60	?
Vegetative Bacteria		
Streptococcus (grp. A)	500,000	175
Yersinia enterocolitica	20,000	2–3
Staphylococcus aureus	8,900,000	5–6
Salmonella (non-typhi)	3,000,000	2,000
Campylobacter spp.	2,100,000	2,100
Shigella spp.	300,000	600
Escherichia coli (pathogenic)	200,000	400
Brucella spp.	50,000	0.1
Vibrio cholerae/vulnificus	13,000	1–2
Vibrio (noncholera)	30,000	300–900
Salmonella typhi	600	36
Listeria monocytogenes	25,000	1,000
Miscellaneous Microorganisms	107,000	11
VIRUSES		
Hepatitis A	48,000	150
Norwalk virus	181,000	0
Other viruses	6,000,000	6
PARASITES		
Trichinella spiralis	100,000	1,000
Giardia lamblia	7,000	0
Toxoplasma gondii	2,300,000	450
Taenia spp.	1,000	10
Fish parasites	1,000	0
CHEMICALS/TOXINS		
Ciguatera toxin	27,000	2.1
Chemical poisons	96,000	5.4
Plant poisons	7,000	5.9
Scrombroid toxin	31,000	0
Paralytic shellfish poison	260	0.3
HARD FOREIGN OBJECTS	?	?
TOTAL	24,779,020	8,870

Source: Data adapted from Bennett et al. (6), Roberts and van Ravenswaay (74), and Todd (84).

Most health inspectors look only for visual indications of cleanliness of the facility and equipment and defects in food holding temperatures. They do not have the education or time to look for defects in food processing. The key to hazard control is to design processes that will reduce the incoming pathogen level to a safe level, validate that they are effective, and then train and coach employees to strive for zero performance mistakes.

IV. PERVASIVENESS OF PATHOGENS IN FOOD

Table 2, compiled by the Council for Agricultural Science and Technology (CAST) (10), reports the pervasiveness of pathogens in food.

V. NEED FOR INTERNATIONAL SAFE AND HAZARDOUS LEVEL GUIDELINES

At the present time there are few worldwide food safety guidelines for upper or lower control limits of potential hazards in foods. Standards, when they exist, may be inappropriate (e.g., specification of the numbers of coliforms in milk and shellfish waters), or may be undesirable or unattainable (e.g., a zero level of both *Salmonella* spp. and *L. monocytogenes* in food). There is no zero level in food safety. There is a point at which measurements cannot be made with any degree of statistical reliability, and this point is frequently taken as zero. For example, processed food is assumed to be safe from *Salmonella* contamination if there are no detectable salmonellae in a sample using the method of analysis described, e.g., by the *Bacteriological Analytical Manual* (24). However, as laboratory methods improve, standards for safe levels of pathogenic material in food may be resolved. When safety standards or guidelines are developed for microbiological, chemical, and physical hazards in food, the standards or guidelines must be based on the risk of causing injury or illness to consumers, not what the processing industry is capable of achieving or what scientific technology is capable of measuring.

VI. THE UNIT AS A FOOD PROCESS SYSTEM

What is included in the retail food system definition? Figure 1 provides process control nomenclature and a systematic way to look at the processes in order to make flow diagrams of them. At the top of the system is *management* as the first element of an HACCP-based design, providing resources, leadership, enforcement, and a good environment in which people can function.

The next step is consideration of what consumers need or desire and the hazard threshold levels necessary for consumer product safety. While there are few government standards, the industry must have guidelines in order to develop acceptably safe food. Guidelines are necessary for the input of the process in order to design the food process *environment, facilities, equipment, employees*, and *supplies* that will give a consumer safe output. The *process* takes the input elements and makes them safe for the *consumer*.

Highly contaminated raw food can be made safe if retail food operators know the level of contamination. In this way they can institute a proper HACCP program to assure that the food not only nourishes the consumer but is safe for consumption. For example, contaminated ground beef and chicken can be sold to a consumer in a retail food market as long as the consumer is informed and aware of handling and cooking procedures that must be accomplished in order to make the food safe to eat [e.g., ground beef: cook to 155°F (68.3°C) food temperature, hold at 155°F (68.3°C) for 15s for a 5D reduction of salmonellae], and how to clean the food contact surfaces that the food touches.

Table 2 Selected Illustrations of the Prevalence of Pathogens in Foods in the United States and Other Countries

Organism	Food	Percent positive	Ref.
Salmonella serovars	Beef	0–2.6	Lammerding et al., 1988 (51); Ternstrom and Molin, 1987 (83)
	Veal carcasses	4.1	Lammerding et al., 1988 (51)
	Pork	0–18	Lammerding et al., 1988 (51); Ternstrom and Molin, 1987 (83); Madden et al., 1986a (55)
	Pork products	3–20	Duitschaever and Buteau, 1979 (19); Farber et al., 1988 (22)
	Turkey carcasses	69	Lammerding et al., 1988 (51)
	Turkey sausage	100	Duitschaever and Buteau, 1979 (19)
	Chicken	0–100	Duitschaever and Buteau, 1979 (19); Izat et al., 1989 (42)
	Shellfish	3.7–33	Colburn et al., 1989 (14); Fraisier and Koburger, 1984 (27)
	Raw milk	0.5–4.7	McEwen et al., 1988 (57); McManus and Lanier, 1987 (58)
Bacillus cereus	Pork	4–7	Konuma et al., 1988 (50); Ternstrom and Molin, 1987 (83)
	Beef	11–63	Konuma et al., 1988 (50); Ternstrom and Molin, 1987 (83)
	Rice	100	Bryan et al., 1981 (7)
	Meat additives	39	Konuma et al., 1988 (50)
	Dairy products	0–63	Ahmed et al., 1983 (3); Mosso et al., 1989 (59); Rodriguez and Barret, 1986 (75)
	Milk, pasteurized	35	Ahmed et al., 1983 (3)
Campylobacter jejuni	Pork	0–24	Mafu et al., 1989 (56); Stern et al., 1984 (80); Ternstrom and Molin, 1987 (83)
	Beef carcasses	50	Garcia et al., 1985 (29)
	Lamb	1–20	Stern et al., 1984 (80)
	Turkey	56–64	Rayes et al., 1983 (70)
	Chicken	8–89	Christopher et al., 1982 (11); Kinde et al., 1983 (48); Norberg, 1981 (67); Shanker et al., 1982 (77); Stern et al., 1984 (80); Ternstrom and Molin, 1987 (83)
Clostridium botulinum	Bacon	0.1	Hauschild and Hilsheimer, 1980 (36)
	Liver sausage	2	Hauschild and Hilsheimer, 1983 (37)
	Corn syrup	20	Kautter et al., 1982 (46)
	Honey	2	Kautter et al., 1982 (46)

Organism	Food	%	References
Enterovirus	Shellfish	0–47.8	Ellender et al., 1980 (20), Gerba and Goyal, 1978 (32); Goyal et al., 1979 (33); Khalifa et al., 1986 (47); Vaughn et al., 1980 (86); Wait et al., 1983 (87)
Clostridium perfringens	Pork	0–39	Bauer et al., 1981 (5); Ternstrom and Molin, 1987 (83)
	Cooked pork	45	Kokubo et al., 1986 (49)
	Beef	22	Ternstrom and Molin, 1987 (83)
	Chicken	0–54	Lillard et al., 1984 (53); Ternstrom and Molin, 1987 (83)
	Seafood	2.4	Abeyta, 1983 (1)
Yersinia enterocolitica	Beef	2	Ternstrom and Molin, 1987 (83)
	Pork	2.5–49	Schiemann, 1980 (76); Ternstrom and Molin, 1987 (83)
	Processed pork products	7–37	Delmas and Vidon, 1985 (17); Ternstrom and Molin, 1987 (83)
	Chicken	11–25	Norberg, 1981 (69); Ternstrom and Molin, 1987 (83)
	Raw milk	2.7–48	Davidson et al., 1989 (16); McManus and Lanier, 1987 (58); Moustafa et al., 1983 (60)
	Ice cream	22	Delmas and Vidon, 1985 (17)
	Raw vegetables	46	Delmas and Vidon, 1985 (17))
Vibrio cholerae	Shellfish	7.4–33	Colburn et al., 1989 (14)
Vibrio parahaemolyticus	Seafood	2.8–46	Abeyta, 1983 (1), Hackney et al., 1980 (35)
Listeria monocytogenes	Raw red meats	0–43	Buchanan et al., 1989 (51); Ternstrom and Molin, 1987 (83)
	Ground beef	77	Farber et al., 1989 (23)
	Ground pork	95	Farber et al., 1989 (23)
	Ground veal	100	Farber et al., 1989 (23)
	Chicken	13–56	Bailey et al., 1989 (4); Farber et al., 1988 (22); Genigeorgis et al. 1989 (31)
	Turkey	12–18	Genigeorgis et al., 1990 (30)
	Cured meats and fermented sausages	0–20	Buchanan et al., 1989 (8); Farber et al., 1989 (23) Trussel et al., 1989 (85)
	Seafood	11–26	Buchanan et al., 1989 (8); Weagant et al., 1988 (88)
	Raw milk	1.6–4.2	Davidson et al., 1989 (16); Farber and Peterkin, 1991 (21); Liewen and Plautz, 1988 (52); Lovett et al., 1987 (54)
	Ice cream	0.25	Farber et al., 1989 (23)

Table 2 Continued

Organism	Food	Percent positive	Ref.
Staphylococcus aureus	Raw beef	16	Ternstrom and Molin, 1987 (83)
	Raw pork	13	Ternstrom and Molin, 1987 (83)
	Pork sausage	33	Farber et al., 1988 (22)
	Raw chicken	41–73	Lillard et al., 1984 (53); Ternstrom and Molin, 1987 (83)
	Seafood	38	Abeyta, 1983 (1)
	Bakery items	9.8	Sumner et al., 1993 (81)
Aeromonas hydrophila	Seafood	19–100	Abeyta, 1983 (1); Abeyta et al., 1989 (2); Colburn et al., 1989 (14); Fricker and Tomsett, 1989 (28); Palumbo et al., 1985 (69)
	Raw milk	33	Palumbo et al., 1985 (69)
	Poultry	16–100	Fricker and Tompsett, 1989 (28); Palumbo et al., 1985 (69); Ternstrom and Molin, 1987 (83)
	Red meat	100	Palumbo et al., 1985 (69)
	Cooked meats	10	Fricker and Tompsett, 1989 (28)
	Produce	95	Callister and Agger, 1987 (9)
Escherichia coli	Beef	3.7	Doyle and Shoeni, 1987 (18)
O157:H7	Pork	1.5	Doyle and Shoeni, 1987 (18)
	Poultry	1.5	Doyle and Shoeni, 1987 (18)
	Lamb	2	Doyle and Shoeni, 1987 (18)
Escherichia coli (verotoxigenic)	Raw beef	17	Willshaw et al., 1993 (89)
	Ground beef	36.4	Read et al., 1990 (71)
	Ground pork	10.6	Read et al., 1990 (71)

Source: Adapted from CAST (Council for Agricultural Science and Technology) 1994. Foodborne pathogens: risks and consequences. Task Force Report No. 122.

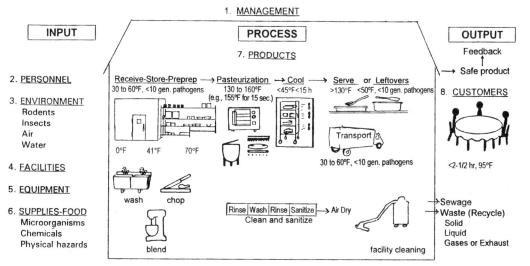

HAZARDS: Microorganisms [bacteria (vegetative cells and spores), viruses, parasites]; chemicals; hard foreign objects.

CONTROLS: Management involvement; hazard analysis and control; written procedures; employee training and empowerment; process measurement, control, and improvement; discipline and consequences.

Figure 1   The unit as a food process system.

VII.   TYPES OF ESTABLISHMENTS INCLUDED IN THE SCOPE OF RETAIL FOOD OPERATIONS

Table 3 is a list of the kinds of establishments included in the scope of retail food operations. The same basic hazard analysis and hazard controls, if applied correctly by each type of operation, will assure that food sold or given to the consumer is safe.

VIII.   HAZARD CLASSIFICATION

The World Health Organization (WHO) Division of Food and Nutrition (91) identifies a hazard as a biological, chemical, or physical agent or condition in food with the potential

Table 3   Types of Establishments Included in the Scope of Retail Food Operations

Food markets, restaurants, food service units, home food preparation	Nursing homes
	Hospitals
Roadside stands and rural meat markets	Cruise ships
Home catering	Military field feeding
Bakeries	In-flight feeding
Convenience stores	Feeding athletes at games
Vending, feeding the public at sports events	Religious institution feeding
Temporary food establishments (food carts, fairs, festivals)	Feeding people on camping trips in the wilderness
Prisons	

to cause harm. In addition to these factors, lack of adequate nutritional requirements necessary for the prevention of disease must also be considered a hazard.

Hazard analysis is the process of collecting and interpreting information to assess the risk and severity of potential hazards (40). It is appropriate to classify hazards as *high concern*, *low concern*, and *no concern*. The definitions for hazards in food are as follows:

1. *High concern*: without control there is a life-threatening risk.
2. *Low concern*: a threat that must be controlled but is not life threatening and requires no government intervention.
3. *No concern*: no threat to the consumer (normally quality and cleanliness issues).

For example, spoilage bacteria, at more than 50 million per gram, are of no known safety concern. Coliform bacteria include both spoilage and pathogenic microorganisms. Therefore, a coliform count of 1000 CFU/g is of no concern until specified levels in the coliform group are identified. Pathogens and pathogenic substances ingested above threshold levels can cause illness, disease, even death.

Also of low concern is the presence of 1000 *Staphylococcus aureus* cells, *Bacillus cereus* spores, or *Clostridium perfringens* spores per gram of food. These organisms are not hazardous until they reach 100,000 to 1 million vegetative cells per gram. A hazard of high concern, however, is the presence of 1 *Escherichia coli* O157:H7 per gram and 10 *Salmonella* spp. per gram of raw ground beef, or 10 *Campylobacter jejuni* per gram of raw poultry. Foodborne illness or disease can occur if there is cross-contamination of these pathogens from these raw meat items to other food items, and if these items are cooked inadequately before being consumed. Even healthy people can become ill if they consume food containing these pathogens at these levels of contamination when food preparers have failed to use procedures that assure safety. To assure that products are safe to consume, raw meat, fish, and poultry products must be heated to temperatures for sufficient periods of time to reduce pathogenic microbial hazards to a safe level. Care must also be taken to prevent cross-contamination of other foods or pasteurized food products when these products are handled.

IX. THE PROCESS OF HAZARD IDENTIFICATION

In order to develop a process or operation capable of protecting public health, which in turn will minimize liability costs, it is essential that a logical process for hazard identification be followed. The following criteria should be included in the hazard analysis:

1. Evidence of a hazard to health based on epidemiological data or a hazard analysis.
2. Type and kind of the natural and commonly acquired microflora of the ingredient or food, and the ability of the food to support microbial growth.
3. The effect of processing on the microflora of the food.
4. Potential for microbial contamination (or recontamination) and/or growth during processing, handling, storage, and distribution.
5. Types of consumers at risk.
6. The state in which food is distributed (e.g., frozen, refrigerated, heat-processed, etc.).
7. Potential for abuse at the consumer level.

8. The existence of GMPs.
9. The manner in which the food is prepared for ultimate consumption (i.e., heated or not).
10. Reliability of methods available to detect and/or quantify the microorganism(s) and toxin(s) of concern.
11. The costs/benefits associated with the application of items 1–10 (as listed above).

Coupling hazard analysis with correct hazard control and operating procedures enables the retail food operator to demonstrate a high degree of ''due diligence'' in the prevention of problems.

X. FDA DEFINITION OF POTENTIALLY HAZARDOUS FOOD

Retail food regulations developed from the FDA food codes are incomplete because the range of foods that can cause a foodborne illness is limited by the *potentially hazardous food* definition. The FDA 1997 Food Code (25) definition is as follows:

(a) "**Potentially hazardous food**" means a *food* that is natural or synthetic and is in a form capable of supporting:
> (i) The rapid and progressive growth of infectious or toxigenic microorganisms;
> (ii) The growth of toxin production of *Clostridium botulinum*; or
> (iii) In shell eggs, the growth of *Salmonella enteritidis*.

(b) "**Potentially hazardous food**" includes an animal *food* (a *food* of animal origin) that is raw or heat-treated; a *food* of plant origin that is heat-treated or consists of raw seed sprouts; cut melons; and garlic and oil mixtures that are not acidified or otherwise modified at a food processing plant in a way that results in mixtures that do not support growth as specified under Subparagraph (a) of this definition.

(c) "**Potentially hazardous food**" does not include:
> (i) An air-cooled hard-boiled egg with shell intact;
> (ii) A *food* with a *water activity* (a_w) value of 0.85 or less;
> (iii) A *food* with a hydrogen ion concentration (pH) level of 4.6 or below when measured at 24°C (75°F);
> (iv) A *food*, in an unopened *hermetically sealed container*, that is commercially processed to achieve and maintain commercial sterility under conditions of nonrefrigerated storage and distribution; and
> (v) A *food* for which laboratory evidence demonstrates that rapid and progressive growth of infectious and toxigenic microorganisms or growth of *S. enteritidis* in eggs or *C. botulinum* cannot occur, such as a food that has an a_w and a pH that are above levels specified under Subparagraphs [c] (ii) and (iii) of this definition and that may contain a preservative, other barrier to the growth of microorganisms, or a combination of barriers that inhibit the growth of microorganisms.
> (vi) A *food* that may contain an infectious or toxigenic microorganism or chemical or physical contaminant at a level sufficient to cause illness,

but that does not support the growth of microorganism as specified under Subparagraph (a) of this definition.

The FDA definition for potentially hazardous food is limited to the microbiological safety of food. This definition should also include physical and chemical hazards as well as substances in food that may produce allergic reactions in some people.

Stating that "potentially hazardous food does not include foods with a pH of 4.6 or below when measured at 24°C (75°F)" is a point of concern. A pH of 4.6 or below is necessary to control the spore outgrowth of *C. botulinum* in pasteurized food. However,

Table 4   Foods Associated with Foodborne Illness Incidents

Meat: Beef: roasts steaks, stews, pies, liver, tongue, gravy, processed products. Pork: ham, bacon, roasts, chops, spareribs, barbecued and processed pork products Veal; lamb; goat; hamburger and other ground meats; frankfurters and other sausages; luncheon meats; game meats

Marine Foods: Fish: salmon (home canned, processed, eggs), tuna, herring, mackerel, sardines, trout, sole, cod. Shellfish: clams and clam chowder, shrimp, crab, lobster, oysters, scallops, squid, mussels.

Poultry: Chicken: fried, roasted, barbecued, pies, soup, processed products, gravy, prepared dishes. Turkey: roast, pie, loaf, dressing, stews, stuffing, soup, gravy. Cornish hen: duck; goose liver pate; roast goose.

Dairy Foods: Milk (pasteurized, raw, canned, evaporated), milk shakes, egg nog, cream sauce, artificial cream, mousse, cream, butter, yogurt, cheese, ice cream.

Bakery Foods: Pizza, cakes, pastries, pies and tarts, puddings, pasta products, bread and muffins, doughnuts, pancakes and crepes, cereal, tacos, pretzels, cookies, biscuits, and crackers.

Eggs: Omelets, scrambled eggs, Hollandaise sauce, deviled eggs, fried eggs, hard-cooked eggs, prepared products with eggs.

Infant Foods: Canned formulas, formula (container not specified), cereal products, beef products in jars, fruit, vegetables.

Confectionery: Chocolate candy, candy bars, licorice, molasses, and honey.

Vegetables and Fruits: Canned and bottled low-acid products, canned and bottle acid products, potatoes, mushroom, wild mushrooms, soup, corn and corn products, beans, vegetable oil, greens (lettuce, broccoli, etc.), molasses, maple and corn syrup, canned tomato juice, other canned and bottled fruit juices, other fruit juices, bottled acid fruits, nuts and nut products, jams and marmalades, dried and preserved fruit.

Salad: Potato, coleslaw, vegetable salads, chicken, seafood, ham, egg, fish, meat, macaroni, multiple ingredients.

Sandwiches: Beef, ham or ham salad, luncheon meats, turkey or chicken, tuna or tuna salad, other fish or shrimp, cheeseburger, sandwich spreads, egg or egg salad, cream cheese, submarine sandwiches, other multiple ingredient sandwiches.

Beverages: Bottled soft drinks, canned soft drinks, bottled, beer, cider, coffee, tea, spirits, wine, flavoring crystals.

Miscellaneous: Margarine, fats and oils, chili sauce, sloppy joes, gravy, soups, seasoning mix, cider vinegar, fish and chips, macaroni and cheese, custard, other multiple foods, popsicles and slush, spaghetti and meatballs, Mexican food, snails and escargots, tube feeding formulas, dressings and dips, sauces and relishes.

Source: Adapted from Health Protection Branch Canada. 1996. Food and Waterborne Disease in Canada, Annual Disease Summaries 1988 and 1989.

in food that is not pasteurized, vegetative pathogens must be considered. The major vegetative pathogen of concern is *Salmonella* spp. because of its low infective dose. Some *Salmonella* species can multiply down to a pH of 4.1 (44,78). Zhuang et al. (92) showed that *Salmonella montevideo* will multiply in chopped, fully ripe tomatoes with an initial pH of 4.1–4.3 when stored at 20°C (68°F) and 30°C (86°F). This is a serious issue if one considers that retail stores often display tomatoes at 21°C (70°F).

In order to assure destruction of *Salmonella* spp. in salad dressings and mayonnaise, these products are acidified to a pH of 4.1 or less. For example, the aqueous phase of mayonnaise contains 9.0–11.0% salt and 7.0–10.0% sugar. The aqueous phase of salad dressings has 3.0–4.0% salt and 20–30% sugar.

Table 4 shows that when the three classes of hazards—microbiological, chemical, and physical—are considered, no food can be excluded from being considered potentially hazardous. Each step in the production of food from farm to fork must be examined, and the probability of failure of potential hazard controls and hazard development must be analyzed.

XI. FACTORS INCREASING THE RISK OF FOODBORNE ILLNESS

Persons who are considered to have reduced resistance to illness include infants; hospital patients; pregnant women; the frail elderly; malnourished individuals; people with controlled physical or metabolic disorders (e.g., diabetes or high blood pressure); people with AIDS. Table 5 is a list of factors that increase the risk of foodborne infection or severity of illness. Because of these factors, some individuals are very sensitive to low levels of microbiological contaminants and are at greater risk of severe illness and even death from foodborne disease. In hospitals or health care facilities, it is easy to assure that patients receive more thoroughly pasteurized food. In the public feeding arena, operators have no idea of the immune status of their customers. It is very difficult for an operator to control safety when there is no requirement for the consumer to declare his or her susceptibility to disease which may, in fact, vary from day to day. The critical control point in this regard is a properly educated consumer who knows the risks associated with food and asks questions of the operator to determine the risk associated with consuming a food being sold. If the operator cannot provide hazard information, the consumer must choose another, less risky item.

If people are immune-suppressed due to illness or immunosuppressant drug therapy, they must request food that is well cooked or otherwise made safe for their consumption because retail food operators have no way of identifying these individuals.

A food can only be considered safe to eat when it has received a process such as canning, which makes the food safe until the container is opened; or when every step from the food source to the point of consumption is analyzed, and it is determined that hazards are controlled to minimize risk to the person consuming the food. Differences in the immunity of consumers must be considered. An increasing number of people are immune-suppressed due to chemotherapy, transplant operations, and human immunodeficiency virus (HIV). These individuals should never engage in high-risk food behavior such as consuming rare beef or raw seafood, or even raw vegetables.

Table 5 Factors Increasing the Risk of Foodborne Infection or the Severity of Illness

Factors	Reasons
Microbial factors	
Type and strain of pathogen ingested	Some pathogens and strains more virulent than others
Quantity of pathogens ingested	Higher numbers ingested may increase severity of illness and/or shorten onset time
Host factors	
Age less than 5 year	Lack of developed immune systems, smaller infective dose-by-weight required
Age greater than 50 or 60 years (depending on pathogen)	Immune systems failing, weakened by chronic ailments, occurring as easily as 50–60 years of age
Pregnancy	Altered immunity during pregnancy
Hospitalized persons	Immune systems weakened by other diseases or injuries, or at risk of exposure to antibiotic-resistant strains
Concomitant infections	Overloaded or damaged immune systems
Consumption of antibiotics	Alteration of normal intestinal microflora
Excessive iron in blood	Iron in blood serving as nutrient for certain organisms
Reduced liver/kidney function (alcoholism)	Reduced digestion capabilities, altered blood iron concentrations
Possession of certain human antigenic determinants duplicated or easily mimicked by microorganisms	Predisposition to chronic illnesses (sequelae)
Surgical removal of portions of stomach or intestines	Reduction in normal defensive systems against infection
Immunocompromised individual including those on chemotherapy or radiation therapy; recipients of organ transplants taking immunocompromising drugs; persons with leukemia, AIDS, or other illnesses	Immune system inadequate to prevent infection
Stress	Body metabolism changes allowing easier establishment of pathogens, or lower dose of toxin required for illness
Poor hygiene	Increased likelihood of ingestion of pathogens
Diet-related Factors	
Nutritional deficiencies either through poor absorption of food (mostly ill or elderly persons) or unavailability of adequate supply of food (starving persons)	Inadequate strength to build up resistance and/or consumption of poor-quality food ingredients, which may contain pathogens
Consumption of antacids	Increased pH of stomach
Consumption of large volumes of liquids, including water	Dilution of acids in the stomach and rapid transit through the gut
Ingestion of fatty foods (such as chocolate, cheese, hamburger) containing pathogens	Protection of pathogens by the fat against stomach acids
Other Factors	
Geographic location	Likelihood of exposure to endemic virulent strains, limited food and water supply, varied distribution of organisms in water and soil

Adapted from CAST (Council for Agricultural Science and Technology). 1994. Foodborne pathogens: risks and consequences. Task Force Report No. 122. (10).

XII. BEGINNING A HACCP PROGRAM

To begin a HACCP continuous improvement program, management must make a long-term commitment to food safety assurance and form a HACCP development team. Next, the operation's food system must be described in order to identify all of the threats that can become hazards, such as those listed in Table 6.

The total system from production to consumption has three components *input*, *process*, and *output*. The output for the system is food for the consumer that is tasty, safe, and assures repeat customer sales. The health status of the consumer is never known. Some consumers may be healthy, and some may be immune-compromised due to age, illness, antibiotics, or chemotherapy. Consumers may also have a variety of sensitivities to toxic and allergic compounds in the food. If customers do not eat the food at the establishment (e.g., carry-out or take-home food after dining out), they may also time/temperature-abuse the food if they are not properly informed how to handle the take-out food correctly.

After examining the *output* and setting microbiological, chemical, and physical (hard foreign object) hazard safety levels based on the consumer, the *input* of supplies and material is considered. The lower the level of the pathogens, chemicals, and hard foreign objects in the input material, the less the food will need to be processed by the cook in order for the food to be safe. Many foods will be contaminated with various environmental organisms such as *C. botulinum, B. cereus, L. monocytogenes*, and *Yersinia enterocolitica* (10). However, when there are fewer pathogenic microorganisms in the growing environments of food animals, poultry, fish, etc., or fruits and vegetables, there is less likelihood of fecal pathogens such as *Salmonella* spp. and *C. jejuni* being on the food. If people wish to eat raw foods such as oysters, steak tartare, or raw cured products, it is essential that there is supplier HACCP certification and that the ingredients be produced with a pathogen level below the illness threshold level of the consumer.

Depending on the intended use of the food that is being taken out, the food operator can introduce various hurdles such as temperature, time, water activity, oxidation–reduction, chemical additives, and packaging into the product design in order to reduce the risk of illness from multiplication of pathogens in the food after it is given to the customer. For example, the spores of *C. perfringens* survive retail food processing (especially in meat and poultry) and, if given about 10 h of 90–100°F (32.2–37.8°C) temperature abuse, are almost sure to cause illness.

Based on the output hazards and the hazards in the input supplies and material, the *process* criteria for the environment, facilities, equipment, personnel, and food processing are developed. The food process will take the raw ingredients; exclude, eliminate, reduce, or control the threats to human health; and produce finished products that, when eaten by the consumer, will have a safe hazard level that will not cause illness and will nourish the consumer.

XIII. NACMCF HACCP

In 1992, the National Advisory Committee on Microbiological Criteria for Foods (NACMCF) (62), using ideas provided by the National Academy of Sciences (61) and principles well established in food process quality control for many years, elaborated the seven principles of HACCP and guidelines for their application. This NACMCF commit-

Table 6 The System for HACCP-based TQM Management (and Government)

Input	Retail process: food service, markets, vending, home	Safety- and quality-assured output
Supplies and material	*Environment*	*Consumer*

• Environment contamination
    Soil, water, air
    Vegetation, plants, grains
    Wild animals, birds, insects, pests
	• Safe air
	• Insect and rodent control
	• Safe water
	• No soil on shoes
		• Proper balance between pleasurable and safe food
		• Safe levels of hazards for consumer, based on immune threshold

• Supplier contamination
    Pesticides, insecticides
    Mold growth in grains
    Filth contamination of food
    Microorganisms, toxins, poisons, hard foreign objects
    Poor nutritional food profiles due to feed supplies, condition of soil where food is grown
    Hazardous feed additives
    Container contamination of food
    Time, temperature abuse
    Inadequate facilities, equipment, and management
	Facilities that are clean and maintained
	Equipment that controls hazards or warns when it is not functioning correctly; construction from safe materials
	
	• Refrigeration that keeps food at less than 30°F (−1.1°C) and cools to less than 45°F (7.2°C) in less than 15 hours
	• Ovens that cook food from 41°F (5.0°C) to above 130°F (54.4°C) in less than 6 hours
	• Hot holding devices that keep food above 130°F (54.4°C)
		• Consumer abuse control information
		• Nutrition profile and contamination control (i.e., food components) for a long, physically excellent qualtiy of life
		• Hurdles to prevent microbial growth
		    Temperature
		    Water activity (a_w)
		    Oxidation/reduction (Eh)
		    Chemical additive
		    Packaging

- Distribution contamination

  Increase in pathogens, toxins, and poisons through mishandling

  Nutrient loss in shipping

  Food spoilage

  Time, temperature abuse

  Inadequate facilities, equipment, and management

- Wholesale processor contamination

  Fecal contamination during slaughter

  Pathogenic environmental organism contamination

  Spoilage waste

  Hard foreign objects

  Unsafe chemical addition

  Food mislabeling

  Underprocessing

  Overprocessing waste and nutrient loss

  Packaging; container poisons

  Time, temperature abuse

  Inadequate facilities, equipment, and management

Personnel
- Hand washing control of transient organisms

Products and services
- Thawing
- Recipe food time and temperature control
- Proper food temperature measurement
- Food contact surface cleaning and sanitizing
- Unsafe chemicals control; additive
- Control of carcinogens in cooking, as in broiling and grilling
- Nutrient loss minimization
- Food thermal pasteurization
- Food acid pasteurization

Table 7 NACMCF Logic Sequence for the Application of HACCP

| Principle 7: Establish effective record keeping procedures that document the HACCP system. | | | | | | | |
Process step	CCP	Chemical, physical, biological hazards	Critical limits	Monitoring, procedures/ frequency/ person(s) responsible	Corrective action(s)/ person(s) responsible	Verification procedure/ person(s) responsible	HACCP records
1.	Yes or No	1. 2. 3. etc					

Source: Adapted from National Advisory Committee on Microbiological Criteria for Food (NACMCF). 1998. Hazard analysis and critical control point principles and guidelines. J. Food Protect. 61(6): 762–775.

tee reconvened again in 1995 to review the 1992 HACCP document and compare it with current HACCP guidance prepared by the Codex Committee on Food Hygiene (63). Based on this review, the Committee made the HACCP principles more concise; revised and added definitions; included sections on prerequisite programs, education and training, and implementation and maintenance of the HACCP plan; revised and provided a more detailed explanation of the application of HACCP principles; and provided an additional decision tree for identifying critical control points (CCPs). This is presented in compact form in Table 7.

The NACMCF endorses HACCP as an effective and rational means of assuring food safety from harvest to consumption (63). Both the FDA and U.S. Department of Agriculture (USDA) are now mandating that slaughtering operations and food production operations institute HACCP procedures.

To develop a HACCP program, a HACCP team must be formed. The HACCP team should be composed of both management and employees. The HACCP team should then meet to discuss the products, system, GMPs, process, and distribution of the food item(s). The seven principles of an HACCP program must be included as a part of a total system safety management program.

The U.S. Environmental Protection Agency (EPA) and the Occupational Safety and Health Administration (OSHA) concept called process safety management is a more complete management program because it requires a systems and management approach. The components of process safety management are contrasted with the seven principles of HACCP in Table 8. Since process safety management is a much more complete control and improvement process, and food operators must also comply with EPA and OSHA regulations, it is reasonable to incorporate the added features of process safety management with HACCP.

XIV. RISK MANAGEMENT AND HACCP

''Over the past 20 years there has been an increased interest in producing safe food in order to protect consumers from foodborne infections and disease. The application of good manufacturing practices (GMP) coupled with process control that is based on the hazard analysis critical control point (HACCP) concept is valuable. However, GMP/HACCP highlights the inability to make food absolutely safe, and that measures to improve safety have an economical consequence. Therefore, the application of risk assessment within the context of GMP/HACCP is of growing interest'' (68).

Table 8 Components of Food Process Safety Management

---

1. Top-down management commitment to employee empowerment and then, enforcement of procedures and standards
   a. The company's values, vision, focus, and goals
   b. The process for prevention and improvement:
      Check/Measure → Plan → Act to change → Operate → Check/Measure
2. The HACCP team (HACCP prerequisite)
3. The operating system (HACCP prerequisite)
   a. The products, distribution, and service
   b. The intended product use and consumers
   c. The processes (flow chart)
4. The organization chart (HACCP prerequisite)
   a. For each person, describe responsibilities
5. Process change management
6. Current process safety technical information, standards, guidelines, codes, and laws
7. Capital review and facilities and equipment design
8. Process risk assessment and documentation
   a. Process hazard analysis (Principles 1, 2, 3[a])
   b. Consequence analysis (Not yet in HACCP)
   c. Risk analysis (Not yet in HACCP)
9. Preventive control policies, procedures, and standards for safe, zero-defect work practices
10. Employee empowerment for zero-defect performance (HACCP prerequisite)
    a. Training—human factors—for mastery
       1) Operations
       2) Maintenance and cleaning
       3) Suppliers
       4) Management
    b. Behavior assurance (antecedents and consequences)
11. Environment, facilities, and equipment preoperation performance assurance (HACCP prerequisite)
12. Formal startup safety review each morning and during operations for performance assurance (HACCP prerequisite)
13. Operation; measurement of performance; control within safe limits (Principles 4, 6[a])
14. Emergency response plans (Principle 5[a])
15. Investigation of incidents; feed forward to the reduction of risk of deviation/defect in the next cycle (Principle 5[a])
16. Periodic evaluations of the (safety) quality assurance process to find opportunities for improvement. (Principle 6[a])/reduction of risk
17. Revision and improvement of food process safety management program (Not formally in HACCP)

---

[a] Related to NACMCF-HACCP.
Source: Adapted from OSHA and EPA Press Safety Management.

To establish any national and international standards for microbiological, chemical, or physical hazards in food, it is essential to understand that there will probably always be a risk factor associated with some contamination of food. If food is grown in the ground or is associated with soil, there will always be the likelihood that the food will be contaminated with pathogens such as *Bacillus cereus, Clostridium botulinum*, and *Listeria monocytogenes*, as well as molds, parasites, and viruses. Rodents, insects, and wild birds are known to carry infections to farm animals, poultry, and fish. Meat and poultry products produced in a typical farm environment may be contaminated with *Salmonella* spp, *Campylobacter jejuni, Escherichia coli*, and other pathogens. Fish and shellfish are likely to be contaminated with *Vibrio* spp. Some contamination from the natural environment will always be present.

Development of vaccines to be used for meat animals and poultry so that they are not colonized with pathogens may be one method of controlling pathogens in these food items in the future. Another method of control is to raise animals and poultry in pathogen-reduced environments. However, raising animals without some degree of environmental contamination will probably never be possible. This means that pathogenic substances in foods must be reduced to low-risk levels by pasteurization, cooking, addition of acid, fermentation, etc., or by reduction in population accomplished by washing food items. These controls must be applied so that the risk of making consumers ill depends on the functioning immune system of each individual.

Risk (as related to food consumption) is the likelihood that individuals (or a population) will incur an increased incidence of adverse effects such as foodborne illness, disease, or death as a result of consuming a food (65). The risk can be further defined by defining the specific cause of illness or disease, e.g., the risk of salmonellosis due to consumption of underheated shell eggs, or the risk of foodborne illness due to *E. coli* O157:H7 in hamburger.

To estimate how much illness or injury can be expected from exposure(s) to a given risk agent and to assist in judging whether these consequences are great enough to require increased management or regulation, a *risk assessment* or analysis should be conducted. A risk assessment can be used as a tool to determine sources of the worst hazards and reduce the presence of these hazards. A risk assessment can also be used to ensure that operational risk management decisions are rational and are based on the best available science (64).

After a risk assessment, risk can be expressed in quantitative probability terms or in qualitative terms. An example of the use of *quantitative probability* terms is the number of illnesses over a lifetime in a population of 1 million exposed people. A risk of 1 illness in 10,000 is described as "10^{-4} risk," while 1 illness in 1 million is described as a "10^{-6} risk." Historically, risks of less than 10^{-6} in magnitude have not been the object of concern. Risk can also be expressed in *qualitative* terms of "low," "medium," and "high." These terms are used when quantification is not feasible or is unnecessary.

Table 9 is an organizational chart for integrated risk management. Integrated risk management can be divided into risk analysis, management risk control, and risk communication. Once the process is specified, there are three components of risk analysis. They are as follows:

1. *Hazard analysis* is based on quantitative epidemiology, expert knowledge, data, or research evidence demonstrating that an agent associated with the consumption of a particular food may cause human illness or injury.

2. *Risk assessment* evaluates the chance that a hazard will be in the food. This type of assessment includes consumers, what they eat, and in what type of environments. For example: Is the product take-out food? How do consumers handle the food? Is there consumer abuse? What is the amount, frequency, and source of the hazard in food? It is also important to assess the effectiveness of the controls of the hazard from food production to consumption.

Table 9 Integrated Risk Management

1. OPERATIONS Process step description	2. RISK ANALYSIS			3. MANAGEMENT RISK CONTROL	4. RISK COMMUNICATION
	A. Hazard analysis	B. Risk assessment	C. Hazard control assessment		
Employee procedures and controls	Hazard identification	Exposure assessment	• Analysis of effectiveness of current unit controls	• If the risk is acceptable, certify the step	• Consumer is informed how to control remaining risk
Get ready	Hazard quantification	Dose response assessment	• Analysis of consumer control	• If the risk is not acceptable, improve control	
Do _____	Critical limits	Risk characterization	•Failure mode effect analysis		
Until _____		($)	•Expected cost per year from failure of this step		
Check _____					
If _____					
Then _____					
Else _____					
Record if _____					
Clean up					
Put away					

A part of the assessment is the ''dose–response'' or the probability of consumers becoming ill at various dose intakes. For example, it should be known what the level is that normally healthy people can consume and remain healthy (and possibly even develop some immunity), as well as the levels known to cause probable illness or injury.

Risk characterization is the severity and cost of a hazard. The development of listeri-orsis in pregnant women and their fetuses is a severe and costly hazard, whereas the incidence of illness due to consumption of *C. perfringens* in improperly cooled food will only cause an illness of short duration and consequence, and at virtually no cost.

3. *Hazard control assessment* is the third component of risk analysis. This type of assessment is generated by the HACCP team through evaluation of the effectiveness of the current process, consumer controls, and estimates of the probability of failure. The HACCP team also estimates the probable cost to the business in the event of possible incidents related to an expected failure of control at this step.

A. Management Risk Control

After the risk has been assessed, and if the risk of financial loss and human suffering is unacceptable, then measures must be taken by management to control or reduce the dollar risk to an acceptable level. This can be accomplished through application of GMPs and the use of a HACCP-based total quality management program in improvement of prerequisite programs and the production of food products.

B. Risk Communication

Risk is never zero. For example, microbial spores in food survive food preparation processes in retail operations. Consumers must be aware of this risk in foods they take out of the food operation to consume at a later time. Also, many people desire undercooked/raw food such as beef, eggs, and fish. While the operators can buy these kinds of items from highly reliable suppliers with HACCP programs, there is still risk. Producers and operators of retail food establishments should communicate the amount of residual risk associated with food items to consumers through labels on containers or other various media presentations. Communicating the risk and consumer control responsibilities (after it leaves management operations control) does not relieve management of all liability, but it does help in a due diligence defense.

XV. FOOD OPERATIONS RISK MANAGEMENT BASED ON COMBINED HAZARD ANALYSIS RISK ASSESSMENT AND FAILURE MODE EFFECT ANALYSIS

Risk management of food operations procedures begins with assuring that prerequisite programs are effective (see the beginning of Table 7). These programs are the basis for assuring that the process controls will be effective. Prerequisite programs assure that there is a control system. Every employee, in addition to HACCP process control, must be able to perform systems control at his or her workplace or work site.

When a process authority or government official certifies a process as safe, he or she must review the process, using the principles of FMEA (failure mode effect analysis). An FMEA is a systematic method of identifying and preventing process and product problems before they occur. FMEAs are focused on preventing process deviations and hence enhancing safety. Ideally, FMEAs are conducted in food product (recipe) design or process development stages. However, conducting an FMEA on existing products and processes can also yield benefits. The objective of a food safety FMEA is to look at a HACCP plan for all ways in which a process can fail and thus produce an unsafe product. Even the simplest foods have many opportunities for failure in terms of microbiological, chemical, and particulate risks. For example, spores can outgrow in retail food, the growth of spoilage microorganisms can produce histamine in scombroid fish, chemical sanitizers can contaminate food, and metal or plastic package clips can fall into open containers of food. The production process must be tested to find out how the controls can fail. In this way, the hazards, standards, and controls at each step can be set to achieve an acceptable risk, and the process can be certified as acceptably safe.

Process capability (C_p) indices can also be used in food processes to evaluate the ability of a stable process to produce food within specification limits (45). For example, all hamburgers will be cooked to at least 155°F (68.3°C) and not over 165°F (73.9°C). The formula for calculating a simple process capability index is:

$$C_p = \frac{\text{allowable process spread}}{\text{actual process spread}}$$

$$= \frac{\text{upper specification limit} - \text{lower specification limit}}{6 \text{ standard deviations } (\sigma) \text{ of the process}}$$

When the difference in the upper safety limit and the lower safety limit is the same as 6σ, then the $C_p = 1$, and the process is considered to be very marginal in capability. For example, when cooking hamburgers, the allowable process spread is 10°F. If 6σ were 10°F, then the C_p would be 1 and the process should be improved because it would produce hamburgers beyond specification limits about 0.14% of the time. For the same process spread of 10°F, if 6σ were 6.25°F, then the capability would be 1.6, and the random occurrence that the process would produce an out-of-the-specification hamburger would be 0.79 ppm, or about 1 in 1 million hamburgers. This is a much more desirable performance standard.

XVI. INTEGRATION OF HACCP AND PROCESS SAFETY MANAGEMENT WITH ISO 9000 AND THE MALCOLM BALDRIGE AWARD

How does food HACCP-based process safety management interface with the ISO 9000 series as well as the Malcolm Baldrige criteria? Quality, as defined by Crosby (15), is the *needs, wants, and expectations of the customer*. Quality is in the beholder's eyes. Therefore, quality becomes:

1. Safe food and environment expected by the consumer.
2. Pleasurable food and dining experience expected by the customer.

3.   Operating efficiency and effectiveness leading to profits expected by the owners of the business so that they can continually improve the HACCP program.

Actually, ISO 9000 is simply a specification of the forms and processes for the quality control elements of a product-making system. It does not deal with the results of the process. It alludes to management and training, but directs its effort to the documentation of the quality control procedures. It does not use the most modern quality control philosophy of employee behavioral management for quality control, but rather focuses on documentation.

Modern quality management says that quality control should be performed by the trained, motivated, empowered employees striving to perform jobs safely and accurately. When employees produce each portion or item of food, they can provide 100% monitoring of what is produced and served to assure that all critical control standards are met.

Quality control is performance of tasks according to prescribed procedures and guidelines that have been designed and tested by management to meet customer needs, wants, and expectations.

Quality assurance is the management process of designing the process to be as capable as possible of producing products and services that meet customer quality needs, wants, and expectations with a high probability of zero deviations. This management process includes training and coaching employees so that all personnel cooperate to achieve quality control.

Quality improvement is the top management process of planning, organizing, directing, and controlling the system of management and technology in order to maintain and improve the quality of products.

Each time a task is performed, information is gathered and fed back to the design process so that during the next product and service cycle of the system the processes can be adjusted as necessary. In this way there is less likelihood of a deviation in the process performance from a target standard, and less risk of a product and service defect. Quality improvement also includes "benchmarking" against competitors' products in order to continually devise acceptable or higher quality products and services. Most importantly, quality improvement processes are intended to continually improve the safety and quality of the products the customer receives.

Total quality management (TQM) is a term currently applied to the integration of employee quality control with management quality assurance and quality improvement. For example, employees are encouraged to become members of a quality management team. Their input and involvement insure the production of safe, high-quality products and services. TQM teams are groups of people doing similar work, such as employees and managers, who work together to identify operational errors and problems; determine causes; analyze solutions; and specify policies, procedures, and standards or guidelines that emphasize and assure the quality of both food and service.

TQM teams often cross organization boundaries. The team may consist of five to eight employees such as a cook, bartender, food server, maintenance person, bus person, and cashier. Each shift should be represented. Initially, the leader is usually a supervisor. The TQM team analyzes problems observed by both management and employees. By setting objectives that support policies and standards or guidelines, they move one step at a time toward achievement of the specified objectives. Regular meetings are arranged so that the TQM teams can review procedures, set objectives, and send them forward for resolution. Formal but brief documentation of meetings is recorded in order to evaluate progress, note agreed-upon procedures, state objectives, and record successes.

XVII. HACCP-BASED TQM UNIT PROCESSES SYSTEM

It is important to understand the processes in a food production system. Figure 2 shows how the system is divided into processes for analysis. It is convenient to use typical systems terminology. The total group of activities will be called a *system*. The inputs, processes, and outputs define a system. In any food processing system there are eight components that must be addressed to assure the production of a safe, quality product. These are *management*; *consumer*; *environment*; *facilities*; *equipment*; *personnel*; *supplies*; and *product processes*. The processes of all of these system components function to provide food at an acceptable risk. Each of the eight components has processes.

A process can then be broken down into *tasks* and *steps*. This logic is fully compatible with computer-structured language programming. Thus, using the procedure for writing structured computer language, it is possible to systematically analyze and write a recipe or cleaning procedure that controls or minimizes process defects. A simple example is as follows on page 450.

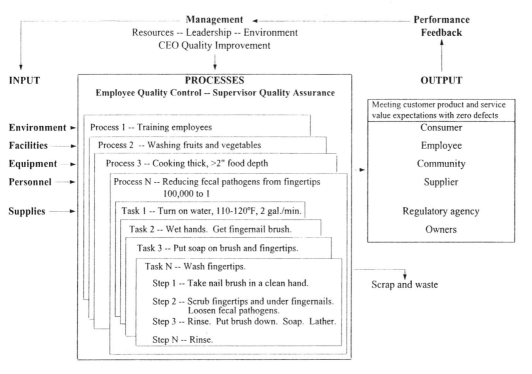

Figure 2 HACCP-based TOM unit processes.

Begin
↓
Get tools and material
↓
Do until critical level process standard is met
↓
Check:
      If not complete, go back and continue the process
      If complete, go on
↓
Clean up
↓
Put away
↓
End

Every task and action that takes place in the facility, from cleaning the floor to cooking meat, to changing light bulbs, is a process with hazards that require controls to assure that safety/quality guidelines are met and not exceeded. Control limits that are exceeded are a source of foodborne illness incidents, loss of customers, and waste. Each process can always be described in terms of input, processes (procedures), and output. Four examples of processes are given: training employees, washing fruits and vegetables, cooking thick foods, and handwashing to remove fecal pathogens from fingertips.

The process, then, is further divided into tasks. For example, for Process N (Fig. 2; removing soil and fecal pathogens from fingertips), first the water must be potable and then the water must be turned on. Second, the employee must wet the hands and get the fingernail brush. Third, soap is put on the brush and on fingertips, etc.

Tasks are finally divided into *steps*, which are the individual actions that control the process. In the case of the task of washing fingertips, the employee first takes the fingernail brush in the clean hand. The second step is to scrub under the fingernails to loosen soil. Step 3 is to scrub the nails and fingers on both hands. Step N is to rinse the hands in 110–120°F (43.3–48.9°C) water flowing at 2 gal/min. Until every process in the food system is safety certified to the step level, the system is not assured to produce a safe output of products.

XVIII. SYSTEM DOCUMENTATION

Operating procedures must be written in such a way that they can be taught, reviewed, and improved. To accomplish this, a policies, procedures, and standards or guidelines manual for food systems operation is needed. Table 10 is an illustration of the table of contents of a food system operating procedures manual (79). The manual combines the principles of hazard analysis, quality control, quality assurance, continuous quality improvement, and the principles of ISO 9000 and process safety management into components that constitute the basis for a company's HACCP-TQM program. Employees are empowered to do their best to perform processes according to the rules, policies, and standards or guidelines specified in the manual. Thus, there is a greater probability of ensuring the production and service of safe food products of designated quality.

Table 10 Policies, Procedures, and Standards Manual, Table of Contents

I. Food Safety Policy
   A. Owner's food safety policy
II. Organization for HACCP-based TQM
   A. Organization chart
   B. HACCP team
   C. Roster by job responsibilities
III. System Description
   A. Description of clientele to be served
   B. Process capacity analysis
   C. Environment surrounding the facility
   D. Facility plan
   E. Facility food flows
   F. Major equipment items
   G. Food storage space; freeze, refrigerated, dry
   H. Menu and menu items categorized by production categories/production centers
IV. Good Manufacturing Practices
   A. Management
   B. Personnel
   C. Environment
   D. Facilities
   E. Equipment
   F. Supplies and materials
   G. Food production
   H. Consumer
V. Supplier HACCP
   A. Supplier HACCP qualification standards
   B. Supplier HACCP/QA qualification list by ingredients purchased
   C. Ingredient specifications
VI. Recipe HACCP
   A. Menu item HACCP process summary
   B. Quality-assured recipe procedures
   C. Product specifications
VII. Cleaning and Sanitizing Schedule and Instructions
   A. Cleaning and sanitizing schedule and equipment
   B. Sanitation procedures and standards
   C. Cleaning verification
   D. Chemicals list and material safety data sheets

VIII. Maintenance Schedule and Instructions
   A. Maintenance schedule and instructions
   B. Maintenance procedures and standards
   C. Equipment maintenance manuals
   D. Maintenance verification
IX. Pest Control Schedule and Instructions
   A. Pest control schedule and equipment
   B. Pest control verification
X. HACCP-TQM Employee Training Program and Record
   A. New employee training record
      1. Continuing education training record
   B. Quality improvement training outline
   C. Employee improvement worksheet
XI. Self-inspection, Continuous Quality Improvment
   A. New product/process development: description, HACCP, testing/validation, implementation
   B. Product process monitoring/sampling plan: microbiological, chemical, physical
   C. Daily self-inspection
   D. Weekly, yearly self-inspection
   E. Corrective action log and preventive action schedule
   F. Employee—customer quality and hazard control problem report and problem removal
   G. Foodborne illness information and action report
   H. Foodborne illness symptoms table
XII. Food Safety Program Verification and Certification
   A. Performance verification and capability certification

The manual is key to the zero defect goal. Since all process tasks and steps are described in the manual, it is the basis for training employees. Thus, the manual becomes the basis for process evaluation and improvement. Each employee is given a chance to suggest ways to improve the system. All suggestions are referenced, so that the operations manual can be improved and the new way of doing a process can be integrated into operations.

Of course, there will always be some defects because of the natural variation of ingredients and supplies to the process, as well as normal variations due to the people running the equipment that produce the products. The closer the input is to design standards or guidelines for supplies, ingredients, personnel procedures, etc., and the closer the processes perform to design targets, then the more stable the process is and the more likely it is that a standard product will be produced within upper and lower control limits.

REFERENCES

1. Abeyta, C. 1983. Bacteriological quality of fresh seafood products from Seattle retail markets. J. Food Protect. 46:901–909.
2. Abeyta, C., Weagant, S. D., Kaysner, C. A., Wekell, M. M., Stott, R. F., Krane, M. H., and Peeler, J. T. 1989. *Aeromonas hydrophila* in shellfish growing waters: incidence and media evaluation. J. Food Protect. 52:7–12.
3. Ahmed, A. A-H., Moustafa, M. K., and Marth, E. H. 1983. Incidence of *Bacillus cereus* in milk and some milk products. J. Food Protect. 46:126–128.
4. Bailey, J. S., Fletcher, D. L., and Cox, N. A. 1989. Recovery and serotype distribution of *Listeria monocytogenes* from broiler chickens in the southeastern United States. J. Food Protect. 52:148–150.
5. Bauer, F. T., Carpenter, J. A., and Reagan, J. O. 1981. Prevalence of *Clostridium perfringens* in pork during processing. J. Food Protect. 44:279–283.
6. Bennett, J. V., Scott, D. H., Rogers, M. F., and Solomon, S. L. 1987. Infectious and parasitic diseases. In the Burden of Unnecessary Illness. Amler, R. W. and Dull, H. B. Oxford Univ. Press, New York.
7. Bryan, F. L., Bartelson, C. A., and Christopherson, N. 1981. Hazard analyses, in reference to *Bacillus cereus*, of boiled and fried rice in Cantonese-style restaurants. J. Food Protect. 44:500–512.
8. Buchanan, R. L., Stahl, H. G., Bencivengo, M. M., and Del Coral, F. 1989. Comparison of lithium chloridephenylethano-moxalactam and modified Vogel Johnson agars for detection of *Listeria* spp. in retail-level meats, poultry, and seafood. Appl. Environ. Microbiol. 55:559–603.
9. Callister, A. M., and Agger, W. A. 1987. Enumeration and characterization of *Aeromonas hydrophila* and *Aeromonas caviae* isolated from grocery store produce. Appl. Environ. Microbiol. 53:249–253.
10. Council for Agricultural Science and Technology (CAST) 1994. Foodborne pathogens: risks and consequences. Task Force Report No. 122. CAST, 4420 West Lincoln Way, Ames, IA.
11. Christopher, F. M., Smith, G. C., and Vanderzant, C. 1982. Examination of poultry giblets, raw milk and meat for *Campylobacter fetus* subsp. *jejuni*. J. Food Protect. 45:260–262.
12. Code of Federal Regulations (CFR). 1995. Title 21. Food and Drugs. Parts 100 to 169. Superintendent of Documents. U.S. Govt. Printing Office, Washington, D. C.
13. Codex Alimentarius Commission. 1981 Procedural Manual, 5th edition. Food and Agriculture Organization of the United Nations/World Health Organization, Rome.

14. Colburn, K. G., Kaysner, C. A., Wekel, M. M., Matches, J. R., Abeyta, C. and Stott, R. F. 1989. Microbiological quality of oysters (*Crassotrea gigas*) and water of live holding tanks in Seattle, WA markets. J. Food Protect. 52:100–104.

15. Crosby, P. B. 1979. Quality Is Free. McGraw-Hill, New York.

16. Davidson, R. L., Sprung, D. W., Park, C. E., and Rayman, M. K. 1989. Occurrence of *Listeria monocytogenes*, *Campylobacter* spp., and *Yersinia enterocolitica* in Manitoba raw milk. Can. Inst. Food Sci. Technol. J. 22:70–74.

17. Delmas, C. L., and Vidon, D. J. M. 1985. Isolation of *Yersinia enterocolitica* and related species from foods in France. Appl. Environ. Microbiol. 50:767–771.

18. Doyle, M. P. and Schoeni, J. L. 1987. Isolation of *Escherichia coli* O157:H7 from retail fresh meats and poultry. Appl. Environ. Microbiol. 53:2394–2396.

19. Duitschaever, C. L. and Buteau, C. 1979 Incidence of *Salmonella* in pork and poultry products. J. Food Protect. 42:662–663.

20. Ellender, R. D., Mapp, J. P., Middlebrooks, B. L., Cook, D. W., and Cake, E. W. 1980. Natural enterovirus and fecal coliform contamination of Gulf Coast oysters. J. Food Protect. 43:105–110.

21. Farber, J. M., and Peterkin, P. I. 1991. *Listeria moncytogenes*: a foodborne pathogen. Microbiol. Rev. 55:476–511.

22. Farber, J. M., Malcom, S. A., Weiss, K. F., and Johnston, M. A. 1988. Microbiological quality of fresh and frozen breakfast type sausages sold in Canada. J. Food Protect. 51:397–401.

23. Farber, J. M., Sanders, G. W., and Johnston, M. A., 1989. A survey of various foods for the presence of *Listeria* species. J. Food Protect. 52:456–458.

24. Food and Drug Administration. 1992. Bacteriological Analytical Manual, 7th ed. AOAC International, Arlington, VA.

25. Food and Drug Administration. 1997. Food Code. U.S. Public Health Service, U.S. Dept. of Health and Human Services. Pub. No. PB97-141204. Washington, D.C.

26. Food Research Institute. 1995. Food Safety 1995. Marcel Dekker, New York.

27. Frasier, M. B. and Koburger, J. A. 1984. Incidence of *Salmonella* in clams and oysters, crabs and mullet. J. Food Protect. 47:343–345.

28. Fricker, C. R., and Tompsett, S. 1989. *Aeromonas* spp. in foods: a significant cause of food poisoning? Int. J. Food Microbiol. 9:17–23.

29. Garcia, M. M., Lior, H., Stewart, R. B., Ruckerbauer, G. M., Trudel, J. R. R., and Skliarevski, A. 1985. Isolation, characterization and serotyping of *Campylobacter jejuni* and *Campylobacter coli* from slaughter cattle. Appl. Environ. Microbiol. 49:667–672.

30. Genegeorgis, C. A., Dutulescu, D., and Garayzabal, J. F. 1990. Prevalence of *Listeria* spp. in turkey meat at the supermarket and slaughter level. J. Food Protect. 53:282–288.

31. Genegeorgis, C. A., Oanca, P., and Dutulescu, D. 1989. Prevalence of *Listeria* spp. in poultry meat at the supermarket and slaughter level. J. Food Protect. 52:618–624.

32. Gerba, C. P., and Goyal, S. M. 1978. Detection and occurrence of enteric viruses in shellfish: A review. J. Food Protect. 41:743–754.

33. Goyal, S. M., Gerba, C. P., and Melnick, J. L. 1979. Human enteroviruses in oysters and their overlying water. Appl. Environ. Microbiol. 37:572–581.

34. Griffiths, J. K. and Keusch, G. T. 1992. *Shigella*, infection and immunity. In: Encyclopedia of Immunology. I. M. Roitt and Delves, P. J., eds. Academic Press, San Diego, pp. 1371–1373.

35. Hackney, C. R., Ray, B., and Speck, M. L. 1980. Incidence of *Vibrio parahaemolyticus* in and the microbiological quality of seafood in North Carolina. J. Food Protect. 43:769–773.

36. Hauschild, A. H. W. and Hilsheimer, R. 1980. Incidence of *Clostridium botulinum* in commercial bacon. J. Food Protect. 43:564–565.

37. Hauschild, A. H. W. and Hilsheimer, R. 1983. Prevalence of *Clostridium botulinum* in commercial liver sausage. J. Food Protect. 46:242–244.

38. Health Protection Branch Canada. 1996. Food and Waterborne Disease in Canada, Annual Disease Summaries 1988 and 1989. Health Canada, Ontario, Canada.

39. Hormaeche, C. E. 1992. Salmonella, infection and immunity. In: Encylopedia of Immunology. I. M. Roitt and Delves, P. J., eds. Academic Press, San Diego, pp. 1350–1352.

40. International Commission of Microbiological Specifications for Foods (ICMSF) 1988. Microorganisms in Foods 4. Application of the hazard analysis critical control point (HACCP) system to ensure microbiological safety and quality. Blackwell Scientific, Oxford.

41. Irving, W. I. 1992. Hepatitis A virus, infection and immunity. In Encylopedia of Immunology. I. M. Roitt and Delves, P. J. eds. Academic Press, San Diego, pp. 660–661.

42. Izat, A. L., Druggers, C. D., Colberg, M., Reiber, M. A., and Adams, M. H. 1989. Comparison of the DNA probe to culture methods for the detection of *Salmonella* on poultry carcasses and processing waters. J. Food Protect. 52:564–570.

43. Jay, J. M. 1995. Foods with low numbers of microorganisms may not be the safest foods or, why did human listerioris and hemorrahagic colitis become foodborne diseases? Dairy Food Environ. Sanit. 15:674–677.

44. Jay, J. M. 1996. Modern Food Microbiology, 5th ed., Chapman & Hall, New York.

45. Kane, V. E. 1989. Defect Prevention—Use of Simple Statistical Tools. Marcel Dekker, New York, Chap. 7.

46. Kautter, D. A., Lilly, T., Solomon, H. M., and Lynt, R. K. 1982. *Clostridium botulinum* spores in infant foods: A survey. J. Food Protect. 45:1028–1029.

47. Khalifa, K. I., Werner, B., and Timperi, R. 1986. Non-detection of enteroviruses in shellfish collected from legal shellfish beds in Massachusetts. J. Food Protect. 49:971–973.

48. Kinde, H., Genigeorgis, C. A., and Pappaioanou, M. 1983. Prevalence of *Campylobacter jejuni* on chicken wings. Appl. Environ. Microbiol. 45:1116–1118.

49. Kokubo, Y. and Matsumoto, M., Terada, A., Saito, M., Shinagawa, K., Konuma, H., and Kurata, H. 1986. The incidence of clostridia in cooked meat products in Japan. J. Food Protect. 49:864–867.

50. Konuma, H. K., Shinagawa, K., Tokumaru, M., Onoue, Y., Konno, J. S., Fujino, N., Shgahisa, T., Kurata, H., Kuwabara, Y., and Lopes, C. A. M. 1988. Occurrence of *Bacillus cereus* in meat products, raw meat and meat product additives. J. Food Protect. 51:324–326.

51. Lammerding, A. M., Garcia, M. M., Mann, E. D., Robinson, Y., Dorward, W. J., Truscott, R. B., and Tittiger, F. 1988. Prevalence of *Salmonella* and thermophilic *Campylobacter* in fresh pork, beef, veal and poultry in Canada. J. Food Protect. 51:47–52.

52. Liewen, M. B. and Plautz, M. W. 1988. Occurrence of *Listeria monocytogenes* in raw milk in Nebraska. J. Food Protect. 51:840–841.

53. Lillard, H. S., Hamm, D., and Thomson, J. E. 1984. Effect of reduced processing on recovery of foodborne pathogens from hot-boned broiler meat and skin. J. Food Protect. 47:209–212.

54. Lovett, J., Francis, D. W., and Hunt, J. W. 1987. *Listeria monocytogenes* in raw milk: Detection, incidence, and pathogenicity. J. Food Protect. 50:188–192.

55. Madden, R. H., Hough, B., and Gillespie, C. W. 1986. Occurrence of *Salmonella* in porcine liver in Northern Ireland. J. Food Protect. 49:893–894.

56. Mafu, A. A., Higgins, R., Nadeau, M., and Cousineau, G. 1989. The incidence of *Campylobacter jejuni* and *Yersinia enterocolitica* in swine carcasses and the slaughter house environment. J. Food Protect. 52:642–645.

57. McEwen, S. A., Martin, S. W., Clarke, R. C., and Tamblyn, S. E. 1988. A prevalence survey of *Salmonella* in raw milk in Ontario, 1986–1987. J. Food Protect. 51:963–965.

58. McManus, C. and Lanier, J. M. 1987. *Salmonella, Campylobacter jejuni*, and *Yersinia enterocolitica* in raw milk. J. Food Protect. 50:51–55.

59. Mosso, A., Arribas, L. G., Cuena, J. A., and de la Rosa, C. 1989. Enumeration of *Bacillus* and *Bacillus cereus* spores in food from Spain. J. Food Protect. 52:184–188.

60. Moustafa, M. K., Ahmed, A. A-H., and Marth, E. H. 1983. Occurrence of *Yersinia enterocolitica* in raw and pastuerized milk. J. Food Protect. 46:276–278.

61. National Academy of Science. 1985. An Evaluation of the Role of Microbiological Criteria for

Foods and Food Ingredients. Food Protection Committee, Subcommittee on Microbiological Criteria, National Research Council. National Academic Press, Washington, DC.

62. National Advisory Committee on Microbiological Criteria for Foods (NACMCF). 1992. Hazard analysis and critical control point system. USDA-FSIS, Washington, DC.

63. National Advisory Committee on Microbiological Criteria for Foods (NACMCF). 1998. Hazard analysis and critical control point principles and application guidelines. J. Food Protect. 61(6):762–775.

64. National Advisory Committee on Microbiological Criteria for Foods (NACMCF). 1998. Potential application of risk assessment techniques to microbiological issues related to international trade in food and food products. J. Food Protect. 61(8):1075–1086.

65. National Advisory Committee on Microbiological Criteria for Foods (NACMCF). 1998. Principles of risk assessment for illness caused by foodborne biological agents. J. Food Protect. 61(8):1071–1074.

66. Newell, D. 1992. Campylobacter, infection and immunity. In: Encylopedia of Immunology. I. M. Roitt and Delves, P. J., eds. Academic Press, San Diego, pp. 272–273.

67. Norberg, P. 1981. Enteropathogenic bacteria in frozen chicken. Appl. Environ. Microbiol. 42: 32–34.

68. Notermans, S., and Batt, C. A. 1998. A risk assessment approach for food-borne *Bacillus cereus* and its toxins. J. Appl. Microbiol. Symp. Suppl. 84:51S–61S.

69. Palumbo, S. A., Maxion, F., Williams, A. C., Buchanan, R. L., and Thayer, D. W. 1985. Starch-ampicillin agar for the quantitative detection of *Aeromonas hydrophila*. Appl. Environ. Microbiol. 50:1027–1030.

70. Rayes, H. M., Genigeorgis, C. A., and Farver, T. B. 1983. Prevalence of *Campylobacter jejuni* in vacuum packaged processed turkey. J. Food Protect. 46:292–294.

71. Read, S. C., Gyles, C. L., Clarke, R. C, Lior, H., and McEwen, S. 1990. Prevalence of vercytotoxigenic *Escherichia coli* in ground beef, pork, and chicken in southwestern Ontario. Epidemiol. Infect. 105:11–20.

72. Reymond, D., Johnson, R. P., Karmali, M. A., Petric, M., Winkler, M., Johnson, S., Rahn, K., Renwick, S., Wilson, J., Clarke, R. C., and Spika, J. 1996. Neutralizing antibodies to *Escherichia coli* vero cytotoxin 1 and antibodies to O157 lipopolysaccharide in healthy farm family members and urban residents.

73. Roberts, T. 1990. 1987 Bacterial foodborne illness costs in the USA. Food Laboratory News, 19 (March):53

74. Roberts, T., and van Ravenswaay, E. 1989. The economics of safe guarding the U.S. food supply. USDA Ag. Info. Bull. No. 566:1–8.

75. Rodriguez, M. H. and Barrett, E. L. 1986. Changes in microbial population and growth of *Bacillus cereus* during storage of reconstituted dry milk. J. Food Protect. 49:680–686.

76. Schiemann, D. A. 1980. Isolation of toxigenic *Yersinia enterocolitica* from retail pork products. J. Food Protect. 43:360–365.

77. Shanker, S., Rosenfield, J. A., Davey, G. R., and Sorrel, T. C. 1982. *Campylobacter jejuni*: incidence in processed broilers and biotype distribution in human and broiler isolates. Appl. Environ. Microbiol. 43:1219–1220.

78. Silliker, J. H. 1982. *Salmonella* foodborne illness. In: Microbiological Safety of Foods in Feeding Systems. ABMPS Report 125, pp. 22–31.

79. Snyder, O. P. 1998. Food Safety Through Quality Assurance Policies, Procedures, and Standards. Hospitality Institute of Technology and Management (HITM), St. Paul, MN.

80. Stern, N. J., Green, S. S., Thaker, N., Krout, D. J., and Chiu, J. 1984. Recovery of *Campylobacter jejuni* from fresh and frozen meat and poultry collected at slaughter. J. Food Protect. 47:372–374.

81. Sumner, S. S., Albrecht, J. A., and Peters, D. L. 1993. Occurrence of enterotoxigenic strains of *Staphylococcus aureus* and enterotoxin production in bakery products. J. Food Protect. 56: 722–724.

82. Svennerholm, A. 1992. *Escherichia coli*, infection and immunity. In: Encylopedia of Immunology. I. M. Roitt and Delves, P. J. eds. Academic Press, San Diego, pp. 526–527.

83. Ternstrom, A. and Molin, G. 1987. Incidence of potential pathogens on raw pork, beef and chicken in Sweden, with special reference to *Erysipelothrix rhusiopathiae*. J. Food Protect. 50:141–146.

84. Todd, E. C. D. 1989. Preliminary estimates of costs of foodborne illness in the United States. J. Food Protect. 52(8):595–601.

85. Trussel, M. 1989. Zum Vorkommen von Listerien bei der Produktion von Bunderfleisch, Salami, Mettwurst. Schweiz. Arch. Tierheilk. 131:409–421.

86. Vaughn, J. M., Landry, E. F., Thomas, M. A., Vicale, T. J., and Panello, W. F. 1990. Isolation of naturally occurring enteroviruses from a variety of shellfish species residing in Long Island and New Jersey marine embayments. J. Food Protect. 43:95–98.

87. Wait, D. A., Hackney, C. R., Carrick, R. J., Lovelace, G., and Sobsey, M. D. 1983. Enteric bacterial and viral pathogens and indicator bacteria in hard shell clams. J. Food Protect. 46: 493–496.

88. Weagant, S. D., Sado, P. N., Colburn, K. G., Thorkelson, J. D., Stanley, F. A., Krane, M. H., Shields, S. C., and Thayer, C. F. 1988. The incidence of *Listeria* species in frozen seafood products. J. Food Protect. 51:655–657.

89. Willshaw, G. W., Smith, H. R., Roberts, D., Thirwell, J., Cheasty, T., and Rowe, B. 1993. Examination of raw beef products for the presence of verotoxin producing *Escherichia coli*, particularly those of serogroup O157:H7. J. Appl. Bacteriol. 75:420–426.

90. Wilson, J. B., Clarke, R. C., Renwick, S. A., Rahn, K., Johnson, R. P., Karmali, M. A., Lior, H., Alves, D., Gyles, C. L., Sandhu, K. S., McEwen, S. A., and Spika, J. S. 1996. Vero cytoxigenic *Escherichia coli* infection in dairy farm families. J. Infect. Dis. 174:1021–1027.

91. World Health Organization (WHO) Division of Food and Nutrition. 1993. Training considerations for the application of the hazard analysis critical control point system to food processing and manufacturing. FOS/93.3. WHO, Geneva.

92. Zhuang, R. Y., Beuchat, L. R., and Angulo, F. J. 1995. Fate of Salmonella montevideo on and in raw tomatoes as affected by temperature and treatment with chlorine. Appl. Environ. Microbiol. 61:2127–2131.

20

Foodservice Operations: HACCP Control Programs

O. Peter Snyder, Jr.
Hospitality Institute of Technology and Management, St. Paul, Minnesota

I. EMPLOYEE HYGIENE

Each day, a small number of people coming to work are potential spreaders of pathogenic microorganisms, even though they seemingly have no symptoms of illness. The main threat of transfer of these pathogens is in salad and cold-food preparation. This problem exists because these menu items are composed of many ingredients that are not cooked or heated sufficiently to inactivate pathogenic microorganisms and are not held at temperatures that prevent the multiplication of pathogens (12,23).

   To control for any possibility of fecal pathogen transfer (viruses, bacteria, and parasites) to food, it should be assumed that every employee is ill. Therefore all employees must be trained and required to use adequate methods for washing fingertips and hands.

A. Handwashing

There are as many as 10^9 salmonellae per gram in the feces of people carrying this pathogen (26). If toilet tissue slips just a little, infected individuals may easily get 0.001 g feces or 10^6 pathogens on their fingertips. Without reduction of these pathogens to a safe level on hands and fingertips, the food, particularly salads, handled by food preparers can become hazardous and make consumers ill. The following double hand/fingertip wash procedure has been shown in a laboratory study by the editor (unpublished data) to reduce pathogens by 10^{-5}.

1. Turn on the water at a temperature of 110–120°F (43.3–48.9°C) at 2 gal/min. (Well-designed sinks should have thermostats set to the proper temperature, and the water flows at 2 gal/min when fully turned on.) A lot of water must be used to wash the detergent with microorganisms from the fingertips and hands. The principal control is dilution. Wet the hands and brush.
2. Put 2.5 to 5 mL of plain, unmedicated hand soap or detergent on a fingernail brush. (Antibacterial soaps destroy beneficial resident microorganisms on the skin.)
3. Under the water, produce lather by using the tips of the fingernail brush gently on the fingertips. Use the fingernail brush to scrub the fingernails. Special attention must be made to the fingertips that held the toilet paper. The purpose of using the fingernail brush is to ensure safe reduction of any fecal pathogens and any other material that harbors pathogens from the fingertips and under the fingernails.
4. Rinsing is a critical step. The microorganisms in the lather are not dead; they are just loosened from the skin and fingertips and are suspended in the lather. Rinsing in flowing water removes the lather and produces a 10^{-3} microbial reduction (34). Rinse the fingernail brush and put it down, placing the bristles up to dry. Do not place the brush in a sanitizer solution because the residual detergent and organic material on the brush will neutralize the sanitizer, and microorganisms can then multiply in the neutralized solution.
5. Again, apply 2.5–5 mL of detergent to the hands.
6. Lather the hands and skin of arms up to the tips of sleeves.
7. Thoroughly rinse the lather from the hands and arms in warm, flowing water.
8. Dry hands thoroughly with clean paper towels. The second handwashing produces another 10^{-2} microbial reduction, and the paper towel perhaps a 10^{-2} reduction (34).

To prevent transfer of pathogenic bacteria and viruses, employees should be banned from carrying all forms of nose wipes in order to prevent inadvertent contamination of hands. If employees must blow their noses, they must be trained to go to the handwashing sink, get a facial tissue, and use it to wipe or blow the nose. After discarding the facial tissue, employees should then wash their hands using the single hand wash method. If employees need to sneeze or cough, they should step back from the food preparation area and sneeze into their shoulder.

When employees have cuts or infections on their hands, the cut or infection must be cleaned, bandaged if necessary, and covered with a glove. The glove is used to keep the bandage on the hand and prevent it from falling into food, not to control the infection, since the cut must be verified by the supervisor as not being infected. (If the cut is infected, the employee should not work in food production or foodservice). The glove must be changed as often as handwashing is necessary for safe food preparation.

Food handlers should stay home when sick, especially with diarrhea, but they do not. It is the responsibility of the supervisor to watch for people who are using the toilet frequently. If these individuals seem ill, they should be asked if they are sick and, if so, sent home.

II. FOOD CONTACT SURFACE CLEANING AND SANITIZING

A. Clean

''Clean'' means free of dirt and soil such as grease. Detergents, hot water, acid cleaners, and wetting agents are used to dissolve and remove grease and soil from a surface. Cleaning must be done with warm to hot water to be effective. Cleaning prior to sanitizing is a critical step. The fact that high proportions of both soil and bacteria (up to 99.8%) can be removed by simple cleaning and dilution with a water rinse has demonstrated that detergency is quantitatively more important than sanitizing (9). If a surface is not clean, the organic matter neutralizes the sanitizer, and the sanitizer becomes ineffective and does not inactivate any microorganisms.

In recent years, it has also been shown that there is no difference between using wooden or plastic cutting boards. Cutting boards made from both types of material get equally soiled, and both can be cleaned adequately (1,2).

B. Sanitized

''Sanitized'' means the reduction of disease-producing organisms by a factor of 10^{-5} or by 99.999%. Unfortunately, this standard applies only to a laboratory test of a few types of pathogens, with fresh sanitizer solutions in test tubes on a scrupulously clean stainless steel disk. The U.S. Public Health Service suggests that a sanitized surface must have fewer than 100 nonspecific aerobic organisms per 8 in.^{2} of surface (e.g., cutting board, dish, tabletop, etc.) or 100 organisms per utensil (e.g., spoon) (41). This is an adequate standard that is used as a sanitation guideline in the Grade A Pasteurized Milk Ordinance (10). Visual cleanliness is not a reliable indicator that a surface is sanitized. Dirty-looking surfaces that have been hot and are dry, such as an oven or grill surface, may have few microorganisms. Clean-looking plastic, Formica, or stainless steel surfaces that have been wiped with a contaminated towel will have high levels of microorganisms adhering to them. For example, it is probably common to find 1000–5000 microorganisms on 8 in.^{2} of plastic cutting boards in retail food operations having no evidence of a food safety problem (22). It is really not so much the question of final count as it is the reduction. If the microorganisms are reduced from 1000 to 1, then it is highly likely that the actual low counts of pathogens on the surface from contaminated food such as chicken will be reduced to a safe level.

C. Critical Controls for the Five-Step Surface Sanitizing Process

Use a clean, hot [110–120°F (43.3 to 48.9°C)] detergent solution and a scrub brush to wash and clean surfaces. Rinse surfaces with a lot of hot [110–120°F (43.3–48.9°C)] water. The temperature of the sanitizing solution must be 75–100°F (23.9–37.8°C) or above. Common household bleach may be used to prepare a 50 ppm chlorine sanitizing solution [1 teaspoon of bleach per gallon of water (pH about 6.5)].

Effective sanitizing involves five basic steps.

1. *Scrape and rinse* the surface with 110–120°F (43.3–48.9°C) water. Use a scrub brush, if possible. This rinse gives at least a 1000:1 reduction per cm^2 (35). If this is not done with cutting boards and knives that have been used on raw food, the pathogens will contaminate the detergent wash water and will multiply about every 20 min at 95°F (35°C).
2. Place item in a wash sink containing a sufficient amount of clean, hot, detergent solution. *Wash and scrub* the item with a brush to loosen and dissolve debris on the surface.
3. Transfer the item to the rinse sink and *rinse and float off* any remaining debris. At this point, the surface must be clean and free of soil and grease. [Steps 2 and 3 reduce counts another 1000 to 1 per cm^2 (35).]
4. *Sanitize.* Use a 50 ppm fresh chlorine solution or other sanitizer solution of equivalent effectiveness. Dispense this solution from a squirt bottle and wipe it across the surface to be sanitized with a clean paper towel. The chlorine sanitizing solution should be made fresh each morning. (A bucket of chlorine solution and a cleaning cloth should not be used because the soil from the cleaning cloth neutralizes the free chlorine after about three or four rinses in the chlorine water.) Paper test strips for determining the effectiveness of chlorine sanitizing solutions do not accurately indicate the amount of free chlorine (the active sanitizing agent) in the solution. If a bucket of sanitizing solution is used, the oxidation–reduction potential of the solution should be measured (as is done for swimming pools) and a level of more than +800 mV maintained by the addition of more chlorine solution (e.g., household bleach). Note that there is no operational evidence that the solution is effective or necessary in retail operations. Therefore, this author recommends that the use of chlorine be discontinued. Safety is achieved by prerinsing items to remove debris and pathogens from the surface and then by the dilution reduction of the pathogens from the surface of items in a large volume of clean detergent water.
5. *Air-dry.* This is a critical step. There will always be some remaining microorganisms because of their ability to adhere to surfaces in cracks and embedded organic material. If the surface remains wet, bacteria can multiply from 1 to 1000

overnight at 75°F (23.9°C). Surfaces must be allowed to air-dry thoroughly. Dry surfaces having a water activity of less than 0.95 do not support a hazardous multiplication of most bacteria.

III. CONTROLLING MICROBIAL GROWTH: ALTERING ENVIRONMENTAL CONDITIONS

Favorable environmental conditions of temperature, nutrient, pH, water activity (a_w), and oxidation–reduction conditions over a period of time promote the multiplication of microorganisms. By altering these conditions, the multiplication of microorganisms can be controlled and/or their destruction achieved.

A. Water Activity (a_w)

Microorganisms require moisture to multiply. Multiplication of microorganisms is restricted in an environment where water is not available, or where water is bound by other food components such as salt, sugar, and glycerol. Foods high in moisture, such as fresh fruits and vegetables, meat, fish, poultry, etc., permit rapid multiplication of microorganisms. The water in the structural system of these foods is available for the metabolic functions of microorganisms. When water is removed to a sufficiently low level (e.g., cereals, dried fruits, and vegetables), the multiplication of microorganisms is suppressed or stopped. Table 1 lists the water activity range of some common foods in relation to water activity for microbial growth. Table 2 lists the minimal water activity necessary for the growth of some pathogenic bacteria.

It is frequently much more difficult to inactivate surviving microorganisms in lower water activity foods, starch-thickened sauces, and desserts containing substantial amounts of sugar. Higher temperatures for longer periods of time are required to ensure destruction. A practical application of this knowledge is the addition of salt and sugar (if used in significant amounts) to a food product only when it has reached a 5D pasteurization temperature of, for example 150°F (65.6°C) for 52 s. Both sugar and salt reduce the water activity (a_w), making it more difficult to inactivate microorganisms. If there is a considerable amount of salt or sugar in a food item, the standard pasteurization value should be increased 10°F (6°C) to achieve the same degree of destruction.

B. Nutrients and Acids (pH)

When the supply of nutrients is low or not optimum, the multiplication of microorganisms is slower and the population declines. The addition of lemon juice, vinegar, or wine to food items lowers the pH of the food products, thus inhibiting bacterial growth and contributing to the destruction of some bacterial pathogens. If the pH of food is less than 4.6, the food will be safe from *C. botulinum* spore outgrowth and multiplication. However, most *Salmonella* spp. will multiply down to 4.1 pH and *Listeria monocytogenes* can multiply slowly in foods below this pH value. It is assumed, by government regulations (6), in the commercial preparation of salad dressing and mayonnaise that ingredients such as the egg yolks are contaminated with *Salmonella* spp. Therefore, these products must be produced to have a pH of 3.8 or less. This is an acetic acid concentration of less than

Table 1 Principal Groups of Food Based on Water Activity

1. 0.98 a_w and above

Fresh meats and fish	Canned vegetables in brine
Fresh fruits and vegetables	Canned fruit in light syrup
Milk and other beverage	

These are very moist foods, including those containing less than 3.5% sodium chloride or 26% sucrose in the aqueous (water) phase. Foodborne pathogenic bacteria and common spoilage microorganisms (bacteria, yeast, and molds), with the exception of extreme xerophiles and halophiles, grow almost unimpeded at levels of a_w within this range.

2. Below 0.98 to 0.93 a_w

Evaporated milk	Processed cheese
Tomato paste	Gouda cheese
Lightly salted fish, pork, beef products	Canned fruits in heavy syrup
Canned cured meats	Bread
Fermented sausages (not dried)	High-moisture prunes
Cooked sausages	

Maximum concentration of salt or sugar in the aqueous phase of the foods will be near 10% and 50%, respectively. All known foodborne pathogenic bacteria can grow in the upper part of this range.

3. Below 0.93 to 0.85 a_w

Dry or fermented sausage (Hungarian, Italian types)	Aged cheddar cheese
Dried beef	Sweetened condensed milk
Raw ham	

This group includes foods up to 17% salt or saturated sucrose in the aqueous phase. Only one bacterial pathogen, *Staphylococcus aureus* can grow in this a_w range. However, many molds that produce mycotoxins can grow in this range.

4. Below 0.85 to 60 a_w

Intermediate moisture food	Molasses
Dried fruit	Heavily salted fish
Flour	Meat extracts
Cereals	Some aged cheeses
Jams and jellies	Nuts

No pathogenic bacteria grow within this range. However, spoilage can occur from growth of xerophilic, osmophilic, or halophilic yeasts and molds.

5. Below 0.60 a_w

Confectionery	Biscuits
Chocolate	Crackers
Honey	Potato chips
Noodles	Dried eggs, milk, and vegetables

Microorganisms do not multiply below 0.60 a_w, but can remain viable for long periods of time.

Source: ICMSF 1980. Microbial Ecology of Foods, Vol. 1, Factors Affecting Life and Death of Microorganisms. Academic Press, New York.

Table 2 Water Activity for Some Pathogenic Microorganisms

Pathogen	Minimal a_w for growth[a]
Bacillus cereus	0.92
Campylobacter jejuni	0.987
Clostridium botulinum type A, proteolytic B and F	0.935
Clostridium botulinum type E, nonproteolytic B and F	0.97
Clostridium perfringens	0.93
Pathogenic strains of *Escherichia coli*	0.95
Listeria monocytogenes	0.92
Salmonella spp.	0.94
Shigella spp.	0.96
Staphylococcus aureus—growth	0.83
Staphylococcus aureus—toxin	0.85
Vibrio cholera	0.97
Vibrio parahaemolyticus	0.94
Vibrio vulnificus	0.96
Yersinia enterocolitica	0.945

[a] Below these levels of water activity, specified bacterial growth is halted.
Source: Adapted from U.S. Department of Health and Human Services. 1998. Fish and Fisheries Products Hazards and Controls Guide, 2nd ed. U.S. Public Health Service, FDA, Center for Food Safety and Applied Nutrition, Office of Seafood, Washington, D.C.

1.4%. At this pH, the salmonellae not only do not multiply but actually die in a period of 5 min to a few hours at room temperature (33).

In recent years, it was determined that fresh, raw tomatoes (whole or cut up) at a pH of 4.2–4.4 may be contaminated with *Salmonella*. To assure that the salmonellae do not multiply, fresh tomato products should be maintained at a temperature of less than 50°F (10.0°C) (3,42).

C. Preservatives

Some retail food operations make their own products such as sausage, which entails the addition of preservatives such as nitrite. This should be done in accordance with USDA regulations. Chemical compounds added to food as preservatives in the United States must be added at levels that are GRAS (Generally Recognized As Safe) (6–8). Common preservatives include nitrites used to prevent growth of *C. botulinum* in meat, butylated hydroxyanisol (BHA) and butylated hydroxytoluene (BHT) used to prevent oxidation of lipids, and sulfites used to preserve color in dried fruits and vegetables. Used in correct amounts, these compounds can be safely added to food. Excessive addition can lead to illness (18). If preservatives are added to foods, food preparers must be taught to use acceptable amounts according to the Code of Federal Regulations (6–8).

D. Oxidation-Reduction Potential

The redox (oxidation–reduction) potential, or E_h, is known to be an important selective factor in all environments, including food, and influences the types of microorganisms found in the food and their metabolism (16). Aerobic microorganisms require positive E_h values (oxidized) for multiplication, while anaerobes require negative E_h values (reduced). Anaerobic bacteria such as *Clostridium* spp. require reduced conditions for multiplication initiation (E_h of less than -200 mV). Aerobic bacteria such as *Bacillus* spp. require oxidizing conditions for multiplication.

By controlling the atmosphere (redox potential) of a food system, it is possible to inhibit or prevent the growth of *C. botulinum* in foods (specifically, to prevent growth of nonproteolytic types of *C. botulinum* in fish, and proteolytic types in fruits and vegetables). If many vegetables, some fruits, and pasteurized smoked fish are packaged anaerobically (at less than 2% oxygen concentration) and left at room temperature for 1–2 days, it has been shown that there is the likelihood of *C. botulinum* toxin production (5,39). This hazard can be controlled by raising the amount of oxygen in the package to more than 4%. [When foods are kept below 38°F (3.3°C), *C. botulinum* type E and nonproteolytic B and F will not multiply. If foods are kept below 50°F (10°C), *C. botulinum* type A and proteolytic B and F will not multiply.] Controlled atmosphere packaging can also be used to prevent fermentation and spoilage due to growth of other anaerobic microorganisms.

IV. TIME AND TEMPERATURE CONTROL: SAFE FOOD HOLDING TIMES AT SPECIFIED TEMPERATURES

A. Pathogen Growth During Processing and Food Handling

If hazardous temperatures for storing and holding food are considered to be that temperature range which allows growth of pathogenic bacteria, the temperature range is 30°F (-1.1°C) to 130°F (54.4°C). However, there is very little refrigerated food in the retail food sector stored below 41°F (5.0°C). Therefore, both time and temperature must be used to control the acceptable limits of growth during processing, holding, and storage of food in retail operations.

A review of all of the infective bacterial pathogen growth leads to the conclusion that *Y. enterocolitica* and *L. monocytogenes* can be selected as the low-temperature process control organisms. These pathogenic bacteria begin to multiply at 29.3°F (-1.5°C), as does *Aeromonas hydrophila* (14). They are the "organisms of choice" for control up to approximately 70°F (21.1°C) because of the severity of illness and speed of multiplication. Disease or illness caused by *L. monocytogenes* is estimated to be fatal to immune-compromised individuals about 27–28% of the time (17). The 1997 FDA Food Code (11) allows a 7-day holding at 41°F (5.0°C) and 4 days at 45°F (7.2°C). This time falls in between the growth rates of *Y. enterocolitica* and *L. monocytogenes* and will allow approximately 10 multiplications (1 microorganism becomes 1024) of *L. monocytogenes* and more of *Y. enterocolitica*.

For temperatures ranging from 70 to 112°F (21.1 to 44.4°C), *Salmonella* spp. is the chosen pathogen because it is commonly present in many foods and can cause serious illness. *Salmonella* spp. is known to multiply about once every 25 min at 104°F (40.0°C) in Chinese barbecued chicken (29).

If the FDA's 4-h food holding standard is applied at 112°F (44.4°C), this will allow the same 10 multiplications of a pathogen. From 112 to 126.1°F (44.4 to 52.3°C), the control microorganism of choice is *C. perfringens*. It multiplies every 8–15 min in the temperature range of 105–120°F (40.6–48.9°C) and is the pathogen that grows at the highest temperature, i.e., 126.1°F (52.3°C) (32). This sets the upper temperature growth limit at slightly less than 130°F (54.4°C). While vegetative cells of *C. perfringens* multiply rapidly, there is about a 2-h lag for outgrowth of the spores. Therefore, 2 h of lag and 10 multiplications again is compatible with allowing a 4-h period of holding at the most dangerous temperature of about 110°F (43.3°C).

Extensive data have been accumulated from scientific literature concerning the growth rates of pathogenic bacteria in food (36,38). Using this information, the growth rates of pathogenic bacteria can be predicted over the entire temperature range by using the formula of Ratkowsky et al. (30).

$$\sqrt{r} = b(T - T_{min})\{1 - \exp[c(T - T_{max})]\}$$

While the original formula specifies the use of temperature in degrees Kelvin, degrees Celsius can be used because a difference is being calculated and the data are normally expressed in degrees Celsius. T_{min} and T_{max} are the minimum and maximum temperatures, respectively, at which the rate of growth is zero. The parameter b is the regression coefficient of the square root of growth rate constant vs. degrees Kelvin/Celsius for temperatures below the optimal temperature, whereas c is an additional parameter that enables the model to fit the data for temperatures above the optimal temperature.

Using the FDA time–temperature standards of 41°F (5°C) for 7 days and 45°F (7.2°C) for 4 days, and assigning 4 h (time food can be held if it is out of "temperature control") to the most rapid multiplication temperature of about 112°F (44.4°C), a regression line that represents FDA time–temperature standards can be developed (see Fig. 1). The formula for the regression line that reasonably fits the FDA safety constraints is

$$y = 0.032 \, [\text{temp} - (-2.924)] \, \{1 - \text{Exp}[0.444 \, (\text{temp} - 52.553)]\}$$

It can be noted that the spoilage microorganisms multiply almost twice as fast as the predicted FDA hazard control times over the temperature range 30–126.1°F (0–52.3°C). The predicted *Y. enterocolitica* growth is slightly faster than the FDA hazard control standard below 50°F (10°C). However, the safety of the food is little compromised because the infective dose needed for *Y. enterocolitica* to cause illness is quite high. The danger of illness due to the presence of a sufficient number of *Clostridium perfringens* organisms in food is controlled when lag is factored in. The FDA 1997 Food Code–recommended temperatures and times for holding food (11) can be used to assure food safety (e.g., that the growth pathogens in food at these times and temperatures, including *C. botulinum*, is not sufficient to cause a hazard or illness).

In an actual kitchen or food production area, food does not remain at any one temperature for long. Therefore, the question becomes: "How long can food be left at other equivalent growth temperatures between 30 and 126.1°F (0 and 52.3°C)?" The formula can be used to calculate the times at specified temperatures. The results are shown in Table 3. Utilizing the set of guidelines given in the table allows the use of refrigeration units that exist in most retail food operations today. Most existing refrigeration units hold food between 40 and 55°F (4.4 and 12.8°C) during normal operations because the doors are opened so often. The temperature of many cold foods, such as those items found in salad bars, is 50–55°F (10–12.8°C) (23). The equivalent growth/generation time calculated from data in Table 3 can be used to calculate and estimate microbial growth and safety of food items. For example, if freshly prepared food is placed on a salad bar at 10:00 a.m. and held at 55°F (12.8°C) until 10:00 p.m., there would be fewer than the allowed 10 multiplications of pathogens, and the food would be safe for consumption during this time period. The food should be discarded at the end of this 12-h holding

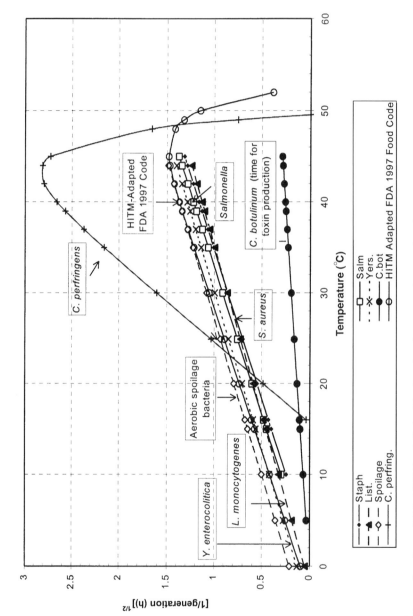

Figure 1 Generation time—FDA 1997 Food Code (Centigrade). Summary of generation times, growth rate data of pathogenic bacteria in a variety of foods compared to the 1997 FDA Food Code Holding/Storage Recommendations (11).

Table 3   Safe Food Holding Times at Specified Temperatures

	°F	Safety (based on 10 multiplications of pathogens)
↑	130	Safe
From	125	31.0 h
130 to	120	5.6 h
	115	4.6 h
|	110	4.7 h
If Heat	105	5.2 h
<6 h	100	5.9 h
	95	6.8 h
	90	7.9 h
	85	9.3 h
	80	11.2 h
	75	13.6 h
If Cool	70	16.9 h
<15 h	65	21.6 h
	60	28.6 h
	55	39.6 h
	50	2.4 d
↓	45	4.0 d
	41	6.5 d
	40	7.5 d
	35	19.3 d
	30	123.8 d
	<30	Safe chilled food holding
	28	Meat, poultry, fish thaw
	23	Spoilage bacteria begin to multiply
	14	Yeasts and molds begin to multiply

period because with overnight storage there would not be enough "safe time" remaining for another day. These data also indicate why it is hazardous to combine leftover cold, ready-to-eat foods with fresh products.

Similarly, within the rapid growth temperature range of 90–115°F (32.2–46.1°C), for which the generation time of pathogenic bacteria such as *Salmonella* spp. and *S. aureus* is approximately every 24 min, there is about 4 h of safe time. If food is on a buffet line, the temperature of the food must be monitored and if the temperature is 90–115°F (32.2–46.1°C) for 4 h, the food must be discarded.

There is also the question of what should be done about customer hot take-out food, when this type of food often cools to below 90°F (32.2°C) within a few minutes after being given to the customer. If the customer does not plan to eat the food for a couple of days, and if the food is refrigerated in less than 2 h, at a depth no greater than 2 in., the food will cool rapidly enough in a home refrigerator and will still be safe to eat after a day or two. The reason for this is that at a 40°F (4.4°C) refrigerated holding temperature there will have been fewer than 10 generations of pathogen multiplication.

V. FOOD PASTEURIZATION STANDARDS

In some countries, the destruction of *Enterococcus faecalis* in food is used as the pasteurization standard. Since it is doubtful that *E. faecalis* is actually a pathogen, it is more appropriate to use destruction of the known pathogenic species, *Salmonella*, for the standard. *Salmonella* spp. cause 30% of documented foodborne illnesses and deaths in the United States (4). Some species of *Salmonella* are rather difficult to inactivate. The resistance of salmonellae to inactivation also varies with the physical and chemical properties of each food. The *D* value necessary to destroy salmonellae in food containing a high amount of sugar or in food with a lower water activity is much larger than in food in which the water activity is sufficient for the growth of these organisms (31). If the *D* values and *z* values described by the USDA for the destruction of *Salmonella* in beef based on research by Goodfellow and Brown (13) are used as the basis for pasteurization control, there will be adequate destruction of other vegetative pathogens such as *E. coli* O157:H7 (19,20).

A. Times at Specified Temperatures for 5D *Salmonella* Inactivation

The 5D inactivation times shown in Fig. 2 will reduce the population of *Salmonella* on meat by a factor of 10^5. The *D* values (times needed to reduce a pathogens by a factor of 10) are derived from standards for hamburger (8,13).

Note that *Salmonella* spp. in this USDA standard has a *z* value (temperature difference for a 10-fold increase in rate of kill) of 10°F (5.6°C). For every 10°F (5.6°C) increase,

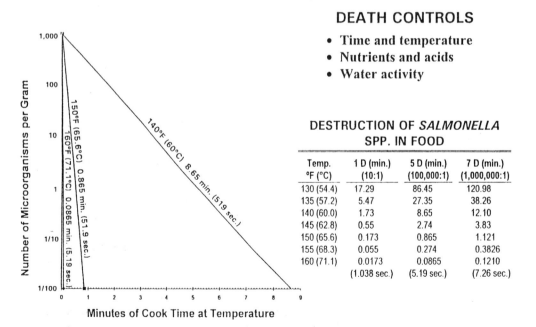

DEATH CONTROLS

- **Time and temperature**
- **Nutrients and acids**
- **Water activity**

DESTRUCTION OF *SALMONELLA* SPP. IN FOOD

Temp. °F (°C)	1 D (min.) (10:1)	5 D (min.) (100,000:1)	7 D (min.) (1,000,000:1)
130 (54.4)	17.29	86.45	120.98
135 (57.2)	5.47	27.35	38.26
140 (60.0)	1.73	8.65	12.10
145 (62.8)	0.55	2.74	3.83
150 (65.6)	0.173	0.865	1.121
155 (68.3)	0.055	0.274	0.3826
160 (71.1)	0.0173	0.0865	0.1210
	(1.038 sec.)	(5.19 sec.)	(7.26 sec.)

Figure 2 Destruction of *Salmonella* in food.

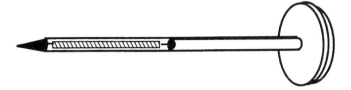

Figure 3 Bimetallic coil thermometer.

Salmonella bacteria are destroyed 10 times faster. This means that a very precise temperature measuring device [$\pm$ 1°F (0.56°C)] must be used for measuring pasteurization temperatures, or there can be serious errors. For example, the time necessary for inactivation of a specified number of pathogens is doubled if the food temperature is 3°F (1.7°C) lower than that specified. If the food pasteurization temperature is not measured accurately, enteric pathogens (*Salmonella* spp. and *E. coli* O157:H7) can survive in the food to cause illness.

B. Thermometers

Currently throughout the world, the bimetallic coil thermometer is commonly used to measure food temperature. Figure 3 is an illustration of the construction of a typical bimetallic coil thermometer. The bimetallic temperature-sensing coil extends from the tip through the length of the stem of the thermometer [approximately 3 in. (7.62 cm)] and averages food contact temperature over this distance. *This device is not tip-sensitive.* Hence, it is impossible for this instrument to determine the coldest temperature spot in hamburgers, chicken breasts, or any other thin, small-volume food items, and accurately validate that the food has been adequately pasteurized.

To ensure correct food temperature measurement, which is essential for correct pasteurization of contaminated food that comes from the wholesale system, all cooks must use tip-sensitive thermocouple thermometers. These thermocouple thermometers should have a tip diameter of 0.040 in. (1.0 mm) or less and an accuracy of $\pm$1°F ($\pm$0.56°C) over the range 0–400°F ($-$17.8–204.4°C) to verify food pasteurization. An example of an accurate tip-sensitive device is the Atkins 33040 (Atkins Technical Inc., Gainesville, FL) thermocouple with 0.040 hemitipped probe. If the food is 1 in. or more in depth, and a time span of 20 s is allowed for an accurate temperature reading, a tip-sensitive, thermistor meter such as the UEI PDT 300 (Universal Enterprises, Beaverton, OR) can be used.

VI. FOOD COOLING

Every foodservice operator has the responsibility to be certain that refrigerators have the capacity to cool foods safely. Spores of *C. botulinum*, *B. cereus*, and *C. perfringens* survive pasteurization. The 1997 FDA Food Code (11) states that food should be cooled from 140°F to 70°F (60°C to 21.1°C) in 2 h and from 70°F to 41°F (21.1°C to 5.0°C) in 4 h. This guideline implies that cooling food is a two-step process. Actually, the FDA has not interpreted cooling technology correctly. It is not a two-segment cooling process but rather a straight-line continuous-cooling process from 140°F to 41°F (60°C to 5°C) in 6 h.

Juneja et al. (21) have shown that the critical cooling time to control the outgrowth and multiplication of surviving *C. perfringens* spores is 15 h, if food is continuously cooled from 130°F to 45°F (54.4°C to 7.2°C) in a 38°F (3.3°C) cooling environment. This is the normal time needed to cool 2 in. of hot food in a 2¹/₂-in. steam table pan, from 130°F to 45°F (54.4°C to 7.2°C) in a 50-fpm (feet per minute) air flow, which is characteristic of commercial walk-in and reach-in refrigerators (36,37). This means that normal storage refrigerators with 1/4-horsepower compressors can cool food overnight if the food mass is less than 25 pounds with a starting temperature of 150°F (65.6°C) and air flow at 50 fpm across the bottom of the pan.

Foodservice refrigerators are designed to hold or store food at cool temperatures and are not designed to cool food (25). Refrigeration units used for food cooling must have additional horsepower (Btu/min) to provide the increased refrigeration capacity required for lower chilling temperatures of 35–25°F (1.7 to −3.9°C). Temperatures below 25°F (−3.9°C) should not be used because the surface of food begins to freeze, and center cooling is no faster. The ideal temperature for food storage for maximum shelf life is about 30°F (−1.1°C). National Sanitation Foundation refrigerators are designed to operate above 35°F (1.7°C) [normally 36–40°F (2.2–4.4°C)]. At less than 35°F (1.7°C), moisture from the air condenses on the refrigerator cooling coils, and ice forms and accumulates to block air circulation. It is possible to purchase freezers with thermostats that allow the units to operate at 25–35°F (−3.9–1.7°C)]. Thus, the problem is solved. Because the freezer coils have an electric or hot gas defrost cycle, the ice can not build up on the coil at 30°F (−1.1°C) and the cooling system functions efficiently.

Blast chilling units that are used to cool foods to 41°F (5.0°C) in 6 h or less must have a rapid flow of air at 1000 fpm (5.08 m/s) blowing directly across the container of food. Air turbulence around the container of food influences the cooling rate. For example, air at 1000 fpm cools food 3 times faster than typical refrigeration units with air flows of 50 fpm (27). If cooling is done in turbulent water, then the heat transfer at the surface of the food is improved above air cooling at 50 fpm by a factor of about 5. Figure 4 illustrates a typical food cooling process. Note that the process follows a semilog relationship. When the ΔT (temperature difference) of the food center temperature vs. air temperature is plotted on the y axis, the cooling curve is a straight line (27,28).

Since 75% of the heat is removed through the bottom of flat pans of food because of the direct contact of food with the bottom surface, pans of food should rest on an open or wire rack where there is no blockage of air flow across the bottom of the pan. Research has shown that the maximum food thickness that can be cooled to 41°F (5.0°C) in 6 h in a 30°F (−1.1°C) high-velocity stream of air is about 2 in. (5 cm). The type of container material, either stainless steel or plastic, has little effect on the cooling rate.

Food should be covered to prevent possible contamination of its food surface with mold spores and other microorganisms from the air circulating through the evaporator coil, which is never cleaned. Covering acid foods, such as spaghetti sauce or chili, with aluminum foil is not recommended. The acid in these foods combines with the aluminum causing pinholes to develop in the foil cover, as well as causing the foods to discolor or darken. However, if plastic wrap is used to cover food, it may blow off. The best solution

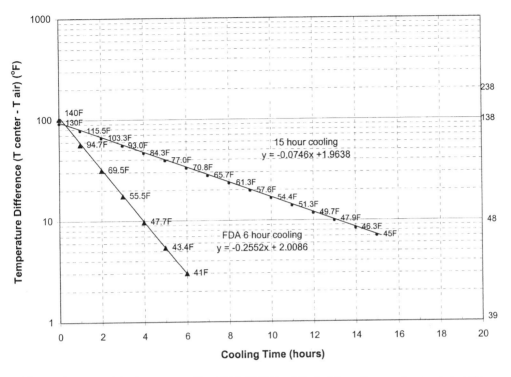

Figure 4 Typical food system cooling. FDA 1997 Food Code 6-h cooling vs. Juneja et al., 1994 15-h cooling. Food center temperature is shown in °F beside plot points.

is to put a layer of silicone paper or plastic wrap over the food surface prior to covering with aluminum foil.

In ordinary retail food operations that do not prepare large volumes of food ahead of time that require cooling, a high-capacity fan is not necessary. Justification for this statement is based on the research by Juneja et al. (21), which has shown that 15-h continuous cooling to 45°F (7.2°C) is safe, and that food in an ordinary refrigerator will cool in this time span if the food mass is 2 in. deep or less, in a covered pan (37). Figure 4 compares the FDA-recommended continuous cooling curve with this 15-h continuous cooling curve.

A. Validation of Cooling

The ability of refrigerators and cold-holding units in food operations to cool food can be validated in a simple manner. Instant mashed potatoes can be rehydrated in a food pan by adding 7% by weight potato flakes to water, until the potato–water mixture in the pan thickens. For cooling tests, the initial temperature of this potato–water mixture should be 140–150°F (60–65.6°C). A thermocouple or thermistor thermometer can be mounted on the pan to measure the temperature in the geometrical middle of the mixture. The temperature of the potato–water mixture is measured and recorded at the time it is placed in the refrigeration unit and thereafter at 30-min intervals. To assure that data are accurate, the thermometer should not be moved during cooling. The temperature data can then be plotted as shown in Fig. 4. When the temperature data are plotted, there will be a linear cooling relationship.

Holding temperatures in salad bars can be checked in the same manner, except that the potato–water mixture is prepared with cold water and the mixture should be 39–40°F (3.9–4.4°C). The mixture should hold a temperature of less than 41°F (5.0°C) [but above 32°F (0°C) to prevent freezing] throughout the container as long as it is in a salad bar, cold prep table, or cold display.

B. Effective Methods for Cooling Food to 41°F (5°C) Rapidly

While Juneja et al. (21) suggest that food can be cooled safely from 130°F to 45°F (54°C to 7.2°C) with 15 h continuous cooling, the FDA 1997 Food Code (11) recommends cooling food to 41°F in 6 h [from 140°F (60°C) to 70°F (21°C) within 2 h and 70°F (21°C) to 41°F (5°C) or below within 4 h]. This requires rapid cooling with either blast coolers, which can cost $10,000, or continuous labor stirring the food in sinks of ice and water. Figure 5 shows the simple ways to cool food rapidly to 41°F (5°C).

Blast chilling. The simplest way to cool food to 41°F (5.0), e.g., in the FDA-recommended 6 h or less, is to use a blast chill refrigeration system. Experiments have shown that the key factors in chilling are:

1.   The velocity of the air across the bottom of the pan (greater than 1000 fpm)
2.   The thickness of the food (less than 2 in.)
3.   The temperature of the cooling air system [less than 35°F (1.7°C)]

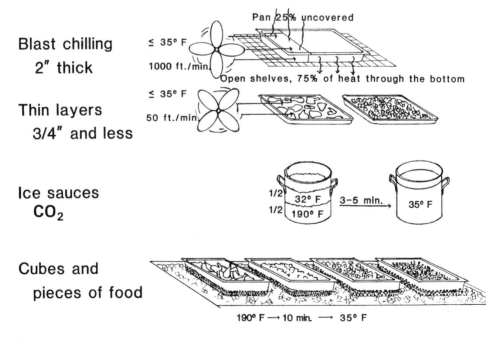

Figure 5   Cooling methods—6 h. to 41°F, pan 25% uncovered.

Whether the pan containing food is stainless steel, aluminum, or plastic makes little difference in cooling rates. However, the air velocity across the pan of food must be 1000 fpm or greater. Most fans, if placed 4 in. or less from the side of the pan, can provide this amount of air velocity.

Small-piece cooling. Another way to prevent cooling problems is to cook ingredients individually and make the sauce separately so that the parts can be cooled easily and the ingredients combined when both are cold. If foods are in small pieces, in thin layers on sheet pans with many sides exposed (e.g., cubed beef, vegetables for stew), the foods cool very rapidly. These types of food will cool to 41°F (5.0°C) in approximately 90 min in a 35°F (1.7°C) refrigeration unit with air blowing across the pan even at 50 fpm. This is an effective way to cool ingredients before they are combined into a salad, sauce, or casserole-type dish. Combining ingredients previously chilled to below 41°F (5.0°C) eliminates the problem of outgrowth of *S. aureus* contamination from the hands used to mix the salad because *S. aureus* and *C. botulinum* (types A and proteolytic B) do not produce toxin below 50°F (10.0°C). It is true that uncovered food cools about twice as fast. However, it will become contaminated with mold, yeast, and other spoilage microorganisms from the refrigerator blower coils, which will grow in the food and cause it to spoil in about 5 days. If the food is to be used within 2–3 days, it can be cooled uncovered. However, if a longer storage period is desired, the food should be covered during cooling.

Gravies and sauces can be cooled quickly if half of the liquid is omitted during preparation of these items. When the gravy/sauce has the correct flavor, it can be thickened to a double-thick consistency with a roux or starch. Cooling is accomplished rapidly (3–6 min) by adding the other half of the liquid as ice (or frozen milk) to chill the gravy or sauce to 35°F (1.7°C). At the same time, the product is diluted to the correct strength and viscosity. To prepare a stew or casserole, the food preparer simply gets the correct amount of cooked, refrigerated cubed beef and vegetables and the correct amount of sauce. The items are combined with the sauce or gravy that is then reheated in a convection oven or microwave oven in 15 min or less. This system eliminates the hazardous procedure of inadequate cooling and hot holding.

Solid or liquid carbon dioxide (e.g., dry ice) can also be used to cool sauces. Carbon dioxide (CO_2) is particularly effective because it does not dilute the liquid. It also has an excellent inhibiting effect on pathogens because it forms carbonic acid with water and the pH of the product declines. The carbon dioxide may give a sauce or gravy a slightly carbonated flavor when cooled. However, when the products are reheated the CO_2 vaporizes and the carbonation is no longer evident. [Note: CO_2 should always be used in a ventilated area to prevent asphyxiation.]

Ice bath cooling. When refrigeration capacity is limited, solid chunks of hot foods such as chicken, beef cubes, pot roast, turkey, potato cubes, macaroni, or rice in perforated pans can be cooled first by being placed in a stream of cold tap water and then cooled further, if necessary, in an ice bath. Ice that is used to cool food in this manner must be prepared from potable water. Most cubed products that are less than 1 in. thick can be removed from the ice bath in about 30 min when the temperature of the food has dropped to below 35°F (1.7°C). Large items such as a whole cooked turkey and roast beef may require 4 h to cool in an ice bath. Cold food should be stored covered, in any size container in the refrigerator at 32–35°F (0 to 1.7°C). Products prepared in this manner have a much longer shelf life because the rapid cooling process suppresses the growth of spoilage microorganisms.

It is very important that the sink or large container containing the slush ice be sanitized prior to using this method in order to prevent cross-contamination of products.

Salad preparation. All ingredients used to prepare cold foods such as salads, which will not receive any further heat treatment, should be cooled to less than 41°F (5.0°C) before ingredients are combined. An easy way to cool freshly cooked salad ingredients such as macaroni or potatoes is to use cold water to cool the items initially and then further cool the items with ice. If all ingredients (including salad dressing) are less than 41°F (5.0°C), it is possible to mix a salad and return it to refrigerator storage before the salad temperature reaches 50°F (10°C). This prevents toxin formation from the multiplication of *S. aureus* and proteolytic types of *C. botulinum*.

VII. FOOD OPERATIONS HAZARD ANALYSIS AND PROCESS CONTROL

Figure 6 presents a summary of the common critical control points in a foodservice operation. Note that consumer nutrition, consumer allergies, and consumer abuse of the food are included, since they are problems that have been identified in lawsuits. Figure 7 summarizes the standards for the control of chilled-food processes. First, employees must wash their hands and fingertips using the double handwash with a fingernail brush. Water must be obtained from a potable source. Insects and rodents must be kept out of the facility, and all food contact surfaces must be cleaned and sanitized between contacts with different potentially hazardous raw products. Assuming that raw food is contaminated, there must be just-in-time delivery of fresh product to assure that there is minimal pathogen multiplication in the wholesale supply system. Food must be stored before the temperature rises more than 5°F (2.8°C). Damaged packages should be returned, and moldy food should be destroyed. Food should be stored for less than 7 days so that *L. monocytogenes*, which is capable of multiplying once per 17 h at 41°F (5.0°C), will not multiply more than 10 generations.

Preparation of the food should begin within 24 h of the time it will be consumed. The food temperature must be kept to less than 50°F (10.0°C) during prepreparation. This way, total multiplication of *L. monocytogenes* will be less than five generations from the time the food is received until it is cooked or consumed. Heating the food to above 130°F (54.4°C) in less than 6 h will control the multiplication of *C. perfringens*. The food should then be pasteurized to reduce *Salmonella* 100,000 to 1, according to the standards shown. Leftovers of cooked/pasteurized food should be cooled to 45°F (7.2°C) in less than 15 h in order to control the multiplication of *C. perfringens*. If salads are made, they should be mixed at a temperature of less than 50°F (10.0°C) in order to control the toxin production by growth of *S. aureus* and *C. botulinum*.

Food held hot on a steam table should be served in less than 30 min in order to retain the thermally sensitive nutrients (e.g., ascorbic acid, niacin, and thiamin). Cross-contamination must be controlled. If the hot take-out food is given to the consumer, then the consumer must be instructed to eat it within 2 h, or to begin to cool it immediately to 41°F (5.0°C) within 2 h. If the consumer cannot do this, then he or she should purchase cold food at 41°F (5.0°C) to take home. Food should be held either below 30°F (−1.1°C) or used within the time limits given in Table 3.

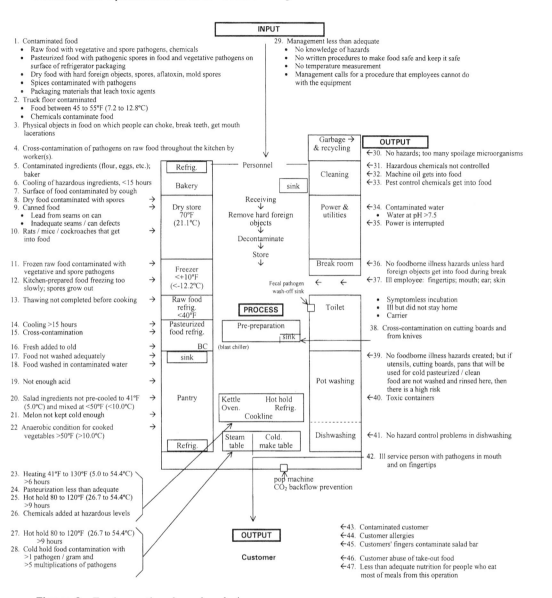

Figure 6 Food operations hazard analysis.

Reheating of food should not be used as a critical hazard control process. If toxins of *S. aureus, B. cereus*, and *C. botulinum* have been produced, they cannot be inactivated by reheating food to 165°F (73.9°C).

The following process standards assure that toxins will not be produced after food is cooked. Keeping food at less than 38°F (3.3°C) controls the growth of nonproteolytic types of *C. botulinum* and *B. cereus*. However, this temperature will allow the growth of *L. monocytogenes* and *Yersinia enterocolitica*. Therefore, to assure the safety of the food it should be used within the time-temperature constraints of Table 3.

INPUT

- Employee hand washing ....*Shigella* spp. and hepatitis A reduction by 10^{-5} by use of fingernail brush and double hand wash.
- Water...Pathogen control by water supplier.
- Insects and rodents.. Exclusion through cleanliness and construction.
- Food contact surfaces ....Surfaces cleaned and sanitized 10^{-5} to ≤ 2 CFU/cm^2.

1. **Expected threat level** in raw food to be controlled:		- Control of hard foreign objects.
Salmonella spp	<10/g	
Listeria monocytogenes	<1/g	
Staphylococcus aureus	<100/g	
Clostridium perfringens	<100/g	
Clostridium botulinum	<0.01/g	
Bacillus cereus	<100/g	

1. **Expected threat level** in raw food to be controlled:
   - *Salmonella* spp — <10/g
   - *Listeria monocytogenes* — <1/g
   - *Staphylococcus aureus* — <100/g
   - *Clostridium perfringens* — <100/g
   - *Clostridium botulinum* — <0.01/g
   - *Bacillus cereus* — <100/g

- Control of hard foreign objects.
- If meat, fish, or poultry is to be eaten rare or raw, the supplier assures and certifies safe pathogen levels.
- The producers / suppliers provide standard plate count data that prove they have a stable, HACCP-controlled process.
- Just-in-time delivery at <0°F (<-18°C) or <41°F (<5.0°C) maximizes freshness and minimizes pathogen multiplication. If food is maintained at <30°F (<-1.1°C), there will be no *Listeria monocytogenes* multiplication.

PROCESS

2. **Receiving:** Some food and beverages will be contaminated and must be checked, sorted, trimmed.

- Food must be stored before temperature reaches 45°F (7.2°C) or 5°F (-15°C). Label with date.
- Damaged packages and cans of food are returned.
- Infested packages and defective products (moldy or spoiled foods) are returned.

3. **Storage:** <41°F (<5.0°C) <4 days. Limit surface mold by controlling a$_w$.

- At 41°F (<5.0°C) and <4 days, *L. monocytogenes* will be controlled to an acceptable increase of <1:16 (4 generations). Cut, chop <24 hours before use.

4. **Pre-preparation and staging** for production <24 hours before use. Cut, chop, wash fruits and vegetables. Weigh and measure. Keep temperatures <50°F (<10°C).

- Clean-as-you-go prevents pathogen cross-contamination. Single hand wash 10^{-2} reduction.
- Control multiplication of *L. monocytogenes* to <1 additional generation. [Total *L. monocytogenes* multiplication is <1:32 (5 generations).]
- Fruits and vegetables are double washed to remove surface filth and reduce pathogens >100:1 (10^{-2}).

5. **Preparation:**
   Cook
   <41 to 130°F (<5.0 to >54.4°C)
   <6 hours
   Pasteurize
   Reduce *Salmonella* spp. 10^{-5}
   Control *C. botulinum*, pH 4.6
   Control salmonellae, pH 4.1
   Cool
   >130 to 45°F (>54.4 to 7.2°C)
   <15 hours
   Mix salads to maintain <50°F (<10°C)

- Heat from <41 to >130°F (5.0C to >54.4°C) <6 hours to prevent multiplication of *Clostridium perfringens*.
- Food pasteurization for 100,000:1 (10-5) *Salmonella* spp. reduction by temperature: 130°F (54.4°C)--86.45 min.; 140°F (60°C)--8.65 min.; 150°F (65.6°C)--0.865 min.; 160°F (71.1°C)--0.0865 min. **or** by addition of sufficient organic acid to decrease the pH below 4.1 with a 2-day hold at 75°F (23.9°C).
- Cool food from 130 to 45°F (54.4 to 7.2°C) in <15 hours to control multiplication of *C. perfringens*.
- Prevent toxin production of *Staphylococcus aureus* when mixing salads by pre-cooling ingredients to 40°F (4.4°C) before mixing and then keeping the ingredients <50°F (<10°C) during mixing and use.

OUTPUT

6. **Finish production:**
   Serve at 150°F (65.6°C) in <30 min.
   or
   Package, chill, and distribute at <41°F (<5.0°C).

- Retain nutrients at 150°F (65.6°C) by serving in <30 minutes.
- Prevent cross-contamination.
- Prevent customer abuse. Label: Consume within 2 hours, or begin cooling within 2 hours and cool to 41°F (5.0°C) in <10 hours.
- Warn consumers with a label that lists ingredients which might cause a food allergic reaction.

7. **Food holding, leftovers, reuse:**
   Do not add fresh to old.

- If stored at 30 to 41°F (-1.1 to 5.0°C), use in a time to control possible post-processing *L. monocytogenes* contamination to <5 generations.
- If stored at <38°F (<3.3°C), pasteurized food can be held until spoiled.
- Since post-cooking contamination is controlled, reheating is not required as a hazard control.

Figure 7 Pasteurized-chilled food process hazard control flow diagram.

VIII. RECIPE HACCP

A. The Seven Recipe Processes

All recipes can be clustered into seven process styles (Fig. 8) in terms of vegetative cell destruction and toxin/spore control. These are

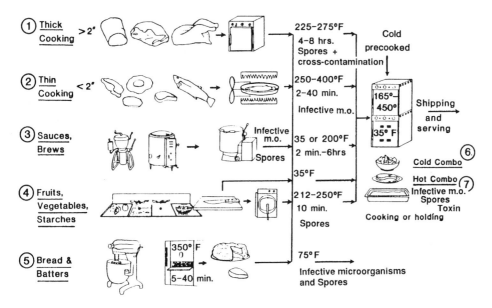

Figure 8 The seven recipe processes. Design for control of infective microorganisms, toxin-producing microorganisms, and spores.

1. Thick foods, greater than 2 in.
2. Thin foods, less than 2 in.
3. Sauces and brews, hot or cold
4. Fruits, vegetables, and starches
5. Bread and batters
6. Cold combinations
7. Hot combinations

Thick foods, such as large pieces of meat and poultry, may have high levels of surface contamination. Usually, these products are cooked or roasted slowly in an oven. During this time, vegetative pathogens on the surface of roasts or poultry are inactivated. The hazard control point for a large piece of meat occurs after it has been cooked, when it is left for some period of time for carving and serving. Often during serving, the meat is at a temperature of 100–110°F (37.8–43.3°C). Therefore, if it is not served and consumed in less than 4 h, there is a serious risk that the *C. perfringens* in the meat will have enough time to multiply to an illness-causing level. The critical control point for any thick food is after it has been cooked, and the control is using the foods within the time/temperature constraints given in Table 3.

Thin foods. The problem with thin foods is that the cooking process can be so fast that the infective microorganisms in the center of the food (e.g., *Salmonella* in eggs, *Trichinella spiralis* in pork, *E. coli* O157:H7 in hamburger, etc.) may not be inactivated with heat. There is also the problem of nonuniform heating (as occurs in a microwave oven), which allows vegetative cells to survive. After the food is cooked, since thin foods are usually individual portions, they probably will be eaten almost immediately, and spores will have no chance to outgrow, nor will *S. aureus* have time to produce a toxin. Thus, the critical control point is correct and uniform heating of the food to temperatures for times that assure adequate pasteurization.

Hot sauces and food products such as soups and gravies are heated sufficiently during preparation or production to easily destroy vegetative cells. However, spores survive. Therefore, if the soups and sauces are not kept hot, above 130°F (54.4°C) after cooking, spores of *C. perfringens* and *B. cereus* can grow out, multiply, and make people ill if the food is not cooled continuously for times and temperatures necessary to prevent their growth. For sauces heated to low temperatures, such as Hollandaise sauce made with raw egg, a combination of low-temperature heating and acid addition accomplish pasteurization. If the pH of the sauce is less than 4.1, salmonellae will be inactivated.

Fruits, vegetables, and starches are contaminated with both vegetative pathogens and spores, and perhaps chemicals. All raw fruits and vegetables must be double-washed in a clean, sanitized vegetable preparation sink. Each washing will reduce the microorganisms by approximately 10 to 1 (24). Therefore, overall, there will be a pathogen reduction of about 10^2 on the surface of the fruits and vegetables. This is the only control for contamination of fruits and vegetables that are to be consumed raw. Hence, washing in a clean, sanitized sink is an extremely important critical control procedure. If the fruits or vegetables are cooked and have a pH above 4.6, *C. botulinum* and *B. cereus* can become a serious hazard if the products are not cooled and stored properly. Hot fruits and vegetables must be kept above 130°F (54.4°C) to prevent spore outgrowth, or cooled continuously to 45°F (7.2°C) within 15 h, and held at 41°F (5.0°C) or less.

Breads and batters are inherently safe because these products are cooked to high temperatures over 180°F (82.2°C), for a period of time that is sufficient to destroy any infective microorganisms that may have contaminated and multiplied in the product. However, baked products that are iced, filled, and manipulated after baking must be handled with care. Care must be taken not to introduce pathogenic microorganisms into icings and fillings that will make the baked products hazardous. Handwashing becomes an important critical control point, as well as correctly pasteurizing and cooling hazardous fillings such as egg and milk/cream-containing fillings and custards.

Cold combination products are usually protein food items mixed with a sauce such as mayonnaise and a starchy food item such as macaroni (e.g., macaroni salad with tuna, egg, or cheese). Critical control procedures make certain that no microorganisms grow during the ingredient cooling step and prevent cross-contamination during mixing. If all ingredients are cooled to 41°F (5.0°C) and kept below 50°F (10.0°C) during mixing, there is no hazard from growth and toxin production by *S. aureus* or proteolytic *C. botulinum*. If salads are stored at less than 41°F (5.0°C) and used in less than 10 days, the hazard of pathogen multiplication is controlled.

Hot combination products, such as casseroles, are safer than cold combinations because if there is some infective microorganism contamination during the mixing step, the organisms will be inactivated during reheating. However, there will be spore and *S. aureus* contamination when the ingredients are mixed. If *S. aureus, B. cereus*, or *C. botulinum* is allowed to multiply due to careless food handling after cooking, the toxins produced by these pathogens will not be inactivated in the reheating step. For example, casseroles should be kept below 50°F (10.0°C) until heating, and then heated to above 130°F (54.4°C) in less than 6 h to prevent multiplication of *C. perfringens* during cooking.

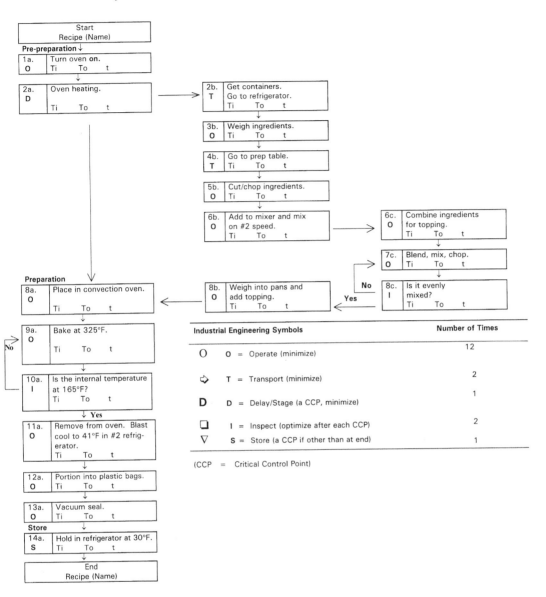

Figure 9   Food process flow charting the cooking and branching of a recipe process flow.

B.   Recipe Flow Charting

The first step in applying HACCP principles to recipes is to do a process flow chart. Figure 9 is a simple illustration of a recipe flow chart. There are three columns identified as *a*, *b*, and *c* columns. The *b* column shows subprocesses as to what is done at a given point in the *a* column. The *c* column shows subprocesses at a given point in the *b* column. Each step is numbered sequentially from 1 to the end. Food processes such as making beef stew for lunch in a institutional kitchen can easily require 150 steps.

One problem with most recipes is that the sequence when first diagrammed is illogical and inefficient. By flow-diagramming the recipe, the proper order in which steps should be accomplished in order to minimize labor and maximize safety becomes obvious. Fundamentally, all raw food is prepared, chopped, cubed, and made ready, as in ingredient preparation.

First, the vegetables are cut or chopped, because these food times change the least in quality waiting to be combined. Next, sauces are made because they can be put into a *bain marie* and held hot without much deterioration in quality. Next, the meat is cooked to the point at which the pre-prepared vegetables, which take less cooking time, can be added. Then, the sauce is combined at the correct point, and the product is finished, panned or bagged, and served. An important point is *not* to cook the meat first and let it remain at ambient temperatures while sauces and vegetables are prepared. Clearly, this procedure could lead to food safety problems. At each step in the flow, the step is identified by one of the five industrial engineering symbols: **Operate, Transport, Delay, Inspect**, and **Store**. The use of these five terms is very important to recipe analysis because it allows comparing the safety and efficiency of one recipe with another. The optimum recipe has a minimum of Operate and Transport steps to achieve the desired sensory properties for the finished product. Ideally, it should have no Delay steps. It has an optimum number of Inspect steps whereby the employee is given specific instructions to verify the quality and safety of the product. Finally, there is only one Store step at the end. Any unnecessary delay in Store steps represent hazard control points.

Each step has provisions for "temperature in" (Ti) of the food, the "temperature out" (To) of the food, and the time (t) it will take to complete the step. When this information is used in combination with the growth and death temperatures and times previously listed, it is possible to validate the control of vegetative and spore pathogens in a process.

Example of QA Recipe Flow—Barley Soup. As previously described, all tasks that are done in the food facility are parts of the seven recipe processes. Thus, only each of the seven recipe processes for each style of product (thick foods; thin foods; sauces, brews; fruits; vegetables, starches; bread, batters; cold combinations; hot combinations) needs to be flow-diagrammed. The processes of each recipe style can be applied to all of the recipes categorized under each style.

All processes can be divided into prepreparation (getting ready), preparation (doing), chill-store/transport-holding, serving, and leftovers. Each process step is numbered so that it can be referenced. Each step is identified as: **O** for **Operation; I** for **Inspect; T** for **transport; D** for **Delay;** or **S** for **Store**. The object is to have a minimum number of delay steps, only one store step at the end, a minimum number of transport steps, and just enough operating steps, controlled by inspecting steps, to safely produce the product.

Each block has a brief description of the step. Since the critical controls in pasteurized food processes are temperature and time, these factors are indicated in the block so that by using the microbiological rules, which have been previously described, one can verify that the process is safe. Conventional computer logic is used in material flow diagramming. The symbols are used by industrial engineers and to optimize processes. When two process methods are compared, the one with the fewest operations that gives the desired product, the least amount of transport and delay, only one store step at the end, and adequate inspect steps, is the best process. Note the inclusion of the inspect step. This is one of the most important elements of hazard control. It emphasizes that the process designer and management must also describe precisely how the employee is to check that he or she has performed a step correctly and what standard(s) must be met.

Figure 10 is an example of a quality-assured recipe flow for barley soup.

C. Quality-Assured HACCP Recipe Procedures, the Critical Hazard Control Document

Since a flow diagram is extremely useful for analysis but very difficult for a foodservice worker to read, the information must be transformed to a conventional recipe format for use by the cook. Figure 11 illustrates such a format for the barley soup recipe. Note that the first page of the recipe is divided into the two traditional sections familiar to cooks. The top section lists ingredients. The cook buys ingredients or assembles ingredients based on this section. The next section details the steps taken to convert the ingredients to products. The second page continues the steps and contains comments.

The recipe is the conventional process control document that has been used for centuries to control the quality and uniformity of food production. The problem is that food time–temperature rules have not been provided for pathogen control. Using the simple rules presented in this paper, a food safety process authority can read the recipe steps and verify the safety of the recipe. For example, with step 10 [Add barley, onions, Worcestershire sauce (40°F) to the kettle (150°F, t = 1 min)] there is no problem because the time of 1 min is too brief. Step 11 [Set temperature to 212°F and bring to a boil (212°F, t = 15 min)] is, again, too short a time for any microbiological multiplication. In addition, when the ingredients reach 212°F (100°C), all of the infective vegetative microorganisms have been destroyed. Step 12 [Return dirty containers to pot and pan washing area. Get clean refrigerator storage racks and bring back to the kettle] assumes that the containers will be properly washed, rinsed, and sanitized before the food dries on them. Step 13 [Reduce kettle temperature to 200°F and simmer for 45 min or until barley is done *al dente*] assures the destruction of all infective microorganisms and nonproteolytic *C. botulinum* in this product. However, spores of proteolytic *C. botulinum*, *B. cereus*, and *C. perfringens* will survive.

At the end of each line, a clock time and initials can be entered by the employee. This records the time that the employee completed the step. In a typical restaurant, this information would not be entered because the work is repetitive. However, in a hospital chilled-food foodservice system, where there is a requirement for documented process control for liability reasons, the written information provides the documentation that can be given to the government to show that this process was done according to government-approved standards.

On the second page, at the bottom of the form, additional information is given regarding the reconstitution instructions, items with which it might be served, and plating instructions. On the second page, an instruction for handling *leftovers* is also given. Use of leftovers is grossly abused in most foodservice operations. There is either a tremendous waste, and they are discarded, or they are reheated multiple times in an attempt to get rid of them. In a good food operation, there is an accurate head count, and there are virtually no leftovers. The few leftovers remaining often can be fed to the employees. The HACCP analysis must include controls to ensure the safety of food from production to consumption or disposal.

Finally, there is a *comments* section, which complements the first side of the recipe sheet, where special comments can be made regarding the specific preparation procedure.

Pre-Preparation (Begin Barley Soup)

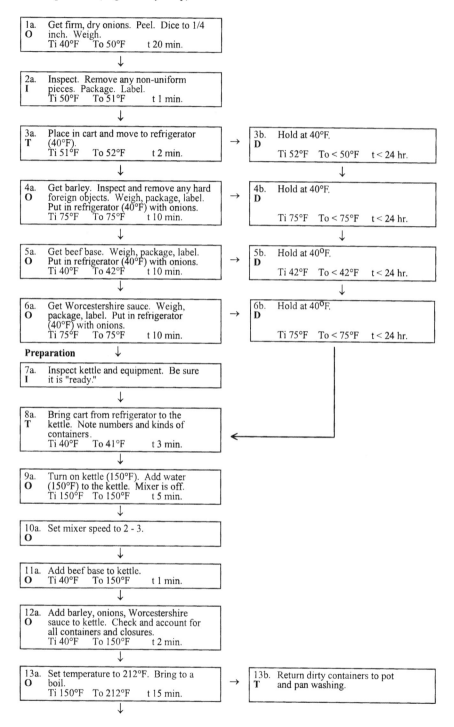

1a. O	Get firm, dry onions. Peel. Dice to 1/4 inch. Weigh. Ti 40°F To 50°F t 20 min.

↓

2a. I	Inspect. Remove any non-uniform pieces. Package. Label. Ti 50°F To 51°F t 1 min.

↓

| 3a.
T | Place in cart and move to refrigerator (40°F).
Ti 51°F   To 52°F   t 2 min. | → | 3b.
D | Hold at 40°F.

Ti 52°F   To < 50°F   t < 24 hr. |

↓

| 4a.
O | Get barley. Inspect and remove any hard foreign objects. Weigh, package, label. Put in refrigerator (40°F) with onions.
Ti 75°F   To 75°F   t 10 min. | → | 4b.
D | Hold at 40°F.

Ti 75°F   To < 75°F   t < 24 hr. |

↓

| 5a.
O | Get beef base. Weigh, package, label. Put in refrigerator (40°F) with onions.
Ti 40°F   To 42°F   t 10 min. | → | 5b.
D | Hold at 40°F.
Ti 42°F   To < 42°F   t < 24 hr. |

↓

| 6a.
O | Get Worcestershire sauce. Weigh, package, label. Put in refrigerator (40°F) with onions.
Ti 75°F   To 75°F   t 10 min. | → | 6b.
D | Hold at 40°F.

Ti 75°F   To < 75°F   t < 24 hr. |

Preparation ↓

7a. I	Inspect kettle and equipment. Be sure it is "ready."

↓

| 8a.
T | Bring cart from refrigerator to the kettle. Note numbers and kinds of containers.
Ti 40°F   To 41°F   t 3 min. | ← |

↓

9a. O	Turn on kettle (150°F). Add water (150°F) to the kettle. Mixer is off. Ti 150°F To 150°F t 5 min.

↓

10a. O	Set mixer speed to 2 - 3.

↓

11a. O	Add beef base to kettle. Ti 40°F To 150°F t 1 min.

↓

12a. O	Add barley, onions, Worcestershire sauce to kettle. Check and account for all containers and closures. Ti 40°F To 150°F t 2 min.

↓

| 13a.
O | Set temperature to 212°F. Bring to a boil.
Ti 150°F   To 212°F   t 15 min. | → | 13b.
T | Return dirty containers to pot and pan washing. |

↓

Figure 10 Barley soup quality-assured recipe flow.

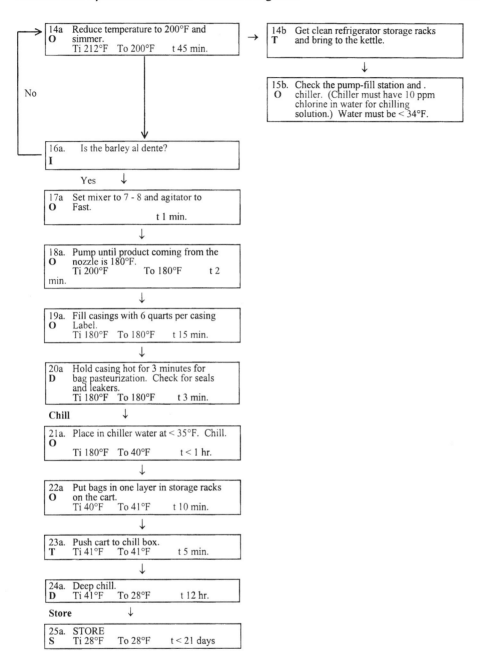

14a O Reduce temperature to 200°F and simmer.
Ti 212°F   To 200°F   t 45 min.

14b T Get clean refrigerator storage racks and bring to the kettle.

15b. O Check the pump-fill station and . chiller. (Chiller must have 10 ppm chlorine in water for chilling solution.) Water must be < 34°F.

No

16a. I Is the barley al dente?

Yes

17a O Set mixer to 7 - 8 and agitator to Fast.
t 1 min.

18a. O Pump until product coming from the nozzle is 180°F.
Ti 200°F   To 180°F   t 2 min.

19a. O Fill casings with 6 quarts per casing Label.
Ti 180°F   To 180°F   t 15 min.

20a D Hold casing hot for 3 minutes for bag pasteurization. Check for seals and leakers.
Ti 180°F   To 180°F   t 3 min.

Chill

21a. O Place in chiller water at < 35°F. Chill.
Ti 180°F   To 40°F   t < 1 hr.

22a O Put bags in one layer in storage racks on the cart.
Ti 40°F   To 41°F   t 10 min.

23a. T Push cart to chill box.
Ti 41°F   To 41°F   t 5 min.

24a. D Deep chill.
Ti 41°F   To 28°F   t 12 hr.

Store

25a. S STORE
Ti 28°F   To 28°F   t < 21 days

Figure 10 Continued

This is a place to note irregularities, so that if there are customer comments that the recipe is worse or better than normal, one can refer to the comments to find out what was done differently and then execute a quality control change in the recipe procedure if appropriate.

Beef Stew. There are many forms that the recipe can take. The recipe for beef stew (Fig. 12) is specifically written for regulatory analysis. The ingredients are listed in stand-

Recipe Name: **Barley Soup**   Portion size (vol./wt.): 7 oz.   Preparation time: 4 hours
Recipe #: **100**   Number of portions: 477   To be prepared by: P. Snyder
Production style: **Soup/Sauce**   Final yield (AS): 25 gal. (208.35 lb.)   Supervisor:
Written by: D. Poland Date: 3/96   Yield: 90 %:   SA/QA by: J. Campbell   Date: 4/96

Gp. #	Ingred. #	Ingredients and Specifications	EP Weight %	Edible Portion (EP) (weight or volume)		User Rec. (wt./vol.)	Nutrition Ref. #
I	1	Barley	11.65	27.0 lb.	(12,258.0 g)		
	2	Onions, chopped 1/4 inch dice	3.46	8.0 lb.	(3,632.0 g)		
	3	Worcestershire sauce	0.22	.5 lb.	(227.0 g)		
II	4	Beef base	5.62	13.0 lb.	(5,902.0 g)		
III	5	Water (22 gal)	79.05	183.0 lb.	(83,082.0 g)		
		Total	100.00	231.5 lb.	(105,101.0 g)		
		Approx. gallons		27.8 gal			

                                                                      Time/ Initials

Pre-preparation   11/19/98 *Begin*

1. Get clean dry onions (45°F). Dice to 1/4 inch and weigh. Inspect for uniformity.   1600 OPS
    Package and label. Place on a cart and move to the refrigerator . (50°F, 23 min.)
2. Get barley (75°F). Weigh, package, and label. Put in refrigerator with onions. (75°F,   1630 OPS
    10 min.)
3. Get beef base (41°F). Weigh, package, and label. Put into refrigerator with onions.   1640 OPS
    (42°F, 10 min.)
4. Get Worcestershire sauce (75°F). Weigh, package, and label. Put in the refrigerator with
    onions. (75°F, 10 min.)   1650 OPS
5. Take cart to kettle preparation area or hold < 41°F until time for preparation (< 3 days).   1700 OPS

Preparation   11/20/98 *Begin*   0630 OPS

6. Inspect kettle and equipment to be sure it is ready.   0635 OPS
7. Bring the cart with food supplies (41°F) from refrigerator to the kettle. (41°F, 3 min.)   0640 OPS
8. With the mixer off, turn on the kettle to 150°F and add tap water. (150°F, 5 min.)
9. Add beef base to the kettle. (150°F, 1 min.)   0645 OPS
10. Add barley, onions, Worcestershire sauce to the kettle. (150°F, 2 min.)   0646 OPS
11. Set temperature to 212°F and bring to a boil. (212°F, 15 min.)   0648 OPS
12. Return dirty containers to pot and pan washing area. Get clean refrigerator storage racks   0659 OPS
    and bring back to the kettle.
13. Reduce kettle temperature to 200°F and simmer until barley is *al dente* (45 min.).   0740 OPS
Pumping
14. Check the pump-fill station. Check that the water chiller station has 5 ppm chlorine for   0742 OPS
    chilling.
15. When barley is *al dente,* set mixer to # 7 and agitator to Fast.   0744 OPS
16. Pump enough product into a bag to get the product temperature coming from the pump to   A
    180°F. (2 min.)
                                                                      B
17. Fill casings with 6 quarts of soup (180°F) per casing. Label. Hold each bag for 3   0801 OPS
    minutes to allow for inside bag surface pasteurization. Check seal before adding the bag
    to the chiller. Put a special label on the last bag so this bag can be found in the chiller.
    The product temperature of this bag is measured to assess cooling of contents of all bags
    before they are removed from the chiller. (15 min.)

Figure 11   Quality-assured recipe procedures—barley soup.

Chilling	**Time / Initials**

Chilling
18. Place in chiller water (< 35°). Chill < 1 hour. When temperature of control bag is < 40°F, all bags can be removed.
19. Put bags (<40°F) one layer deep in storage racks on the cart. (<40°F 10 min.)
20. Push cart to chill box. (40°F, 5 min.)
21. Allow food to deep chill to 29°F (± 1°F) in < 12 hours.

0905 OPS

0910 OPS

0915 OPS

C

Store
22. Hold at 29°F (± 1°F) and use within < 21 days.

Reheating, Plating, and Serving
23. When needed, heat soup to 165°F (no higher for quality) by an appropriate method.
24. Hold covered at 165°F and serve within 30 minutes so that food is >150°F when consumed.

Leftovers
25. Dispose of leftovers, or cool to 40°F in < 4 hours and serve within 24 hours.

Comments:

A   QC took a sample of the soup from the first bag pumped for microbial. analysis.

B   Got 15 1/2 - 6 quart bags, rather than 16 1/2 bags.   OK - J.C.

C   Checked bags at 1745. Temperature was 30°F   OPS

11/24/98   The APC was 55/ml. - L. Dugan

Figure 11   Continued

ard form, along with process steps. There is a column for *hazards*, in which the various hazards associated with a given step are identified. The last column provides a *hazard control analysis*, which includes, when appropriate, the effectiveness of the policies, procedures, and standards or guidelines manual to control these threats.

Quality-Assured HACCP Recipe Procedures—Beef Stew. Another version of the beef stew recipe (Fig. 13) is written in the more traditional format like the barley soup recipe that lists ingredients at the top of the page.

Ingredients are listed by weight rather than volume because this is the most accurate way to measure ingredients in food preparation. Ingredients are listed in both pounds/ounces and grams. The measurement system used is dependent on what system (English or metric) the kitchen is utilizing.

It is particularly important to convert weights to an edible portion weight percent. When this is done, it is possible to look at an ingredient, such as black pepper at approximately 0.004%, to determine whether or not the customer will be able to taste it. If salt or monosodium glutamate is added to a recipe, the edible portion weight percent becomes extremely important because it is easy to determine overuse of an ingredient.

The second part of the quality-assured HACCP recipe procedure is the written instructions. A HACCP recipe is different from a typical recipe because each step must include food time and temperature. Because of these food times and temperatures in each step, it is possible to read the recipe and almost instantly verify that the times and temperatures are adequate to prevent a microbiological problem in the food. Unusual or incomplete steps can and should be discussed and modified until corrected.

BEEF STEW   Portion: 6 oz ladle (8 oz by weight)

Yield Beef Cube Base Vegetables, cooked Total	24 Servings (1 pan) 3 1/2 qt 6 1/2 lb. 6 qt	96 Servings (4 pans) 3 1/2 gal 26 lb. 6 gal	Process Steps: Policies, Procedures, and Guidelines	Hazard	Hazard Control Analysis (Effectiveness of the Policies, Procedures, and Standards
Ingredients: Beef Cube Base: Beef cubes, 1 inch	 4 1/2 lb.	 18 lb.	**Pre-Preparation:** • Take beef cubes from the refrigerator (40°F) (15 min.) and brown in kettle or heavy skillet (190°F) (15 min.).	Vegetative pathogens. Spores.	Cooking at 190°F kills vegetative pathogens. Spores survive but temperature is too high for growth.
Tomatoes, ground (No. 10 can) Bay leaf Pepper, black Soy sauce Beef base Water	2 C ---- 1 ea 1/4 t 3 T 1 1/2 oz 1 1/2 qt	2 qt (2/3 can) 4 ea 1 t 3/4 C 6 oz 1 1/2 gal	**Preparation:** • Add tomatoes, bay leaf in cheese- cloth bag, black pepper, soy sauce, beef base, and water (70°F) (15 min.). • Bring to boil (212°F). Cover and simmer (190°F) for approximately 1 hour or until meat is tender.	Bay leaf: HFO. Black pepper has high pathogen counts. Tomatoes reduce pH. Beef base only has spores.	Cheese cloth bag controls HFO. 1 hour kills vegetative pathogens. Spores that survive will not multiply .because temperature is >130°F.
Beef Stew: Flour Water, cold	 4 oz 3/4 C	 1 lb. 3 C	• Combine flour and room temperature water, add while stirring, to boiling beef base. • Simmer (195°F) for 5 minutes to thicken while vegetables are prepared, <50°F.	Flour and water have pathogens. Possibility of pesticide or herbicide residue on vegetables. Vegetables contain both vegetative pathogens and spores.	Vegetables must be washed and rinsed. Simmering inactivates vegetative cells. Heating in the next step will kill vegetative pathogens.
Carrots, 1/4 inch slice Celery, 1/4 inch slice Onions, 3/4 inch dice Potatoes, 3/4 inch dice	1 1/2 lb. 1 1/2 lb. 1 1/2 lb. 2 lb.	6 lb. 6 lb. 6 lb. 8 lb.	• Boil prepared carrots, celery, onions, and potatoes 8 to 10 minutes in a minimum amount of water or steam for 5 to 8 minutes (vegetables >212°F). Drain vegetables if necessary. • Add vegetables to thickened meat mixture.	Vegetative and spore microorganisms.	Vegetative cells are killed. Spores are controlled by temperature >130°F.
			• Bring combination to >185°F. • Pour 6 qt of Beef Stew into each steam table pan and cover with film. Do this in 10 minutes and maintain food temperature of >165°F. • Transfer immediately to a 165°F hot holding box. Use in 1 hour. • Leftover stew should be cooled covered 2 inches deep in a 2 1/2- inch pan in a 35°F refrigerator at an air flow of >1,000 fpm to 41°F in <6 hours.	Spores survive. Spores survive. Spores survive. Spore outgrowth.	Vegetative cells are dead. Temperature >165°F, vegetative cells killed. No time for spore outgrowth. Temperature >165°F, all pathogens controlled. Use in <1 hour controls nutrient loss. No vegetative cells introduced. Spores controlled by cooling to 45°F in <15 hours.

Figure 12   Recipe for beef stew.

At the bottom of Fig. 13 is the format for each line entry in the most stringent hazard-controlled process. The line should be identified by:

1. Process step number
2. The starting food center temperature
3. The thickest food dimension
4. Container size
5. Whether or not the food is covered
6. The temperature on or around the food
7. The ending food center temperature
8. The time (in hours and minutes) it takes to complete the process step

Recipe Name: **Beef Stew**
Recipe #:
Production style: **Hot combination**
Written by: Cleveland Range   Date: 3/96

Portion size (vol./wt.):
Number of portions:
Final yield (AS): **90 gal.**
Yield: **88 %**:

Preparation time:
To be prepared by: **P. Snyder**
Supervisor:
SA/QA by: P. Snyder   Date: 4/96

Gp. #	Ingred. #	Ingredients and Specifications	EP Wt. %	Edible Portion (EP) (weight or volume)		User Rec. (wt./vol.)	Nutrition Ref. #
I	1	Cooking oil	0.12	1.06 lb.	(480.0g)		
II	2	Beef, boneless, raw, cut in 1" cubes	33.45	288.00 lb.	(130,752.0 g)		
	3	Onions, chopped	7.78	67.00 lb.	(30,418.0 g)		
III	4	Beef stock (20 gallons)	19.40	167.0 lb.	(75,818.0 g)		
IV	5	Carrots diced 1/2 inch	7.78	67.00 lb.	(30,418.0 g)		
	6	Celery	7.78	67.00 lb.	(30,418.0 g)		
	7	Potatoes	20.91	180.00 lb.	(81,720.0 g)		
V	8	Flour	1.16	10.00 lb.	(4,540.0 g)		
	9	Water, cold (5 quarts)	1.21	10.44 lb.	(4,739.8 g)		
VI	10	Salt	.35	3.00 lb.	(1,362.0 g)		
	11	Pepper	.02	0.19 lb.	(86.3 g)		
	12	Garlic powder	.02	0.19 lb.	(86.3 g)		
		Total	99.98	860.88 lb.	(390,838.4 g)		
		Approx. gallons		103 gal.			

Pre-preparation   **Time/ Initials**

1. Inspect the weight and condition of all ingredients.

Preparation

2. Pour oil (72°F) into kettle, or use a lecithin spray and omit the oil.
3. Set mixer on #3 setting. Add 25% of the cubed beef (41°F). Turn heat **On**. Add remaining meat. Brown the meat. (225°F, 10 min.)
4. Add beef stock and onions (41°F). Heat to 200°F.
5. Simmer at 190°F until meat is tender (approx. 1.5 hours).
6. Add carrots, celery and potatoes (41°F). Heat to 190°F. Simmer 10 min.
7. While kettle is heating, mix flour (72°F, room temperature) and water (70°F). (10 min.).
8. Add flour / water mixture to kettle mixture. Add salt, pepper, and garlic powder. Cook until thickened. (200°F, 10 min.)

Pumping

9. Set mixer speed at #4 and agitator speed on **slow to medium**.
10. Fill casings with 6 quarts of soup (180°F) per casing. Label. Hold each bag for 3 minutes to allow for inside bag surface pasteurization. Check seal before adding the bag to the chiller. Put a special label on the last bag so that this bag can be found in the chiller. The product temperature of this bag is measured to assess cooling of contents of all bags before they are removed from the chiller. (15 min.)

Chilling

11. Place casings in chiller water (< 35°). (Make sure the tumble chiller water has 5 ppm chlorine or equivalent chemical.) Chill to <41°F in < 1 hour. When temperature of control bag is < 41°F, all bags can be removed.
12. Put bags (< 41°F) one layer deep in storage racks on the cart. Push cart to chill box (< 41°F, 15 min.).
13. Cool food to 29°F (± 1°F) in < 12 hours.

Store

14. Hold at 29°F (± 1°F) and use within < 21 days.

Reheating, Plating, and Serving

15. When needed, heat beef stew to 165°F (no higher for quality) by an appropriate method.
16. Hold covered at 165°F and serve within 30 minutes so that food is >150°F when consumed.

Leftovers

17. Dispose of leftovers, or cool to 41°F in < 4 hours and serve within 24 hours.

Ingredients that could produce possible adverse or allergic reactions: Flour

Process step #	Start food ctr. temp., °F	Thickest food dimension (in.)	Container size H x W x L (in.)	Cover Yes/No	Temp. on/ around food	End food ctr. temp., °F	Process step time, hr./min.

Figure 13   Quality-assured HACCP recipe procedures—beef stew.

If these variables are indicated for each process step, then any process control authority can read this recipe and certify whether or not it is safe and determine the competency of the person who wrote the HACCP recipe.

Also on the recipe form there is a line for *Ingredients that could produce possible allergic reactions.* Some individuals have serious, life-threatening allergies to certain foods. For example, on this recipe for beef stew, flour is listed as an ingredient that could cause an adverse reaction (notably for people with gluten intolerance).

IX. MANAGEMENT CONTROL

The most important element of an effective HACCP-based food safety program is the management process, which precontrols the system to assure that food products are safe. Government's responsibility is only to verify that the operation is following agreed-to HACCP policies, procedures, and guidelines or standards to produce safe food products for human consumption. A major reason that most food safety regulations at the present time are ineffective is that they do not require unit management to demonstrate how food products are safely produced, stored, served, and sold. Rather, they are based on the false premise that government inspection is control. Government inspection of food operations does not assure food safety. Employees on the line who handle the food can check each item and ensure the safety and quality of food. Management's precontrol responsibility is to educate and train employees in safety-validated food handling techniques and provide adequate resources for continuing employee training and process improvement.

X. COMPONENTS OF AN EFFECTIVE HACCP PROGRAM

The following are the six components of an effective HACCP program. These must be in place to assure virtual zero-defect control of the hazards.

1. **Management responsibility for food safety**
    a. Demonstrate management commitment through food safety promotion actions (e.g., safety committees, incentives, awards, etc.).
        * Set challenging, measurable, and attainable improvement safety goals.
        * Set the safety example in all activities.
        * Interview employees, during walkaround, to hear and respond to their suggestions for process improvement.
    b. Allocate sufficient resources to accomplish food safety goals.
    c. Establish an organizational chart that shows assigned line and staff responsibility for control of food hazards.
    d. Establish a system to measure the economic saving of doing the right tasks correctly the first time.
    e. Develop food safety committees.
    f. Establish accountability measures for meeting food safety responsibilities.
    g. Hold regular staff meetings to reinforce the safety principles and to listen to employee improvement suggestions.
    h. Implement ongoing inspection and monitoring programs to identify and improve controls of changing workplace hazards.
    i. Take action to maintain and improve process control and stability.

2. **Hazard analysis and control**
   a. Identify and analyze food hazards through food safety audits, environmental monitoring, and self-inspections to *identify hazards in each job* that could lead to food-borne illness.
   b. Examine each job for hazards and list the following:
      * Sequence of job process steps identifying ingredients, time, temperature, and equipment essential to hazard control.
      * How hazards can lead to illness.
      * Procedures and guidelines that, when used, will control the hazards.
   c. Make arrangements for a hazard control, quality assurance manager to be available on all work shifts.
   d. Evaluate the safety performance of new equipment, supplies, and materials before purchase, and processes before implementation to assure production of safe food.

3. **Written program** with clearly stated goals and objectives for food safety assurance that promotes safe and sanitary working conditions and has a clearly stated plan for meeting the goals and objectives.
   Owner(s)/manager(s) must:
   a. Write a food safety policy statement concerning hazards and controls specific to the workplace.
   b. Write a food safety action plan and program clearly describing how food safety assurance and safety improvement goals will be met.
   c. Develop and implement written food safety procedures and guidelines.
   d. Write plans for conducting and documenting at least an annual review of the program effectiveness and then for improving the program based on the findings.
   e. Institute appropriate equipment programs to cover the calibration, use, cleaning, maintenance, and eventual replacement of equipment.

4. **Communication and training**
   Owner(s)/manager(s) must:
   a. Communicate the food safety program to all employees.
   b. Allow for employee input in bringing hazardous food operating conditions to management's attention.
   c. Provide training prior to all new job assignments, including training on specific hazards and controls.
   d. Update training at least annually or as work processes and ingredients change.
   e. Maintain records of training (date/topic/content/attendance).
   f. Train supervisors in pertinent food safety matters, food safety leadership, coaching, and employee empowerment to take action at any time to prevent a problem.
   g. Evaluate training needs to determine specialized training and retraining. Use supervisors and employees to give feedback as to how to improve training.

5. **Process control problem investigation and corrective action**
   Owner(s)/Manager(s) must:
   a. Develop procedures for process control problem reporting, problem investigation, corrective action, and follow-up.
   b. Conduct workplace prevention inspection of facilities and equipment (e.g., refrigeration, cooking and hot holding devices, pot and dish washing and sanitizing, insect and rodent control).
   c. Write reports following process control problems showing what preventive/corrective action is being taken to prevent similar problems, e.g.,

Table 4 Food Operation Evaluation

Operation ————————————————————— Date ————————————————

Director of SA/QA ————————————————————————————————

Criteria	Capable (0–5)[a]	Perform[b]	Haz. Ctrl. Y/N[c]	Comments
1. Management Safety commitment and resources Safety leadership Safety enforcement Action Log				
2. Quality Control, Assure, Improve Improvement audits Continuous improvement program Hazard analysis Hazard control policies, procedures, and stand- ards or guidelines Quality costs Schedules for maintenance, cleaning, training Operational safety self-evaluation System changes are controlled Safety controlled by the line employee(s) Recipes are HACCP'd				
3. Personnel Everyone knows his/her responsibilities Selection and qualification Training and performance certification Coaching and improvement by supervisors Personal hygiene and dress, handwashing Knowledge of hazards and controls				
4. Facilities Sufficient capacity Maintenance Cleanliness and sanitation Insect and rodent control Waste and trash control				
5. Equipment Adequate capacity Maintenance Cleaning and sanitizing Backflow CO_2 prevention Calibration of control and temperature measur- ing thermometers				
6. Supplies Specifications Supplier food safety certification Chemical material safety data sheets and con- trols				

Table 4 Continued

Operation ————————————————— Date ————————————

Director of SA/QA ——————————————————————

Criteria	Capable (0–5)[a]	Perform[b]	Haz. Ctrl. Y/N[c]	Comments
7. Receiving and storage Package and container damage control Spoilage culling, cleaning of fruits and vegetables Rapid and correct storage; food covered, off floor Product dated; FIFO; <41°F (5.0°C), use in <7 days.				
8. Production Pre-prep temperatures <50°F (10°C) Cross-contamination control; clean-as-you-go Salad ingredients cooled to <41°F (5.0°C) before mixing; kept at <50°F (10°C) during mixing Heat food to >130°F (54.4°C) in <6 h and then heat to get a 5D *Salmonella* pasteurization: 130°F (54.4°C)—86.45 min 140°F (60°C)—8.65 min 150°F (65.6°C)—0.865 min 160°F (71.1°C)—0.0865 min (5.19 s)				
9. Finished product and service Hot hold >140°F (60°C); serve in <30 min for maximum nutrition Cool food, continuously to <45°F (7.2°C) in <15 h [FDA reg.: 140°F (5.0°C) to 70°F (21°C) in 2 h, 70°F (21°C) to 41°F (5.0°C) in 4 h] Cold holding <41°F (5.0°C) <7 days (allows less than 10 generations of *L. monocytogenes*) Service persons know possible allergenic ingredients in recipes and can communicate to customers Leftovers not mixed with fresh Leftovers kept <41°F (5.0°C) (Reheating to 165°F (73.9°C), not used for critical control) Consumer abuse of food is controlled				

[a] **Capability**

  0–No measurable activity in the performance area.

  1–Partial but inadequate activity in the performance area.

  2–An informal program that meets minimal criteria of the performance area.

  3–A formal program that has not been fully developed and implemented in the performance area.

  4–A formal program that meets all basic criteria of the performance area and is backed by management.

  5–A formal written program that meets or exceeds all performance criteria and is fully implemented, communicated and reviewed annually.

[b] **Performance**: Quality defects per 1,000 items.

[c] **Hazard Control**: Explain negative responses.

- Equipment modified
- Work method modified
- Equipment changed or added
- Employee retrained or special needs accommodated.

   d. Maintain, summarize, and analyze foodborne illness data (e.g., first reports of illness) to determine tasks and operations where incidents have occurred. Take action to prevent recurrence.

6. Program enforcement

Owner(s)/Manager(s) must:

   a. Write an enforcement statement on safe food operation practices, food safety rules, and standard operating procedures.

   b. Maintain records of disciplinary actions and warnings.

   c. Develop policies that hold all personnel accountable for fulfillment of food safety responsibilities. (Safety is behavioral control. Everyone is responsible for his or her safety behavior.)

All of this information should be assembled in a *Food Safety through Quality Assurance: Policies, Procedures, and Standards or Guidelines Manual.* Organization of this manual was discussed in the previous chapter, "Foodservice Operations: HACCP Principles."

XI. FOOD OPERATION EVALUATION

There is a fundamental management principle that says: "If a standard is not measured, it will not be enforced." The food operation evaluation form (Table 4) enables the operator to validate the operation to assure that standards are followed and that the processes are stable. Stability is the basis for continuous quality improvement and reliably safe products. The results of this evaluation should never be used to punish employees but rather to coach them and to assure improved performance.

XII. SUMMARY

In summary, the actual rules for food safety HACCP-based self-control are simple.

1. All raw food is assumed to be contaminated.
2. Raw food is used as soon as possible.
3. Employees wash their hands and fingertips correctly, at times required, to prevent transfer of pathogens to food.
4. Utensils, work surfaces, cutting boards, etc. are prerinsed, washed, and sanitized correctly.
5. Infective pathogens in food are reduced to safe levels by sufficient heating of food, (pasteurization or cooking), washing it, or acidifying it. Food should be heated in less than 6 h to over 130°F (54.4°C).
6. After preparation, food is served and consumed as soon as possible, or is cooled continuously to 45°F (7.2°C) in less than 15 h [FDA: 140–41°F (60–5.0°C) in 6 h or less] and stored at less than 41°F (5.0°C). If the food temperature is not below 30°F (−1.1°C), above 130°F (54.4°C), or pathogen growth is controlled

in some other way, then it must be used in a time that limits the growth of *L. monocytogenes* or *Salmonella* spp. to fewer than 10 generations.

7. The presence of chemicals and hard foreign objects in food is monitored and controlled.

8. Ingredients in all food products are known so that consumers who have food allergies receive accurate answers regarding composition of food items.

The key to performing these simple food safety tasks is for all food preparers and servers to be trained and to follow these procedures 100% of the time until the rules are officially changed. With documentation of all recipe procedures and policies on handwashing, cutting board cleaning, etc., the only remaining factor is behavioral control. Management leadership is crucial. Effective management trains and coaches all employees to use correct procedures for handling, producing, and serving food before they are allowed to perform their tasks. Employees are given positive reinforcement when jobs are completed satisfactorily, so that the goal of producing and serving safe food through hazard-controlled processes becomes achievable.

REFERENCES

1. Ak, N. O., Cliver, D. O., and Kaspar, C. W. 1994. Cutting boards of plastic and wood contaminated experimentally with bacteria. J. Food Protect. 57: 16–22.
2. Ak, N. O., Cliver, D. O. and Kaspar, C. W. 1994. Decontamination of plastic and wooden cutting boards for kitchen use. J. Food Protect. 57: 23–30.
3. Asplund, K., and Nurmi, E. 1991. The growth of *Salmonellae* in tomatoes. Int. J. Food Microbiol. 12: 177–182.
4. Bean, N. H., Griffin, P. M., Goulding, J. S., and Ivey, C. B. 1990. Foodborne disease outbreaks, 5-year summary, 1983–1987. MMWR 39 (SS-1): 15.
5. Cann, D. C., Wilson, B. B., Hobbs, G., and Shewan, J. M. 1965. The growth and toxin production of *Clostridium botulinum* Type E in certain vacuum packed fish. J. Appl. Bact. 28(3): 431–436.
6. Code of Federal Regulations (CFR). 1995. Title 21. Food and Drugs. Parts 100 to 169. Superintendent of Documents, U.S. Gov. Printing Office, Washington, D.C.
7. Code of Federal Regulations (CFR). 1995. Title 21. Food and Drugs. Parts 170 to 199. Superintendent of Documents, U.S. Gov. Printing Office, Washington, D.C.
8. Code of Federal Regulations (CFR). 1995. Title 9. Animal and Animal Products. 200 to end. 318.17 Superintendent of Documents, U.S. Gov. Printing Office, Washington, D.C.
9. Dunsmore, D. G., Twomey, A., Whittlesone, W. G., and Morgan, H. W. 1981. Design and performance of systems for cleaning product-contact surface of food equipment: a review. J. Food Protect. 44: 230–240.
10. Food and Drug Administration/Public Health Service. 1995. Grade A Pasteurized Milk Ordinance. Pub. No. 229, Washington, D.C.
11. Food and Drug Administration. 1997. Food Code. U.S. Public Health Service, U.S. Dept. of Health and Human Services. Pub. No. PB97-141204, Washington, D.C.
12. Frazer, A. M., and Matthews, M. E. 1990. Effectiveness of cold-holding methods in foodservice operations. J. Food Protect. 53: 336–340.
13. Goodfellow, S. J., and Brown, W. L. 1978. Fate of *Salmonella* inoculated into beef for cooking. J. Food Protect. 41: 598–685.
14. Hudson, J. A., Mott, S. J., and Penney, N. 1994. Growth of *Listeria monocytogenes, Aeromonas hydrophila, Yersinia enterocolitica* on vacuum and saturated carbon dioxide controlled atmosphere-packaged sliced roast beef. J. Food Protect. 57(3): 204–208.

15. International Commission of Microbiological Specifications for Foods (ICMSF) 1980. Microbial Ecology of Foods, Vol. 1, Factors Affecting the Life and Death of Microorganisms. Academic Press, New York.

16. International Commission of Microbiological Specifications for Foods (ICMSF) 1988. Microorganisms in Foods 4. Application of the Hazard Analysis Critical Control Point (HACCP) System to Ensure Microbiological Safety and Quality. Blackwell Scientific, Oxford.

17. Jay, J. M. 1992. Microbiological food safety. Crit. Rev. Food Sci. Nutr. 31(3): 177–190.

18. Jones, J. M. 1992. Food Additives in Food Safety. Eagan Press, St. Paul, MN, pp. 203–258.

19. Juneja, V. K., Snyder. O. P., and Marmer, B. S. 1996. Thermal destruction of *Escherichia coli* O157:H7 in beef and chicken: determination of D- and z-values. Int. J. Food Microbiol. (in press).

20. Juneja, V. K., Snyder, O. P., Williams, A. C., and Marmer, B. S. 1997. Thermal destruction of *Escherichia coli* O157:H7 in hamburger. J. Food Protect. 60(10):1163–1166.

21. Juneja, V. K., Snyder, O. P., and Cygnarowicz-Provost, M. 1994. Influence of cooling rate on outgrowth of *Clostridium perfringens* spores in cooked ground beef. J. Food Protect. 57(12): 1063–1067.

22. Jung, M., Stomberg, L., Redondo Fernandez, D., Crespo, I., and Snyder, O. P. 1995. A food safety survey of eleven food service operations in the Minneapolis/St. Paul, Minnesota metropolitan area. Hospitality Institute of Technology and Management, St. Paul, MN.

23. Matthews, M. E. 1991. Safety issues related to use of take-out food. J. Foodserv. Syst. 6: 41–59.

24. Nagle, N. 1997. Dole Inc., private communication.

25. National Science Foundation. 1994. Food Service Refrigerators and Storage Freezers. No. 7. NSF, Ann Arbor, MI.

26. Pether, J. V. S., and Scott, R. J. D. 1982. *Salmonella* carriers: are they dangerous? A study to identify finger contamination with salmonellae by convalescent carriers. J. Infect. 5: 81–88.

27. Pflug, I. J., and Blaisdell, J. L. 1963. Methods of analysis of pre-cooling data. ASHRAE J. 5(11): 133.

28. Pflug, I. J., Blaisdell, J. L., and Kopelman, I. J. 1965. Developing temperature-time curves for objects that can be approximated by sphere, infinite plate, or infinite cylinder. ASHRAE Trans. 1. 71: 238.

29. Pivnick, H., Erdman, I. E., Manzatiuk, S., and Pommier, E. 1968. Growth of food poisoning bacteria on barbecued chicken. J. Milk Food Technol. 31: 198–201.

30. Ratkowsky, D. A., Lowry, R. K., McMeekin, T. A., Stokes, A. N., and Chandler, R. E. 1983. Model for bacterial culture growth rate throughout the entire biokinetic temperature range. J. Bacteriol. 154(3): 1222–1226.

31. Riemann, H. 1968. Effect of water activity on the heat resistance of *Salmonella* in ''dry'' materials. Appl. Microbiol. 16: 1621–1622.

32. Shoemaker, S. P. and Pierson, M. D. 1976. ''Phoenix phenomenon'' in the growth of *Clostridium perfringens.* Appl. Microbiol. 32(6): 803–807.

33. Smittle, R. B. 1977. Microbiology of mayonnaise and salad dressing: a review. J. Food Protect. 40(6): 415–422.

34. Snyder, O. P. 1997. A ''Safe Hands'' Hand Wash Program for Retail Food Operations. Hospitality Institute of Technology and Management, St. Paul, MN.

35. Snyder, O. P. 1997. The Microbiology of Cleaning and Sanitizing a Cutting Board. Hospitality Institute of Technology and Management, St. Paul, MN.

36. Snyder, O. P. 1994. HACCP-Based Safety and Quality Assured Pasteurized-Chilled Food Systems. Hospitality Institute of Technology and Management (HITM), St. Paul, MN.

37. Snyder, O. P. 1997. Two-Inch and four inch food cooling in a commercial walk-in refrigerator. Dairy Food Environ. Sanitat. 17(7): 398–404.

38. Snyder, O. P. 1996. Use of time and temperature specifications for holding and storing food in retail food operations. Dairy Food Environ. Sanitat. 16(6): 374–378.

39. Solomon, H. M., Kautter, D. A., Lilly, T., and Rhodehamel, E. J. 1990. Outgrowth of *Clostridium botulinum* in shredded cabbage at room temperature under a modified atmosphere. J. Food Protect. 53(10): 831–833.

40. U.S. Department of Health and Human Services. 1998. Fish and Fisheries Products Hazards and Controls Guide, 2nd ed. U.S. Public Health Service, FDA, Center for Food Safety and Applied Nutrition, Office of Seafood, Washington, D.C.

41. U.S. Public Health Service. 1967. Standard Methods for the Examination of Dairy Products: Microbiological and Chemical, 11th ed. USPHS Pub. 1631, Washington, D.C.

42. Zhuang, R. Y., Beuchat, L. R., and Angulo, F. J. 1995. Fate of *Salmonella montevideo* on and in raw tomatoes as affected by temperature and treatment with chlorine. Appl. Environ. Microbiol. 61: 2127–2131.

Index